Advances in Marine Genomics

Volume 2

Series Editor
J. Mark Cock
CNRS, UMR 8227, Integrative Biology of Marine Models,
Station Biologique de Roscoff, Sorbonne Universités,
UPMC Univ Paris 06, CS 90074,
F-29688 Roscoff cedex, France

More information about this series at http://www.springer.com/series/8684

Iñaki Ruiz-Trillo • Aurora M. Nedelcu
Editors

Evolutionary Transitions to Multicellular Life

Principles and Mechanisms

Springer

Editors
Iñaki Ruiz-Trillo
Institut de Biologia Evolutiva (UPF-CSIC)
Barcelona
Spain

Aurora M. Nedelcu
Department of Biology
University of New Brunswick
Fredericton
New Brunswick
Canada

ISSN 1879-8012 ISSN 1879-8020 (electronic)
Advances in Marine Genomics
ISBN 978-94-024-0359-6 ISBN 978-94-017-9642-2 (eBook)
DOI 10.1007/978-94-017-9642-2
Springer Dordrecht Heidelberg New York London

Printed on acid-free paper

Springer Netherlands is part of Springer Science+Business Media (www.springer.com)

Foreword

As a young child growing up in Florida, I would set out on "treasure hunts" to look for fossils. The discovery of a Megalodon tooth or a mysterious fossilized bone would inspire thoughts about the lives of long-extinct creatures in a world before humans. And while my musings started with the fossils I could find and hold, with time I became curious about organisms and events from increasingly ancient times. How did tetrapods evolve? What about their ancestors, early aquatic vertebrates? How did developmental patterning evolve in the first bilaterians? And before that? What did the first animals look like? And from what did they evolve?

How animals evolved from their single celled ancestors is one of the great mysteries in evolution and was likely set in motion by the origin of multicellularity. A remarkable process, one so striking that it is considered one of only eight "Major Transitions" in evolutionary history (Maynard Smith, John; Szathmáry, Eörs (1995). *The Major Transitions in Evolution*. Oxford, England: Oxford University Press. ISBN 0-19-850294-X), the transition to multicellularity occurred not just once, but repeatedly in diverse lineages. From relatively inconspicuous beginnings—sister cells that remained attached following division rather than going it alone—evolved the many multicellular life forms that fill our visible world today: brown algae, red algae, green algae, land plants, fungi, and animals.

Despite the inherent interest surrounding the origins of multicellularity, relatively little is known about when, how, and why multicellularity evolved in each lineage. The potential barriers to reconstructing ancient transitions to multicellularity are diverse. For most multicellular lineages, the transition to multicellularity occurred hundreds of millions of years ago. The unicellular progenitors of each multicellular lineage were unlikely to have been preserved in the fossil record, and even if they were, how would we recognize them? Moreover, the identities of the closest living unicellular relatives of most multicellular lineages were, until recently, unknown. This has meant that the powerful comparative and experimental approaches used to reconstruct the evolution of animal and plant developmental patterning have been difficult to apply to questions regarding the evolution of multicellularity. Finally, it has been unclear whether there are common evolutionary themes underlying the many independent transitions to multicellularity.

But much has changed over the last decade or so. Molecular phylogenetic analyses have revealed the connections among multicellular lineages and their closest living single celled relatives. Genome sequences from diverse unicellular eukaryotes have begun to reveal the deep evolutionary histories of many gene families whose functions were previously thought to be restricted to multicellular organisms. And the development of new model organisms with increasing experimental tractability is beginning to offer insights into the molecular mechanisms that may have contributed to the transition to multicellularity. At the same time, a rich tradition of theoretical work is increasingly informing experimental approaches to the question of multicellularity, while also benefitting from the growing collection of data being generated by experimentalists.

This book captures the excitement of our field at a critical juncture, as findings from these diverse approaches are yielding meaningful insights into some of the biggest questions surrounding multicellularity. I am reminded of the excitement that I first felt while gazing at fossils from long ago. Despite the long passage of time since the earliest origins of multicellularity, it is now possible to envision the biology, morphology, and even the genome contents of the organisms from which diverse macroscopic life forms first evolved.

University of California
Berkeley
Department of Molecular and Cell Biology

Nicole King

Preface

The transition to multicellular life is associated with some of the most important events in the history of life, such as the evolution of animals, land plants, multicellular fungi and many groups of macroscopic algae, including seaweeds. Yet, despite its importance, we are far from fully understanding the forces driving this transition as well as the genetic and molecular basis for the evolution of multicellularity. However, in recent years, emerging data from various fields are providing new insight into the factors and processes underlying the evolution of multicellularity in different lineages.

The aim of this book is to offer a synthesis of the current issues and research into this fundamental biological question, by providing several complementary perspectives (both theoretical and experimental) and using examples from various lineages in which multicellularity evolved. The ultimate goal of the book is to integrate our understanding of how and why such transitions occurred and to facilitate the identification of general principles and mechanisms. We believe this book is a timely contribution to the field, given the research efforts, dedicated meetings and workshops that are being now committed to this question. The book will be of great interest to all researchers working in the field as well as to young scientists generally interested in evolutionary questions.

The chapters in this book have been written by leading researchers in their respective fields. Each chapter provides a review of the current state of the field and/or offers new perspectives for future research. We have tried to provide a balance of topics, model-systems and approaches; however, due to space limits, many interesting and important subjects had to be omitted.

The book starts out with a forward by Nicole King, a prominent scientist whose research program is devoted to understanding the early evolution of multicellularity in the animal lineages. The chapters are organized around five themes corresponding to the five parts of the book. Part 1—Multicellularity in the tree of life—emphasizes general issues and questions relevant to the evolution of multicellularity and provides an overview of the transition to multicellularity in the context of the history of life (when, why, how many times). Part 2—Model-systems—features several systems currently used to investigate independent origins of multicellularity in distinct taxonomic groups (e.g., metazoans and their unicellular relatives, fungi, green algae, brown algae). Part 3—Theoretical approaches—highlights several distinct approaches (using mathematical modeling, computer simulations) and frameworks (based on cooperation and conflict, phenotypic plasticity, or physics) that have been recently developed to address the forces driving the transition to multicellularity and identify the main factors. Part 4—Genomics insights—stresses the use of genomic approaches to provides insights into the genetic basis for the evolution of multicellularity in distinct lineages, including the evolution of the metazoan developmental toolkit, the evolution of morphological complexity in the green plant lineage, the independent emergence of complex multicellularity in brown and red algae, and of aggregative multicellularity in social amoebae. Part 5—Molecular mechanisms—covers several molecular mechanistic issues related to the emergence of multicellularity in various lineages, including changes in transcriptional regulation, cell-cell signaling and changes in signaling patterns (e.g., evolution of developmental signaling), allorecognition, and co-option of cellular pathways already present in the unicellular ancestors.

Each chapter was reviewed by one or two external reviewers or chapter contributors, and at least one of the editors. We are extremely grateful to all reviewers for taking their time to read the submitted contributions and to provide very thoughtful, critical and helpful suggestions for improving the chapters. We are also indebted to Nicole King for contributing the foreword to this book.

Last but not least, our greatest thanks to all the contributors, for their dedication to this field and their willingness to share their knowledge, data and views as part of this book; without their enthusiasm, timely contributions, and patience during the entire process this book would not have been possible!

The editors acknowledge funding support from NSERC (to AMN) and MINECO and ERC (to IRT).

(Barcelona) Iñaki Ruiz-Trillo
(Fredericton) Aurora M. Nedelcu
July 2014

Contents

Part V Molecular Mechanisms

Contributors

Maja Adamska Sars International Centre for Marine Molecular Biology, University of Bergen, Bergen, Norway

Claudio Aguilar Department of Microbiology, Institute of Plant Biology, University of Zurich, Zurich, Switzerland

Homayoun C Bagheri Institute of Evolutionary Biology and Environmental Studies, University of Zurich, Zurich, Switzerland

Matthew W. Brown Centre for Comparative Genomics and Evolutionary Bioinformatics, Department of Biochemistry and Molecular Biology, Dalhousie University, Halifax, Nova Scotia, Canada
Department of Biological Sciences, Mississippi State University, Mississippi, MS, USA

Zhi-hui Chen College of Life Sciences, University of Dundee, Dundee, UK

J Mark Cock CNRS, UMR 8227, Integrative Biology of Marine Models, Station Biologique de Roscoff, Sorbonne Universités, UPMC Univ Paris 06, Roscoff cedex, France

UMR 8227, Integrative Biology of Marine Models, Station Biologique de Roscoff, Sorbonne Universités, UPMC Univ Paris 06, Roscoff cedex, France

CNRS, UMR 8227, Integrative Biology of Marine Models, Station Biologique de Roscoff, Roscoff cedex, France

Susana M Coelho CNRS, UMR 8227, Integrative Biology of Marine Models, Station Biologique de Roscoff, Sorbonne Universités, UPMC Univ Paris 06, Roscoff cedex, France

Jonas Collén UMR 8227, Integrative Biology of Marine Models, Station Biologique de Roscoff, Sorbonne Universités, UPMC Univ Paris 06, Roscoff cedex, France

CNRS, UMR 8227, Integrative Biology of Marine Models, Station Biologique de Roscoff, Roscoff cedex, France

Alex de Mendoza Institut de Biologia Evolutiva, CSIC-Universitat Pompeu Fabra, Barcelona, Spain

Bernard M. Degnan School of Biological Sciences, University of Queensland, Brisbane, QLD, Australia

Qingyou Du College of Life Sciences, University of Dundee, Dundee, UK

Salva Duran-Nebreda ICREA-Complex Systems Lab, Institut de Biologia Evolutiva-UPF-CSIC, Barcelona, Spain

Leo Eberl Department of Microbiology, Institute of Plant Biology, University of Zurich, Zurich, Switzerland

Catherine Eichwald Institute of Virology, University of Zurich, Zurich, Switzerland

Laura Eme Centre for Comparative Genomics and Evolutionary Bioinformatics, Department of Biochemistry and Molecular Biology, Dalhousie University, Halifax, Nova Scotia, Canada

Douglas H. Erwin Department of Paleobiology, National Museum of Natural History, Washington, DC, USA

Stephen R. Fairclough Department of Energy, Joint Genome Institute, Walnut Creek, CA, USA

Garry Farnham CNRS, UMR 8227, Integrative Biology of Marine Models, Station Biologique de Roscoff, Sorbonne Universités, UPMC Univ Paris 06, Roscoff cedex, France

The Marine Biological Association of the United Kingdom, The Laboratory, Plymouth, Devon, UK

Plymouth Marine Laboratory, Plymouth, UK

Vanina J. Galzenati Laboratorio de Biología Comparada de Protistas, IBBEA, CONICET-UBA, Universidad de Buenos Aires, Buenos Aires, Argentina

Gernot Glöckner Leibniz-Institute of Freshwater Ecology and Inland Fisheries, IGB, Berlin, Germany

Institute for Biochemistry I Medical Faculty, University of Cologne, Köln, Germany

Olivier Godfroy CNRS, UMR 8227, Integrative Biology of Marine Models, Station Biologique de Roscoff, Sorbonne Universités, UPMC Univ Paris 06, Roscoff cedex, France

Laura F. Grice School of Biological Sciences, University of Queensland, Brisbane, QLD, Australia

Erik R. Hanschen Department of Ecology and Evolutionary Biology, University of Arizona, Tucson, AZ, USA

Matthew D. Herron Division of Biological Sciences, University of Montana, Missoula, MT, USA

Dale Kaiser Departments of Biochemistry and of Developmental Biology, Stanford University School of Medicine Stanford, Stanford, CA, USA

Yoshinori Kawabe College of Life Sciences, University of Dundee, Dundee, UK

John O. Kessler Department of Physics, University of Arizona, Tucson, AZ, USA

Physics Department, University of Arizona, Tucson, AZ, USA

Daniel Lang Plant-Biotechnology, Faculty of Biology, University of Freiburg, Freiburg, Germany

Hajara Lawal College of Life Sciences, University of Dundee, Dundee, UK

Charles R. Marshall Museum of Paleontology and Department of Integrative Biology, University of California, Berkeley, CA, USA

Richard E. Michod Department of Ecology and Evolutionary Biology, University of Arizona, Tucson, AZ, USA

Aurora M. Nedelcu Department of Biology, University of New Brunswick, Fredericton, NB, Canada

Akira F Peters CNRS, UMR 8227, Integrative Biology of Marine Models, Station Biologique de Roscoff, Sorbonne Universités, UPMC Univ Paris 06, Roscoff cedex, France

Bezhin Rosko, Santec, France

Stefan A. Rensing Faculty of Biology, University of Marburg, Marburg, Germany

Andrew J. Roger Centre for Comparative Genomics and Evolutionary Bioinformatics, Department of Biochemistry and Molecular Biology, Dalhousie University, Halifax, Nova Scotia, Canada

Valentina Rossetti Institute of Evolutionary Biology and Environmental Studies, University of Zurich, Zurich, Switzerland

Anthropological Institute and Museum, University of Zurich, Zurich, Switzerland

Iñaki Ruiz-Trillo Institut de Biologia Evolutiva (CSIC-Universitat Pompeu Fabra), Barcelona, Spain

Departament de Genètica, Universitat de Barcelona, Barcelona, Spain

Institució Catalana de Recerca i Estudis Avançats (ICREA), Barcelona, Spain

Pauline Schaap College of Life Sciences, University of Dundee, Dundee, UK

Christina Schilde College of Life Sciences, University of Dundee, Dundee, UK

Delphine Scornet CNRS, UMR 8227, Integrative Biology of Marine Models, Station Biologique de Roscoff, Sorbonne Universités, UPMC Univ Paris 06, Roscoff cedex, France

Arnau Sebé-Pedrós Institut de Biologia Evolutiva, CSIC-Universitat Pompeu Fabra, Barcelona, Spain

Susan C. Sharpe Centre for Comparative Genomics and Evolutionary Bioinformatics, Department of Biochemistry and Molecular Biology, Dalhousie University, Halifax, Nova Scotia, Canada

Deborah E. Shelton Department of Ecology and Evolutionary Biology, University of Arizona, Tucson, AZ, USA

Ricard V. Solé Santa Fe Institute, Santa Fe, NM, USA

ICREA-Complex Systems Lab, Institut de Biologia Evolutiva-UPF-CSIC, Barcelona, Spain

Cristian A. Solari Laboratorio de Biología Comparada de Protistas, IBBEA, CONICET-UBA, Universidad de Buenos Aires, Buenos Aires, Argentina

Mansi Srivastava Whitehead Institute for Biomedical Research, Cambridge, MA, USA

Martina Strittmatter CNRS, UMR 8227, Integrative Biology of Marine Models, Station Biologique de Roscoff, Sorbonne Universités, UPMC Univ Paris 06, Roscoff cedex, France

Hiroshi Suga Faculty of Life and Environmental Sciences, Prefectural University of Hiroshima, Shobara, Hiroshima, Japan

Sarah M. Tweedt Department of Paleobiology, National Museum of Natural History, Washington, DC, USA

Behavior, Ecology, Evolution, & Systematics, University of Maryland, College Park, 20742 MD, USA

Toshiki Uji CNRS, UMR 8227, Integrative Biology of Marine Models, Station Biologique de Roscoff, Sorbonne Universités, UPMC Univ Paris 06, Roscoff cedex, France

James W. Valentine Museum of Paleontology and Department of Integrative Biology, University of California, Berkeley, CA, USA

Part I
Multicellularity in the Tree of Life

Timing the Origins of Multicellular Eukaryotes Through Phylogenomics and Relaxed Molecular Clock Analyses

Susan C. Sharpe*, **Laura Eme***, **Matthew W. Brown and Andrew J. Roger**

Abstract Multicellularity has evolved many times during eukaryote evolution. Deciphering the evolutionary transitions to multicellularity requires a robust deep phylogeny of eukaryotes to clarify the relationships amongst multicellular groups and determine their closest unicellular relatives. Here we review progress in understanding the phylogenetic relationships amongst multicellular and unicellular eukaryotes, as well as estimates of the ages of multicellular groups based on relaxed molecular clock (RMC) analyses. In addition, we present an RMC analysis of a large phylogenomic dataset to estimate the divergence dates of select major eukaryotic multicellular groups. Our analyses (and other recent studies) tentatively suggest that multicellular eukaryotes such as Metazoa, Fungi and two of the major multicellular red algal taxa first emerged in the mid-Neoproterozoic, whereas the dictyostelids arose in the Paleozoic. We also hypothesize that the first multicellular organisms emerged within 300–600 Myr after the Last Eukaryotic Common Ancestor. The age of land plants is less clear and is highly dependent on methodology, the genes analyzed, and the nature of fossil constraints. In general, there is great variability in all these age estimates, and their credible intervals frequently span hundreds of millions of years. These estimates are highly sensitive to both the models and methods of RMC analysis, as well as the manner in which fossil calibrations are treated in these analyses. As paleontological investigations continue

---

* Susan C. Sharpe and Laura Eme have contributed equally.

A. J. Roger (✉) · L. Eme · M. W. Brown · S. C. Sharpe
Centre for Comparative Genomics and Evolutionary Bioinformatics,
Department of Biochemistry and Molecular Biology, Dalhousie University,
Halifax, Nova Scotia, Canada
e-mail: Andrew.Roger@DAL.CA

L. Eme
e-mail: laura.eme@dal.ca

M. W. Brown
Department of Biological Sciences, Mississippi State University, Mississippi, MS, USA
e-mail: protist@live.com

S. C. Sharpe
e-mail: susansharpe@dal.ca

I. Ruiz-Trillo, A. M. Nedelcu (eds.), *Evolutionary Transitions to Multicellular Life*,
Advances in Marine Genomics 2, DOI 10.1007/978-94-017-9642-2_1

to fill out the Proterozoic fossil record, genomic data is gathered from a greater diversity of eukaryotes and RMC methodology improves, we may converge on more precise estimates of the ages of multicellular eukaryotes that can be correlated with Earth's ancient geochemical record.

Keywords Multicellularity · Relaxed molecular clock · Origins of complex multicellularity · Bayesian framework · Microfossils · Eukaryote phylogeny

Eukaryote Phylogeny

As our knowledge of the deepest relationships between all extant eukaryotes improves, we are gaining a better understanding of how multicellularity has developed several times in distantly related eukaryote lineages. Early phylogenies of eukaryotes based on small subunit ribosomal RNA genes (SSU rDNA) showed the animals (Metazoa), plants, fungi and many protistan groups emerging from an unresolved radiation. This radiation was preceded by the divergence of a series of protistan lineages, with anaerobic 'amitochondriate' protists emerging as the earliest branches next to the prokaryotic outgroup (for example see Cavalier-Smith and Chao 1996). This supported the idea that eukaryotic evolution proceeded by a gradual increase in complexity, from simple cells without mitochondria, through more complex unicellular organisms, to complex multicellular organisms. This understanding of eukaryotic evolution appeared to be founded on methodological artifacts (Roger 1999; Roger and Hug 2006) and on an incorrect notion of the nature of so-called 'amitochondriate' eukaryotic lineages. It is now clear that the latter all possess homologs of mitochondria in the form of mitochondrion-related organelles (MRO) (Tsaousis et al. 2012). Furthermore, because their SSU rDNA sequences have evolved more rapidly than other eukaryotes, their deep-branching position in the eukaryote tree is likely a result of the infamous long branch attraction (LBA) artifact whereby they are artificially clustering with the long branches leading to the prokaryotic outgroup (Roger and Hug 2006). With more data from multiple genes and better analytical methods it has become clear that the apparently 'deep-branching' lineages on early SSU rDNA trees in fact emerge in multiple distinct places in the eukaryote tree (Roger 1999; Roger and Hug 2006; Keeling et al. 2005; Baldauf et al. 2000) as do multicellular groups that each show affinities to distinct ancestral unicellular protistan lineages (Burki et al. 2012).

Recent advances in our understanding of deep eukaryotic phylogeny have come from analyses of large sets of concatenated genes that provide more information on ancient nodes (Parfrey et al. 2010; Brown et al. 2012a; Burki et al. 2007). While the deep branching order of eukaryotic lineages is still controversial (e.g., see Zhao et al. 2012), a number of relatively 'stable' eukaryotic supergroups have been identified. One of the earliest recognized higher-level groupings of eukaryotes are animals, fungi and their unicellular relatives (Wainright et al. 1993) collectively known as the Opisthokonta. In unrooted phylogenies of eukaryotes, opisthokonts are adjacent to the Amoebozoa, a group that includes a wide variety of unicellular amoebae, anaerobic species previously thought to be basal eukaryotes, and the Eumycetozoa (social amoebae, myxogastrids and relatives). Opisthokonts, Amoebozoa, in addition to Breviata, Apusomonadida and a number of unicellular organisms of unclear phylogenetic affiliation (e.g., Ancyromonadida and *Mantamonas*: (Kim et al. 2006; Cavalier-Smith and Chao 2010; Brown et al. 2013)) form the major division Amorphea (Adl et al. 2012), roughly equivalent to "unikonts" of Stechmann and Cavalier-Smith (2003) and Richards and Cavalier-Smith (2005). Excavata is a possibly paraphyletic supergroup including many long branches that contain several of the lineages originally thought to be basal to other eukaryotes, but are known to be united by ultrastructural and molecular characteristics (Simpson 2003; Simpson et al. 2006; Hampl et al. 2009). The remainder of eukaryotic diversity is encompassed in a grouping referred to as Diaphoretickes that contains most of the photosynthetic lineages of eukaryotes (Adl et al. 2012). Within Diaphoretickes, the Archaeplastida encompasses eukaryotes with a primary plastid, including glaucophytes, rhodophytes (red algae) and green algae (containing land plants). Stramenopiles, Alveolata and Rhizaria form an assemblage known as SAR. These supergroups are generally well supported in phylogenomic analyses (Brown et al. 2013; Hampl et al. 2009; Burki et al. 2012), including the 159-gene analysis shown in Fig. 1. Finally, other lineages have more uncertain placements in the eukaryote tree; these include the Haptophytes, Cryptophytes, telonemids, and collodictyonids (Zhao et al. 2012; Burki et al. 2012).

It is uncertain how Diaphoretickes, Amorphea and Excavates are related, as this depends on the location of the root of the tree of all eukaryotes. Since the topology recovered in early rDNA analyses has been discredited as an artifact of LBA, several other possibilities for the root have emerged, and rare genomic changes have been used to define its location. For example, a fusion between dihydrofolate reductase (DHFR) and thymidylate synthase (TS) (Stechmann and Cavalier-Smith 2002) and the distribution of myosin II suggested that the root might lie between Opisthokonts + Amoebozoa (Amorphea) and all other eukaryotes (Richards and Cavalier-Smith 2005). More recently, it has become clear that the distribution of these features in various eukaryote lineages could not be simply explained by 'single gain' scenarios and that so-called 'rare' changes may have occurred more frequently than was once thought (Kim et al. 2006; Roger and Simpson 2009). Using a molecular phylogenetic approach, Derelle and Lang (2012) analysed a collection of mitochondrion-derived genes and found support for a root between Amorphea and all other eukaryotes. However, analyses of another dataset with better taxonomic sampling (He et al. 2014) instead supported a eukaryotic root between Excavata and all other eukaryotes. Other root positions have been recovered using alternative types of data. For example, an approach minimizing gene family duplication and loss apparently supported a root between Opisthokonts and other eukaryotes (Katz et al. 2012). In contrast, Cavalier-Smith has suggested that the root lies between the Euglenozoa and all other eukaryotes because the former lack a number of molecular and morphological features that are conserved in most eukaryotes (Cavalier-Smith 2012). Clearly, there is no consensus on the position of the eukaryote root and many candidate positions are plausible given the current evidence.

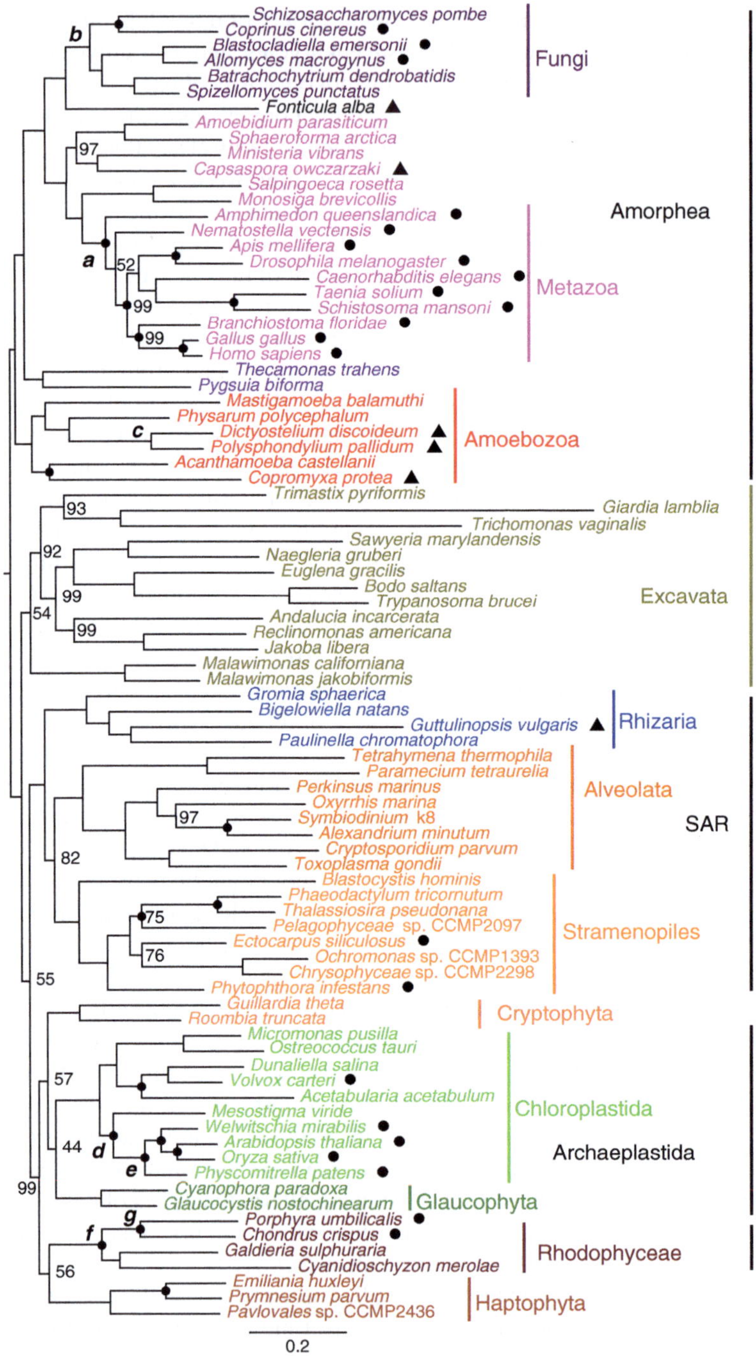

Fig. 1 Phylogenetic tree of eukaryotes based on a phylogenomic dataset. Additional taxa were added to a 159 gene dataset (Brown et al. 2013) to maximize available fossil calibrations (total of 85 taxa, 43099 sites). A maximum likelihood (ML) phylogenetic tree was obtained from 60 heuristic searches

Placing Multicellular Groups on the Tree of Eukaryotes

Multicellularity represents a spectrum of organizations ranging from simple groups of cells (colonies) to complex differentiated multicellular organisms (Bonner 1998). Colonial growth is common throughout eukaryotes and varies in complexity and the degree of intercellular integration. Examples include diatoms, which can interlock their silica shells to form chains, and the ciliates of the genus *Zoothamnium*, which have a sessile colonial form that can contract when exposed to stimuli (Bonner 1998). In this chapter, we do not consider simple colonial organisms; instead we focus on more complex multicellular taxa, including Metazoa, Fungi and land plants (Embryophyta) as well as less well-known forms such as red algae and dictyostelids. Many of these possess sophisticated mechanisms for communication and transport of nutrients between cells. It is generally accepted that most of these multicellular groups evolved independently from distinct unicellular ancestors (Knoll 2011; Brown and Silberman 2013).

The largest and/or most familiar multicellular organisms develop through cells dividing but not separating, followed by the differentiation of cell lineages into different specialized types. This kind of 'multicellularity-by-division' is employed in both simple organisms with a few cell types (e.g., the green algal genus *Volvox*; see Chapter "Volvocine Algae: From Simple to Complex Multicellularity") and in complex organisms with hundreds of cell types (e.g., Metazoa) (Rokas 2008). Alternatively, multicellular organisms known informally as 'cellular slime moulds' can develop through the aggregation of single cells (Brown and Silberman 2013; see

←

Fig. 1 employing RAxML version 7.2.6 (Stamatakis 2013) under the LG + Γ + F amino acid substitution model. Numbers indicate bootstrap support (BS) for splits estimated from 500 pseudoreplicates. Split with BS < 100 % are shown (all others are 100 %). Fossil calibrated nodes are indicated by a filled *circle*, and nodes representing the divergence of multicellular groups are indicated by letters (**a**: Metazoa; **b**: Fungi; **c**: dictyostelids; **d**: Streptophyta; **e**: Stomatophyta; **f**: Rhodophyceae; **g**: Bangiales/Florideophyceae). Taxa that are multicellular by division are indicated by a filled *circle*, while aggregative multicellular taxa are indicated by a *triangle*. Tree is shown arbitrarily rooted at the base of Amorphea. Four chains of Multiple Markov Chain Monte Carlo (MCMC) were run with Phylobayes 3.2 (Lartillot et al. 2009) for each of the CAT-GTR, CAT-POISSON and catfix C60-Poisson substitution models. The four MCMC chains did not converge, although the post-burnin consensus tree was identical to the ML tree except for an unresolved multifurcation at the base of Excavata. This phylogeny was used for relaxed clock molecular dating analyses with Phylobayes 3.2 (Lartillot et al. 2009). For all analyses, a birth-death tree prior was applied. Two chains were run until diagnostic stastistics indicated convergence or until estimated dates on nodes of interest for the two chains were < 10 % different. Fossil calibrations were taken from (Parfrey et al. 2011) with the following adjustments: the Gonyaulacales, spirotrich, Foraminifera and euglenid calibrations were removed due to insufficient gene coverage; the 'Ciliate' tetrahymenol-based calibration (Summons and Walter 1990), was removed as it is not ciliate specific (Takishita et al. 2012). The coccolithophorid calibration was adjusted to an uninformative maximum (3000 Ma) because of lack of genomic sampling from relevant haptophytes. The oldest known cestode (tapeworm) fossil (Dentzien-Dias et al. 2013) was added to calibrate platyhelminths. The minimum age (250 Ma) was taken from the youngest possible age of the fossil (lower boundary of the Permian) and the upper bound was taken from the limit on the next-oldest calibrated node (Bilateria)

Chapters "Social Amoebae and Their Genomes: On the Brink to True Multicellularity" and "The Evolution of Developmental Signalling in Dictyostelia from an Amoebozoan Stress Response"). In this form of simple multicellularity, organisms exist as single cells for part of their life cycle, but come together to form specialized structures for the dispersal of spores. While the same basic challenges of cell adhesion and communication need to be solved for both types of multicellularity, they appear to have evolved in different types of environments. Aggregative multicellularity has evolved in lineages that live predominantly in terrestrial environments, while multicellularity-by-division has apparently evolved in lineages that were originally aquatic (Bonner 1998).

Multicellular organisms face several challenges that are not relevant to strictly unicellular organisms, including cell-to-cell communication and adhesion. The various distinct lineages of multicellular organisms have solved these challenges in different ways. In order to make inferences about the transition to multicellularity we must understand both how these multicellular groups are related to each other and to unicellular organisms, and how the genes involved in their multicellularity have evolved (see Chapters "A Comparative Genomics Perspective on the Origin of Multicellularity and Early Animal Evolution", "The Evolution of Transcriptional Regulation in the Viridiplantae and Its Correlation with Morphological Complexity", "Independent Emergence of Complex Multicellularity in the Brown and Red Algae", "Social Amoebae and their Genomes: On the Brink to True Multicellularity"). For example, Dickinson and colleagues have argued that biochemical and morphological similarities between the epithelial tissues of one group of aggregative multicellular organisms (dictyostelids) and the Metazoa indicated their common amorphean ancestor was multicellular (Dickinson et al. 2012). However, Parfrey and Lahr conducted more detailed evolutionary bioinformatic analyses that show the proteins involved are paralogs that have evolved convergent functions (Parfrey and Lahr 2013). Dictyostelid and metazoan epithelia are therefore unlikely to be homologous (Parfrey and Lahr 2013). In general, there is currently little reason to suppose that any of the molecular mechanisms underpinning multicellularity in the various lineages we discuss are homologous.

Multicellularity-by-division is exhibited by multiple lineages within the tree of eukaryotes. Within Archaeplastida (Fig. 1) this form of multicellularity has evolved both within the red algae and green algae. Both Florideophyceae and Bangiales within the red algae are multicellular, and a fossil *Bangia*-like organism, *Bangiomorpha*, apparently provides the earliest fossil evidence of eukaryotic multicellularity (i.e., dated at 1198 +/− 24 Ma) (Butterfield 2000). Within the green algae, multicellularity has developed multiple times giving rise to a wide variety of forms, for example the nets of *Hydrodictyon* (Bonner 1998) or the ball-like volvocine algae (Herron et al. 2009). *Volvox* is a well-studied example of the latter, which is useful in a comparative genomics context because many close relatives with varying degrees of complexity exist (Herron 2009; Kirk 2005; and Chapter "Volvocine Algae: From Simple to Complex Multicellularity"). Finally, the Embryophyta (land plants) are the best-known and most conspicuous multicellular organisms within the Archaeplastida (see Chapter "The Evolution of Transcriptional Regulation in the Viridiplantae and Its Correlation with Morphological Complexity").

Animals (Metazoa) include the most complex of all multicellular organisms, and range from simple sponges to mammals with elaborate nervous systems. Besides animals, the supergroup Opisthokonta includes another charismatic multicellular-by-division group, the Fungi. Fungi often display filamentous 'hyphal' growth, and complex multicellularity with tissue differentiation occurs in the fruiting bodies of several lineages that evolved separately in the Basidiomycota and Ascomycota (Knoll 2011; Stajich et al. 2009). Other fungal lineages have apparently experienced reductive evolution to unicellularity (e.g., yeasts) (Stajich et al. 2009). Distantly-related eukaryotes, the Oomycetes (a stramenopile lineage) have converged on a similar lifestyle to Fungi and show filamentous growth (Beakes et al. 2012). Finally, brown algae, another stramenopile group include large kelps and the model organism *Ectocarpus* (Cock et al. 2010 and Chapter "Emergence of Ectocarpus as a Model System to Study the Evolution of Complex Multicellularity in the Brown Algae").

Among organisms showing the second main type of multicellularity (i.e., aggregative multicellularity), the best-studied are the dictyostelids (Chapters "Social Amoebae and Their Genomes: On the Brink to True Multicellularity" and "The Evolution of Developmental Signalling in Dictyostelia from an Amoebozoan Stress Response"). Many other organisms that display aggregation have been discovered over the years, but only recently have they been placed in the tree of eukaryotes using molecular data (Brown and Silberman 2013). The dictyostelids belong to the supergroup Amoebozoa, and are closely related to the plasmodial slime molds, which achieve a macroscopic form by growing into multinucleate plasmodia (Schilde and Schaap 2013). Other aggregative protists include *Copromyxa*, another amoebozoan (Brown et al. 2011); *Capsaspora owczarzaki*, a relative of Metazoa (Sebe-Pedros et al. 2013); and *Fonticula alba*, an amoeba that groups as a sister lineage to Fungi (Brown et al. 2009). Within the SAR clade, there are examples of aggregative multicellularity within each of the three main groups: *Guttulinopsis* within Rhizaria (Brown et al. 2012a), *Sorodiplophrys* in Stramenopiles (Dykstra and Olive 1975), and the ciliate *Sorogena* within Alveolates (Olive and Blanton 1980). Finally, the acrasid amoebae within Excavates are aggregative (Brown et al. 2012b), leaving Archeaplastida as the only supergroup without an aggregative multicellular representative.

Using Molecular Data to Date the Emergence of Lineages

In addition to being useful for the elucidation of phylogenetic relationships, molecular data can be used to estimate the date of divergence between organisms. From molecular sequence data it is possible to calculate an evolutionary distance, which is the product of rate of substitution (i.e., fixed nucleotide or amino acid changes) and time. Consequently, if the date of divergence for two taxa is known, an average rate of substitution can be inferred. If the rates of substitution are equal across all branches on a tree (i.e., a 'molecular clock' holds), this rate can then be used to convert branch lengths from the rest of the tree to dates (Zuckerkandl and Pauling 1965). Given that evolutionary rates often vary across subgroups of the tree of life (for example, see discussion of LBA above), methods using strict molecular clocks are not appropriate unless explicit tests are conducted to prove that their use is justified (Welch and Bromham 2005; Takezaki et al. 1995).

Once it became clear that a strict molecular clock does not generally hold (Langley and Fitch 1974), many efforts were made to 'relax' the molecular clock (Sanderson 1997; Sanderson 2002; Kishino et al. 2001; Thorne et al. 1998; Lepage et al. 2007). Strategies ranged from assigning subsections of the tree to evolve at different rates (Yoder and Yang 2000), to employing a complex Bayesian framework to model branch-specific rates, along with other parameters associated with the tree (Yang 2006). Substitution rates depend on biological processes such as mutation rate and generation time (Ho 2009), and so may be correlated on neighbouring branches. Whether the correlation is applicable between branches that cover a large phylogenetic distance, such as the entire breadth of eukaryotic diversity, is less clear. Correlated models (such as the lognormal (LogN) (Kishino et al. 2001)) and Cox-Ingersoll-Ross (CIR) (Lepage et al. 2007) used in our analyses below) draw the rate on a particular branch from a probability distribution of rates centered on the rate of the 'parent' branch. Uncorrelated models (for example the uncorrelated gamma (UGam) model in our analyses) draw rates for each branch from one global probability distribution, not taking into consideration the rates on adjacent branches (Drummond et al. 2006).

Another difficulty with molecular clock analysis concerns the assignment of fossil-based time calibrations. There is only indirect evidence for the date of divergence of two species: groups under study must have existed for some time before the right conditions occurred for preservation in the geological record. Moreover, even if an organism fossilizes, estimates of its geological age also have associated uncertainty. In addition, identification of fossils is often controversial (especially in the case of simple 'soft-bodied' multicellular or unicellular organisms), as they often lack characteristic features that would allow them to be definitively assigned to an extant group. Furthermore, fossil assignments depend on whether a fossil is a member of the 'crown group' (i.e., it descends from the most recent common ancestor (MRCA) of all extant species in the group), or instead represent a 'stem group' lineage, (i.e., it diverged prior to the MRCA of extant members of the group, and thus does not possess any currently living descendant). Unfortunately, it is sometimes difficult to determine whether a given fossil corresponds to an extinct crown group lineage or is a stem group organism. Clearly, this distinction is extremely important as each of these types of fossils conveys different information about the ages of nodes on molecular phylogenies of extant organisms. For example, stem lineage fossils may be older or younger than the node defining the MRCA of the crown group on phylogenies, whereas crown group fossils must be younger than this node (and can therefore provide a lower bound on its age).

In addition to the difficulties associated with assigning fossil constraints to specific nodes on phylogenetic trees, another problem relates to the fact that crown group fossil dates must always be treated as minimum possible ages (as the MRCA of the group must predate the appearance of the first fossil). While a few groups of protists have a continuous fossil record that can be fairly reliably translated into minimum and maximum dates (Berney and Pawlowski 2006), most organisms on the tree of life are only sporadically fossilized, and so a fossil represents only a minimum restriction on the age of that group, which is not particularly informative for subsequent analyses.

Finally, there has been much debate over how fossil dates should be treated during RMC analyses, notably because of the uncertainty associated with the dating of the rocks in which the fossils are found (for a detailed review on this question, see Parham et al. 2012 and references therein). There are several ways of applying fossil constraints. Treating calibrations as 'hard bounds' implies that the estimated age of a constrained node will necessarily fall within the fixed time intervals specified by the paleontological evidence (Kishino et al. 2001). In contrast, the 'soft bound' approach uses a probabilistic treatment of fossil age data, and therefore can accommodate potential error in calibrations. The simplest of these is the use of a uniform prior probability distribution for the node falling within the bounds defined by the fossil age data (Yang and Rannala 2006), leaving some probability of the node falling outside the bounds (e.g., a smoothly decreasing probability distribution on each side of the bounds, as implemented in Phylobayes (Lartillot et al. 2009) and used for our analyses below). Alternatively, fossil evidence can also be represented as parametric probability distributions such as lognormal or gamma, with a 'mix' of a hard lower bound and soft upper bound (e.g., as implemented in the software BEAST (Drummond et al. 2006)). The manner in which fossil dates are treated by the software that is estimating dates can have a large impact on the final results (Inoue et al. 2010).

Translating fossil and other geological evidence to a range or distribution of dates is often subjective and has attracted much criticism (Shaul and Graur 2002). Recently, efforts have focused on formalizing strategies for interpreting fossil data (Parham et al. 2012) and selecting age distributions (Nowak 2013); however, these have yet to be widely applied.

Dating the Emergence of Multicellular Eukaryotic Lineages with Relaxed Molecular Clocks

In the last few decades, many attempts have been made to date deep divergences within the tree of life, including many of the multicellular eukaryote lineages. Here, we review the most recent attempts that employ relaxed molecular clock methods. In addition, we present an analysis of the age of various multicellular groups based on a phylogenomic dataset with molecular data and fossil calibrations from representatives of the full breadth of eukaryotic diversity. We focus specifically on multicellular groups that are well represented in our own analyses (Fig. 1 and below) including the Metazoa, Fungi, embryophytes (land plants), the red algae and dictyostelids.

For our estimation of the ages of multicellular groups (nodes indicated by letters in Fig. 1), we used the Bayesian implementation of RMC models in Phylobayes

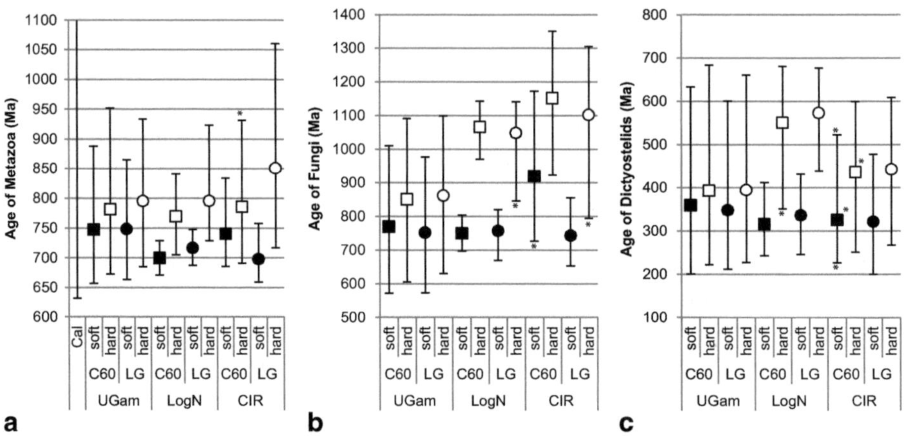

Fig. 2 Estimates of the age of the most recent common ancestor of Metazoa (**a**), Fungi (**b**) and Dictyostelids (**c**). Results are shown for the combination of three different RMC models (UGam, LogN and CIR) with two substitution models (C60, *squares*; LG, *circles*), and with soft (*filled shapes*) or hard (*open shapes*) bounds. *Error bars* indicate 95 % credible intervals. For each node with calibration, the *error bar above* 'Cal' shows the age range used to calibrate the node. Where *upper* bound is not shown, it was equal to 3000 Ma. Asterisks (*) indicate estimates where the two chains showed more than a 10 % difference in either the mean, the *upper* 95 % limit or the *lower* 95 % limit

(Lartillot et al. 2009) to analyze a large phylogenomic dataset (159 proteins, 85 taxa). The phylogenetic tree was obtained using maximum likelihood analysis and was calibrated by 19 fossil calibrations from diverse eukaryote groups, the majority of them taken from Parfrey et al. (2011) (for details, see Fig. 1). Our analyses included three different RMC models (the uncorrelated UGam model, and the correlated LogN and CIR models), two substitution matrices (the site-heterogeneous empirical profile mixture model C60-Poisson (C60) (Le et al. 2008) and the more classical site-homogeneous LG substitution matrix (Le and Gascuel 2008)), and two ways of treating fossil calibrations ("soft" and "hard" bounds).

A few general trends can be noted in our results. First, soft bounds yielded younger date estimates than hard bounds when other parameters (i.e., relaxed clock model, substitution matrix) were identical (Figs. 2 and 3). As the use of hard bounds is hardly justified given the uncertainties discussed above, we will mainly focus on results obtained with soft bounds. Second, the uncorrelated UGam relaxed molecular clock model gave larger credible intervals (so that the range for soft and hard bounds largely overlapped), with little change based on the substitution model. Third, analyses with the LogN and CIR rate evolution models showed more variation, both in terms of estimated ages of nodes, and of size of the confidence interval. This variation was often only seen when both the type of calibration used (hard or soft bounds) and the substitution matrix (C60 or LG) was changed, underlying the complex interaction between the various features of RMC analyses in influencing the

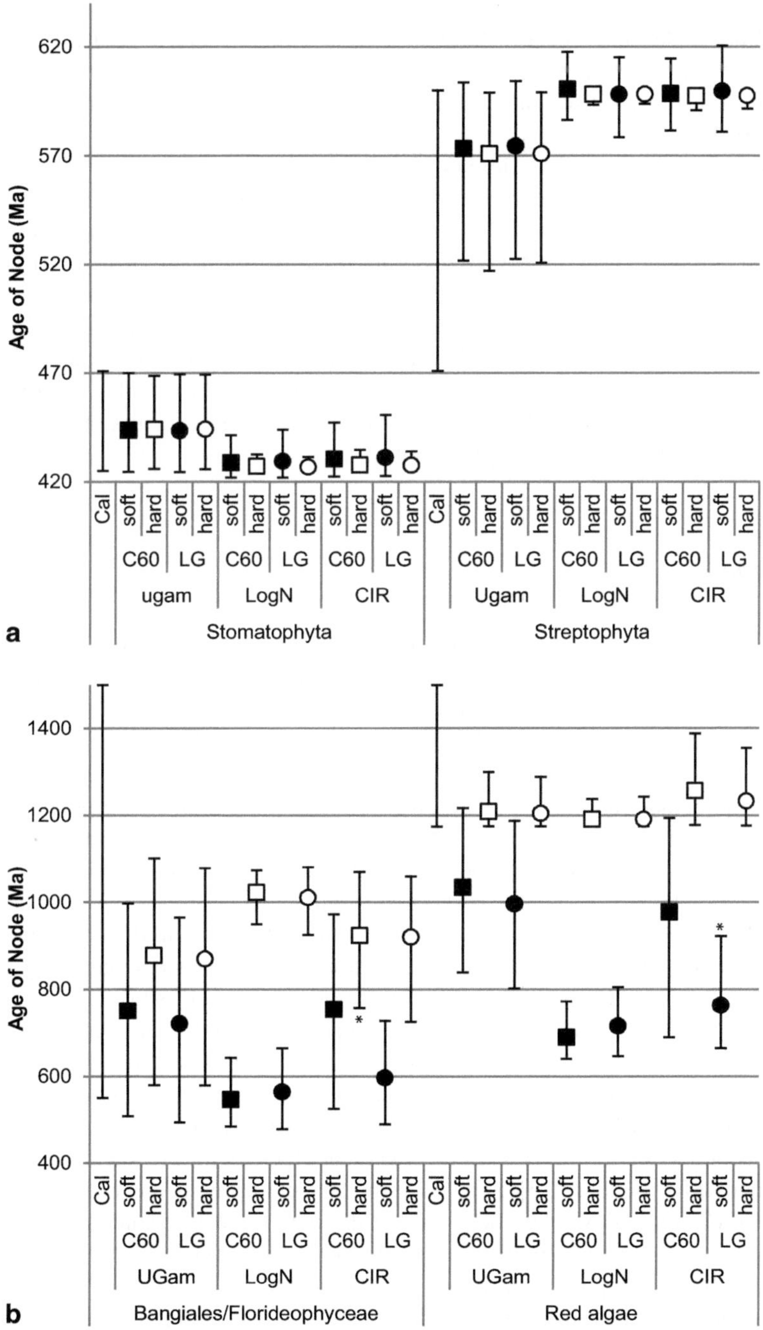

Fig. 3 Estimates of the age of the most recent common ancestor of Stomatophyta and Stretophyta (**a**), and Bangiales/Florideophyceae and Rhodophyceae (**b**). See Fig. 2 for details

age estimates obtained. Fourth, in general, the choice of the substitution matrix often did not make much of a difference (Fig. 2, compare black squares for a given clock model), although more substitution matrix-based changes were seen in CIR analyses. For further discussion on the impact of these various parameters, see Eme et al. (2014).

Dating these ancient evolutionary events requires the specification of the position of the root of the eukaryotic tree. For this reason, we tested the impact of three alternative positions for the root: (i) between Amorphea and the remainder of eukaryotes (i.e., the classical "unikont-bikont" rooting (Derelle and Lang 2012); (ii) between the Obazoa (Opisthokonta + apusomonads + breviates) and all other taxa (similar to Katz et al. 2012); and (iii) between Excavata and all other eukaryotes (He et al. 2014). Interestingly, these three alternative placements had little impact on the estimated ages of the various groups considered here (not shown, also see Eme et al. 2014). Therefore we only discuss our RMC age estimates for major multicellular groups obtained with the Amorphea root.

Metazoa

There have been a large number of molecular clock-based studies attempting to date groups within Metazoa, in particular the Bilateria. These have often sought to illuminate the Cambrian explosion, an apparently sudden emergence of many divergent bilaterian phyla in the fossil record at the start of the Cambrian Period (530 Ma) (Levinton 2008). While much has been made about this change in the fossil record, molecular clock analyses often date the first Bilaterian divergence as much older (for example, Hedges et al. 2004; Erwin et al. 2011). Basal metazoans such as cnidarians, placozoans and sponges existed for many years previous to this, as sponge biomarkers (Love et al. 2009) and fossil evidence for sponges (Cohen et al. 2009) dates to 632 Ma. The transition from unicellularity to multicellularity would have happened before the last common ancestor of all extant Metazoans (Fig. 1, node a) and we focus on the estimated age of this node in previous studies.

Hedges et al. (2004) estimated the Porifera-Animalia divergence at 1,351 Ma $\pm$ 120 (1116–1586), more than twice as old as the first fossil evidence for Metazoa. Their 17–19 protein gene dataset (depending on the method applied) was calibrated using the bird-mammal divergence as a primary calibration, and three additional secondary calibrations; various molecular clock methods were used, including Bayesian RMC modeling. The use of secondary calibrations has been vigourously criticized (Graur and Martin 2004) for failing to add new information to the dating analyses, propagating earlier biases from the analyses that generated the calibrations, and for failing to properly 'carry forward' the uncertainty associated with them.

A much younger estimate was obtained with analyses using nine primary invertebrate fossil calibrations and a dataset of seven genes (Peterson and Butterfield 2005). These molecular clock analyses (using the software r8s, see Sanderson 2003) yielded different estimates depending on the phylogenetic method used to construct the tree.

The minimum evolution tree yielded an estimate of 664 Ma whereas the ML tree gave 867 Ma. Later re-analyses of this dataset (Roger and Hug 2006) found substantial variation in the estimates (approximately 750–1300 Ma) and extremely broad confidence intervals depending on the substitution models, relaxed molecular clock method/models and the manner in which fossil constraints were applied.

Using 22 Phanerozoic minimum calibrations with only four maximum calibrations based on fossils from a wide diversity of eukaryotic groups, Berney and Pawlowski (2006) performed Bayesian RMC analyses of SSU rDNA sequences to date the divergence of major eukaryotic groups. They estimated that the primary split within Metazoa occurred 812 Ma (671–985).

In contrast to Berney and Pawlowski's analyses of one gene with many fossil calibrations, Lartillot et al. (2009) analyzed a many-gene dataset (68 genes) from 52 Holozoans (Metazoa plus unicellular opisthokont relatives) using only two fossil calibrations (within the Metazoa) as soft bounds. Dates were estimated using the PhyloBayes 3 software, with a CAT-GTR substitution model and lognormal autocorrelated RMC model. An age of $\sim$ 800 Ma (700–900) was obtained for the MRCA of Metazoa.

With the goal of estimating the age of eukaryotes and their supergroups, Parfrey et al. (2011) analyzed a 15-gene dataset with rich taxon sampling and 22 fossil calibrations associated with different nodes throughout eukaryotic diversity. Using the uncorrelated lognormal RMC model in BEAST (Drummond et al. 2006), they found the age of the last common ancestor of all Metazoans to be 780 Ma (782–820). The estimate was about 60 Myr younger when the seven Proterozoic fossil calibrations (including *Bangiomorpha*) were excluded.

In a taxon-rich analysis focused on Metazoan evolution, Erwin et al. (2011) analyzed a 7-gene dataset with 118 metazoan taxa and 8 other taxa from Amorphea, with 24 fossil calibrations. Their Phylobayes analyses estimated the common ancestral Metazoan node to be between 747 (690–825) and 1093 (962–1260) Ma, depending on tree topology, RMC model (CIR or UGam), prior probability on the age of the root and the degree of soft bound relaxation (i.e., ranging from 5–50 % prior probability of the node falling outside of the interval defined by the fossil dates).

While the geological evidence supports a bound on the youngest possible age for animals at 632 Ma, there is lack of clear evidence for a plausible upper bound. Therefore, for our analyses, we employed an uninformative maximum age for this node at 3000 Ma, following Parfrey and colleagues (Parfrey et al. 2011). All of our analyses placed the last common ancestor of Metazoa within 220 Myr of the lower calibration, with age estimates between 698 Ma and 851 Ma, depending on the parameters used (Fig. 2a). These estimates fall within the range of age estimates of most of the other recent studies discussed above; these span roughly 650–850 Ma (e.g., Peterson and Butterfield 2005; Berney and Pawlowski 2006; Lartillot et al. 2009; Parfrey et al. 2011). The exceptions are the estimates obtained by Erwin et al. (2011). They recovered slightly older estimates (i.e., $\sim$ 750–1100 Ma) that appeared to be fairly strongly influenced by the root age prior placed on the Holozoa-Fungi split.

Fungi

Complex multicellularity with tissue differentiation in fruiting bodies has evolved independently within two subgroups of the Fungi: the ascomycete lineage Pezizomycotina and the basidiomycete lineage Agaricomycotina (Stajich et al. 2009). Here we will focus specifically on the origin of simple filamentous 'hyphal' growth, which was likely a feature of the common ancestor of most extant Fungi (with the possible exception of Cryptomycota and Microsporidia) (Stajich et al. 2009). Although the deepest branching order of the fungal tree is still controversial, we will provisionally consider the Chytridiomycota and Blastocladiomycota to form a clade that split basally from all other Fungi (Ebersberger et al. 2012; James et al. 2013, see Fig. 1), excluding the Cryptomycota and Microsporidia.

There have been few efforts to apply molecular clock methods to the Fungi as a whole, or even to specific groups of fungi, largely because of the lack of fungal fossils, and the difficulty of classifying the fossils that do exist. Using the Langley-Fitch method from the r8s software, Taylor and Berbee (2006) showed that age of the node uniting *Rhizopus* (Mucoromycotina) and the Ascomycota + Basidiomycota group was highly uncertain (estimates ranged from 435–1979 Ma depending on the calibration used) (Taylor and Berbee 2006). When three calibrations (one each within Fungi, Metazoa and plants) were used, an age of 792 Ma was estimated. More recently, they used Bayesian analyses implemented in the BEAST software program to analyze a 50-gene dataset with three primary calibrations and a prior on the age of opisthokonts (Berbee and Taylor 2010). These analyses led to large confidence intervals for the majority of nodes, including the *Rhizopus*/Ascomycota + Basidiomycota split which they dated at ~750 Ma. Global eukaryotic analyses dated the deepest split in extant Fungi (i.e., basal Chytridiomycota versus all other Fungi, a deeper divergence than the one considered by Berbee and Taylor (2010)) at 798 Ma (634–1003) (Berney and Pawlowski 2006) and 1070 Ma (980–1220) (Parfrey et al. 2011).

In our analyses, the CIR model showed the greatest range of age estimates for the split of the Chytridiomycota + Blastocladiomycota group from all other fungi (Fig. 1, node b). It yielded both the youngest age estimate (744 Ma, employing LG + soft bounds), and the oldest (1152 Ma, employing C60 + hard bounds, Fig. 2b). This illustrates the difficulty of separating the synergistic effects of different parameters in molecular clock analysis. CIR also recovered wide credible intervals on the age, whereas the LogN model had the smallest (and non-overlapping) intervals, with soft bound estimates significantly older than the hard bound estimates. Soft bound estimates for the age of Fungi were similar (~760 Ma) over all combinations of substitution and RMC models, with the exception of the CIR + C60 combination, which was notably older (920 Ma).

Our estimates are consistent with those of Berney and Pawlowski (2006). Surprisingly, Parfrey et al. (2011) estimated an older age (980–1,220 Ma), despite the fact most of their calibrations were identical to ours. The paucity of fossil constraints within Fungi (i.e., only one) is likely largely responsible for this discrepancy; because this node is not strongly constrained by ancestral or descendant nodes, it is more susceptible to age estimate variation depending on the dataset and methodologies used.

Dictyostelids

Using either a six-protein dataset or an rDNA tree, Fiz-Palacios et al. (2013) estimated the divergence of dictyostelids using a combination of four fossil calibrations from opisthokonts and land plants. Variation in estimates was found between those obtained by the Bayesian package BEAST (employing the uncorrelated lognormal RMC model) and MCMCtree (using independent rates RMC model), with most estimates centering between 570–730 Ma. The one lower estimate obtained by Fiz-Palacios and colleagues (341 Ma (247–699)) was obtained when only the two opisthokont fossils were used and among site rate variation (ASRV) was modeled using a gamma distribution (modeling ASRV is essential for model realism). Their analyses showed that trees calibrated with only the two plant fossils failed to return appropriate dates for the opisthokont fossils using fossil cross-validation (Near and Sanderson 2004). Land plants are phylogenetically distant from Amoebozoa, and so in absence of better sampling of fossil calibrations across the tree, may heavily distort the results.

Our estimates for the age of the last common ancestor of dictyostelids were around 330 Ma (Fig. 2c) and remained similar across most clock models and substitution matrices (with soft bounds). However, even when restricting attention to soft bounds analysis, confidence intervals tended to be very wide, spanning from 160–430 Myr. In contrast, age estimates obtained by Fiz-Palacios et al. (2013) were much greater ($\sim$ 530–730 Ma), except in the single case described above where their estimate was closer to ours ($\sim$ 340 Ma).

Embryophytes, Stomatophytes and Streptophytes

Embryophytes, the 'land plants', emerge from within the Streptophyta clade that also includes a number of green algal lineages with unicellular, colonial/filamentous and multicellular forms (Leliaert et al. 2012). Multicellularity seems to have evolved multiple times within the streptophytes, and reversion to simpler unicellular forms has probably occurred as well (Becker and Marin 2009; Leliaert et al. 2012). It is therefore difficult to pinpoint the origins of multicellularity within this group, although a key aspect of 'complex multicellularity' associated with embryophytes is the development of a diploid zygote into a multicellular diploid sporophyte (Becker and Marin 2009). Here, we will focus on the crown embryophyte node (uniting liverworts with all other land plants), which post-dates the origin of embryophyte-type multicellularity and the crown streptophyte node which predates it. Note that many recent molecular clock studies (including ours below) have not included liverworts (the deepest branching embryophytes), so the crown 'land plant' node referred to in these studies is the somewhat younger common ancestor of the Stomatophyta (Fig. 1, node e) that includes mosses and all other land-plants (Clarke et al. 2011). Fossil constraints that are commonly included in molecular clock analyses correspond to the first appearance of crown stomatophyte and embryophyte spores at ~ 420 Ma and ~ 449 Ma and to the oldest unassignable (i.e., possibly stem or crown) embryophyte spores at ~ 472 Ma (see Clarke et al. 2011; Magallón et al. 2013, and references therein).

Of the global eukaryotic analyses, the study by Hedges et al. (2004) recovered the oldest estimate for the age of the MRCA of stomatophytes: 707 Ma (515–899) (no streptophyte estimate was given). In contrast, Berney and Pawlowski's (2006) estimate for the embryophytes node was much younger 510 Ma (431–645), with stomatophytes dated at ~ 442 Ma (again, no crown streptophyte age was available). Parfrey et al. (2011) estimated a similar age for stomatophytes (~ 460 Ma with very small credible intervals) with streptophytes dated at ~ 745 Ma (~ 625–850)

Although there have been many molecular clock analyses in the last two decades that focus specifically on the streptophytes, here we discuss only two of the most recent studies. Clark et al. (2011) analyzed seven plastid genes and 17 fossil calibrations within the embryophytes using a Bayesian RMC method (MCMCtree, independent rates model). They examined fossil calibration consistency using cross-validation (Near et al. 2005) as well as extensively explored the impact of uniform versus Cauchy distribution priors on fossil calibrations. Using a uniform prior with hard lower bound and a soft upper bound for all constraints, their 'best' estimates for stomatophyte and embryophyte crown nodes were 632 (548–750) and 670 Ma (568–815), respectively. However, they found that the estimated ages of these deepest nodes were greatly influenced by the upper calibration bound, as well as the nature of the prior distribution (i.e., uniform versus Cauchy) on the fossil bounds. For the above estimates they used a uniform prior with an extremely high upper bound of 1042 Ma for all three nodes (with the lower bound defined by the first fossil evidence for each group).

Magallón et al. (2013) estimated the age of crown stomatophytes as ~ 458 Ma (446–469) and the ancestral embryophyte node at 475 Ma (471–480) using the uncorrelated lognormal RMC model implemented in BEAST. While this dataset was comprehensive in terms of taxon sampling (80 taxa) and fossil calibrations (26), there used only five plastid genes. In contrast to the analyses by Clark and colleagues, Magallón and colleagues set a strong narrow lognormal prior distribution on the age of the MRCA of embyrophytes centred at 472 Ma (Magallón et al. 2013) which likely explains the narrow credible region from the posterior distribution they obtained for this node (spanning only 9 Myr).

Our taxon sampling does not include the liverworts, the earliest branching lineage within Embryophyta (Magallon 2013). We therefore focus on the age of the common ancestor of living stomatophytes (the split between mosses and all other land plants; Fig. 1 node e) as a lower bound on the age of embryophytes. We can estimate the upper bound on the age of complex embryophyte-type multicellularity (i.e., diploid multicellular sporophytes) by estimating the age of the basal streptophyte split (*Mesostigma* versus all other land plants; Fig. 1, node d). This divergence (i.e., Streptophyta, the upper bracket) was calibrated between 471 Ma (the oldest fossilized embryophyte spores (Rubinstein et al. 2010)), and 600 Ma (following Parfrey and colleagues (2011)). All of our analyses estimated the age of this node to be close to the upper bound. In fact, the youngest estimates were obtained under the UGam model (~ 570 Ma) and were associated with very wide confidence intervals spanning more than half the calibration range. In contrast, with LogN and CIR models estimates were older, approaching the upper bound of the calibration, and in one case, just outside of it (LogN + C60, with soft bounds). The lower bracket (i.e., last common ancestor of stomatophytes (Fig. 1, node e)) shows the opposite trend: CIR and LogN estimates are grouped near the lower calibration bound for that node (425 Ma).

Our estimate of ~430–450 Ma for Stomatophyta is in agreement with results from Berney and Pawlowski (2006) (~442 Ma), Parfrey et al. (2011) (~460 Ma), and Magallón et al. (2013) (~458 Ma). This congruence is interesting given that these analyses used fairly different age constraints on nodes from this region of the tree. In contrast, Clarke et al. (2011) estimated this node to be much older (~630 Ma). However, the results obtained by these authors seem to have been heavily influenced by the root prior, which usually reflects that the data itself contains little information (Felsenstein 2004; Rannala 2002; Zwickl and Holder 2004).

Red Algae

The first evidence of florideophyte red algae comes from fossils from the Doushantuo Formation dated at 550–600 Ma (Xiao et al. 2004). These fossils have been used by a number of molecular clock studies to provide a lower bound on the split of two of the multicellular red algal lineages: the Bangiales and the Florideophyceae (Yoon et al. 2004; Berney and Pawlowski 2006; Parfrey et al. 2011). The second putative red algal fossil often used is the *Bangia*-like *Bangiomorpha pubescens* from the Hunting Formation dated at 1198 +/− 24 Ma (Butterfield 2000). As Bangiales and Florideophyceae are a clade, and a number of red algal lineages (some of which are multicellular) branch off before them within the Rhodophyceae (Yoon et al. 2004; Saunders and Hommersand 2004), this fossil is often used as a bound indicating the first the appearance of multicellular red algae (Yoon et al. 2004; Berney and Pawlowski 2006; Parfrey et al. 2011), although there has been some argument about its attribution (e.g. see Cavalier-Smith 2002).

In attempting to date the origin of major eukaryotic photosynthetic groups, Yoon et al. (2004) used two rate-smoothing RMC approaches in r8s (Sanderson 2003) to analyze five plastid genes from a large diversity of photosynthetic eukaryote lineages and seven fossil age constraints (two within the red algae, four within the embryophytes and a maximum age on the root). Because their methods assumed hard bound constraints (and their bounds were narrowly defined around fossil dates), they could only obtain an unconstrained age estimate for the basal red algal split between the bangiophyte + florideophyte clade and the unicellular Cyanidiales. Their estimate for this node was 1,370 Ma (1,350–1,416). This age was much older than comparable estimates by Berney and Pawlowski (2006)—they estimated an age of 780 Ma (680–950) for this split, while the MRCA of the Floridiophyceae/Bangiales divergence was estimated to be 740 Ma (600–929). Note that Berney and Pawlowski did not include the *Bangiomorpha* calibration in their analyses. Parfrey and colleagues (2011) obtained widely varying estimates for the basal red algal split depending on whether or not they included the *Bangiomorpha* and other Proterozoic fossil constraints. Without Proterozoic fossils, they obtained a 95 % credible age range for crown red algae of 625–959 Ma, whereas, with them, they estimated the range to be 1180–1285 Ma. Their estimate for the florideophyte/Bangiales split was ~765 Ma (~630–915) when Proterozoic fossil constraints were included, and ~620 Ma (~495–700) when excluded.

We estimated the age of the Bangiales/Florideophyceae common ancestor as well as the basal divergence between the unicellular Cyanidiales (*Cyanidioschyzon* and *Galdieria*) and the Bangiales/Flordeophyceae clade. The Bangiales/Florideophyceae divergence (Fig. 1, node g) was calibrated by the first appearance of florideophytes in the fossil record (Xiao et al. 2004) with a younger age bound of 550 Ma and an uninformative maximum bound at 3000 Ma. All estimates of the age of this node were within 210 Myr of the lower bound (with soft bounds). It is worth noting that the estimates for this node were likely considerably affected by the estimated age of its immediate ancestral node on our tree. The latter represents the basal divergence in the red algae (Fig. 1, node f) and was calibrated using the fossil *Bangiomorpha pubescens* (Butterfield 2000) as a lower bound at 1174 Ma. (Note that while these fossils resemble some Bangiales (Butterfield 2000), here we have treated it as if it were a stem multicellular red algal lineage, following Parfrey et al. 2011). The effect of this fossil is clear when comparing hard and soft bounds: all the hard bound estimates were within 85 Myr of the minimum calibration, with small confidence intervals constrained by this calibration. However, the soft bound analyses yielded dramatically younger estimates: LogN + C60 returned an age of 690 Ma, which is 484 Myr younger that the minimum bound set by the fossil; the *Bangiomorpha* fossils seem at odds with other calibrations in our analyses, and seem to be much older than any of the dates estimated from molecular clock data. This discrepancy can, in theory, be explained in a number of ways. For example, either the identification of *Bangiomorpha* fossils as red algae, or the estimated age of the rocks in which they are found, could be in error (although the latter is thought to be unlikely, see discussion in (Parfrey et al. 2011; Knoll et al. 2006). Alternatively, the changes in the rates of evolution within some lineages, including this one, might be poorly captured by the currently available RMC models.

While our estimates of the most recent common ancestor of Bangiales/Florideophyceae red algae are quite variable (~550–720 Ma, with credible intervals spanning up to 500 Myr), they are relatively consistent with those obtained by Berney and Pawlowski (2006) (~700 Ma), and by Parfrey et al. (2011) (~620 Ma) when the latter excluded all Proterozoic fossil constraints (including *Bangiomorpha*).

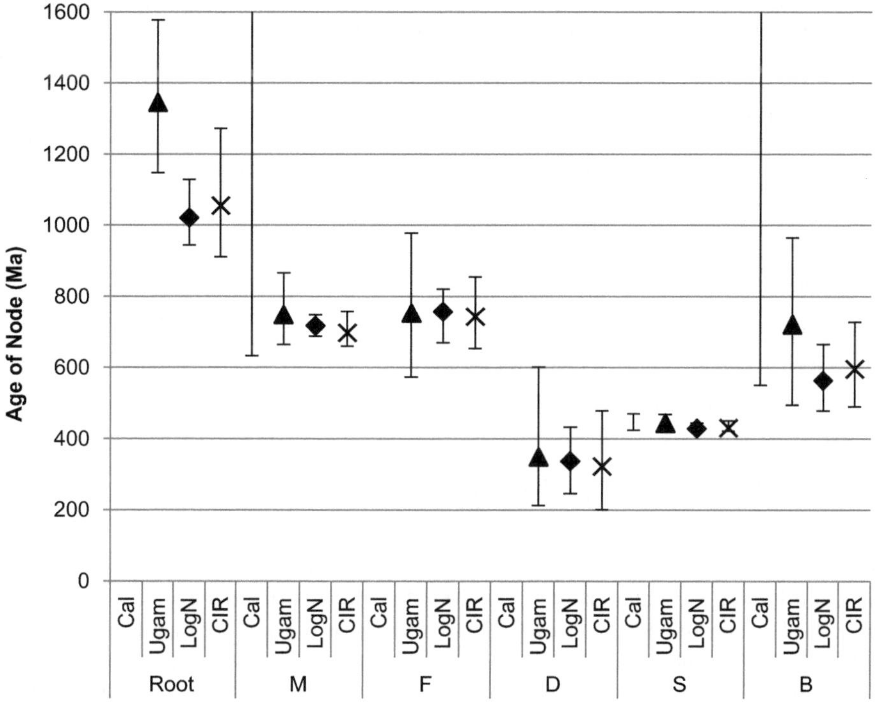

Fig. 4 Age of the LECA compared to multicellular eukaryotic groups. All estimates were calculated with the LG substitution model and soft bounds. Results are shown for the UGam (*triangles*), LogN (*diamonds*), and CIR (*crosses*) relaxed molecular clock models. *Error bars* indicate 95 % credible intervals. For each nodes with calibration, the *error bar above* 'Cal' shows the age range used to calibrate the node. Where *upper* bound is not shown, it was equal to 3000 Ma. *M* Metazoa, *F* Fungi, *D* Dictyostelids, *S* Stomatophyta, *B* Bangiales/Florideophyceae

However, Parfrey and colleagues' estimates for this node were much older ($\sim$820 Ma) when Proterozoic fossils were included. The basal split between Cyanidiales and the Bangiales/Florideophyceae clade showed much greater differences. Our estimates vary greatly (by up to 500 Myr, with credible intervals even more extreme) depending on the model and style of constraints (Fig. 3b) and similar degrees of variation in age estimates for this node were found by Parfrey et al. (2011) depending on the inclusion/exclusion of Proterozoic fossils constraints. Berney and Pawlowski estimated younger ages (680–950 Ma), which were similar to the lower ranges we obtained (e.g., 639–771 Ma for LogN + C60) and Parfrey and colleagues (625–959 Ma without Proterozoic fossils). The sensitivity of estimates associated with this and the previously discussed node seems to be largely correlated with how 'strongly' the *Bangiomorpha* dates are imposed on it as a constraint. The weaker the constraint (soft bounds in our analysis or removal of the calibration by Parfrey and colleagues), the younger the age estimate for this node.

Table 1 Limits on the age of emergence of multicellularity in different eukaryotic groups. Lower Limit represents the date of the most recent common ancestor of all sampled members of the multicellular group. The upper limit column refers to one node deeper on the rooted tree, i.e., the date of divergence of the group from its most closely related unicellular organism (in parentheses is the node letter(s) from Fig. 1, where there are two, the second letter represents the upper limit). 95 % credible intervals are indicated within the parentheses. Analyses shown were performed using soft bounds, an LG substitution matrix and the UGam relaxed clock model

	Lower limit (Ma)	Upper limit (Ma)
Metazoa (a)	748 (663–865)	872 (758–1024)
Fungi (b)	752 (573–977)	927 (738–1145)
Dictyostelids (c)	348 (212–601)	793 (542–1062)
Embryophytes (e, d)	574 (522–604)	852 (654–1067)
Bangiales/Florideophyceae (g, f)	721 (494–964)	996 (801–1186)

Relative Ages of Multicellular Groups

According to our analysis, it appears that the Metazoa and Fungi are the oldest groups, with similar age estimates around 700–800 Ma when restricted to soft bound analyses (Fig. 4). The credible intervals on the age of Metazoa are smaller than Fungi presumably because this node, as well as many of its descendant nodes, is well constrained by fossil data. It is worth noticing the relatively small difference (~300–600 Myr, depending on the RMC model used) between these two nodes and the age estimates for LECA (~1,000 and ~1,350 Ma). This suggests that the first multicellular organisms emerged relatively rapidly after LECA. The Bangiales/Florideophyceae split is estimated to be the next oldest group in soft bound analyses at ~550–720 Ma (hard bound analyses yielded substantially older ages because of the fossil *Bangiomorpha*, as discussed above). Stomatophyte age estimates are somewhat younger (430–450 Ma) and overlap with those for the dictyostelids, which appear to be the youngest multicellular group we considered here (~330 Ma). It should be noted that all of the age estimates (with the exception of the Stomatophyta) are associated with large 95 % credible intervals spanning hundreds of millions of years. Thus, despite the large size of our data set, considerable uncertainty in the age estimates persists.

It is important to keep in mind that the dates shown on Fig. 4 represent age estimates for the most recent common ancestor of each multicellular group, which, without doubt, was not the first multicellular organism in this lineage. The time of first emergence of multicellularity for the lineages we have discussed can be bracketed by considering not only the age of the most recent common ancestor of the group, but also the date of divergence of this lineage from its closest unicellular relative (as we discussed for embryophytes and red algae in the previous section). These bracketed age estimate ranges are given in Table 1; only UGam results are shown as they encompass the range of estimates given by the other two models. Note that the precision of these 'brackets' on the age of the emergence of multicellularity depends critically on the taxonomic sampling of the groups. Sparse sampling of the closest unicellular relatives of groups, or the presence of long stem lineages, will artificially inflate the age ranges.

Conclusions

Here, we highlighted some of the progress and difficulties in estimating the age of multicellular eukaryotic lineages using sophisticated 'relaxed molecular clock' (RMC) methods. Such estimates are critical if we are to understand how these major evolutionary transitions correlate with major geological events, such as the oxygen rise in the atmosphere and oceans, or the 'snowball Earth' glaciation periods in the Cryogenian. In fact, the association between the origins of animals and oxygen has long been discussed (see Lenton et al. 2014; Knoll and Sperling 2014 and references therein). While the emergence of Metazoa is in broad synchrony with an increase in atmospheric oxygen and the Sturtian glaciation period, determining which of these events happened first is not a simple task. It was long thought that a rise in atmospheric oxygen concentrations led to the oxygenation of the ocean, and thus triggered metazoan evolution. The alternative hypothesis would be that the Ediacaran oxygen transition was a consequence rather than cause of animal diversification: surface oceans of dense bacterial populations would have been consumed by filter-feeding animals, while fecal pellets from planktonic bilaterians would have been rapidly sinking from the surface, lessening the oxygen demand. To further complicate this issue, there is no evidence for a significant rise in atmospheric or oceanic oxygen levels before the Cryogenian glaciations (including the Sturtian ($\sim$670–730 Ma) and Marinoan ($\sim$635–650 Ma) "Snowball Earth" events), and it was thought that these glaciations posed a major barrier to the survival of eukaryotic life. Yet, the fossil record indicates that numerous eukaryote lineages continued through the glaciations. In fact, our results (and those of others we discuss) postulate that the last common ancestor of metazoans and Fungi existed $\sim$700–800 Ma, possibly before the Cryogenian glaciations. Indeed, it has been argued that sponges were present during, and perhaps, even before these glaciations (Love et al. 2009). If confirmed, this would suggest that the origins of the first metazoans could not have occurred as a response to increasing ocean oxygenation.

This example emphasizes the importance of being able to pinpoint precisely the timing of the origin of multicellular lineages and correlate them with respect to the ancient geochemical record. However, our results clearly show that the obtained molecular dating estimates depend heavily on the models and methods used, and on the nature and treatment of fossil calibrations. This, combined with the significant uncertainty associated with most of the age estimates (often spanning many hundreds of millions of years) suggests that our conclusions remain quite tentative. Hopefully, as further investigations fill out the Proterozoic fossil record associated with protistan eukaryotes, more genomic data is gather from diverse eukaryote lineages and improvements are made in relaxed molecular clock modeling and methods, we will obtain more precise estimates of the ages of multicellular groups.

Summary

1. Multicellularity is found in all major eukaryotic groups and represents a spectrum of organization ranging from simple colonies of cells to complex differentiated multicellular organisms There is currently little reason to suppose that the molecular mechanisms underpinning multicellularity in the various lineages are homologous.
2. Multicellularity has emerged several times over the evolution of eukaryotes. Uncertainty about several regions of the eukaryotic tree makes it difficult to date the origin of multicellular groups.
3. Relaxed Molecular Clock (RMC) methods allow the estimation of unknown divergence dates from phylogenetic trees with fossil calibrations, accounting for variation in rate of sequence evolution among lineages. There are several ways to model this variation across phylogenetic trees.
4. The fossils used to calibrate RMC models add another level of uncertainty to the dating of the emergence of multicellular groups, as the phylogenetic placement and age of fossils are uncertain. In addition, the best way to incorporate this uncertainty within the framework of RMC modeling is not clear.
5. Our RMC analysis of a 159-gene, 85-taxa dataset calibrated with 19 fossils and biomarkers showed considerable variation depending on the RMC model and substitution matrices employed and the manner in which fossil calibrations were applied.
6. Considering previous work and the analysis presented here, we conclude Metazoa, Fungi and two of the major multicellular red algal taxa emerged in the mid-Neoproterozoic, while the Dictyostelids emerged in the Paleozoic.

Acknowledgments This work and MWB's postdoctoral fellowship were supported by a Discovery grant (227085-11) and an Accelerator grant from the Natural Sciences and Engineering Research Council of Canada (NSERC) awarded to AJR. LE is supported by a Centre for Comparative Genomics and Evolutionary Bioinformatics postdoctoral fellowship from the Tula Foundation; SCS is supported by a graduate scholarships from NSERC and Killam trusts. AJR acknowledges the Canadian Institute for Advanced Research, Program in Integrated Microbial Biodiversity and the Canada Research Chairs program. Computations were partially performed on the supercomputers at the SciNet HPC Consortium. SciNet is funded by the Canada Foundation for Innovation under the auspices of Compute Canada; the Government of Ontario; Ontario Research Fund—Research Excellence; and the University of Toronto (Loken et al. 2010).

References

Adl SM, Simpson AGB, Heiss A, et al (2012) The revised classification of eukaryotes. J Eukaryot Microbiol 59(5):429–493. doi:10.1111/j.1550-7408.2012.00644.x

Baldauf SL, Roger AJ, Wenk-Siefert I, et al (2000) A kingdom-level phylogeny of eukaryotes based on combined protein data. Science 290(5493):972–977. doi:10.1126/science.290.5493.972

Beakes GW, Glockling SL, Sekimoto S (2012) The evolutionary phylogeny of the oomycete "fungi". Protoplasma 249(1):3–19. doi:10.1007/s00709-011-0269-2

Becker B, Marin B (2009) Streptophyte algae and the origin of embryophytes. Ann Bot 103(7):999–1004 doi:10.1093/aob/mcp044

Berbee ML, Taylor JW (2010) Dating the molecular clock in fungi—how close are we? Fungal Biol Rev 24(1–2):1–16. doi:10.1016/j.fbr.2010.03.001

Berney C, Pawlowski J (2006) A molecular time-scale for eukaryote evolution recalibrated with the continuous microfossil record. Proc Biol Sci 273(1596):1867–1872 doi:10.1098/rspb.2006.3537

Bonner JT (1998) The origins of multicellularity. Integr Biol: Issues News Rev 1(1):2736. doi:10.1002/(SICI)1520-6602(1998)1:1<27::AID-INBI4>3.0.CO;2-6

Brown MW, Silberman JD (2013) The non-dictyostelid sorocarpic amoebae. In: Romeralo M, Baldauf S, Escalante R (eds) Dictyostelids. Springer, pp 219–242

Brown MW, Spiegel FW, Silberman JD (2009) Phylogeny of the "forgotten" cellular slime mold, Fonticula alba, reveals a key evolutionary branch within Opisthokonta. Mol Biol Evolut 26(12):2699–2709. doi:10.1093/molbev/msp185

Brown MW, Silberman JD, Spiegel FW (2011) "Slime Molds" among the Tubulinea (Amoebozoa): molecular systematics and taxonomy of *Copromyxa*. Protist 162(2):277–287. doi:10.1016/j.protis.2010.09.003

Brown MW, Kolisko M, Silberman JD, et al (2012a) Aggregative multicellularity evolved independently in the eukaryotic supergroup Rhizaria. Curr Biol 22(12):1123–1127. doi:10.1016/j.cub.2012.04.021

Brown MW, Silberman JD, Spiegel FW (2012b) A contemporary evaluation of the acrasids (Acrasidae, Heterolobosea, Excavata). Eur J Protistol 48(2):103–123. doi:10.1016/j.ejop.2011.10.001

Brown MW, Sharpe SC, Silberman JD, et al (2013) Phylogenomics demonstrates that breviate flagellates are related to opisthokonts and apusomonads. Proc Biol Sci 280(1769):20131755. doi:10.1098/rspb.2013.1755

Burki F, Shalchian-Tabrizi K, Minge M, et al (2007) Phylogenomics Reshuffles the Eukaryotic supergroups. PLoS ONE 2(8). doi:10.1371/journal.pone.0000790

Burki F, Okamoto N, Pombert JF, et al (2012) The evolutionary history of haptophytes and cryptophytes: phylogenomic evidence for separate origins. Proc Biol Sci 1736(279):2246–2254. doi:10.1098/rspb.2011.2301

Butterfield NJ (2000) *Bangiomorpha pubescens* n. gen., n. sp.: Implications for the evolution of sex, multicellularity, and the Mesoproterozoic/Neoproterozoic radiation of eukaryotes. Paleobiology 26(3): 386–404

Cavalier-Smith T (2002) The neomuran origin of archaebacteria, the negibacterial root of the universal tree and bacterial megaclassification. Int J Syst Evol Microbiol 52(Pt 1):7–76

Cavalier-Smith T (2012) Early evolution of eukaryote feeding modes, cell structural diversity, and classification of the protozoan phyla Loukozoa, Sulcozoa, and Choanozoa. Eur J Protistol 49(2):115–178. doi:10.1016/j.ejop.2012.06.001

Cavalier-Smith T, Chao EE (1996) Molecular phylogeny of the free-living archezoan *Trepomonas agilis* and the nature of the first eukaryote. J Mol Evol 43(6):551–562. doi:10.1007/BF02202103

Cavalier-Smith T, Chao EE (2010) Phylogeny and evolution of apusomonadida (protozoa: apusozoa): new genera and species. Protist 161(4):549–576. doi:10.1016/j.protis.2010.04.002

Clarke JT, Warnock R, Donoghue PC (2011) Establishing a time-scale for plant evolution. New Phytol 192(1):266–301. doi:10.1111/j.1469-8137.2011.03794.x

Cock JM, Sterck L, Rouzé P, et al (2010) The *Ectocarpus* genome and the independent evolution of multicellularity in brown algae. Nature 465(7298):617–621. doi:10.1038/nature09016

Cohen PA, Knoll AH, Kodner RB (2009) Large spinose microfossils in Ediacaran rocks as resting stages of early animals. Proc Natl Acad Sci USA 106(16):6519–6524. doi:10.1073/pnas.0902322106

Dentzien-Dias PC, Poinar G, Jr, de Figueiredo AE, et al (2013) Tapeworm eggs in a 270 million-year-old shark coprolite. PLoS One 8(1):e55007. doi:10.1371/journal.pone.0055007

Derelle R, Lang BF (2012) Rooting the eukaryotic tree with mitochondrial and bacterial proteins. Mol Biol Evol 29(4):1277–1289. doi:10.1093/molbev/msr295

Dickinson DJ, Nelson WJ, Weis WI (2012) An epithelial tissue in *Dictyostelium* challenges the traditional origin of metazoan multicellularity. BioEssays 34(10):833–840. doi:10.1002/bies.201100187

Drummond AJ, Ho SYW, Phillips MJ, et al (2006) Relaxed phylogenetics and dating with confidence. PLoS Biol 4(5). doi:10.1371/journal.pbio.0040088

Dykstra MJ, Olive LS (1975) Sorodiplophrys: an unusual Sorocarp-producing protist. Mycologia 67(4):873–879. doi:10.2307/3758346

Ebersberger I, de Matos Simoes R, Kupczok A, et al (2012) A consistent phylogenetic backbone for the fungi. Mol Biol Evol 29(5):1319–1334. doi:10.1093/molbev/msr285

Eme L, Sharpe SC, Brown MW, Roger AJ (2014) On the age of eukaryotes: evaluating evidence from fossils and molecular clocks. Cold Spring Harb Perspect Biol 6(8):165–180, a016139

Erwin DH, Laflamme M, Tweedt SM, et al (2011) The Cambrian conundrum: early divergence and later ecological success in the early history of animals. Science 334(6059):1091–1097. doi:10.1126/science.1206375

Felsenstein J (2004) Inferring phylogenies. Sinauer Associates Sunderland, Mass

Fiz-Palacios O, Romeralo M, Ahmadzadeh A, et al (2013) Did terrestrial diversification of amoebas (amoebozoa) occur in synchrony with land plants? PLoS one 8(9):e74374. doi:10.1371/journal.pone.0074374

Graur D, Martin W (2004) Reading the entrails of chickens: molecular timescales of evolution and the illusion of precision. Trends Genet 20(2):80–86. doi:10.1016/j.tig.2003.12.003

Hampl V, Hug L, Leigh JW, et al (2009) Phylogenomic analyses support the monophyly of excavata and resolve relationships among eukaryotic "supergroups". Proc Nat Acad Sci 106(10):3859–3864. doi:10.1073/pnas.0807880106

He D, Fiz-Palacios O, Fu C, et al (2014) An alternative root for the eukaryote tree of life. Curr Biol 24(4):465–470. doi:10.1016/j.cub.2014.01.036

Hedges SB, Blair JE, Venturi ML, et al (2004) A molecular timescale of eukaryote evolution and the rise of complex multicellular life. BMC Evol Biol 4:2. doi:10.1186/1471-2148-4-2

Herron M (2009) Many from one: lessons from the volvocine algae on the evolution of multicellularity. Commun integr biol 2(4):368–370. doi:10.4161/cib.2.4.8611

Herron MD, Hackett JD, Aylward FO, et al (2009) Triassic origin and early radiation of multicellular volvocine algae. Proc Natl Acad Sci U S A 106(9):3254–3258. doi:10.1073/pnas.0811205106

Ho SYW (2009) An examination of phylogenetic models of substitution rate variation among lineages. Biol Lett 5(3):421–424. doi:10.1098/rsbl.2008.0729

Inoue J, Yang Z, Donoghue PCJ (2010) The impact of the representation of fossil calibrations on bayesian estimation of species divergence times. Syst Biol 59(1):74–89. doi:10.1093/sysbio/syp078

James TY, Pelin A, Bonen L, et al (2013) Shared signatures of parasitism and phylogenomics unite cryptomycota and microsporidia. Curr Biol 23(16):1548–1553. doi:10.1016/j.cub.2013.06.057

Katz LA, Grant JR, Parfrey LW, et al (2012) Turning the crown upside down: gene tree parsimony roots the eukaryotic tree of life. Syst Biol 61(4):653–660. doi:10.1093/sysbio/sys026

Keeling PJ, Burger G, Durnford DG, et al (2005) The tree of eukaryotes. Trends Ecol Evol 20(12):670–676. doi:10.1016/j.tree.2005.09.005

Kim E, Simpson AG, Graham LE (2006) Evolutionary relationships of apusomonads inferred from taxon-rich analyses of 6 nuclear encoded genes. Mol Biol Evol 23(12):2455–2466. doi:10.1093/molbev/msl120

Kirk DL (2005) A twelve-step program for evolving multicellularity and a division of labor. Bioessays 27(3):299–310. doi:10.1002/bies.20197

Kishino H, Thorne JL, Bruno WJ (2001) Performance of a divergence time estimation method under a probabilistic model of rate evolution. Mol Biol Evol 18(3):352–361

Knoll AH (2011) The multiple origins of complex multicellularity. Annu Rev Earth Planet Sci 39:217–239. doi:10.1146/annurev.earth.031208.100209

Knoll AH, Sperling EA (2014) Oxygen and animals in earth history. Proc Natl Acad Sci U S A 111(11):3907–3908. doi:10.1073/pnas.1401745111

Knoll AH, Javaux EJ, Hewitt D, et al (2006) Eukaryotic organisms in Proterozoic oceans. Philos Trans R Soc Lond B Biol Sci 361(1470):1023–1038. doi:10.1098/rstb.2006.1843

Langley CH, Fitch WM (1974) An examination of the constancy of the rate of molecular evolution. J Mol Evol 3(3):161–177. doi:10.1007/BF01797451

Lartillot N, Lepage T, Blanquart S (2009) PhyloBayes 3: a Bayesian software package for phylogenetic reconstruction and molecular dating. Bioinformatics 125(17):2286–2288. doi:10.1093/bioinformatics/btp368

Le SQ, Gascuel O (2008) An improved general amino acid replacement matrix. Mol Biol Evol 25(7):1307–1320. doi:10.1093/molbev/msn067

Le SQ, Gascuel O, Lartillot N (2008) Empirical profile mixture models for phylogenetic reconstruction. Bioinformatics 24(20):2317–2323. doi:10.1093/bioinformatics/btn445

Leliaert F, Smith DR, Moreau H, et al (2012) Phylogeny and molecular evolution of the green algae. Crit Rev Plant Sci 31:1–46. doi:10.1080/07352689.2011.615705

Lenton TM, Boyle RA, Poulton SW, et al (2014) Co-evolution of eukaryotes and ocean oxygenation in the Neoproterozoic era. Nat Geosci 7 257–265. doi:10.1038/ngeo2108

Lepage T, Bryant D, Philippe H, et al (2007) A general comparison of relaxed molecular clock models. Mol Biol Evol 24(12):2669–2680. doi:10.1093/molbev/msm193

Levinton JS (2008) The Cambrian explosion: how do we use the evidence. Bioscience 58(9):855–864. doi:10.1641/B580912

Loken C, Gruner D, Groer L, et al (2010) SciNet: lessons learned from building a power-efficient top-20 system and data centre. J Phys: Conf Ser 256(1). doi:10.1088/1742-6596/256/1/012026

Love GD, Grosjean E, Stalvies C, et al (2009) Fossil steroids record the appearance of Demospongiae during the Cryogenian period. Nature 457(7230):718–721. doi:10.1038/nature07673

Magallon S, Hilu KW, Quandt D (2013) Land plant evolutionary timeline: gene effects are secondary to fossil constraints in relaxed clock estimation of age and substitution rates. Am J Bot 100(3):556–573. doi:10.3732/ajb.1200416

Near TJ, Sanderson MJ (2004) Assessing the quality of molecular divergence time estimates by fossil calibrations and fossil-based model selection. Philos Trans R Soc Lond B Biol Sci 359(1450):1477–1483. doi:10.1098/rstb.2004.1523

Near TJ, Meylan PA, Shaffer HB (2005) Assessing concordance of fossil calibration points in molecular clock studies: an example using turtles. Am Nat 165(2):137–146. doi:10.1086/427734

Nowak MD (2013) A simple method for estimating informative node age priors for the fossil calibration of molecular divergence time analyses. PLoS ONE 8(6):e66254. doi:10.1371/journal.pone.0066245

Olive L, Blanton R (1980) Aerial sorocarp development by the aggregative ciliate, *Sorogena stoianovitchae**. J Eukaryot Microbiol 27(3):293–299. doi:10.1111/j.1550-7408.1980.tb04260.x

Parfrey LW, Lahr DJG (2013) Multicellularity arose several times in the evolution of eukaryotes (Response to DOI 10.1002/bies.201100187). Bioessays 35(4):339–347. doi:10.1002/bies.201200143

Parfrey LW, Katz LA, Grant J, et al (2010) Broadly sampled multi-gene analyses yield a well-resolved eukaryotic tree of life. Syst Biol 59(5):518–533. doi:10.1093/sysbio/syq037

Parfrey LW, Lahr DJG, Katz LA, et al (2011) Estimating the timing of early eukaryotic diversification with multigene molecular clocks. Proc Natl Acad Sci U S A 108(33):13624–13629. doi:10.1073/pnas.1110633108

Parham JF, Donoghue PC, Bell CJ, et al (2012) Best practices for justifying fossil calibrations. Syst Biol 61(2):346–359. doi:10.1093/sysbio/syr107

Peterson KJ, Butterfield NJ (2005) Origin of the Eumetazoa: testing ecological predictions of molecular clocks against the Proterozoic fossil record. Proc Natl Acad Sci U S A 102(27):9547–9552. doi:10.1073/pnas.0503660102

Rannala B (2002) Identifiability of parameters in MCMC Bayesian inference of phylogeny. Syst Biol 51(5):754–760. doi:10.1080/10635150290102429

Richards TA, Cavalier-Smith T (2005) Myosin domain evolution and the primary divergence of eukaryotes. Nature 436(7054):1113–1118. doi:10.1038/nature03949
Roger AJ (1999) Reconstructing early events in eukaryotic evolution. Am Nat 154(S4):S146–S163. doi:10.1086/303290
Roger AJ, Hug LA (2006) The origin and diversification of eukaryotes: problems with molecular phylogenetics and molecular clock estimation. Philos Trans R Soc B Biol Sci 1470(361):1039–1054. doi:10.1098/rstb.2006.1845
Roger AJ, Simpson AGB (2009) Evolution: revisiting the root of the eukaryote tree. Curr Biol 19(4):R165–R167. doi:10.1016/j.cub.2008.12.032
Rokas A (2008) The origins of multicellularity and the early history of the genetic toolkit for animal development. Annu Rev of Genet 42:235–251. doi:10.1146/annurev.genet.42.110807.091513
Rubinstein CV, Gerrienne P, de la Puente GS, et al (2010) Early middle ordovician evidence for land plants in Argentina (eastern Gondwana). New Phytol 188(2):365–369. doi:10.1111/j.1469-8137.2010.03433.x
Sanderson MJ (1997) A nonparametric approach to estimating divergence times in the absence of rate constancy. Mol Biol Evol 14(12):1218–1231
Sanderson MJ (2002) Estimating absolute rates of molecular evolution and divergence times: a penalized likelihood approach. Mol Biol Evol 19(1):101–109
Sanderson MJ (2003) r8s: inferring absolute rates of molecular evolution and divergence times in the absence of a molecular clock. Bioinformatics 19(2):301–302. doi:10.1093/bioinformatics/19.2.301
Saunders GW, Hommersand MH (2004) Assessing red algal supraordinal diversity and taxonomy in the context of contemporary systematic data. Am J Bot 91(10):1494–1507. doi:10.3732/ajb.91.10.1494
Schilde C, Schaap P (2013) The amoebozoa. In: Eichinger L, Rivero F (eds) *Dictyostelium discoideum* Protocols. Methods in molecular biology, vol 983. Springer, pp 1–15
Sebe-Pedros A, Irimia M, del Campo J, et al (2013) Regulated aggregative multicellularity in a close unicellular relative of metazoa. eLife 2:e01287–e01287. doi:10.7554/eLife.01287
Shaul S, Graur D (2002) Playing chicken (*Gallus gallus*): methodological inconsistencies of molecular divergence date estimates due to secondary calibration points. Gene 300(1–2):59–61. doi:10.1016/S0378-1119(02)00851-X
Simpson AG (2003) Cytoskeletal organization, phylogenetic affinities and systematics in the contentious taxon Excavata (Eukaryota). Int J Syst Evol Microbiol 53(Pt 6):1759–1777. doi:10.1099/ijs.0.02578-0
Simpson AG, Inagaki Y, Roger AJ (2006) Comprehensive multi-gene phylogenies of excavate protists reveal the evolutionary positions of "primitive" eukaryotes. Mol Biol Evol 23(3):615–625. doi:10.1093/molbev/msj068
Stajich JE, Berbee ML, Blackwell M, et al (2009) The fungi. Curr Biol 19(18):R840–845. doi:10.1016/j.cub.2009.07.004
Stamatakis A (2006) RAxML-VI-HPC: maximum likelihood-based phylogenetic analyses with thousands of taxa and mixed models. Bioinformatics 22(21):2688–2690
Stechmann A, Cavalier-Smith T (2002) Rooting the eukaryote tree by using a derived gene fusion. Science 297(5578):89–91. doi:10.1126/science.1071196
Stechmann A, Cavalier-Smith T (2003) The root of the eukaryote tree pinpointed. Curr Biol 13(17):R665–R666
Summons RE, Walter MR (1990) Molecular fossils and microfossils of prokaryotes and protists from Proterozoic sediments. American Journal of Science 290A:212–244
Takezaki N, Rzhetsky A, Nei M (1995) Phylogenetic test of the molecular clock and linearized trees. Mol Biol Evol 12(5):823–833
Takishita K, Chikaraishi Y, Leger MM, et al (2012) Lateral transfer of tetrahymanol-synthesizing genes has allowed multiple diverse eukaryote lineages to independently adapt to environments without oxygen. Biol Direct 7:5. doi:10.1186/1745-6150-7-5

Taylor JW, Berbee ML (2006) Dating divergences in the Fungal tree of life: review and new analyses. Mycologia 98(6). doi:10.3852/mycologia.98.6.838

Thorne JL, Kishino H, Painter IS (1998) Estimating the rate of evolution of the rate of molecular evolution. Mol Biol Evol 15(12):1647–1657

Tsaousis AD, Leger MM, Stairs CA, Roger AJ (2012) The biochemical adaptations of mitochondrion-related organelles of parasitic and free-living microbial eukaryotes to low oxygen environments. In: Altenbach A, Bernhard JM, Seckbach J (eds) Anoxia. Cellular origin, life in extreme habitats and astrobiology, vol 21. Springer, pp 51–81

Wainright PO, Hinkle G, Sogin ML, et al (1993) Monophyletic origins of the metazoa: an evolutionary link with fungi. Science 260(5106):340–342. doi:10.1126/science.8469985

Welch JJ, Bromham L (2005) Molecular dating when rates vary. Trends ecol evol 20:320–327. doi:10.1016/j.tree.2005.02.007

Xiao S, Knoll AH, Yuan X, et al (2004) Phosphatized multicellular algae in the Neoproterozoic Doushantuo Formation, China, and the early evolution of florideophyte red algae. Am J Bot 91(2):214–227. doi:10.3732/ajb.91.2.214

Yang Z (2006) Computational molecular evolution. Oxford University Press, Oxford

Yang Z, Rannala B (2006) Bayesian estimation of species divergence times under a molecular clock using multiple fossil calibrations with soft bounds. Mol Biol Evol 23(1):212–226. doi:10.1093/molbev/msj024

Yoder AD, Yang Z (2000) Estimation of primate speciation dates using local molecular clocks. Mol Biol Evol 17(7):1081–1090

Yoon HS, Hackett JD, Ciniglia C, et al (2004) A molecular timeline for the origin of photosynthetic eukaryotes. Mol Biol Evol 21(5):809–818. doi:10.1093/molbev/msh075

Zhao S, Burki F, Bråte J, et al (2012) *Collodictyon*—an ancient lineage in the tree of eukaryotes. Mol Biol Evol 29(6):1557–1568. doi:10.1093/molbev/mss001

Zuckerkandl E, Pauling L (1965) Evolutionary divergence and convergence in proteins. In: Bryson V, Vogel HJ (eds) Evolving genes and proteins: A symposium (on evolving genes and proteins), held at the institute of microbiology of rutgers. Academic Press, pp 97–166

Zwickl D, Holder M (2004) Model parameterization, prior distributions, and the general time-reversible model in Bayesian phylogenetics. Syst Biol 53(6):877–888. doi:10.1080/10635150490522584

Fossil and Transcriptomic Perspectives on the Origins and Success of Metazoan Multicellularity

James W. Valentine and Charles R. Marshall

Abstract Multicellularity independently evolved numbers of times—many estimates are in the mid to high twenties—but it is within only two clades, Embryophyta and especially Metazoa, that the multicellular condition led to the evolution of richly diverse and morphologically disparate taxa that have so transformed the biosphere over the last half billion years. Here we first examine the fossil record of metazoans for clues to this morphological profligacy. Part of the reason for their success appears to lie in their early macroevolutionary pattern of rapid invasions of newly accessible adaptive zones followed by exploitation of the morphological possibilities inherent in their new adaptive capacities—their bodyplans–which led to the hierarchical pattern exploited by Linnaean taxonomy. The recent ability to investigate genomically the initial morphological radiation of the phyla through the comparison of ontogenetically dissected transcriptomes has revealed a genomic signature of the phylotypic stage. This suggests, in combination with the paleontologic pattern, that the phylum-level radiation of the metazoans involved the radiation of the phylotypic stages of the phyla. Further transcriptome data offer the possibility of testing evolutionary hypotheses, such as proposed heterochronies, which may be associated with the origin of major morphological novelties; for example, the possibility that the eumetazoan phylotypes descended from sponge larval developmental modules, rather than from those of adult sponges. Finally, we suggest that the morphological disparity of the metazoans (and embryophytes) may be due to their developmental architecture, which includes a mid-embryonic morphological conservatism and transcriptional complexity, and the ability to modify the transcriptome at any developmental stage.

Keywords Cambrian explosion · Metazoa · Phyla · Phylotypic stage · Transcriptome modules · Paedomorphosis · Heterochrony · Disparity · Macroevolution

J. W. Valentine (✉) · C. R. Marshall
Museum of Paleontology and Department of Integrative Biology
University of California, Berkeley, CA 94720, USA
e-mail: jwvsossi@berkeley.edu

C. R. Marshall
e-mail: crmarshall@berkeley.edu

I. Ruiz-Trillo, A. M. Nedelcu (eds.), *Evolutionary Transitions to Multicellular Life*,
Advances in Marine Genomics 2, DOI 10.1007/978-94-017-9642-2_2

Introduction

Among the major clades that have composed Earth's biosphere, Metazoa stand out for the richness of their forms, habits and behaviors, and for the breadth of their disparity and diversity. These attributes are clearly derived from their multicellular construction. Although unicellular clades continue to dominate the major biogeochemical processes of our planet, metazoans exemplify the vast morphological distances that organic evolution is capable of traversing. Here, we use the metazoan fossil record to ask how metazoans have managed to accomplish those feats, and why their particular approach to exploiting the multicellular condition has been so transformative of the biosphere. As paleontologists we have naturally turned to morphologies in the early fossil record for clues, but it is in concert with the exploration of the genome and developmental processes that answers to these questions may eventually be found.

The fossil record has been largely silent on the origins of multicellularity, and although molecular evidence has tentatively identified the clades most closely related to the metazoans (and to the embryophytes and many algal groups; see Butterfield 2009), these groups lack or have very poor fossil records during the periods when their multicellular sister groups arose. Details of earliest metazoan morphologies are also not recorded, although they can be inferred from a morphological comparison of the most basal metazoans (the sponges or possibly the ctenophores (Dunn et al. 2008; Hejnol et al. 2009), although these unusual trees seem to be the result of insufficient taxon sampling and long branch attraction (Pick et al. 2010)) with their putative sister group (the choanoflagellates). Clues to the earliest metazoans can also be garnered from non-morphological fossils such as traces left by their activities, by chemical fossils (biomarkers), and by paleontological dating of branching events in molecular phylogenies, which suggest that stem metazoans arose about 780 Ma (million years ago) (Erwin et al. 2011; also see Chapter "Timing the Origins of Multicellular Eukaryotes Through Phylogenomics and Relaxed Molecular Clock Analyses"). However, nodes on molecular phylogenies simply record the divergences between clades, and not the origin of the significant morphological novelties that those clades may come to possess (Marshall and Valentine 2010). Thus, the metazoan stem lineage may have consisted of species that were morphologically at a choanoflagellate grade of organization for tens of millions of years after the origin of the last common ancestor (LCA) of metazoans and crown choanoflagellates (Fig. 1)—although it might sound counter-intuitive, the first stem group metazoans were not sponges nor were they multicellular (Fig. 1). It is thus useful to supplement the molecular clock estimates for the time of origin of morphologically defined taxa, such as phyla and classes, with morphologies mapped onto phylogenies (rather than cladograms—see Fig. 1), as well as examine the fossil record of the early appearances of those morphologies, and evaluate the patterns of disparity that are found over time.

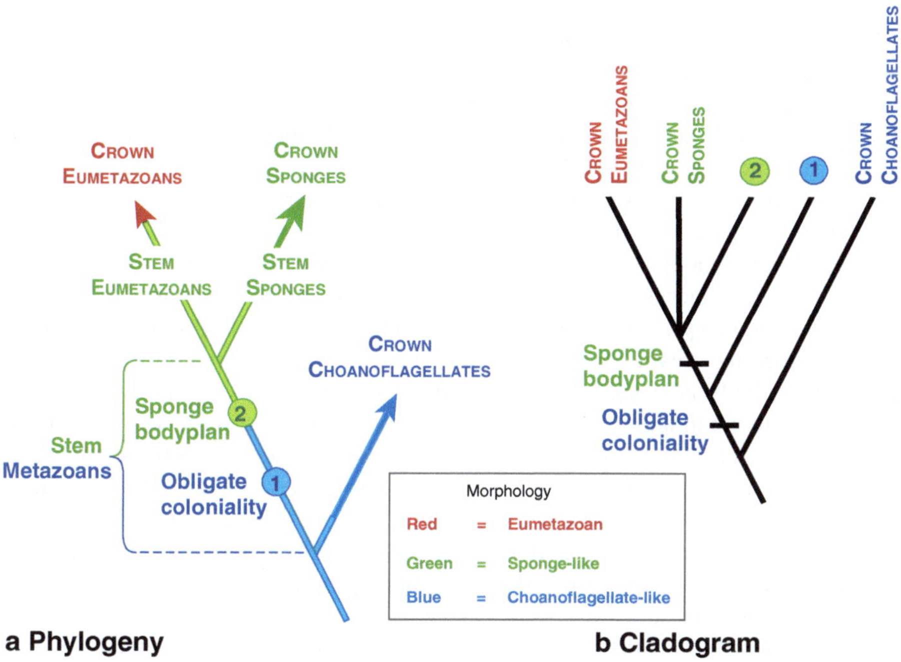

Fig. 1 Morphologies associated with the emergence of the metazoans. **a** Morphologies (*colors*) and morphological changes within a phylogeny depicting the emergence of the metazoans. Lineages with choanoflagellate-like morphologies (*blue*) split to produce crown choanoflagellates and the lineage that gave rise to living sponges and metazoans. At some point the ancestors of metazoans, which were biologically choanoflagellates, evolved obligate coloniality (*1*), which in turn gave rise to a differentiated multicellular organism with a sponge-like bodyplan (*green*, at *2*). Later, a lineage of these sponges, which cladistically were stem metazoans, split, with one lineage giving rise to the eumetazoans, the other to living sponges. If living sponges are paraphyletic (Sperling et al. 2009, 2010) then there will have been more than one branch leading to living sponges. Thus, the earliest eumetazoans (*red*) had a sponge-like bodyplan. **b** The same events captured in a cladogram. Note that the cladogram, in lacking the explicit temporal perspective, does not adequately capture the sequence of morphological changes associated with the emergence of the metazoans

Metazoa Before the Cambrian Explosion

Choanoflagellates to Sponges

The ancestry of sponges in Choanoflagellata (Chapter "Choanoflagellates: Perspective on the Origin of Animal Multicellularity") is suggested by morphology, i.e., the similarity of sponge choanocytes to choanoflagellates, e.g. Laval (1971), although there are differences in the cell structure and cytoskeleton between the two (Karpov and Leadbeater 1998). Choanoflagellate ancestry of sponges is also suggested by molecular evidence (e.g. Lang et al. 2002; Medina et al. 2003; King 2004), but we note that there has been important gene loss in choanoflagellates with respect to the metazoan developmental toolkit (Sebé-Pedrós et al. 2010, 2011; and Chapter

"Transcription Factors and the Origin of Animal Multicellularity"), making the continuity between choanoflagellates and metazoans less clear. That the last common ancestor (LCA) of choanoflagellates and sponges was colonial is possible but not certain, although obligate coloniality was almost certainly present in the metazoan stem lineage as an early step in the evolution of multicellularity (Fig. 1). The exact mechanism by which the colonial form evolved is uncertain, but recent evidence from ichthyosporeans suggests it may have been via cellularization of a multinucleate syncytium (Suga and Ruiz-Trillo 2013; Chapter "Filastereans and Ichthyosporeans: Models to Understand the Origin of Metazoan Multicellularity"). Assuming a choanoflagellate ancestry, feeding chambers were produced as sponges evolved by clustering choanocytes within a scaffold of supporting cells to provide for a more powerful feeding stream, clearing a larger volume of water and permitting the rise of larger individuals. New specialized cell types also appeared as the sponge bodyplan was elaborated, although this almost certainly predated the divergence between the lineage that led to living sponges and the eumetazoans (Fig. 1). As the sponges use food items, largely bacteria, that are quite similar to those of choanoflagellates, they would have fitted easily into the benthic trophic web of the time; no novel conditions of trophic ecology seem required for their emergence. The adaptive advantages of becoming a sponge probably involved feeding and reproductive efficiencies associated with larger body sizes.

The earliest fossils that most likely represent metazoans are indeed possible sponges, that are found in rocks deposited earlier than the Ediacaran Period (c. 630–542 Ma), perhaps significantly before 635 Ma (Maloof et al. 2010), while biomarkers that characterize demosponges today also appear earlier than 635 Ma (Love et al. 2009). Hexactinellids seem to be closely allied with and perhaps basal to demosponges, but most early records of their spicules are suspect, and their crown groups may have diversified as late as the early Cambrian (Dohrmann et al. 2013). It has been postulated that the appearance of large sponge populations filtering the water column influenced the oxygenation of the oceans (Laflamme et al. 2009) and could have played an important role in the eventual appearance of larger, active eumetazoans.

The morphological bridge between sponge and eumetazoan bodyplans is unknown. Many early phylogenies postulated that sponges branched from a protistan lineage while eumetazoans arose independently from another (probably colonial) protistan (reviews in Clark 1964; Willmer 1990). The alternative, a route to eumetazoans through the sponge body plan (see Sperling et al. 2009, 2010 for the case based on the paraphyly of the living sponges) seems more difficult, because the sponge bodyplan (although not the genome) must then be lost. In the past, a common solution has been to suggest that eumetazoans descended from sponge larvae via paedomorphosis (e.g. Salvini-Plawen 1978), shifting reproduction into a worm-like larva (Fig. 2). Crown sponge larvae include somewhat elongate forms with anteroposterior axes that might suggest how bilaterality was foreshadowed (e.g. the trichimella of hexactinellidans and the cinctoblastula of homoscleromorphs; Boury-Esnault et al. 2003; Leys and Ereskovsky 2006), although the fact that adult calcareans and hexactinellids have polarized bodyplans leaves the door open to the possibility that

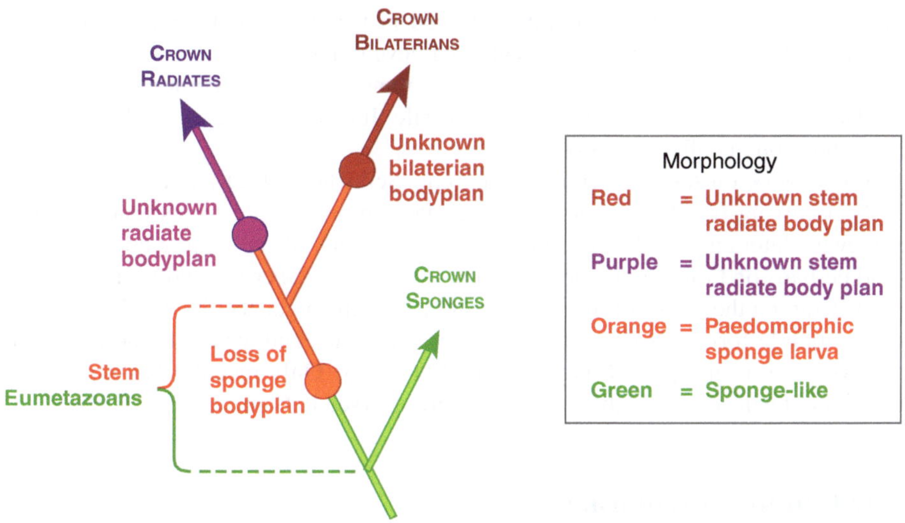

Fig. 2 Morphologies associated with the emergence of the eumetazoans. Morphologies (*colors*) within a phylogeny depicting the emergence of the eumetazoans. Lineages with sponge morphologies (*green*) split to produce crown sponges (or more than one crown group if they are paraphyletic) and the lineage that gave rise to eumetazoans. Under the paedomorphic theory for the origin of eumetazoans, their ancestors, which were biologically sponges, lost the sponge bodyplan with the transfer of reproduction to the larval stage (*orange*). Later, a split occurred with one lineage (or perhaps more than one) that went on to become the radiates, while the other gave rise to the bilaterians. At present, we do not know what the associated bodyplans looked like before these superphyla differentiated

eumetazoans passed through a sponge-like organism that had a limited adult morphology in comparison to the adult morphologies seen in the living sponges. At any rate, the fact that some cell lineages in sponges may readily be transformed from one cell type to another (T. L. Simpson 1984)—for example, archaeocytes into gametes—suggests that paedomorphosis might be more accessible to sponges than to most metazoans. Nevertheless, the issue of whether eumetazoans descended from sponge larvae remains unsettled.

The Ediacarans

A group of large-bodied taxa that are commonly interpreted as early metazoans appeared near 578 Ma. These are the "ediacarans", which seem to have been multicellular and appeared later than the first sponges but a bit earlier than the first convincing evidence for bilaterians (see Xiao and Laflamme 2009). They persisted at least to about the Ediacaran/Cambrian boundary interval (c. 542 Ma). They have been allied with a wide variety of taxa, including various unicellular forms, lichens, or fungi, or as multicellular forms that arose independently of metazoans, or, commonly, as early branches of epithelial metazoans themselves. When they first appear their bodies are highly patterned in frond-like architectures, many of which are fractal, indicating well-regulated developmental processes. If these first ediacarans were metazoans they must have branched along the eumetazoan stem. All lack organs and appear to have been sessile. It has been suggested that they fed by osmotrophy (Laflamme et al. 2009).

Early work on the ediacaran faunas (e.g. Seilacher 1992) tended to lump the early occurring frondose forms with a fauna of more mobile forms that appeared somewhat later and lived on and under microbial mats, likely feeding on those mats (and perhaps on biofilms) in the fashion of Placozoans (Sperling and Vinther 2010), or by scraping the mats or films in the fashion of some algal-feeding mollusks. Most of the mobile forms also have highly patterned bodies and appear to include more than one major group, and may include bilaterians. Only recently has a practical taxonomy been set up for these ediacaran groups (see Laflamme et al. 2012).

The Cambrian Explosion

Patterns of Preservation

The fossil record improves dramatically during the Cambrian (c. 542–489 Ma), especially with the appearance of the rich, chiefly benthic assemblages of the Chengjiang fauna of south China (c. 520 Ma) and the later Burgess-shale type faunas from British Columbia, Canada (c. 510 Ma; see Zhu et al. 2006). These fossils are so exceptionally well preserved that they include many details of their soft-part anatomies, even though some of them appear to lack stiffened organic or mineralized integuments. These assemblages provide a window into an important fraction of Cambrian faunas that would otherwise be lost. Another mode of preservation that appears in the early Cambrian is represented by "small shelly fossils", largely phosphatized skeletons, chiefly under 2 mm in their largest dimension, commonly of sclerites (individual plates, spines, etc., typically components of multi-element skeletons or scleritomes). And the most common Phanerozoic mode of benthic marine fossil preservation, as mineralized or otherwise stiffened organic skeletons, is well represented in the Cambrian as well. Thus, there are three important modes of preservation represented in Cambrian faunas, two of which are unusual.

The highest taxonomic levels, for example the Linnaean phyla and classes, are well represented in the Cambrian fossil faunas, while the faunas seem underrepresented by taxa at lower levels, such as species and genera, at least by modern standards. For example, there are 20 Phanerozoic echinoderm taxa that are deemed distinctive enough to be assigned to class level by Linnaean criteria (Sprinkle and Kier 1987), but many of the first appearing ones are extinct, have very short durations compared with living classes, and are represented by relatively few lower taxa; eight of them are known from fewer than 10 genera each. It is possible to interpret the low generic (and specific) diversities recorded for these classes as owing to a fossil record so spotty that it fails to sample most lower-level taxa, while still leaving a relatively complete record at the higher taxonomic levels (given that it takes only one record of one genus or species to establish the presence of its class). And the spottier the record, the fewer lower taxa will be captured, and the shorter the apparent temporal range of the phylum or class will be. Thus one can ask, is the Cambrian pattern of high rates of appearance of often short-lived higher taxa, and low rates of appearance of lower taxa simply owing to a record biased in this way? Or does the fossil record portray the true underlying macroevolutionary dynamics of the Cambrian explosion—high rates of major morphological innovation with low rates of lower taxonomic innovation? The latter possibility is supported by the fact that the Cambrian is characterized by the three major modes of high quality preservation as described above, and by the observation that peak rates of origination shift towards lower taxonomic levels through time (Valentine 1969; Erwin et al. 1987), as if evolution was running out of novel gambits. That is, the higher the taxon, the earlier its peak rate of diversification: phylum-level diversification peaks during the early Cambrian, class-level diversification peaks during the late Cambrian to early Ordovician, orders in the mid Ordovician, while families peaked later still.

Building a Macroevolutionary Case for Early Metazoan Diversification Patterns

Morphometric studies can, in principle, shed light on the relationship between the observed patterns of Cambrian diversification and the processes that produced them. However, there are methodological problems that make such studies difficult. Two are particularly discouraging: the lack of reasonably large numbers of lower-level taxa, required for statistical significance; and, the lack of common morphological landmarks among the very disparate Cambrian higher taxa, many of which are stem groups at the level of phyla or superphyla (Budd and Jensen 2000). Trilobites are a major exception, for they are represented by more species than all other major invertebrate clades combined during the Cambrian, although at its appearance this clade is already distinctive and its branching position among arthropods is difficult to establish. But extensive studies of trilobite lineages find nothing in their developmental pattern that is unique among arthropods (review by Hughes 2005)—the emergence and initial radiation of trilobites does not seem to be the result of an unusual mode of development or macroevolution, although their success may owe in part to their earlier acquisition of more easily preserved (chiefly mineralized) skeletons than other contemporaneous clades.

Perhaps a better place to test for a similar macroevolutionary patterns is among Ordovician faunas (c. 488–444 Ma); they are better preserved and much more diverse than those of the Cambrian thanks to the "great Ordovician biodiversity event" (see Webby et al. 2004). That diversification raised overall family diversity by a factor of about 2.5 (Sepkoski 1981) to a standing diversity that held reasonably steady until the end-Permian extinctions, despite some extinction spikes. The numbers of lower taxa available within many of the more common phyla and classes became large enough to support morphometric treatments, which were successfully pursued among Echinodermata in a series of important studies by Foote (especially 1992, 1994,

1995). These studies are particularly interesting because the Cambrian appearance of echinoderms is emblematic of the low-diversity, high-disparity pattern described above, and some classes continued that pattern during the Ordovician. Foote's studies included Paleozoic blastozoans (Foote 1992) and crinoids from the Ordovician to Devonian (Foote 1994) and also from the Carboniferous to Permian (Foote 1995).

The morphological pattern that Foote found to be the most common in his strato-morphometric studies was that of the early radiation of higher taxa relative to lower ones. The earliest members of the diversifying taxon were highly disparate, more or less defining the full morphological space that was to become more densely occupied as the clade reached its peak diversity—the later appearing members tend to fill in the morphospace between the founders. Thus, the founders were more distinctive from each other (and usually assigned to a higher taxonomic level) than were the later appearing members from each other. Within more inclusive clades, their sub-clades typically repeat this pattern, even if late evolving—*their* early branches tending to outline the morphological region that their descendants would come to occupy, and so on. The early morphospace regions staked out by early-appearing novelties became more densely occupied over time as later branches originated. Such a history of the filling of morphospace produces a hierarchy of disparities, which lend themselves easily to classification in the Linnaean fashion. Thus, the dissection of diversity patterns at different levels of the Linnaean hierarchy is highly informative (despite arguments to the contrary, e.g. Smith (1994), and see Foote (1996) for a well-reasoned counter-argument).

This pattern of early rise of disparity is by no means universal, but it occurs in other phyla that were important in early metazoan history for which data are available. Other major clades in which disparities among groups appear abruptly before many lower-level taxa appear include: arthropods, the most dominant fossil group during the Cambrian explosion, which reach a level of disparity in the Cambrian that approximates that of the modern marine arthropod fauna (Briggs et al. 1992; Wills et al. 1994); brachiopods, of which 12 orders appear in the Cambrian (Curry and Brunton 2007); and, also at the ordinal level, early bryozoans (appearing in the early Ordovician—Anstey and Pachut 1995). The generality that emerges is that for taxa in the Cambrian as in the Ordovician, the higher the taxonomic level the earlier they diversified. This pattern holds from the level of phylum down to the family level (see Valentine 1969; Sepkoski 1981; Erwin et al. 1987; Campbell and Marshall 1986 for echinoderms).

Subsequent expansion of morphospace regions certainly does occur within some higher taxa, especially after major extinctions that presumably released spatial and trophic resources utilized by the extinct groups that were then taken up by a surviving group; this may be the case with the origins of some orders following extinctions (Erwin et al. 1987). Especially clear cases of morphospace expansion of established clades involve the invasion of large but previously unoccupied regions of adaptive space, most spectacularly by the invasion of the land, as by insects and by tetrapods. However a group arises, the size of its morphospace is most strongly affected by the presence of large peripheral subgroups (Foote 1993).

As the reality of this "top-down" filling of morphospace has been confirmed for Ordovician taxa, and as a similar (but spottier) pattern can be seen in the much less diverse Cambrian faunas, it seems likely that similar macroevolutionary processes were responsible for the similarities in the record of higher taxa origination during both Periods. This raises two questions: (1) Is it reasonable to assume that the Ediacaran pattern was also similar, even though we may not have a record of a much larger proportion of the taxa of the time; and (2) What caused or permitted the relatively rapid origin of the novel bodyplans that founded the Cambrian explosion, and of the sub-plans that begat the awesome morphological diversity of the Metazoa. We tentatively answer the first question affirmatively since there is no obvious reason that the best-known macroevolutionary pattern among major metazoan morphologies was not inherited from that of earlier times. We cannot answer the second question, but recent advances in the analysis of change in gene expression during development offers an opportunity to better understand the evolutionary processes responsible for the "top-down" filling of morphospace, which we now explore.

The Phylotype, Transcriptomes, and the "Top-Down" Filling of Morphospace

The Phylotypic Stage

A long-standing observation is that metazoan morphological lability during ontogeny is shaped like an hourglass, the neck of which corresponds to a conserved "phylotypic stage" (Slack et al. 1993) with more divergent morphologies developed both earlier and later (Duboule 1994; Raff 1996). Descriptively, morphological lability is said to be constrained at that stage because morphological similarities in the developing embryos among species belonging to the various subgroups within a phylum (or other taxon) resemble each other most closely then. Common features are seen across the subgroups, and these features can be referred to at the phylum level as constituting its underlying bodyplan.

The "top-down" filling of morphospace implies that the phylum-level morphological differences were among the first stages in the evolution of the metazoans, and thus that, in essence, the beginnings of the radiation of metazoans was the radiation of the morphologies that retrospectively we identify as the phylotypes. Thus, given that the genome captures a great deal of historical information in its sequences and its functions, we here posit that the genomic analysis of the phylotypic stages of living taxa should offer insight into the first steps in the diversification of the phyla.

The Search for the Genomic Signatures of the Phylotypic Stage

Recently, phylostratigraphic analyses (Domazet-Loso et al. 2007) of ontogenetically dissected transcriptome data from model organisms, fruitflies (Kalinka et al. 2010) and zebrafish (Domazet-Loso and Tautz 2010), have been used to examine transcriptional lability during development. Kalinka et al. (2010) studied six species of

Drosophila and found that a reduction in temporal lability of transcription occurs in mid-embryogenesis and that this developmental stage is characterized by the relatively enhanced transcription of the regulatory genes that are chiefly responsible for the bodyplan of the phylum—the arthropod phylotypic stage. Furthermore, tests for the strength of stabilizing and directional selection among the species were consistent with the presence of significant stabilizing selection during mid-embryogenesis.

Domazet-Loso and Tautz (2010) used a different type of analysis, asking what is the average phylogenetic age of the genes expressed during each ontogenetic stage of the developing zebrafish. They found that mid-embryonic phylotypic stages were both preceded and followed in ontogeny by the transcription of, on average, phylogenetically younger genes—their data suggested that the hourglass shaped lability of morphology seen in development is mirrored by an hourglass-shaped average age of the genes expressed during ontogeny.

However, re-analysis of Domazet-Loso and Tautz's (2010) data by Piasecka et al. (2013) suggest that rather than the average age of the genes expressed at each developmental stage exhibiting an hourglass shape, the average age of the genes expressed simply gets younger through ontogeny. In that case the phylotype is not reflected in the average age of the genes expressed during ontogeny. Nonetheless, Piasecka et al. (2013) show a significant enrichment in the number of transcription factors expressed during the phylotypic stage, and that those transcription factors have significant concentrations of highly conserved non-coding elements and transposon-free regions. Thus, consistent with Kalinka et al.'s (2010) observations and the ideas of Duboule (1994) and Raff (1996), the zebrafish phylotypic stage is at least characterized by a peak in evolutionary stability.

Further, and perhaps most significantly, Piasecka et al. (2013) showed that there is a significant over-representation of genes common to all bilaterians expressed during the vertebrate phylotypic stage. Unfortunately, they were only able to employ a few model systems in their analysis, so it is unclear how many of those genes have origins that lie deeper in the tree, for example at the origin of the Eumetazoa or Metazoa or even deeper. Nonetheless, Piasecka et al.'s (2013) analysis of Domazet-Loso and Tautz's (2010) data lends considerable weight to the hypothesis that the first stages of the metazoan radiation were characterized by the divergence of the developmental stages that were later to become entrenched in the phylotypic stages of the differentiated phyla (see de Mendoza et al. 2013).

Using Transcriptome Profiles Help to Solve Evolutionary Problems in the Deep Past?

The increasing sophistication of the approaches used to understand the transcriptional history of genes in the early history of the metazoans offers hope that further work will yield solutions to some of the most vexing problems associated with the nature of the morphological innovation among early metazoans. The most obvious major problem for which help may arrive involves the identification of transcriptional shifts among ontogenetic stages that are hypothesized to be associated with heterochronic events during the origin of major novelties that produced new higher taxa. The transition from sponges to eumetazoans, discussed above, provides a good example of the possibilities. If transcriptome profiles of the stages of sponge development can be obtained, they might be used to test the proposition that the adult sponge stages were essentially dropped from the developmental repertoire of eumetazoans, and that early bilaterian developmental stages evolved from early embryonic stages associated with sponge larvae (Fig. 2). A clear solution would solve a very long-standing problem, although one can imagine a wide range of unanticipated outcomes that might ensue. More generally, it is conceivable that heterochronies underlie many important morphological changes in metazoans (e.g. McKinney and McNamara 1991), and thus that access to information on the transcriptional shifts between taxa and ontogenetic stages might well illuminate many major problems in morphogenesis.

The broader questions that concern the controls on the relatively rapid morphological changes implied by the well diversified and disparate explosion faunas and the clear constraints on evolutionary experimentation within the bodyplans that followed might also be usefully probed by comparative studies of transcription within ontogenetic modules. For example, are there modules devoted to class-level bodyplans, which can be nearly as long-lived as their phyla? Surely a class-level developmental subsystem is responsible for the echinoid bodyplan as distinct from crinoids, etc., for example. And if they can be found, what are the relations between the modules devoted to the development of phylum-level versus class-level bodyplans? Are these subsystems integrated with the phylotype and if so how is the class-level divergence (which of course arose after the phylotype) managed? What controls the initial burst of morphological change, and what brings it to a halt? Are genome changes active causes or more passive participants?

There are two main sources of speculation about this last question. One set of hypotheses, the more ecological, derives from the contributions of G. G. Simpson (e.g. 1944; and see Valentine 1980; Marshall 2006; Erwin and Valentine 2013). Simpson envisioned the environment as being subdivided into potential adaptive zones, surrounded by adaptive barriers that could be breached by pre-adapted traits that happened to permit at least tentative entrance to the empty zone. Once into the zone a lineage may rapidly explore various sub-zones (producing novel branches when broaching sub-zonal barriers) that may create a morphologically hierarchical taxon. The "filling" of adaptive space within the subzones eventually saturates the adaptive regions available to the morphologies of this lineage and significant further diversification then depends on extinctions or breakouts into other adaptive zones. The other speculations are based on genome properties (see Davidson 2006; Davidson and Erwin 2006). These workers note that many important developmental functions become protected from change by being woven into complex multigene combinations in which no one gene can be changed without losing the entire function (the kernels of Davidson 2006; Davidson and Erwin 2006); thus selection is strong for both the promotion and maintenance of these combinations. Exploration of changes in modules associated with the development of kernel-like systems might bear on the relative importance of these hypotheses, which however, are not mutually exclusive.

From Where does Metazoan Morphological Success Derive?

If we only knew. It is clearly not multicellularity *per se*, which has not had similar effects within most clades where it is found, tending to be just another adaptation that may have helped in promoting the success of its lineage but that has hardly transformed the biosphere. Embryophyta is the other major multicellular group that has had truly transformative effects, at least terrestrially, where it has certainly supported the success of terrestrial metazoans. Embryophytes and metazoans have the most complex transcription factor repertoires, and much of this complexity arose in their respective unicellular ancestors (Mendoza et al. 2013; Chapter "Transcription Factors and the Origin of Animal Multicellularity"). Embryophytes also have similar developmental architectures and structurally their gene regulatory networks resemble those of metazoans, although mostly using different families of genes from metazoans (de Mendoza et al. 2013; Chapter "The Evolution of Transcriptional Regulation in the Viridiplantae and its Correlation with Morphological Complexity"). And indeed embryophytes are said to display an hourglass-shaped transcriptome during development that can be interpreted as having a sort of "embryotypic" stage with fewer but older genes expressed at the neck as well (Quint et al. 2012). However, Quint et al. (2012) used the same methodology as Domazet-Loso and Tautz (2010) so it is possible that with re-analysis (i.e. see Piesecka et al. 2013) their conclusions may be altered. Embryophytes evidently arose from among streptophytes, freshwater green algae that lack the heteromorphic life history of embryophytes (Becker and Marin 2009). It thus seems that streptophytes or other green algae may possess neither a developmental hourglass nor, presumably, the genomic evolutionary flexibility of embryophyte ontogenies, although there does not seem to be enough molecular data to test this possibility as yet.

Molecular phylostratigraphic data also seem to be lacking on brown algae, although they have independently evolved complex multicellular developmental systems that are now under study (see Charrier et al. 2008; Chapters "Emergence of *Ectocarpus* as a Model System to Study the Evolution of Complex Multicellularity in the Brown Algae" and "Independent Emergence of Complex Multicellularity in the Brown and Red Algae"). Likewise, transcriptional aspects of fungal development are under study. Work reported by Nygren et al. (2012) indicates the presence of heterogeneities in evolutionary rates during fungal ontogenies, but whether this will translate into translational modules similar to metazoans and embryophytes is not yet known.

As the two most morphologically successful groups, metazoans and embryophytes, each with sophisticated and similarly structured gene regulatory networks, it is tempting to choose the evolution of such network architectures as a major candidate for their morphological preeminence. One of the key features of these networks appears to be their abililty to rework the transcriptome at most, or all, developmental stages. If so, it might be suggested that one of the reasons that this developmental architecture has been so successful in producing disparity is its ability to mediate heterochronic changes that permit the morphological elaboration of distinctive body plans and provide the top-down patterns of disparity that we find, and that led to

Linnaean-style classification over 250 years ago. That is, we suggest that the origin and evolvability of even the earliest pre-phylotypic developmental stages in metazoans is a consequence of the remarkable flexibility of the gene regulatory system.

Summary

1. The evolution of multicellularity is not uncommon among major clades, but it did not usually lead to the great morphological disparities that characterize metazoans and underlie their adaptive success.
2. The fossil record shows that the morphologies that define Linnaean taxa precede the morphologies that define lower taxa. This pattern of top-down origin of metazoan body plans is clearest during the Ordovician diversifications within the phyla, and is supported by morphometric analysis of some abundant taxa. In general, body plan and sub-plan diversification precedes species and genus-level diversifications within the major metazoan clades.
3. A similar macroevolutionary pattern can be traced back into the Cambrian explosion but is not well recorded in the preceding fossil record of the Ediacaran, when the bodyplans of metazoan phyla were emerging.
4. Transcriptome studies, backed up by data from time calibrated molecular phylogenies, are consistent with a principally Ediacaran radiation of phylotypes, which characterize the phylum-level bodyplans that were radiating during the explosion.
5. A cis-regulatory system similar to that of metazoans is shown by embryophytes, the only other kingdom that displays important morphological disparity.
6. This style of gene regulatory architecture may thus be an essential element in understanding why we do not see the evolution of exceptional disparity in the majority of multicellular groups.

Acknowledgement We thank John Taylor and Jeffrey Townsend for information on fungal genomes, Brent Mishler and Chelsea Specht for discussions of plant genomes, and especially Inaki Ruiz-Trillo for the invitation to contribute to this volume and for thoughtful editing. Two anonymous reviews were most helpful and their comments led to significant rewriting. This is University of California Museum of Paleontology contribution number 2060.

References

Anstey RL, Pachut JL (1995) Phylogeny, diversity history, and speciation in paleozoic bryozoans. In: Erwin DH, Anstey RL (eds) New approaches to studying speciation in the fossil record. Columbia University Press, New York, p 239–284

Becker B, Marin B (2009) Streptophyte algae and the origin of embryophytes. Ann Bot 103:999–1004

Boury-Esnault N, Ereskovsky A, Bézac C et al (2003) Larval development in the homoscleromorpha (*Porifera, Demospongiae*). Invertebr Biol 122:187–202

Briggs DEG, Fortey RA, Wills MA (1992) Morphological disparity in the Cambrian. Science 256:1670–1673

Budd GE, Jensen S (2000) A critical reappraisal of the fossil record of the bilaterian phyla. Biol Rev 75:253–95

Butterfield NJ (2009) Modes of pre-Ediacaran multicellularity. Precamb Res 173:201–211

Campbell KSW, Marshall CR (1986) Rates of evolution among palaeozoic echinoderms. In: Campbell KSW, Day MF (eds) Rates of evolution. Allen and Unwin, London, p 61–100

Charrier B, Coelho SM, LeBail A et al (2008) Development and physiology of the brown alga *Ectocarpus siliculosus*: two centuries of research. New Phytol 177:319–332

Clark RB (1964) Dynamics in metazoan evolution. Oxford University Press, Oxford

Curry GB, Brunton CHC (2007) Stratigraphic distribution of brachiopods. In: Williams A, Brunton CHC, Carlson SJ et al (eds) Treatise on invertebrate paleontology H:6 (Brachiopoda Revised). Geological Society of America, Boulder and University of Kansas Press, Lawrence, p 2901–2965

Davidson EH (2006) The regulatory genome. Academic Press, San Diego

Davidson EH, Erwin DH (2006) Gene regulatory networks and the evolution of animal body plans. Science 311:796–800

De Mendoza A, Sebé-Pedrós A, Šestak MS et al (2013) Transcription factor evolution in eukaryotes and the assembly of the regulatory toolkit in multicellular lineages. Proc Nat Acad Sci USA doi:10.1073/pnas.1311818110

Dohrmann M, Vargas S, Janussen D et al (2013) Molecular paleobiology of early-branching animals: integrating DNA and fossils elucidates the evolutionary history of hexactainellid sponges. Paleobiology 39:95–108

Domazet-Loso T, Tautz D (2010) A phylogenetically based transcriptome age index mirrors ontogenetic divergence patterns. Nature 468:815–818

Domazet-Loso T, Brajkovic J, Tautz D (2007) A phylostratigraphic approach to uncover the genomic history of major adaptations in metazoan lineages. Trends Genet 23:533–539

Duboule D (1994) Temporal colinearity and the phylotypic progression: a basis for the stability of a vertebrate Bauplan and the evolution of morphologies through heterochrony. Dev Suppl 1994:135–142

Dunn CW, Hejnol A, Matus DQ, et al. (2008) Broad phylogenomic sampling improves resolution of the animal tree of life. Nature 452:745–749

Erwin DH, Valentine JW (2013) The Cambrian explosion: the construction of animal biodiversity. Roberts & Co., Greenwood Village

Erwin DH, Valentine JW, Sepkoski JJ Jr (1987) A comparative study of diversification events: the early paleozoic versus the mesozoic. Evolution 41:1177–1186

Erwin DH, Laflamme M, Tweedt SM et al. (2011) The Cambrian conundrum: early divergence and later ecological success in the early history of animals. Science 334:1091–1097

Foote M (1992) Paleozoic record of morphological diversity in blastozoan echinoderms. Proc Nat Acad Sci U S A 89:7325–7329

Foote M (1993) Contributions of individual taxa to overall morphological disparity. Paleobiology 19:403–419

Foote M (1994) Morphological disparity in Ordovician-Devonian crinoids and the early saturation of morphological space. Paleobiology 20:320–344

Foote M (1995) Morphology of carboniferous and permian crinoids. Contrib Mus Paleontol Univ Mich 29:135–184

Foote M (1996) Perspective: evolutionary patterns in the fossil record. Evolution 50:1–11

Hejnol A, Obst M, Stamatakis A et al. (2009) Assessing the root of bilaterian animals with scalable phylogenomic methods. Proc R Soc Lond B Biol Sci 276:4261–4270

Hughes N (2005) Trilobite construction: building a bridge across the micro- and macroevolutionary divide. In: Briggs DEG (ed) Evolving form and function: fossils and development. Peabody Museum of Natural History, Yale University Special Publication, New Haven, p 139–158

Kalinka AT, Varga KM, Gererard DT et al. (2010). Gene expression divergence recapitulates the developmental hourglass model. Nature 468:811–814

Karpov SA, Leadbeater BSC (1998) Cytoskeleton structure and composition in choanoflagellates. J Eukaryot Microbiol 45:361–367

King N (2004) The unicellular ancestry of animal development. Dev Cell 7:313–325
Laflamme M, Xiao X, Kowalewski M (2009) Osmotrophy in modular Ediacara organisms. Proc Nat Acad Sci USA 106:14438–14443
Laflamme M, Darroch AF, Tweedt SM et al (2012) The end of the Ediacara biota: extinction, biotic replacement, or cheshire cat? Gondwana Res. doi:10.1016/j.gr.2012.11.004
Lang BF, O'Kelly C, Nerad T et al (2002) The closest unicellular relatives of animals. Curr Biol 12:1773–1778
Laval M (1971) Ultrastructure et mode de nutrition du choanoflagellé *Salpingoecia pelagica* sp. nov.; Comparaison avec les choanocytes des spongiaires. Protohistologica 8:325–336
Leys SP, Ereskovsky AV (2006) Embryogenesis and larval differentiation in sponges. Can J Zool 84:262–287
Love GD, Grosjean E, Stalvies DA et al. (2009) Fossil steroids record the appearance of *Demospongiae* during the Cryogenian period. Nature 457:718–721
Maloof AC, Rose CV, Beach R et al. (2010) Possible animal-body fossils in pre-Marinoan limestones from South Australia. Nature Geosci 3:653–659
Marshall CR (2006) Explaining the Cambrian "explosion" of animals. Ann Rev Earth Planet Sci 34:355–384
Marshall CR, Valentine JW (2010) The importance of preadapted genomes in the origin of the animal bodyplans and the Cambrian explosion. Evolution 64:1189–1201
McKinney ML, McNamara KJ (1991) Heterochrony: the evolution of ontogeny. Plenum Press, New York
Medina M, Collins AG, Taylor JW et al (2003) Phylogeny of opisthokonta and the evolution of multicellularity and complexity in fungi and metazoa. Int J Astrobiol 2:203–211
Nygren K, Walberg A, Samils N et al (2012) Analyises of expressed sequence tags in *Neurospor*a reveal rapid evolution of genes associated with the early stages of sexual reproduction in fungi. BMC Evol Biol 12:229. doi:10.1186/1471-2148-12-229
Piasecka B, Lichocki P, Moretti S et al (2013) The hourglass and the early conservation models—coexisting patterns of development constraints in vertebrates. PLOS Genet 9(4):e1003476
Pick KS, Philippe H, Schreiber F et al (2010) Improved phylogenomic taxon sampling noticeably affects nonbilaterian relationships. Mol Biol Evol 27:1983–1987
Quint M, Drost H-G, Gabel A et al (2012) A transcriptome hourglass in plant embryogenesis. Nature 490:98–101
Raff RA (1996) The shape of life. Chicago University Press, Chicago
Salvini-Plawen LV (1978) On the origin and evolution of the lower metazoa. Z Zool Syst Evolutionforsch 16:40–88
Sebé-Pedrós A, Roger AJ, Lang FB et al (2010) Ancient origin of the integrin-mediated adhesion and signaling machinery. Proc Natl Acad Sci USA 107:10142–10147
Sebé-Pedrós A, de Mendoza A, Lang BF et al (2011) Unexpected repertoire of metazoan transcription factors in the unicellular holozoan *Capsaspora owczarzaki*. Mol Biol Evol 28:1241–1254
Seilacher A (1992) Vendobionta and Psammocorallia: lost constructions of Precambian evolution. J Geol Soc London 149:607–613
Sepkoski JJ Jr (1981) A factor analytic description of the phanerozoic marine fossil record. Paleobiology 7:36–53
Simpson GG (1944) Tempo and mode in evolution. Columbia University Press, New York
Simpson TL (1984) The cell biology of sponges. Springer, New York
Slack JM, Holland PW, Graham CF (1993) The zootype and the phylotypic stage. Nature 361: 490–492
Smith AB (1994) Systematics and the fossil record: documenting evolutionary patterns. Blackwell Science, Oxford
Sperling EA, Vinther J (2010) A placozoan affinity for Dickinsonia and the evolution of late Proterozoic feeding modes. Evol Dev 12:201–209
Sperling EA, Peterson KJ, Pisani D (2009) Phylogenetic-signal dissection of nuclear housekeeping genes supports the paraphyly of sponges and the monophyly of eumetazoa. Mol Biol Evol 26:2261–2274

Sperling EA, Robinson JM, Pisani D, Peterson KJ (2010) Where's the glass? Biomarkers, molecular clock, and microRNAs suggest a 200-Myr missing Precambrian fossil record of siliceous sponge spicules. Geobiology 8:24–36
Sprinkle J, Kier PM (1987) Phylum Echinodermata. In: Boardman RS, Cheatham AH, Rowell AJ (eds) Fossil invertebrates. Blackwell Scientific, Palo Alto, p 550–611
Suga H, Ruiz-Trillo I (2013) Development of ichthyosporeans sheds light on the origin of metazoan multicellularity. Dev Biol. doi:10.1016/j.ydbio.2013.01.009
Valentine JW (1969) Patterns of taxonomic and ecological structure of the shelf benthos during Phanerozoic time. Palaeontology 12:684–709
Valentine JW (1980) Determinants of diversity in higher taxonomic categories. Paleobiology 6:444–450
Webby BD, Paris F, Droser ML, Percival IG (eds) (2004) The great Ordovician biodiversification event. Columbia University Press, New York
Willmer P (1990) Invertebrate relationships. Cambridge University Press, Cambridge
Wills MA, Briggs DEG, Fortey RA (1994) Disparity as an evolutionary index: a comparison of Cambrian and recent arthropods. Paleobiology 20:93–130
Xiao SH, Laflamme M (2009) On the eve of animal radiation: phylogeny, ecology and evolution of the Ediacara biota. Trends Ecol Evol 24:31–40
Zhu M-Y, Babcock LE, Peng S-C et al (2006) Advances in Cambrian stratigraphy and paleontology: integrating correlation techniques, paleobiology, taphonomy and paleoenvironmental reconstruction. Paleoworld 15:217–222

Origin of Metazoan Developmental Toolkits and Their Expression in the Fossil Record

Sarah M. Tweedt and Douglas H. Erwin

Abstract Developmental regulatory genes (largely transcription factors and signaling pathways) were once viewed as tightly connected to the origin of the morphological features with which they are associated in bilaterians. With the increased study of basal metazoans (sponges and cnidarians) as well as other eukaryotic clades, it is now clear that many of these highly conserved genes arose much earlier in evolution, and served different biological purposes. This provides a new view of the nature of developmental toolkits associated with the early origin of Metazoa: ancient regulatory genes were only later co-opted for the various developmental roles associated with bilaterian morphology. Here we review the nature of the toolkits at the origin of Metazoa, the Placozoan-Eumetazoan last common ancestor (LCA), the Cnidarian-Bilaterian LCA, and the Protostome-Deuterostome LCA. Integrating this data with recent molecular clock results and data on the fossil record reveals long macroevolutionary lags between the origin of the molecular toolkits and their developmental potential, and the origin of crown group morphologies as documented in the fossil record.

Keywords Metazoa · Phylogeny · Fossil record · Genetic toolkit · Ediacaran · Cambrian · Macroevolution

Introduction

Although multicellularity has evolved in many eukaryotic lineages (Knoll 2011), differentiated cell types and tissues are relatively rare, having evolved in fungi, algae (including land plants) and animals. The production of differentiated body plans in

S. M. Tweedt (✉) · D. H. Erwin
Department of Paleobiology, National Museum of Natural History,
P.O. Box 37012, MRC-121, Washington, DC, 20013–7012, USA
e-mail: TweedtS@si.edu

S. M. Tweedt
Behavior, Ecology, Evolution, & Systematics, University of Maryland,
College Park, 20742 MD, USA

D. H. Erwin
e-mail: ERWIND@si.edu

I. Ruiz-Trillo, A. M. Nedelcu (eds.), *Evolutionary Transitions to Multicellular Life*,
Advances in Marine Genomics 2, DOI 10.1007/978-94-017-9642-2_3

these lineages requires a suite of developmental genes, cell-cell interactions, and dynamic regulatory signaling. These have been best studied among animals, where comparative studies of considerable phylogenetic breadth have been carried out, supplemented more recently by a growing number of whole-genome sequences. Although their phylogenetic coverage is not yet comprehensive, comparative evolutionary developmental studies have revealed deep conservation of key developmental tools. Indeed, recent studies of such basal groups as sponges and cnidarians have uncovered many genes and components of developmental gene regulatory networks (dGRNs) previously thought unique to bilaterians (see Chapter "A Comparative Genomics Perspective on the Origin of Multicellularity and Early Animal Evolution"). The presence of these genes in early-branching metazoan lineages raises important questions about their ancestral role in early metazoans.

In this chapter we discuss the acquisition of metazoan developmental characteristics within a developmental framework, focusing on a series of key nodes, including the metazoan last common ancestor (LCA) and the cnidarian and bilaterian LCAs. At each node we evaluate the evidence based on genomic data, expression data, and conserved pathways or functions, each progressively a more robust basis for inferring ancestral development. The summary here updates earlier reviews (e.g. Rokas 2008; Erwin 2009). Data from the fossil record, particularly from the Ediacaran (635–542 million years ago [Ma]) and the Cambrian (542–488 Ma) provides one critical insight when combined with molecular clock and comparative developmental studies: a long macroevolutionary lag separates the acquisition of the developmental toolkit needed to form bilaterians and their appearance in the fossil record (Erwin et al. 2011).

Phylogenetic Context

A phylogenetic context is required to integrate fossil and developmental evidence to understand the early evolution of Metazoa. In this section we present the current basic framework of metazoan phylogeny provided by molecular sequence data, with fossil data integrated as appropriate. The dating of important nodes is provided by molecular clock analyses. Although molecular clock studies of early metazoan history have yielded highly variable results, there is increasing congruence between different studies (see Chapter "Timing the Origins of Multicellular Eukaryotes Through Phylogenomics and Relaxed Molecular Clock Analyses"). In this section we follow the phylogenetic structure and molecular clock results presented in Erwin et al. (2011) except where indicated. We also highlight where discrepancies persist about the placement of particular clades.

Molecular Phylogeny

Recent studies of metazoan phylogeny have produced discordant results for the base of the tree (e.g. Edgecombe et al. 2011; Philippe et al. 2011b; Erwin et al. 2011; Ryan et al. 2013; Nosenko et al. 2013; see also discussion on the history of metazoan phylogenetic studies in Edgecombe et al. 2011). These studies differ in taxon coverage, sequences, and analytic details, but all reported strongly supported tree topologies. Two recent analyses of these studies evaluated the methodological problems contributing to the discordant results, particularly those associated with short internal branches between nodes caused by rapid diversification and long-branch attraction (Philippe et al. 2011b; Nosenko et al. 2013).

Although we acknowledge areas of continuing uncertainty below, drawing from these studies we have adopted the topology shown in Fig. 1 (based on Erwin et al. 2011), with choanoflagellates as the outgroup to Metazoa and paraphyletic sponges at the base of the tree. Placozoans, represented by *Trichoplax adhaerens,* branch next followed by coelenterates, including both cnidarians and possibly ctenophores (not shown). The position of ctenophores is critical to evaluating the monophyly of triploblastic animals with nerve and muscle cells, and is discussed below. The crown of bilaterians is well supported by molecular studies, and is comprised of the deuterostomes and the two major protostome clades: Lophotrochozoa and Ecdysozoa. Acoel flatworms, historically part of the Platyhelminthes and more recently placed between cnidarians and the bilaterian LCA, have been suggested to group with *Xenoturbella* at the base of the deuterostomes based in part on microRNA (miRNA) data (Philippe et al. 2011a). However, other studies support acoel and nematodermatid flatworms + *Xenoturbella* as more basal metazoans (Edgecombe et al. 2011), and as yet this clade remains unresolved. Regardless, one of the surprising insights of recent studies of bilaterian phylogeny is that the basal clades in each of the three major groups are morphologically simple, and as will be discussed further below, may have been simplified from morphologically more complex ancestors. The most recent metazoan phylogenies suggest that the basal groups are acoels and xenoturbellids among the deuterostomes, priapulids among the ecdysozoans, and platyhelminths among the lophotrochozoans. This pattern has important implications for interpreting the developmental and morphological complexity of the PDA and the LCA of the three major bilaterian clades.

Several aspects of this topology remain uncertain. Some molecular studies indicate that living sponges are paraphyletic, with the demosponges, calcareous sponges and homoscleromorphs all distinct clades (Sperling et al. 2009), a result that receives support from studies of miRNA (Robinson et al. 2013). Yet the paraphyletic nature of sponges, though supported by molecular data, remains contentious and Nosenko et al. (2013) concluded that sponge monophyly vs. paraphyly depends on root placement. There are several groups of fossil sponges that may well represent additional distinct poriferan clades.

Ctenophores have proven difficult to place phylogenetically, and have been suggested to be the sister group to cnidarians as well as both basal to and above cnidarians in the analyses cited above. Less plausibly they have been placed at the base of the metazoan tree (Ryan et al. 2013), probably a consequence of long-branch attraction (Nosenko et al. 2013). Support for this basal position is not strong, and clearly more data is needed to resolve the ctenophore problem. The position of acoels, as discussed above, also remains contentious. Morphologic and developmental evidence

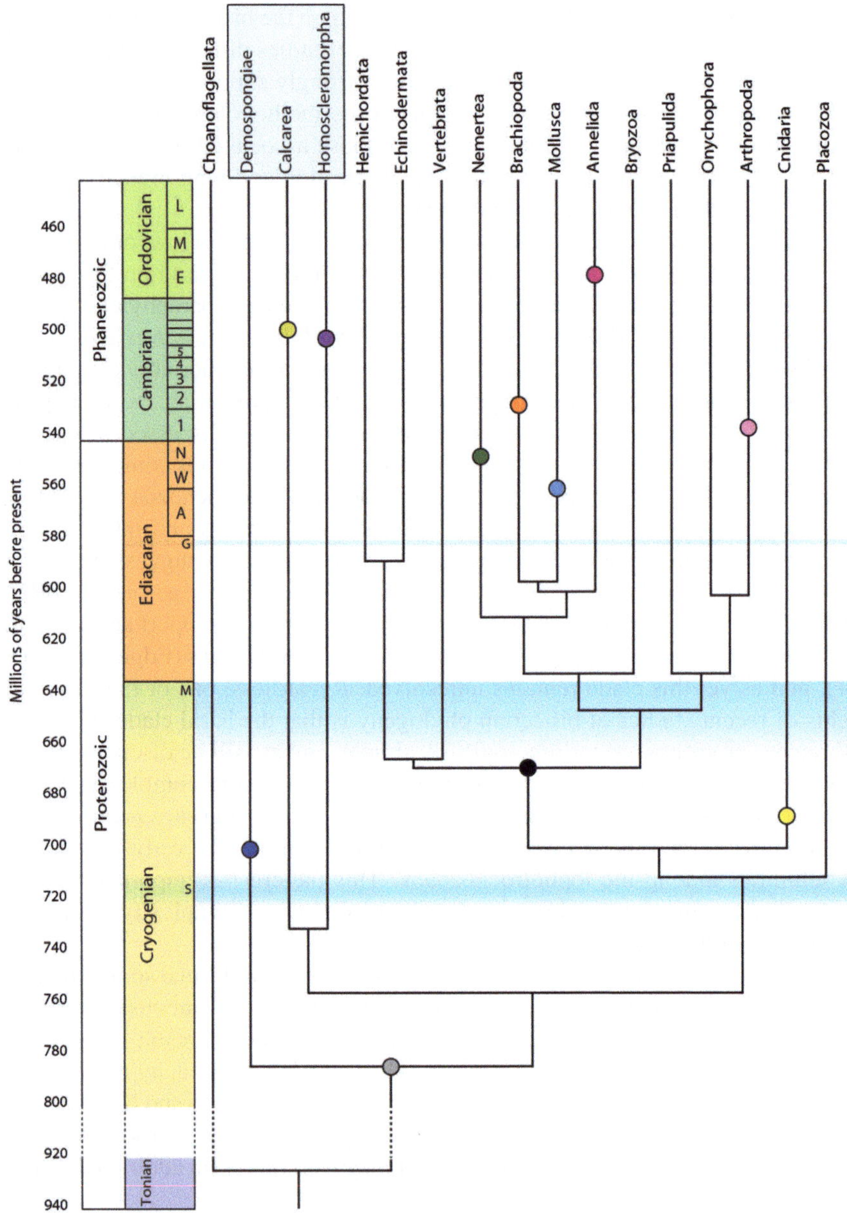

Fig. 1 The origin of animals and major metazoan clades as inferred from molecular clock analyses, based on Erwin et al. (2011). Age estimates (in millions of years [Ma]) for major nodes are indicated by *circles*—*grey* (Metazoa), *black* (Bilateria)—while crown group age estimates are indicated by the *colored circles* on each branch. Sponge paraphyly is indicated by the *blue box* containing the major sponge groups. *Horizontal blue bars* represent glaciations, abbreviated as: *S* Sturtian, *M* Marinoan, *G* Gaskiers. Major Ediacaran temporal/fossil assemblages are abbreviated: *A* Avalon, *W* White Sea, *N* Nama. Early stages of the Cambrian period are labeled (1–5), and Ordovician stages abbreviated as *E* Early, *M* Middle, *L* Late

suggested that they represent descendants of a basal bilaterian clade that branched before the PDA (Edgecombe et al. 2011), however, additional phylogenetic studies suggest a position as basal deuterostomes (Philippe et al. 2011a). Even if the latter hypothesis is correct, it is quite likely that acoelomorph-grade organisms, evidently now extinct, branched between cnidarians and the PDA. In this paper we follow the consensus phylogeny shown in Fig. 1, noting where continuing topological uncertainties would affect our discussion.

Origin of Major Metazoan Clades and Crown Groups

Molecular clocks have been applied to dating early metazoan divergences for several decades, although with considerable spread in the results. The suite of available genes, the methods of analysis, and the variety of fossil calibration points have all improved and consequently different clock studies are beginning to converge. As outlined above, here we adopt a recent analysis that is largely consistent with other recent studies (Erwin et al. 2011). Its results are robust to a variety of tests, including root placement, choice of molecular clock model, and subsampling of the calibration points.

Results place the Metazoan LCA node at about 800 Ma, followed by the rapid divergence of sponge clades, the appearance of cnidarians, and their divergence into the Anthozoa and Hydrozoa by about 700 Ma (Fig. 1). If these results are generally correct, they indicate that all of these divergences, as well as the appearance of the bilaterian LCA, occurred during the Cryogenian Period. As its name implies, the Cryogenian began with the Sturtian glaciation and ended, at about 635 Ma, with the end of the Marinoan glaciation. Each glaciation was global in extent, and each has been associated with possible ‘Snowball Earths’ in which the Earth’s surface became entirely glaciated. Whatever the validity of the Snowball Earth hypothesis, growing geochemical and other environmental evidence confirms considerable environmental turmoil during the Cryogenian.

Bilaterian clades diverged during the Ediacaran Period, with the time of origin of most bilaterian crown groups clustered into a relatively narrow interval from the late Ediacaran into the Cambrian. The overall pattern, on which we elaborate in the discussion, indicates genetic divergence of major metazoan clades during the Cryogenian and Ediacaran, but there is no evidence from the fossil record that representatives of these lineages had yet evolved Phanerozoic morphologies. After some 150–200 million years of evolution the relatively simultaneous appearance of new bilaterian morphologies is documented in the fossil record by the Cambrian explosion, and in the crown group nodes estimated via molecular clock analysis. A new study estimates major ecdysozoan divergences occurring in the Ediacaran, with the pancrustaceans diversifying during the early Cambrian (Rota-Stabelli et al. 2013).

Developmental Toolkits in a Phylogenetic Perspective

Genomic studies of many eukaryotic lineages have demonstrated that the widespread occurrence of multicellularity (Knoll 2011) reflects the broad distribution of many of the developmental tools required for the generation, organization, and maintenance of multicellular structures (Parfrey and Lahr 2013; Rokas 2008; Sebé-Pedrós et al. 2011). These tools include members of a variety of transcription factor families (Hox, Sox, T-box, Pax, ANTP, etc.), components of the major cell signaling pathways such as Wnt, Notch/Delta, Hedgehog, and TGF-β, as well as cell adhesion and structural molecules. Early information about the phylogenetic distribution of the developmental toolkit came from studying expression patterns in transcription factors and signaling systems, which turn out to display greater phylogenetic conservation than expected several decades ago (e.g., Chapters "Transcription Factors and the Origin of Animal Multicellularity" and "Developmental Signalling and Emergence of Animal Multicellularity"). Although these studies initially considered only model animal systems, more recent work has engaged a phylogenetically more diverse array of organisms, providing a stronger basis for inferring the nature of developmental systems in early Metazoa. For our discussion of early metazoan developmental toolkits, four nodes on the tree in Fig. 1 are of particular significance: the last common metazoan ancestor, which documents the complexity of gene regulatory and developmental tools shared across all living animals; the node between placozoans and all higher metazoans; the cnidarian-bilaterian LCA; and the protostome-deuterostome ancestor (often considered the bilaterian LCA as well).

One theme that has emerged recently is that most of the developmental tools responsible for patterning bilaterians have much deeper roots and many developmental genes—and even gene pathways—appear to have achieved their current functions through gene co-option. Current function is thus often a poor guide to ancestral function, and can severely constrain inferences about the nature of ancestral clades (Erwin and Davidson 2002; Davidson and Erwin 2010). Reconstructing developmental tools and capabilities at various points along metazoan phylogeny depends upon an array of different evidence, including genomic data, expression data, and evidence for conserved developmental pathways and functions (in increasing order of the difficulty of obtaining the data). Genomic data (see Table 1) provides a baseline estimate of the shared developmental toolkit, and information about the presence of critical protein-binding domains at most may suggest potential function. Demonstrating spatiotemporal patterns of expression is often a first step in hypothesizing gene developmental roles, however, only data from experimental manipulation verifies conserved pathways and roles for toolkit components (Table 2). The strength of inferences about the nature of LCAs differs depending on the source of data; consequently, in this section we discuss evidence for each specific type of data. Tables 1 and 2 summarize a selection of the genomic and expression/functional evidence (respectively) discussed below. This approach provides a conservative basis for inferring the likely morphology at each key node.

Table 1 Genomic presence/absence of a selection of developmental tools integral to metazoan development. This subset of genes is representative of the evolving molecular capacity for core developmental processes, such as morphogenesis and patterning/differentiation. Summarized from (King et al. 2008; Srivastava et al. 2008; Fahey and Degnan 2010; Srivastava et al. 2010b; Fairclough et al. 2013)

		Choanoflagellatea	Sponges	Placozoa	Cnidaria	Bilateria
Cell-cell adhesion	Integrins	○	•	•	•	•
	Fibrillar collagen	○	•	•	•	•
Cell polarity	Crumbs		○	•	•	•
	Stardust/MPP5/Pals-1				•	•
	Bazooka/Par-3		○	○	•	•
	Par-1	•	•	•	•	•
	Lethal Giant Larva (Lgl)	○	•	•	•	•
	Discs Large (Dlg)	•	•	•	○	•
Adherens junctions	Cadherins	○	•	•	•	•
	Catenins		•	•	•	•
	Patj	•	•	•	•	•
	Vinculin	•	•	•	•	•
Basal lamina	Type IV collagen		○[a]	•	•	•
	Laminin	○	○	○	•	•
	Nidogen			•	•	•
	Perlecan			○	•	•
Tight junctions	Claudin					•
	Occludin					•
Septate junctions	Neurexin			○	○	•
	Neuroglian			•	○	•
	Contactin					•
Signaling pathways	Notch/Delta	○	•	○	○	•
	Hedgehog	○	○	○	•	•
	Wnt		•	•	•	•
	JAK/STAT	○	○	○	○	•
	RTK	○	○	•	•	•
	TGF-β		•	•	•	•

Table 1 (continued)

		Choanoflag-ellatea	Sponges	Placozoa	Cnidaria	Bilateria
Transcription factor families	Sox	•	•	•	•	•
	Fox	•	•	•	•	•
	T-box		•	•	•	•
	POU		•	•	•	•
	LIM		•	•	•	•
	ANTP		•	•	•	•
			•	•	•	•
	Six		•	•	•	•
	PRD		•	•	•	•

Closed circles represent the presence of gene homologs identified in one or more species of the clade. *Open circles* represent the presence of aberrant or incomplete homologous genes. In the case of signaling pathways, *closed circles* indicate homologs of a complete signaling pathway, and *open circles* indicate the absence of core pathway components
[a]indicates the presence of type IV collagen in one sponge group (Homoscleromorpha)

Metazoan LCA

The availability of choanoflagellate and sponge whole-genome sequences (King et al. 2008; Srivastava et al. 2010b; Fairclough et al. 2013) as well as genomic data from several additional sponges has greatly improved our ability to infer the shared genetic features of the metazoan LCA. However, function cannot be confidently assigned by sequence presence/absence data alone, and both gene expression data and experimental reconstruction of functional pathways lag behind that of both the eumetazoan and bilaterian LCAs.

Genomic Data

Choanoflagellates contain genes that encode conserved protein domains found in cadherins, some cell adhesion and polarity proteins, and components of signaling pathways used in cellular differentiation and patterning in metazoans (King et al. 2008; Fairclough et al. 2013; Chapter “Choanoflagellates: Perspective on the Origin of Animal Multicellularity”). Representative domains of some metazoan transcription factor families are present as well, although these are less diverse than in metazoans. Of the seven major metazoan signaling pathways (Wnt, Hh, TGF-β, Notch/Delta, JAK/STAT, NHR, RTK), elements of four (Hh, Notch/Delta, JAK/STAT, RTK) are known from choanoflagellates, although the JAK/STAT pathway is represented by only a single gene (Larroux et al. 2008) (see Chapter “Transcription Factors and the Origin of Animal Multicellularity”).

Table 2 Comparison of functional roles of a subset of metazoan developmental tools. Summary of experimental evidence demonstrating the likely developmental functions of select genes/pathways in extant groups, and by extension, the potentially ancient origin of some of these roles. See text for full discussion

	Sponges	Placozoa	Cnidaria	Bilateria
Canonical Wnt pathway	Early polarized expression in *Amphimedon* larvae suggests axial patterning; disruption affects aquiferous system organization	Pathway components present in genome	"Organizer" role in *Nemtaostella, Clytia, Hydra;* specification of oral/aboral axis, oral regional fate and endoderm fate	"Organizer" in many bilaterians; anterior/posterior axial patterning, specification of posterior fate, gastrulation site, endomesoderm
Notch/Delta pathway	Dynamical expression in *Amphimedon*	Pathway components present but incomplete; lacking key domains	Oriented cell division and shape change, cell fate partitioning in tentacle elongation in *Nematostella*	Many roles, including morphogenesis, boundary formation, cell fate partitioning
Pax/Six	Pax/Six expressed in *Ephydatia* choanoderm active growth zone; involved in aquiferous system organization, cell type differentiation	Six3/6 and second Six family transcription factor present in genome	Six3/6 involved in early identity of aboral region; activates "anterior" Fox gene homologs	Pax/six involved in development of tissues and organs; Six3/6 involved in anterior patterning
FOX	Fox genes present in genome	Ectopic expression of *Trichoplax* FoxJ1 in zebrafish induces expression of ciliogenic target genes	FoxJ1 may be involved in specification of ciliary cells in *Nematostella* apical organ	FoxJ1 regulates ciliogenesis in vertebrates; expressed in sea urchin larvae apical tuft
HOX	True Hox genes absent; likely secondary loss	Expressed in interstitial cells around periphery of animal; disruption halts binary fission/growth	Exhibits staggered expression along *Nematostella* oral/aboral axis	Involved in regional patterning across bilateria
LIM	Lhx3/4(lim-3) and 1/5 (lin-11) expressed in cells specific to *Amphimedon* larval "eye" pigment rings	Lhx 1/5 and other LIM family representatives present in genome; 3 of 6 expressed in cultured *Trichoplax*	Potential involvement in neural cell fate; expressed in neuralized regions	Lim-3, lin-11 involved in organ development, specification of neural cell types

Reconstruction of the evolution of signaling pathway components between choanoflagellates and the metazoan LCA suggests that it largely involved the shuffling of ancient eukaryotic domains with completely new domains. For example, domains of key components of both the Wnt and Hedgehog pathways, although present in choanoflagellate genomes, are not found in the organizations known to be critical for pathway function in metazoans ("Transcription Factors and the Origin of Animal Multicellularity"). Interestingly, although the pathways are incomplete, some signaling components shared between choanoflagellates and metazoans have been found to be disproportionately upregulated in both thecate cells and multi-cell colonies of the choanoflagellate *Salpingoeca rosetta* (Fairclough et al. 2013, see also "Choanoflagellates: Perspective on the Origin of Animal Multicellularity"); perhaps the evolution of these domains may be related to the capacity to form these different cell morphologies and multi-cell associations.

Sponges have multiple cell types with some regional patterning and coherent body architectures. They possess homologs within most of the major transcription factor families, including ANTP, Sox, Fox, T-Box (including Brachyury), and PRD-like (Larroux et al. 2008; Srivastava et al. 2010b). Although T-box genes have not been identified in choanoflagellates, they have been reported from the amoeba *Capsaspora owczarzaki* and other opisthokonts, including a Brachyury homolog (Sebé-Pedrós et al. 2013). Moreover, the T-box family appears to have diversified at the base of Metazoa. The developmental tools necessary to generate basic sponge-grade features are present, but sponges appear to lack many of the tools necessary for the level of developmental and regulatory control found in more derived clades, and most of the transcription factor families listed above have relatively few members. True Hox genes, responsible for regional identity establishment in bilaterians, have not been identified in sponge genomes, although this seems to be due to the secondary loss of Hox and ParaHox loci in the poriferan lineage (Ramos et al. 2012). Adamska et al. (2011) suggest that the Wnt, TGF-β and Notch/Delta signaling pathways were largely complete and probably functioned similarly to those of eumetazoans, while the Hedgehog and tyrosine kinase growth factor pathways were evidently still missing their key signaling ligands (see "Developmental Signalling and Emergence of Animal Multicellularity"). The genome of the homoscleromorph sponge *Oscarella carmela*, as in the demosponge *Amphimedon queenslandica*, has components of six of the seven primary signaling pathways in metazoans (Nichols et al. 2006).

The evolution of epithelia is a critical innovation in metazoans, and there has long been debate over the extent to which sponges can be considered to have true epithelia (Leys and Riesgo 2012; Fahey and Degnan 2010). Part of the controversy reflects different definitions of epithelia, but critical to this subject is the extent to which sponges contain adherens and tight junctions as well as a basement membrane or lamina. The *Amphimedon* genome has evidence for several orthologs of cell polarity genes and adherens junction components, but no evidence for orthologs of genes associated with either tight or septate junctions, or basal lamina (Srivastava et al. 2010b; Fahey and Degnan 2010; Nichols et al. 2012). Consistent with the pattern seen in other proteins, many of the domains that were eventually assembled to form septate and tight junctions and basal laminae are present in the *Amphimedon* genome and were likely present in the metazoan LCA (Fahey and Degnan 2010). In contrast, homoscleromorphs have long been known to have type IV collagen, a key component of basement membranes, and Leys and Riesgo (2012) have identified the sequence for it in a calcareous sponge as well. Leys and Riesgo (2012) also suggest that poriferan claudin-like genes indicate septate junctions were present in the metazoan LCA.

Cadherins are cell surface receptors involved in cell adhesion, polarity and developmental signaling. Although many cadherin protein domains have been identified in choanoflagellates, these do not include the conserved cytoplasmic domain found in metazoan cadherins (King et al. 2008; Fairclough et al. 2013). Analysis of sponges indicates that representatives of at least three cadherin families, lefftyrin, coherin and hedgling (a sponge and cnidarian-specific precursor to Hedgehog), were present in the metazoan LCA and, as discussed below, may have been able to participate in intracellular signaling (Nichols et al. 2012).

An expressed sequence tag (EST) comparison of sponge genes to those of other metazoans showed the greatest similarity between sponges, cnidarians and deuterostomes, with the lowest similarity among protostomes (Harcet et al. 2010). This result is consistent with earlier comparisons between *Nematostella vectensis* and other lineages indicating extensive gene loss on the protostome line (Ogura et al. 2005; Chapman et al. 2010; Forêt et al. 2010).

Expression Data

The Sox genes are a family of HMG box transcription factors involved in the regulation of cell type specification and development in animals, and the family appears to have arisen between choanoflagellates and the metazoan LCA. Fortunato et al. (2012) identified seven Sox and four Sox-like genes in a calcareous sponge, and showed that they were dynamically and differentially expressed in unique patterns during embryonic development, or in specific adult sponge cell types. Similarly, *Amphimedon* embryos dynamically express the genes for a single Notch receptor and five Delta ligands, reflecting the multiple developmental roles of this signaling pathway (Richards and Degnan 2012). In bilaterians, Pax and Six genes are part of a regulatory network that is involved in development of many tissues and organs. Recent work has shown Pax and Six expression in cells located in active growth zone in the developing choanoderm of a freshwater demosponge (Rivera et al. 2013), and additional knock-down experiments (as discussed below) have elucidated some Pax/Six regulatory interactions.

The Wnt signaling pathway is absent in choanoflagellates, and the presence of almost all Wnt pathway components in multiple sponges indicates this to be a metazoan developmental innovation (Petersen and Reddien 2009; Adamska et al. 2010; "Developmental Signalling and Emergence of Animal Multicellularity"). Wnts carry out many functions in bilaterian development, including the establishment and organization of primary body axes. The dynamic expression patterns of Wnt signaling components during *Amphimedon* embryonic development suggest multiple, temporally distinct regulatory roles. Adamska et al. (2010) observed early, polarized expression of Wnt pathway components suggestive of broad anterior-posterior (A/P) axis patterning in addition to later expression localized to a larval organ-like structure. Although these distinct spatio-temporal expression patterns are compelling, further study is needed to demonstrate the precise functional roles of the Wnt pathway in generating these sponge features.

The expression patterns of genes associated with more terminally differentiated cell types have also been investigated in sponges. Although sponges lack mesoderm or muscle cells, Steinmetz et al. (2012) identified orthologs of both striated muscle (ST) and non-muscle (NM) heavy chain myosin (MyHC) in two demosponges. Non-muscle MyHC was detected in many adult cell types, including the pinacocytes responsible for sponge peristalsis-like contraction, while ST MyHC was restricted to the outlet pores (apopyles) which house a sieve-like cell that controls water flow. Steinmetz et al. propose that an early metazoan contractile apparatus existed in the LCA, the components of which independently evolved to form distinct muscle types in the cnidarian and bilaterian lineages (see below). Though sponges lack neurons and photoreceptive opsin pigments, *Amphimedon* larvae possess posterior ciliated pigment rings that are believed to mediate phototaxis (Leys et al. 2002). Rivera et al. (2012) recently discovered two genes encoding cryptochrome (Aq-Cry1 and Aq-Cry2), one of which (Cry2) is expressed in the larval pigment ring eye and maximally absorbs blue light—the same wavelength corresponding to peak larval swimming activity. The LIM homeobox family of transcription factors are involved in the development of many organs in bilaterians, but notably all members of this family have roles in specifying neural cell fates. Lhx3/4 (lim-3), Lhx1/5 (lin-11) and Islet are all expressed in developing *Amphimedon* embryos, including cells specific to the aforementioned eye pigment rings (Srivastava et al. 2010a). These expression studies suggest that molecular tools gained their developmental roles in multifunctional cells prior to the segregation of these functions in additional specialized cell types.

Conserved Pathways and Conserved Functions

Complex developmental GRNs have not yet been identified in sponges, and some workers have suggested that this may limit their morphologic complexity (Adamska et al. 2007; Fahey et al. 2008). Some evidence for conserved functions, however, has been described. In line with the observed polarized expression of Wnt in *Amphimedon* embryos (Adamska et al. 2010), Windsor and Leys (2010) have found that disruption of the canonical Wnt pathway effects aspects of sponge aquiferous system organization, suggesting a role of the Wnt pathway in establishing sponge axial polarity. And, although studies have yet to be done *in vivo*, based on amino acid domain comparisons Nichols et al. (2012) propose that the classical cadherin identified in *Oscarella carmela* may have had the ability to bind β-catenin, thus regulating cell-cell adhesion and contributing to other downstream signaling cascades. The regulatory interactions between Pax and Six genes have been established in the freshwater demosponge *Ephydatia muelleri*, indicating that a component of the network is present and may have been involved in the formation of a multifunctional epithelia (Rivera et al. 2013).

Extinct spongiform clades have been described from the lower Paleozoic fossil record, including archaeocyathids, chancelloriids and other groups. Archaeocyathids include a variety of early Cambrian fossils with a unique, double-walled and heavily calcified skeleton. They have been assigned to a number of groups, from sponges to an extinct phylum, but the discovery of modern sclerosponges has confirmed that archaeocyathids were indeed sponges (Rowland 2001). It now appears likely that both sponge spicules and massive, sclerosponge-like skeletons evolved multiple times among the different sponge lineages; other fossil clades of equivalent rank may also exist (Sperling et al. 2007). Further study of the unique character combinations of some of these groups might shed further light on the evolution of developmental complexity (Erwin and Valentine 2013).

The paraphyletic nature of sponges indicates that the metazoan LCA was likely a simple filter-feeding organism, with genetic elements of what would eventually become elaborate signaling pathways and developmental GRNs. Multiple cell types with some regional body patterning would have been present, but although homologs of most major transcription factor families were present, these were likely utilized in cell-type specification rather than in establishing more sophisticated morphology. Any limited patterning may have included axial polarization/organization via Wnt signaling, for example, as there is little evidence that sponges were capable of more sophisticated regulatory control. Tight junctions and basal laminae may have been present, with cadherins mediating cell adhesion and signaling; thus, the metazoan LCA may have been a proto-epithelial, simply-patterned organism.

Placozoan-Eumetazoan LCA

The complete genome sequence of the placozoan *Trichoplax adhaerens* has provided insights into development in eumetazoans (Srivastava et al. 2008). The presence of epithelia in *Trichoplax* has led to the definition of placozoans plus all eumetazoans as the epitheliozoa (Sperling et al. 2009), suggesting the presence of true epithelia in this basal group. However, while polar cells with zonula adherens are present in placozoans, in addition to genes encoding collagen and laminins, there is no underlying basal lamina and evidence for septate junctions is ambiguous. Thus, true epithelia seems to be absent (Fahey and Degnan 2010). Placozoans also lack nerve or muscle cells (although they do respond to external stimuli), organs, and anterior-posterior differentiation, but share members of the gene families associated with such patterning with eumetazoans. For an excellent discussion of the tribulations of *Trichoplax* since its discovery and an overview of its morphology see Schierwater et al. (2011). Some of the soft-bodied Ediacaran macrofossils may represent either placozoans or related lineages (Sperling and Vinther 2010).

Genomic, Expression and Functional Data

Evidence on the nature of the placozoan developmental toolkit is based largely on genomic data, with only a few papers providing information on gene expression. The genome of *Trichoplax* shows a high degree of conservation of gene structure and synteny, and little of the intron loss observed in other species (Srivastava et al. 2008) (see also "A Comparative Genomics Perspective on the Origin of Multicellularity and Early Animal Evolution"). The whole-genome analysis identified a variety of transcription factors used by eumetazoans in regional patterning and cell type specification, despite the limited evidence for placozoan regional patterning and the presence of only four or five cell types (Srivastava et al. 2008). A total of 23 different transcription factor families have been identified, including homeobox members of the paired and ANTP classes and Brachyury from the T-box family. Within the ANTP class, representatives of the Hox/ParaHox, NKL and extended Hox groups have been identified (Schierwater et al. 2011). A more recent analysis of orthologous genes syntenous to Hox and ParaHox loci indicates that the single Hox-like gene in *Trichoplax* is likely a ParaHox ortholog, with only remnants of a Hox locus remaining after the loss this gene in the lineage (Ramos et al. 2012).

Several signaling pathways are present, including Wnt, TGF-β, Notch and JAK/STAT, although some have lost key components for signal transduction. As some sponges appear to have these components, this suggests they were lost in the placozoan lineage. Some cell types in *Trichoplax* may have retained ancestral multifunctionality, which would explain why the genome contains evidence for ion channels, neurotransmitter and neuropeptide receptors but lacks evidence for the ability to generate either neurotransmitters or synapses. Srivastava et al. (2008) emphasize that the rich diversity of *Trichoplax* developmental control systems is consistent with the suggestion of Erwin and Davidson (2002) that these systems were co-opted in cnidarians and other eumetazoans for regional patterning.

Gene expression and functional data in *Trichoplax* is limited, but hints at a developmental complexity belied by the basic placozoan morphology. Trox-2, the single ParaHox gene described above, is expressed around the entire animal periphery in small cells between the upper and lower epithelial layers (Jakob et al. 2004). Disruption of this gene halts growth and binary fission, suggesting a role in the specification of a stem cell population. The T-box family genes Brachyury and Tbx2/3 are also expressed in *Trichoplax*, and because these genes show distinct localizations uncorrelated with anatomical features, Martinelli and Spring (2008) suggest the presence of undescribed cryptic cell types. The placozoan upper cell layer consists of monociliated cells, and placozoans do possess a homolog to FoxJ1, a transcription factor involved in ciliogenesis. Ectopic expression of this placozoan homolog in zebrafish embryos induces the expression of known ciliogenic target genes, supporting a conserved role for placozoan FoxJ1 in the development of cilia (Vij et al. 2012). The transcription of other key toolkit components has been reported without localization information, via *in situ* hybridization (LIM homeobox genes; Srivastava et al. 2010a) and mass spectrometry of the proteome (Notch/Delta, Wnt and TGF-β pathway components, ECM-related proteins; Ringrose et al. 2013). The

Trichoplax proteome analysis only increases ambiguity concerning the existence of basal lamina, for although many basement membrane proteins are translated there is no evidence for either classical cadherin expression or the associated catenins known to mediate adhesion at zonula adherens junctions (Ringrose et al. 2013).

As Sperling and Vinther (2010) note: "crown-group placozoans likely represent a limited and highly derived subset of [Ediacaran + placozoan clade] diversity" (p. 204). This seems to us quite likely, with *Trichoplax* a highly derived remnant of this clade. Ancestral members of this clade could have been more morphologically sophisticated, given the developmental complexity of elements found in *Trichoplax*, but in a pattern first observed in sponges (and that will recur in cnidarians) the developmental potential of these clades appears to be higher than their realized morphologic complexity.

Cnidarian-Bilaterian LCA

As described in the metazoan phylogeny section, we believe that the weight of the current evidence supports a topology with ctenophores as the sister group to cnidarians (Coelenterata hypothesis), and we view claims that ctenophores are basal to sponges as an artifact of long-branch attraction (Nosenko et al. 2013). The genome of the ctenophore *Mnemiopsis* has been sequenced (Ryan et al. 2013), and recent papers suggest that it lacks a variety of developmental genes found in other eumetazoans. Among the missing elements in *Mnemiopsis* are five of the 11 defined homeodomain classes and several Hox class genes (Ryan et al. 2010), parts of the TGF-β signaling pathway (Pang et al. 2011), and both miRNAs as well as the Drosha and Pasha nuclear proteins required for miRNA processing (Maxwell et al. 2012). This has been interpreted as supporting a position for ctenophores basal to sponges (as in Ryan et al. 2013) and would imply that muscles, elements of the nervous system, and other attributes of ctenophores evolved twice, requiring sponges to be secondarily simplified from a more complex ancestor. Given the well-documented problems with establishing relationships among early-branching metazoans, it seems more plausible that crown ctenophores have secondarily lost many elements.

Cnidarians exhibit polarity along a primary body axis (oral-aboral) and are diploblastic, with two germ layers (ectoderm and endoderm) giving rise to two epithelial layers with myoepithelial cells, a non-centralized nerve net nervous system, and the highly specialized stinging cells (cnidocytes) which define the group. It is now evident that cnidarian body plans, morphologically simple in comparison to the diversity of bilaterian forms, are underlain by complex developmental programs that deploy many of the same molecular tools and regulatory pathways thought to be unique to bilaterians.

Genomic and Expression Data

The genomes of *Nematostella vectensis* and *Hydra magnipapillata* share a surprisingly greater number of linkage groups and conserved synteny with the human genome than what may be expected given the estimated cnidarian-bilaterian divergence of 700 Ma (Putnam et al. 2007; Chapman et al. 2010; Steele et al. 2011; Erwin et al. 2011) (see Chapter "A Comparative Genomics Perspective on the Origin of Multicellularity and Early Animal Evolution"). And, although it has less conserved gene organization than in *Nematostella*, *Hydra* displays genomic conservation far beyond that of the more recently diverged protostomes *Drosophila melanogaster* and *Caenorhabditis elegans* (Chapman et al. 2010). Preliminary analysis of EST data from the hydrozoan *Clytia hemisphaerica* finds it has a higher proportion of unique, taxon-specific genes than either *Nematostella* or *Hydra*, perhaps owing to unique genetic requirements of both polyp and medusa life cycle stages (Forêt et al. 2010). Given the deep split between the cnidarian crown groups Anthozoa and Hydrozoa, the forthcoming *Clytia* whole-genome sequence should be enlightening.

Homologs for members of all of the major developmental transcription factor families seem to be present in cnidarians, with the possible exception of the NF-κB and NFAT families. The single *Nematostella* NFAT-like protein groups ambiguously with other NFAT genes, and the IκB C-terminal domain of all bilaterian NF-κB proteins is absent in the *Nematostella* homolog (but present in different protein sequence) (Sullivan et al. 2006). Sox, Fox, and T-box genes are found in the *Nematostella* genome (Magie et al. 2005; Putnam et al. 2007), and, like their homologs in bilaterians, appear to be associated with processes as diverse as regional patterning, germ layer and cell fate specification, and morphogenesis. Sox genes expressed in the blastopore region of *Nematostella* gastrulae may be functioning to restrict ectoderm vs. endoderm fate, while a repertoire of *Nematostella* Fox genes expressed along the oral-aboral axis may be involved in defining distinct domains along this primary axis (Magie et al. 2005).

Nematostella possesses representatives of all classes of homeobox transcription factors, barring *engrailed*, while the faster rate of evolution in *Hydra* has lead to the loss of many homeobox proteins in its genome (Chapman et al. 2010; Forêt et al. 2010). Members of the homeobox transcription factor LIM family are present in both taxa (e.g. Lhx, Lmx, islet), and as in bilaterians, may work in combination to specify specific neural cell fate. *Nematostella* LIM genes are combinatorially expressed in three major neuralized regions—the planula apical tuft and the polyp oral and pharyngeal nerve rings—but expression of these genes in neurons and similar expression in *Hydra* or *Clytia* have yet to be demonstrated (Srivastava et al. 2010a). And, while the staggered expression of *Nematostella* anterior and posterior-class Hox genes along the oral-aboral axis (Finnerty et al. 2004) is suggestive of a conserved "Hox code" between cnidarians and bilaterians, Hox homolog expression domains in *Clytia* do not demonstrate conservation of such a role across Cnidaria (Chiori et al. 2009), and a "true" Hox patterning system likely postdates the cnidarian-bilaterian split (Schierwater and Kamm 2010).

By the evolution of the cnidarian-bilaterian LCA, genes representing all of the major developmental signaling pathways are present and largely complete (Putnam et al. 2007; Chapman et al. 2010). Components of the non-canonical (PCP) Wnt pathway, absent in sponges (with the exception of atypical cadherin Flamingo in *Oscarella carmela* (Nichols et al. 2012)), are present in the *Nematostella* and *Hydra* genomes (Putnam et al. 2007; Chapman et al. 2010), along with eumetazoan BMP/Chordin signaling components and downstream TGF-β effectors. The first true Hedgehog proteins also arose in the cnidarian-bilaterian LCA lineage, which likely possessed both Hedgling and Hedgehog proteins. Interestingly, because the Hedgehog ligand requires polarized cells and associations with basement membrane-related proteins for processing (Ingham et al. 2011), the evolution of this signaling pathway may have been dependent upon the evolution of true epithelia.

Cnidarian contractile myoepithelial cells comprise both striated and smooth muscle types, both of which express the highly conserved type II myosin heavy chain protein. Striated-type (ST) MyHC is expressed in cnidarian fast-contracting muscles and non-muscle (NM) MyHC broadly in both smooth and non-muscle cells in the endoderm. However, cnidarians lack defining features of bilaterian striated muscle, such as the troponin complex, which is completely missing, and z-disc components, which are expressed either ubiquitously or only in smooth muscle (Steinmetz et al. 2012). Steinmetz et al. (2012) propose that cnidarian and bilaterian striated muscle each evolved independently from an ancient shared contractile machinery based on ST and NM MyHC.

Conserved Pathways and Conserved Functions

Functional studies in cnidarians—primarily *Nematostella*, *Clytia*, and *Hydra*—increasingly point to an ancient conserved role for canonical Wnt signaling as a major developmental "organizer", contributing to axis establishment, germ layer specification, and gastrulation. Throughout Bilateria, early nuclear β-catenin localization via activation of canonical Wnt signaling is a major contributor to patterning the future posterior (and thus A/P axis), specifying endomesoderm cell fate, setting the site of gastrulation, and antagonizing anterior-patterning signals. The Spemann organizer in the frog *Xenopus* provides a classic example: when transplanted to new regions of an embryo, cells that have received these signals induce a secondary axis.

A growing body of evidence indicates a conserved role for the canonical Wnt pathway in determining a similar cnidarian "organizer" and in patterning axial polarity. The cnidarian blastopore forms in the animal hemisphere of the embryo at the oral pole, and cells that ingress form the cnidarian endoderm. As in the formation of bilaterian posterior domains, oral pole fate seems to be determined by known bilaterian "posterior" cues. In *Nematostella* and *Clytia*, β-catenin stabilization specifies the oral-aboral axis by defining presumptive oral territory; inhibition of the canonical Wnt pathway disrupts gastrulation, while ectopic pathway activation favors endoderm formation (Momose and Houliston 2007; Lee et al. 2007; Röttinger et al. 2012). Transplantation of the *Nematostella* blastopore lip even induces the formation of a secondary body axis (Kraus et al. 2007). In *Hydra*, Wnt, downstream effector Tcf, and β-catenin are known to exhibit "head" organizer effects, and maintain adult polarity (Broun et al. 2005). These findings make a compelling case for the molecular homology of the cnidarian oral and the bilaterian posterior poles.

Many deuterostome embryos are "pre-loaded" with maternal transcripts to deploy this Wnt cascade, and a similar mechanism in *Clytia* has been identified (Momose et al. 2008). Maternal transcripts of the *Clytia* Wnt receptor Frizzled (CheFz1 and CheFz3) were found to be localized to the animal and vegetal halves of the egg, respectively, with CheFz1 forming a cytoplasmic gradient from animal to vegetal pole and CheFz3 tightly localized to the vegetal cortex. These receptors regulate β-catenin stabilization positively (CheFz1 in oral pole) and negatively (CheFz3 in aboral pole), effectively establishing polarized Wnt activation (and oral tissue identity) along the oral-aboral axis.

Cnidarian orthologs of bilaterian "anterior" genes also appear to specify the cnidarian aboral pole, adding further evidence for a conserved primary axis patterning regulatory program (Sinigaglia et al. 2013). Six3/6, a key regulator of anterior patterning in all three major bilaterian clades, specifies the early identity of the *Nematostella* aboral region, and activates a suite of anterior-patterning homologs which set more finely-resolved regional boundaries. These include FoxQ2, FGFs (known to antagonize Wnt signaling and thus suppress posterior fate in bilaterians), and FoxJ1, which seems to specify development of ciliary cells in the apical organ (Sinigaglia et al. 2013).

Other cnidarian signaling pathways share more broadly related patterning and morphogenetic roles with bilaterians. BMP2/4/Dpp and Chordin establish endodermal molecular asymmetry along the *Nematostella* directive (perpendicular to oral/aboral) axis (Saina et al. 2009). The bilaterian dorsal/ventral axis is also specified by BMP signaling, but though it is tempting to homologize these pathways, Saina et al. (2009) point out that differences in wiring make it likely that BMPs were first used for axial patterning before being modified in bilaterians for specific D/V axis formation. The non-canonical Wnt pathway (planar cell polarity [PCP] pathway), well-studied in *Drosophila* and vertebrates, is involved in the generation/lateral orientation of cellular structures (including cilia) in epithelial sheets, and through this mechanism, planar-oriented cell divisions. Momose et al. (2012) have demonstrated that the PCP pathway is necessary for ciliogenesis and the oral/aboral orientation of cilia in the *Clytia* planula larva in addition to the oriented cell divisions necessary for embryo elongation, supporting an ancient role in cellular organization. Notch/Delta signaling in *Nematostella* appears involved in the oriented cell division, cell shape change, and cell fate partitioning required for embryonic tentacle elongation (Fritz et al. 2013), and though these reflect a subset of Notch/Delta roles in bilaterians, it is also notable that tentacle outgrowth does not seem to employ a Distal-less/Dlx-based signaling circuits as does many bilaterian outgrowths/appendages.

By the evolution of the cnidarian-bilaterian LCA, metazoans possessed a large and functionally diverse repertoire of developmental signaling pathways and transcription factors. These animals had the capacity to establish both a primary and an orthogonal body axis, as well as organize sensory and other anatomical structures in relation to these axes. A mouth and gut were probably present, in addition to multifunctional cells necessary for navigating and responding to the environment. Developmental components seemed to have participated in roles similar to, but less specific than, those in bilaterians, contributing to broad fate restriction and patterning but not yet to the extensive segregation of function in bilaterian germ layers, organs, and tissues.

Protostome-Deuterostome LCA

Developmental biologists initially had little reason to expect that the developmental pathways, and even the genes controlling development in organisms as different as flies and vertebrates, would be related. The discovery of extensive conservation of both genes and pathways led to early proposals on the nature of the protostome-deuterostome LCA (Arendt and Nubler-Jung 1994; Shenk and Steel 1994; De Robertis and Sasai 1996) and more detailed discussions as additional data accumulated (Knoll and Carroll 1999; Baguñà et al. 2001; Carroll et al. 2001; Valentine et al. 1999). As described in the preceding sections, however, the identification of many of these highly conserved genes in phylogenetically more distant clades called into question many of the earlier inferences about developmental homologies (Erwin and Davidson 2002; Davidson and Erwin 2010).

As the cnidarian-bilaterian LCA was characterized by general body patterning produced by conserved organizing pathways well-described in bilaterians, we will limit our discussion of the PDA toolkit to those components contributing to more urbilaterian-specific features.

Genomic Data

Inferring the structure of the bilaterian LCA genome and developmental toolkit requires comparisons between the three major bilaterian clades: the protostome groups Ecdysozoa and Lophotrochozoa, and Deuterostomia. Early work with model ecdysozoans *Drosophila* (fruit flies) and *Caenorhabditis elegans* (nematode worm) seemed to indicate that Ecdysozoa had undergone extensive gene loss, but with the addition of many more taxa it is clear that gene loss is more a function of divergence time and high rates of evolution regardless of lineage (Wyder et al. 2007; Miller and Ball 2009). Data from lophotrochozoans still lags behind the other clades. A preliminary analysis of three lophotrochozoan genomes—owl limpet *Lottia gigantea*, marine polychaete *Capitella teleta*, and freshwater leech *Helobdella robusta*—finds that these genomes exhibit greater similarity in organization and content to basal deuterostomes than to other protostomes, and the authors suggest that these genomes may more accurately approximate the ancient bilaterian condition (Simakov et al. 2013).

Gene Expression and Conserved Pathways and Functions

As in cnidarians, the Wnt pathway is utilized for axial patterning across Bilateria, from ecdysozoans (e.g. *C. elegans*, insects) and lophotrochozoans (planarians, nemerteans) to both basal deuterostomes (urochordates, echinoderms, cephalochordates) and chordates. Canonical Wnt signaling is utilized by all groups for specifying posterior fate, but it appears that the Hedgehog pathway takes on a more prominent role in antagonizing Wnt signals in the developing bilaterian anterior (Petersen and Reddien 2009; Ingham et al. 2011). Wnt and Hedgehog signaling interactions also establish and maintain A/P segmental boundaries in arthropods, and possibly annelids as well (Dray et al. 2010; Ingham et al. 2011). In *Xenopus*, the endo- and mesodermal fate-specifying signals that set up early organizer and A/P cell fates also induce BMP antagonist expression, initiating the dorsal/ventral patterning pathway (De Robertis and Kuroda 2004), although in *Drosophila* this is achieved by maternal factors localized prior to fertilization, making an ancestral Wnt signal input for this pathway less likely.

Regardless of input, a "true" BMP/BMP antagonist-based D/V patterning circuit (see preceding discussion of cnidarian D/V patterning above) was in place in the PDA. BMP2/4 and antagonist Chordin (Dpp and Sog in *Drosophila*) establish dorsal-ventral polarity in all model organisms (with the exception of *C. elegans*) as well as in the hemichordate *Saccoglossus kowalevskii* (Lowe et al. 2006). The leech *Helobdella* additionally utilizes BMP5–8 and antagonist Gremlin for D/V axis specification, demonstrating both the conservation of BMP D/V signaling in Lophotrochozoa as well as a change in the regulatory logic of this circuit (Kuo and Weisblat 2011).

It increasingly appears that the bilaterian LCA had more finely tuned axial and regional patterning afforded by the pathways discussed above as well as expanded and diversified Hox and ParaHox gene clusters. Bilaterian Hox and ParaHox genes are involved in providing regional specification of cell fate in many developing tissues and structures, from the nervous system and gut to muscles and appendages. These genes are often deployed in spatial arrangements mirroring their genomic arrangement, and this collinear expression (or "Hox code") was likely present in the bilaterian LCA (Chiori et al. 2009). Any deviations from colinearity commonly occur in quickly evolving lineages with disintegrated Hox/ParaHox clusters (Hui et al. 2009; Ikuta 2011). An accounting of the various spatiotemporal expression patterns and functional roles of these genes across the Bilateria is beyond the scope of this chapter, and the determination of specific ancestral function is complicated by their extensive, lineage-specific duplication, loss, break-up, and divergence. The bilaterian LCA may have had seven or eight Hox genes and three ParaHox genes. Although the smaller acoelomorph set has been proposed to approximate the ancestral bilaterian condition (Hejnol and Martindale 2009), the potential placement of acoels at the base of deuterostomes (Philippe et al. 2011a) makes a better case for gene loss in this group. As reviewed by Holland (2013), the taxonomically widespread role of bilaterian Hox genes in A/P patterning of ectoderm/mesoderm and the role of ParaHox genes in the A/P patterning of the gut has fueled the hypothesis that these were their ancestral functions, but more work is needed to resolve this question.

The conserved utilization of the Distalless/Dlx pathway in proximal/distal patterning of vertebrate and arthropod appendages as well as other body outgrowths has lead to the expectation that the PDA possessed, if not appendages, similarly specified body protrusions (e.g. Knoll and Carroll 1999). However, conflicting expression patterns and the lack of Distalless and Dac homologs in the distal region of developing parapodia in the polychaete *Neanthes arenaceodentata* suggests that these tools did not pattern appendages in the ancestral bilaterian (Winchell et al. 2010). Additional work by Winchell and Jacobs (2013) indicates that the LIM homeobox (Lhx) transcription factors *apterous* and *lim1*, required for proximodistal patterning in fruit flies and the flour beetle *Tribolium*, are expressed in developing nervous/sensory structures and parapodial muscle precursor cells and not in a proximodistal fashion. As protostome limb development does not seem to be produced by a conserved molecular pathway, the authors posit that Distalless and Lhx pathways originally patterned sensory outgrowths, and that these were independently co-opted in different lineages for proximodistal patterning.

The presence of Pax6 (or Pax6-like genes) in many bilaterians, and its role in eye specification in both vertebrates and *Drosophila*, has long been regarded as evidence of eyes in the bilaterian LCA. Studies of photoreceptors across bilaterian lineages, however, presents a more complicated picture. Molecular characterization of visual-system related cell types across metazoans indicates that both rhabdomeric and ciliary opsin predates the cnidarian and bilaterian split, and that these were independently modified in the different bilaterian lineages for either photoreceptive or non-visual sensory roles (Arendt 2008). While Pax6 seems to be involved in specifying cephalic eyes generated from brain outgrowths, a recent study (Backfisch et al. 2013) has demonstrated that noncephalic photoreceptors are specified by another Pax system (Pax2/5/8) and may predate the origin of cephalic eyes. Photoreceptive cells in the nerve cord and notopodia of annelids may be related to r-opsin-expressing organs of the amphioxus neural tube, and interestingly, orthologs of annelid r-opsin are expressed in both zebrafish retinal ganglia and neuromasts—the specialized mechanosensory cells of the fish lateral line system. Backfisch et al. (2013) suggest that an ancient type of r-opsin photoreceptive cell participated in many sensory roles in the ancestral bilaterian. This may mean that multifunctional sensory apparatuses were more likely to have existed in the bilaterian LCA than cephalic eyes, again suggesting a pattern of ancient multifunctionality to derived specialization of metazoan cell types (Arendt 2008).

While the structure of the brain/CNS of the bilaterian ancestor is still unclear, there is increasing evidence for conserved molecular pathways in establishing CNS signaling centers in the PDA (e.g. Caestro et al. 2005; Pani et al. 2012); additionally, the discovery of innervated stalked eyes and a presumptive optic lobe in the arthropod *Fuxianhuia* (Ma et al. 2012) suggests that the brain and CNS were highly specialized by the early Cambrian. This may not have been the case for the ancestral bilaterian circulatory pump. A discussion of bilaterian pumping organ homology by Xavier-Neto et al. (2007) posits that rather than a heart or peristaltic pump itself being ancestral to bilaterians, a common tissue specification machinery (which included the "heart" gene tinman/Nkx2–5) was independently modified in the bilaterian lineages to produce different pumping organs. This is yet another demonstration of conserved GRN components not necessarily indicating conserved morphology.

Thus, the ancient bilaterian, as compared to more basal metazoans, was an animal tuned for interaction with its environment. The PDA possessed A/P and D/V axial patterning capacities necessary for arranging morphological structures in a fashion conducive to directed movement, the genetic underpinnings for more specific control of regional patterning, as well as molecular pathways that could establish anatomical outgrowths for exploring the environment. Though the bilaterian LCA may not have possessed recognizable structures for sensation, movement, and feeding, it clearly had the developmental machinery needed for these systems in place, and different lineages may have elaborated these developmental capacities in different ways.

Discussion

The rapid expansion of work on basal metazoans continues to challenge our understanding of the evolution of developmental complexity. Additional whole-genome sequences of sponges and cnidarians are in progress and the first ctenophore genome has very recently been published. Combined with further comparative developmental studies on other clades, these will lead to ongoing revision of the work presented here. Yet in many ways the general pattern now appears unambiguous: (1) much of the developmental toolkit was present among basal metazoan clades and (2) the early origin of these clades was evidently decoupled from the origin of the morphologies that characterize the crown groups. This provides a basis for evaluating the changing views on the nature of the LCAs at critical nodes in metazoan evolution, for integrating with data from the fossil record, and for inferring the implications for evolutionary processes.

Changing Views of the Last Common Ancestors

As comparative sequence data accumulated from *Drosophila*, *Caenorhabditis* and *Mus* during the 1990s, and molecular phylogeny provided a new view of metazoan relationships (Aguinaldo et al. 1997) several authors realized that it could be used to infer the nature of the bilaterian last common ancestor, or *Urbilateria* (Shenk and Steel 1994; De Robertis and Sasai 1996). Previous efforts at ancestral reconstruction were based primarily on morphological comparisons and many pre-dated the spread of phylogenetic approaches as more rigorous analysis of shared characters. Evidence for shared developmental genes, and eventually shared complex developmental pathways, led to numerous discussions of the morphological complexity of the bilaterian LCA (Valentine et al. 1999; Knoll and Carroll 1999; Carroll et al. 2001; Erwin and Davidson 2002).

By 2003 the accumulated studies suggested that the bilaterian LCA was morphologically complex. A minimal reconstruction would include (see Erwin and Davidson 2002; Erwin 2009; Carroll et al. 2001 for citations): seven to eight Hox genes controlling anterior/posterior differentiation; a larger cluster of ANTP-class genes including the ParaHox and NK genes; dorsal-ventral patterning controlled by sog/chordin/dpp/BMP2/4; anterior patterning via ems/Emx and otd/Otx; a tripartite brain; posterior patterning of the developing embryo via evenskipped/evx, caudal/cdx; segmentation controlled through engrailed and Delta-Notch ligands; eye formation controlled by Pax6 and ey; a regionalized gut and endoderm produced through GATA transcription factors, brachyury and goosecoid; a heart with development controlled via Nkx2.5/tinman; and possible appendage formation moderated through distalless. The fossil record posed a significant challenge to this interpretation. Most molecular clock studies of that time suggested that the bilaterian LCA dated to over 750 Ma, and such a complex bilaterian would necessarily have left evidence in the trace fossil record of Cryogenian and Ediacaran-aged rocks (Valentine et al. 1999; Erwin and Davidson 2002).

Yet even by 2003 there were greatly differing views of how to interpret the occurrence of highly conserved developmental control genes across bilaterians. The dominant view was that the conservation of the genes necessarily implied conservation of function, and specifically, the conservation of the morphogenetic pathways documented in extant metazoans. In other words, for example, the presence of Pax6 and ey in flies and mice was evidence that eyes were present in the bilaterian LCA (Callaerts et al. 1997; Arendt and Wittbrodt 2001). But at the time there was little data from basal metazoan clades and seemingly little reason to question such conclusions. Indeed, some authors continue to favor a morphologically and developmentally complex bilaterian LCA (De Robertis 2008; Baguñà et al. 2008). The alternative is that at least some of the bilaterian conserved developmental genes may have been co-opted from other developmental roles, suggesting a potentially much simpler bilaterian LCA. Erwin and Davidson (2002) argued that many of these conserved genes were responsible for cell specification and regional patterning rather than complex morphogenetic pathways, with more precise spatial and developmental control systems intercalated into smaller, less connected simpler networks. As discussed earlier in this chapter, the identification of much of the bilaterian developmental toolkit in more basal metazoan clades basically supports this view. Sponges, placozoans, and cnidarians all possess developmental genes that appear capable of generating more complex morphologies than occur in these clades. The integration of evidence from the fossil record with molecular clock results may help to resolve this apparent conflict.

Fossil Record of Early Metazoa

Although molecular clock estimates suggest that Metazoa originated by about 780 Ma, and the Bilateria by about 660 Ma (Fig. 1; also see Chapter "Timing the Origins of Multicellular Eukaryotes Through Phylogenomics and Relaxed Molecular Cock Analyses"), there is scant fossil evidence of animals before the appearance of the Ediacaran biota after 579 Ma. Despite many reports, the only plausible earlier records are of a putative sponge biomarker (a diagnostic lipid) in rocks older than 635 Ma from Oman (Love et al. 2009) and a possible sponge fossil of about the same age (Maloof et al. 2010). For recent reviews of the fossil record of early metazoan diversification see Erwin et al. (2011) and Erwin and Valentine (2013).

After 579 Ma there is abundant evidence of metazoans, including some body fossils known as the Ediacaran macrobiota, fossil embryos, and a variety of trace fossils (horizontal burrows). The Ediacaran macrobiota (579–542 Ma) encompasses a number of independent clades, only two of which may have been bilaterian (*Kimberella* and the dickinsoniomorphs). The remaining suite of fronds, discs and other entities are almost entirely soft-bodied, and have no apparent mouths, guts, appendages, and with a few exceptions, evidence of motility. The phylogenetic affinities of the Ediacaran macrofossils beyond the two potential bilaterians remains a subject of considerable discussion. The fossil embryos, from the Doushantuo Formation in southern China, probably represent members of basal metazoan clades and possibly some other extinct lineages. Trace fossils of the Ediacaran are largely superficial graces, but increase in complexity and diversity toward the base of the Cambrian.

The Cambrian Explosion *sensu stricto* began about 542 Ma with the appearance of penetrating, vertical burrows, denoting the presence of a coelom in the burrower, and of a diverse array of spines, plates and other skeletal elements known as the small shelly fauna. These fossils are quickly followed by the appearance of a variety of bilaterian lineages, with the order of first appearances largely controlled by preservational issues. By about 520 Ma all the major clades of durably skeletonized marine organisms had appeared with the exception of the Bryozoa (which first appear in the fossil record during the Early Ordovician, although they probably originated earlier). Many other clades, such as lobopods, priapulids, and various early arthropods, appear in extraordinarily well-preserved assemblages such as the Burgess Shale and the Chengjiang faunas.

Molecular clock estimates of divergence times clearly establish a gap of 200–100 million years between the acquisition of many elements of the metazoan developmental toolkit and the appearance of these clades in the fossil record (Erwin et al. 2011). This explosion of bilaterian fossils during the early Cambrian coincides with the origins of bilaterian crown groups based on molecular clock evidence (Erwin et al. 2011). From this decoupling of the origin of the clades and their appearance in the fossil record we have concluded (see Erwin et al. 2011) that the early phase of metazoan evolution involved numerous lineages, probably of small body size, and lacking the morphologies that characterize the later crown groups. This also suggests that the highly conserved developmental tools may have been operating in different ways than later in the history of these clades (in simple patterning and for cell-type specification, rather than in complex regulatory pathways leading to regional patterning) (see Erwin and Davidson 2002; Davidson and Erwin 2010). Thus the emerging evolutionary pattern is one of early divergence of major metazoan clades during the Cryogenian, followed by divergences of metazoan phyla in the latest Ediacaran and

Cambrian. The latter was associated with the class 'Cambrian explosion' as described from the fossil record (Erwin and Valentine 2013).

Importance of Macroevolutionary Lags

The long, macroevolutionary lag between the origin of much of the developmental toolkit and its utilization in a wide array of bilaterian bodyplans is not unusual (Jablonski and Bottjer 1990). Indeed macroevolutionary lags between the origin of a clade and the ecological success of elements of the clade are not infrequently associated with major macroevolutionary innovations. The evolution of grasses, for example, predated the spread of grasslands by some 15–20 myr, during which grasses were virtually invisible in the fossil record (Stromberg 2005). Macroevolutionary lags highlight an important feature of evolutionary dynamics. In contrast to the views of Ernst Mayr (Mayr 1960) and others, evolution is not always highly opportunistic, taking advantage of new possibilities as they arise. Rather, there may be a long delay between the origin of a novelty, and even the diversification of the resulting clade into several lineages, and the time when the clade becomes ecologically and evolutionarily important. This feature of evolution parallels the longstanding distinction among historians of technology between invention (the creation of something new, often as recorded by patents) and innovation (the economic success of an invention). In evolution, changes in the physical environment or in ecological interactions are often required before an innovation may succeed and spread (Erwin and Valentine 2013; Erwin 2008). In the case of the grasses studied by Stromberg, it appears that changes in climate and water availability triggered the spread of grasslands during the Miocene. With the origin of animals a plausible case has been made that increased oxygen levels in the ocean were an important contributory cause to the diversification of Bilateria beginning about 550 Ma (Erwin and Valentine 2013).

What macroevolutionary lags emphasize, however, is that data on the origin of developmental novelties alone is insufficient to fully understand the dynamics of evolutionary innovation. Ultimately the success of developmental novelties such as those discussed here is dependent upon environmental context and ecological opportunities.

Summary

1. Developmental genes originally thought to be exclusive to bilaterians and linked to specific bilaterian features are deeply conserved, and the number of these genes identified in early-branching metazoans, particularly sponges and cnidarians, continues to grow.

2. Although much of the metazoan developmental toolkit was present among basal metazoans, these genes were only later co-opted for the various developmental roles associated with bilaterian morphology.
3. The prevalence of co-option means the current function of developmental genes in living groups is a poor guide to ancestral function. A wide array of comparative developmental data is required to infer the nature of ancestral metazoan development; gene functional data derived from experimental manipulation provides the strongest evidence for these inferences.
4. A current metazoan phylogenetic framework is presented to contextualize and polarize fossil and developmental information related to the evolution of metazoan development.
5. We evaluate current available comparative developmental data to make conservative inferences for likely morphology at each key basal metazoan node.
6. The acquisition of metazoan developmental characteristics likely proceeded with the evolution of more precise spatial and developmental control via the evolution of gene regulatory networks.
7. A long macroevolutionary lag exists between the origin of much of the metazoan developmental toolkit and its utilization in the wide array of bilaterian body plans observed in the fossil record. This decoupling of invention and innovation emphasizes the importance of environmental context and ecological opportunity in explaining the success of developmental genetic potential in the Cambrian explosion.

Acknowledgment SMT and DHE acknowledge financial support from a NASA Astrobiology Institute grant (MIT node; to DHE).

References

Adamska M, Degnan SM, Green KM, Adamski M, Craigie A, Larroux C, Degnan BM (2007) Wnt and TGF-beta expression in the sponge *Amphimedon queenslandica* and the origin of metazoan embryonic patterning. PLoS One 2(10):e1031

Adamska M, Larroux C, Adamski M, Green K, Lovas E, Koop D, Richards GS, Zwafink C, Degnan BM (2010) Structure and expression of conserved Wnt pathway components in the demosponge *Amphimedon queenslandica*. Evol Dev 12(5):494–518

Adamska M, Degnan BM, Green K, Zwafink C (2011) What sponges can tell us about the evolution of developmental processes. Zoology 114(1):1–10

Aguinaldo AMA, Turbeville JM, Linford LS, Rivera MC, Garey JR, Raff RA, Lake JA (1997) Evidence for a clade of nematodes, arthropods and other moulting animals. Nature 387:489–493

Arendt D (2008) The evolution of cell types in animals: emerging principles from molecular studies. Nat Rev Genet 9:868–883

Arendt D, Nubler-Jung K (1994) Inversion of dorso-ventral axis. Nature 371:26

Arendt D, Wittbrodt J (2001) Reconstructing the eyes of Urbilateria. Proc R Soc Lond B Biol Sci 356:1545–1563

Backfisch B, Rajan VBV, Fischer RM, Lohs C, Arboleda E, Tessmar-Raible K, Raible F (2013) Stable transgenesis in the marine annelid *Platynereis dumerilii* sheds new light on photoreceptor evolution. Proc Natl Acad Sci U S A 110(1):193–198

Baguñà J, Ruiz-Trillo I, Paps J, Loukota M, Ribera C, Jondelius U, Riutort M (2001) The first bilaterian organisms: simple or complex? New molecular evidence. Int J Dev Biol 45(S1): S133–S134

Baguñà J, Martinez P, Paps J, Riutort M (2008) Back in time: a new systematic proposal for the Bilateria. Philos Trans R Soc Lond B Biol Sci 363:1481–1491

Broun M, Gee L, Reinhardt B, Bode HR (2005) Formation of the head organizer in hydra involves the canonical Wnt pathway. Development 132:2907–2916

Callaerts P, Halder G, Gehring WJ (1997) Pax-6 in development and evolution. Annu Rev Neurosci 20:483–532

Caestro C, Bassham S, Postlethwait J (2005) Development of the central nervous system in the larvacean *Oikopleura dioica* and the evolution of the chordate brain. Dev Biol 285(2):298–315

Carroll SB, Grenier JK, Weatherbee SD (2001) From DNA to diversity. Blackwell Science, Malden

Chapman JA, Kirkness EF, Simakov O, Hampson SE, Mitros T, Weinmaier T, Rattei T, Balasubramanian PG, Borman J, Busam D, Disbennett K, Pfannkock C, Sumin N, Sutton GG, Devi Viswanathan L, Walenz B, Goodstein DM, Hellsten U, Kawashima T, Prochnik SE, Putnam NH, Shu S, Blumberg B, Dana CE, Gee L, Kibler DF, Law L, Lindgens D, Martinez DE, Peng J, Wigge PA, Bertulat B, Guder C, Nakamura Y, Ozbek S, Watanabe H, Khalturin K, Hemmrich G, Franke A, Augustin R, Fraune S, Hayakawa E, Hayakawa S, Hirose M, Hwang J-S, Ikeo K, Nishimiya-Fujisawa C, Ogura A, Takahashi T, Steinmetz PRH, Zhang X, Aufschnaiter R, Eder M-F, Gorny A-K, Salvenmoser W, Heimberg AM, Wheeler BM, Peterson KJ, Böttger A, Tishler P, Wolf A, Gojobori T, Remington KA, Strausberg RL, Venter JC, Technau U, Hobmayer B, Bosch TCG, Holstein TW, Fujisawa T, Bode HR, David CN, Rokhsar DS, Steele RE (2010) The dynamic genome of Hydra. Nature 464:592–597

Chiori R, Jager M, Denker E, Wincker P, Da Silva C, Le Guyader H, Manuel M, Quéinnec E (2009) Are Hox genes ancestrally involved in axial patterning? Evidence from the Hydrozoan *Clytia hemisphaerica* (Cnidaria). PLoS One 4(1):e4231

Davidson EH, Erwin DH (2010) An integrated view of Precambrian eumetazoan evolution. Cold Spring Harb Symp Quant Biol 79:65–80

De Robertis EM (2008) Evo-Devo: variations on ancestral patterning in bilateria. Nature 380:37–40

De Robertis EM, Kuroda H (2004) Dorsal-ventral patterning and neural induction in *Xenopus* embryos. Annu Rev Cell Dev Biol 20:285–308

De Robertis EM, Sasai Y (1996) A common plan for dorsoventral patterning in bilateria. Nature 380:37–40

Dray N, Tessmar-Raible K, Le Gouar M, Vibert L, Christodoulou F, Schipany K, Guillou A, Zantke J, Snyman H, Béhague J (2010) Hedgehog signaling regulates segment formation in the annelid *Platynereis*. Science 329(5989):339–342

Edgecombe GD, Giribet G, Dunn CW, Hejnol A, Kristensen RM, Neves RC, Rouse GW, Worsaae K, Sørensen MV (2011) Higher-level metazoan relationships: recent progress and remaining questions. Org Divers Evol 11:151–172

Erwin DH (2008) Macroevolution of ecosystem engineering, niche construction, and diversity. Trends Ecol Evol 23:304–310

Erwin, DH (2009) Early origin of the bilaterian developmental toolkit. Philos Trans R Soc Lond B Biol Sci 364:2253–2261

Erwin DH, Davidson EH (2002) The last common bilaterian ancestor. Development 129(13): 3021–3032

Erwin DH, Valentine JW (2013) The cambrian explosion: the construction of animal biodiversity. Roberts & Co., Greenwood

Erwin DH, Laflamme M, Tweedt SM, Sperling EA, Pisani D, Peterson KJ (2011) The Cambrian Conundrum: early divergence and later ecological success in the early history of animals. Science 334:1091–1097

Fahey B, Degnan BM (2010) Origin of animal epithelia: insights from the sponge genome. Evol Dev 12:601–617

Fahey B, Larroux C, Woodcroft BJ, Degnan BM (2008) Does the high gene density in the sponge NK homeobox gene cluster reflect limited regulatory capacity? Biol Bull 214(3):205–217

Fairclough SR, Chen Z, Kramer E, Zeng Q, Young S, Robertson HM, Begovic E, Richter DJ, Russ C, Westbrook MJ, Manning G, Lang BF, Haas B, Nusbaum C, King N (2013) Premetazoan genome evolution and the regulation of cell differentiation in the choanoflagellate *Salpingoeca rosetta*. Genome Biol 14:R15

Finnerty JR, Pang K, Burton P, Paulson D, Martindale MQ (2004) Origins of bilateral symmetry: *Hox* and *Dpp* expression in a sea anemone. Science 304:1335–1337

Forêt S, Knack B, Houliston E, Momose T, Manuel M, Quéinnec E, Hayward DC, Ball E, Miller DJ (2010) New tricks with old genes: the genetic basis of novel cnidarian traits. Trends Genet 26(4):154–158

Fortunato S, Adamski M, Bergum B, Guder C, Jordal S, Leininger S, Zwafink C, Rapp HT, Adamska M (2012) Genome-wide analysis of the sox family in the calcareous sponge *Sycon ciliatum*: multiple genes with unique expression patterns. EvoDevo 3:14

Fritz AE, Ikmi A, Seidel C, Paulson A, Gibson MC (2013) Mechanisms of tentacle morphogenesis in the sea anemone *Nematostella vectensis*. Development 140:2212–2223

Harcet M, Roller M, Cetkovic H, Peina D, Wiens M, Muller WE, Vlahovicek K (2010) Demosponge EST sequencing reveals a complex genetic toolkit of the simplest metazoans. Mol Biol Evol 27(12):2747–2756

Hejnol A, Martindale MQ (2009) Coordinated spatial and temporal expression of *Hox* genes during embryogenesis in the acoel *Convolutriloba longifissura*. BMC Biol 7:65

Holland PWH (2013) Evolution of homeobox genes. WIREs Dev Biol 2:31–45

Hui JHL, Raible F, Korchagina N, Dray N, Samain S, Magdelenat G, Jubin C, Segurens B, Balavoine G, Arendt D, Ferrier DEK (2009) Features of the ancestral bilaterian inferred from *Platynereis dumerilii* ParaHox genes. BMC Biol 7:43

Ikuta T (2011) Evolution of invertebrate deuterostomes and Hox/ParaHox genes. Genomics Proteomics Bioinformatics 9(3):77–96

Ingham P, Nakano Y, Seger C (2011) Mechanisms and functions of Hedgehog signaling across the metazoa. Nat Rev Genet 12:393–406

Jablonski D, Bottjer DJ (1990) The origin and diversification of major groups: environmental patterns and macroevolutionary lags. In: Taylor PD, Larwood GP (eds) Major evolutionary radiations. Systematics association special, vol 42. Clarendon Press, Oxford, pp 17–57

Jakob W, Sagasser S, Dellaporta S, Holland P, Kuhn K, Schierwater B (2004) The *Trox-2* Hox/ParaHox gene of *Trichoplax* (Placozoa) marks an epithelial boundary. Dev Genes Evol 214(4):170–175

King N, Westbrook MJ, Young SL, Kuo A, Abedin M, Chapman J, Fairclough S, Hellsten U, Isogai Y, Letunic I, Marr M, Pincus D, Putnam N, Rokas A, Wright KJ, Zuzow R, Dirks W, Good M, Goodstein D, Lemons D, Li W, Lyons JB, Morris A, Nichols S, Richter DJ, Salamov A, Sequencing JGI, Bork P, Lim WA, Manning G, Miller WT, McGinnis W, Shapiro H, Tjian R, Grigoriev IV, Rokhsar D (2008) The genome of the choanoflagellate *Monosiga brevicollis* and the origin of metazoans. Nature 451(7180):783–788

Knoll AH (2011) The multiple origins of complex multicellularity. Annu Rev Earth Planet Sci 39:217–239

Knoll AH, Carroll SB (1999) Early animal evolution: emerging views from comparative biology and geology. Science 284:2129–2137

Kraus Y, Fritzenwanker JH, Genikhovich G, Technau U (2007) The blastoporal organiser of a sea anemone. Curr Biol 17:R874–876

Kuo D-H, Weisblat DA (2011) A new molecular logic for BMP-mediated dorsoventral patterning in the leech *Helobdella*. Curr Biol 21:1281–1288

Larroux C, Luke GN, Koopman P, Rohksar D, Shimeld SM, Degnan BM (2008) Genesis and expansion of metazoan transcription factor gene classes. Mol Biol Evol 25:980–996

Lee PN, Kumburegama S, Marlow HQ, Martindale MQ, Wikramanayake AH (2007) Asymmetric developmental potential along the animal-vegetal axis in the anthozoan cnidarian, *Nematostella vectensis*, is mediated by Dishevelled. Dev Biol 310:169–186

Leys SP, Riesgo A (2012) Epithelia, an evolutionary novelty of metazoans. J Exp Zool B Mol Dev Evol 318(6):438–447

Leys SP, Cronin TW, Degnan BM, Marshall JN (2002) Spectral sensitivity in a sponge larva. J Comp Physiol A 188:199–202

Love GD, Grosjean E, Stalvie C, Fike DA, Grotzinger JP, Bradley AS, Kelly AE, Bhatia M, Meredith W, Snape CE, Bowring SA, Condon DJ, Summons RE (2009) Fossil steroids record the appearance of Demospongia during the Cryogenian period. Nature 457:718–721

Lowe CJ, Terasaki M, Wu M, Freeman RM Jr, Runft L, Kwan K, Haijo S, Aronowicz J, Lander E, Gruber C, Smith M, Kirschner M, Gerhart J (2006) Dorsoventral patterning in hemichordates: insights into early chordate evolution. PLoS Biol 4(9):e291

Ma X, Hou X, Edgecombe GD, Strausfeld NJ (2012) Complex brain and optic lobes in an early Cambrian arthropod. Nature 490:258–263

Magie CR, Pang K, Martindale MQ (2005) Genomic inventory and expression of *Sox* and *Fox* genes in the cnidarian *Nematostella vectensis*. Dev Genes Evol 215:618–630

Maloof AC, Rose CV, Calmet CC, Beach R, Samuels BM, Erwin DH, Poirier GR, Yao N, Simons FJ (2010) Probable animal body-fossils from pre-Marinoan limestones, South Australia. Nat Geosci 3:653–659

Martinelli C, Spring J (2008) Distinct expression patterns of the two T-box homologues *Brachyury* and *Tbx2/3* in the placozoan *Trichoplax adhaerens*. Dev Genes Evol 213(10):492–499

Maxwell EK, Ryan JF, Schnitzler CE, Browne WE, Baxevanis AD (2012) MicroRNAs and essential components of the microRNA processing machinery are not encoded in the genome of the ctenophore *Mnemiopsis leidyi*. BMC Genomics 13:714

Mayr E (1960) The emergence of novelty. In: Tax S (ed) The evolution of life. University of Chicago Press, Chicago, pp 349–380

Miller DJ, Ball EE (2009) The gene complement of the ancestral bilaterian—was Urbilateria a monster? J Biol 8:89

Momose T, Houliston E (2007) Two oppositely localized frizzled RNAs as axis determinants in a cnidarian embryo. PLoS Biol 5:e70

Momose T, Derelle R, Houliston E (2008) A maternally localized Wnt ligand required for axial patterning in the cnidarian *Clytia hemisphaerica*. Development 135:2105–2113

Momose T, Kraus Y, Houliston E (2012) A conserved function for Strabismus in establishing planar cell polarity in the ciliated ectoderm during cnidarian larval development. Development 139:4375–4382

Nichols SA, Dirks W, Pearse JS, King N (2006) Early evolution of animal cell signaling and adhesion genes. Proc Natl Acad Sci U S A 103(33):12451–12456

Nichols SA, Roberts BW, Richter DJ, Fairclough SR, King N (2012) Origin of metazoan cadherin diversity and the antiquity of the classical cadherin/beta-catenin complex. Proc Natl Acad Sci U S A 109(32):13046–13051

Nosenko T, Schreiber F, Adamska M, Adamski M, Eitel M, Hammel J, Maldonado M, Muller WEG, Nickel M, Schierwater B, Vacelet J, Wiens M, Wörheide G. (2013) Deep metazoan phylogeny: when different genes tell different stories. Mol Phylogenet Evol 67:223–233

Ogura A, Ikeo K, Gojobori T (2005) Estimation of ancestral gene set of bilaterian animals and its implication to dynamic change of gene content in bilaterian evolution. Gene 345(1):65–71

Pang K, Ryan JF, Baxevanis AD, Martindale MQ (2011) Evolution of the TGF-beta signaling pathway and its potential role in the ctenophore, *Mnemiopsis leidyi*. PLoS One 6(9):e24152

Pani AM, Mullarkey EE, Aronowicz J, Assimacopoulos S, Grove EA, Lowe CJ (2012) Ancient deuterostome origins of vertebrate brain signaling centres. Nature 483:289–296

Parfrey LM, Lahr DJG (2013) Multicellularity arose several times in the evolution of eukaryotes. Bioessays 35:339–347

Petersen CP, Reddien PW (2009) Wnt signaling and the polarity of the primary body axis. Cell 139:1056–1068

Philippe H, Brinkmann H, Copley RR, Moroz LL, Nakano H, Poustka AJ, Wallberg A, Peterson KJ, Telford MJ (2011a) Acoelomorph flatworms are deuterostomes related to *Xenoturbella*. Nature 470:255–258

Philippe H, Brinkmann H, Lavrov DV, Littlewood DTJ, Manuel M, Wörheide G, Baurain D (2011b) Resolving difficult phylogenetic questions: why more sequences are not enough. PLoS Biol 9(3):e1000602

Putnam NH, Srivastava M, Hellsten U, Dirks B, Chapman J, Salamov A, Terry A, Shapiro H, Lindquist E, Kapitonov VV, Jurka J, Genikhovich G, Grigoriev IV, Lucas SM, Steele RE, Finnerty JR, Technau U, Martindale MQ, Rokhsar DS (2007) Sea anemone genome reveals ancestral eumetazoan gene repertoire and genomic organization. Science 317:86–96

Ramos OM, Barker D, Ferrier DEK (2012) Ghost loci imply Hox and ParaHox existence in the last common ancestor of animals. Curr Biol 22:1951–1956

Richards GS, Degnan BM (2012) The expression of Delta ligands in the sponge *Amphimedon queenslandica* suggest an ancient role for Notch signaling in metazoan development. EvoDevo 3:1–15

Ringrose JH, van den Toorn HWP, Eitel M, Post H, Neerincx P, Schierwater B, Altelaar AFM, Heck AJR (2013) Deep proteome profiling of *Trichoplax adhaerens* reveals remarkable features at the origin of metazoan multicellularity. Nat Commun 4(1408):1–7

Rivera AS, Ozturk N, Fahey B, Plachetzki DC, Degnan BM, Sancar A, Oakley TH (2012) Blue-light-receptive cryptochrome is expressed in a sponge eye lacking neurons and opsin. J Exp Biol 215:1278–1286

Rivera A, Winters I, Rued A, Ding S, Posfai D, Cieniewicz B, Cameron K, Gentile L, Hill A (2013) The evolution and function of the Pax/Six regulatory network in sponges. Evol Dev 15(3):186–196

Robinson JM, Sperling EA, Bergum B, Adamski M, Nichols SA, Adamska M, Peterson KJ (2013) The identification of microRNAs in Calcisponges: Independent eolution of microRNAs in basal metazoans. J Exp Zool B Mol Dev Evol 320:84–93

Rokas A (2008) The origins of multicellularity and the early history of the genetic toolkit for animal development. Annu Rev Genet 42:235–251

Rota-Stabelli O, Daley AC, Pisani D (2013) Molecular timetrees reveal a Cambrian colonization of land and a new scenario for Ecdysozoan evolution. Curr Biol 23:392–398

Röttinger E, Dahlin P, Martindale MQ (2012) A framework for the establishment of a cnidarian gene regulatory network for "endomesoderm" specification: the inputs of β-catenin/TCF signaling. PLoS Genet 8(12):e1003164

Rowland SJ (2001) Archaeocyaths—a history of phylogenetic interpretation. J Paleontol 75:1065–1078

Ryan JF, Pang K, Mullikin JC, Martindale MQ, Baxevanis AD, Progra NCS (2010) The homeodomain complement of the ctenophore *Mnemiopsis leidyi* suggests that Ctenophora and Porifera diverged prior to the ParaHoxozoa. EvoDevo 1:9.

Ryan JF, Pang K, Schnitzler CE, Nguyen AD, Moreland RT, Simmons DK, Koch BJ, Francis WR, Havlak P, NISC Comparative Sequencing Program, Smith SA, Putnam NH, Haddock SHD, Dunn CW, Wolfsberg TG, Mullikin JC, Martindale MQ, Baxevanis AD (2013) The genome of the ctenophore *Mnemiopsis leidyi* and its implications for cell type evolution. Science 342:124592

Saina M, Genikhovich G, Renfer E, Technau U (2009) BMPs and Chordin regulate patterning of the directive axis in a sea anemone. Proc Natl Acad Sci U S A 106(44):18592–18597

Schierwater B, Kamm K (2010) The early evolution of Hox genes: a battle of believe? In: Deutsch J (ed) Hox Genes' studies from the 20th to the 21st century, advances in experimental medicine and biology, vol 689. Landes Bioscience, New York, pp 81–90

Schierwater B, Eitel M, Osigus HJ, von der Chevallerie K, Bergmann T, Hadrys H, Cramm M, Heck L, Jakob W, Lang MR, DeSalle R (2011) *Trichoplax* and Placozoa: one of the crucial keys to understanding metazoan evolution. In: DeSalle R, Schierwater B (eds) Key transitions in animal evolution. Science Publishers, Enfield, pp 289–326

Sebé-Pedrós A, Ruiz-Trillo I, de Mendoza A, Lang BF, Degnan BM (2011) Unexpected repertoire of metazoan transcription factors in the unicellular holozoan *Capsaspora owczarzaki*. Mol Biol Evol 28(3):1241–1254

Sebé-Pedrós A, Ariza-Cosano A, Weirauch MT, Leininger S, Yang A, Torruella G, Adamski M, Adamska M, Hughes TR, Gómez-Skarmeta JL, Ruiz-Trillo I (2013) Early evolution of the T-box transcription factor family. Proc Natl Acad Sci U S A 110(40):16050–16055

Shenk MA, Steel MA (1994) A molecular snapshot of the metazoan 'Eve'. Trends Biochem Sci 18:459–463

Simakov O, Marletaz F, Cho S-J, Edsinger-Gonzales E, Havlak P, Hellsten U, Kuo D-H, Larsson T, Lv J, Arendt D, Savage R, Osoegawa K, de Jong P, Grimwood J, Chapman JA, Shapiro H, Aerts A, Otillar RP, Terry AY, Boore JL, Grigoriev IV, Lindberg DR, Seaver EC, Weisblat DA, Putnam NH, Rokhsar DS (2013) Insights into bilaterian evolution from three spiralian genomes. Nature 493:526–532

Sinigaglia C, Busengdal H, Leclère L, Technau U, Rentzsch F (2013) The bilaterian head patterning gene *six3/6* controls aboral domain development in a cnidarian. PLoS Biol 11(2):e1001488

Sperling EA, Vinther J (2010) A placozoan affinity for *Dickinsonia* and the evolution of late Proterozoic metazoan feeding modes. Evol Dev 12:201–209

Sperling EA, Pisani D, Peterson KJ (2007) Poriferan paraphyly and its implications for Precambrian palaeobiology. In: Vickers-Rich P, Komarower P (eds) The rise and fall of the Ediacaran biota, geological society special publication, vol 286. Geological Society of London, London, pp 355–368

Sperling EA, Peterson KJ, Pisani D (2009) Phylogenetic-signal dissection of nuclear housekeeping genes supports the paraphyly of sponges and the monophyly of eumetazoa. Mol Biol Evol 26:2261–2274

Srivastava M, Begovic E, Chapman J, Putnam NH, Hellsten U, Kawashima T, Kuo A, Mitros T, Salamov A, Carpenter ML, Signorovitch AY, Moreno MA, Kamm K, Grimwood J, Schmutz J, Shapiro H, Grigoriev IV, Buss LW, Schierwater B, Dellaporta SL, Rokhsar DS (2008) The *Trichoplax* genome and the nature of placozoans. Nature 454(7207):955–960

Srivastava M, Larroux C, Lu DR, Mohanty K, Chapman J, Degnan BM, Rokhsar DS (2010a) Early evolution of the LIM homeobox gene family. BMC Biol 8(4):1–13

Srivastava M, Simakov O, Chapman J, Fahey B, Gauthier MEA, Mitros T, Richards GS, Conaco C, Dacre M, Hellsten U, Larroux C, Putnam NH, Stanke M, Adamska M, Darling A, Degnan SM, Oakley TH, Plachetzki DC, Zhai YF, Adamski M, Calcino A, Cummins SF, Goodstein DM, Harris C, Jackson DJ, Leys SP, Shu SQ, Woodcroft BJ, Vervoort M, Kosik KS, Manning G, Degnan BM, Rokhsar DS (2010b) The *Amphimedon queenslandica* genome and the evolution of animal complexity. Nature 466(7307):720–723

Steele RE, David CN, Technau U (2011) A genomic view of 500 million years of cnidarian evolution. Trends Genet 27(1):7–13

Steinmetz PRH, Kraus JEM, Larroux C, Hammel JU, Amon-Hassenzahl A, Houliston E, Wörheide G, Nickel M, Degnan BM, Technau U (2012) Independent evolution of striated muscles in cnidarians and bilaterians. Nature 487(7406):231–234

Stromberg CAE (2005) Decoupled taxonomic radiation and ecological expansion of open-habitat grasses in the Cenozoic of North America. Proc Natl Acad Sci U S A 102:11980–11984

Sullivan JC, Kalaitzidis D, Gilmore TD, Finnerty JR (2006) Rel homology domain-containing transcription factors in the cnidarian *Nematostella vectensis*. Dev Genes Evol 217:63–72

Valentine JW, Jablonski D, Erwin DH (1999) Fossils, molecules and embryos: new perspectives on the Cambrian explosion. Development 126:851–859

Vij S, Rink JC, Ho HK, Babu D, Eitel M, Narasimhan V, Tiku V, Westbrook J, Schierwater B, Roy S (2012) Evolutionarily ancient association of the FoxJ1 transcription factor with the motile ciliogenic program. PLoS Genet 8(11):e1003019

Winchell CJ, Jacobs DK (2013) Expression of the Lhx genes *apterous* and *lim1* in an errant polychaete: implications for bilaterian appendage evolution, neural development, and muscle diversification. EvoDevo 4:4

Winchell CJ, Valencia JE, Jacobs DK (2010) Expression of *Distal-less*, *dachshund*, and *optomotor blind* in *Neanthes arenaceodentata* Annedlida, Nereididae) does not support homology of appendage-forming mechanisms across the Bilateria. Dev Genes Evol 220:275–295

Windsor PJ, Leys SP (2010) Wnt signaling and induction in the sponge aquiferous system: evidence for an ancient origin of the organizer. Evol Dev 12(5):484–493

Wyder S, Kriventseva EV, Schröder R, Kadowaki T, Zdobnov EM (2007) Quantification of ortholog losses in insects and vertebrates. Genome Biol 8(11):R242

Xavier-Neto J, Castro RA, Sampaio AC, Azambuja AP, Castillo HA, Cravo RM, Simões-Costa MS (2007) Parallel avenues in the evolution of hearts and pumping organs. Cell Mol Life Sci 64:719–734

Multicellularity in Bacteria: From Division of Labor to Biofilm Formation

Claudio Aguilar, Catherine Eichwald and Leo Eberl

Abstract Introduced nearly two decades ago, the concept of multicellularity in bacteria is currently accepted as a general trait of bacterial physiology. The view of bacteria being more than just unicellular, non-organized, selfish organisms is to a large degree based on the findings that division of labor and cell-to-cell communication within bacterial communities are ubiquitous across bacterial species. Bacteria are able to form complex communities in which cells can specialize in a spatiotemporal fashion, using extracellular signals to coordinate the expression of specific genes required for structural development. Despite the enormous progress made by researchers in the field over the past years, knowledge of the molecular mechanisms that govern bacterial multicellularity and biofilm development is scarce and remains a highly interesting field for future research.

Keywords Biofilm · Division of labor · Cell differentiation · Quorum sensing

Introduction

Long before bacterial multicellularity was to be proposed as a fundamental tenet of microbiology (Shapiro 1998, 1988), scientists were locked in the pure culture paradigm initiated by Robert Koch. The significance of Koch's studies in the development of microbiology, especially in the medical field, is undeniable. The Russian microbiologist Sergei Winogradsky was the first who challenged the paradigm of pure culture by appreciating the complex interactions among bacteria in the soil and

C. Aguilar (✉) · L. Eberl
Department of Microbiology, Institute of Plant Biology, University of Zurich, Zollikerstrasse 107, 8008 Zurich, Switzerland
e-mail: claudio_aguilar@access.uzh.ch

C. Eichwald
Institute of Virology, University of Zurich, Winterthurerstrasse 266a, 8057 Zurich, Switzerland
e-mail: ceichwald@vetvir.uzh.ch

L. Eberl
e-mail: leberl@botinst.uzh.ch

I. Ruiz-Trillo, A. M. Nedelcu (eds.), *Evolutionary Transitions to Multicellular Life*, Advances in Marine Genomics 2, DOI 10.1007/978-94-017-9642-2_4

began the systematic study of mixed bacterial populations as living systems (Shapiro and Dworkin 1997). We now acknowledge that bacteria can no longer be considered solitaire creatures but rather, they are organisms capable of cell differentiation and complex behaviors. Multicellularity is one of these behaviors and will be the subject of this chapter.

What are the developmental requirements that bacteria have for multicellular organization? Despite the fact that the molecular mechanisms that bacteria employ to become multicellular may differ between species, we can find several common concepts, including: (i) adhesion, or the ability to stick together by means of secreted substances that can establish cell-cell connections within the consortium; (ii) coordination of activities, by the aid of cell-to-cell communication and (iii) programmed cell death. Even if these characteristics are ubiquitous in the bacterial world, this by no means indicates that multicellularity is obligate. For example, cyanobacteria have diversified through evolution and can be found today in both unicellular or multicellular forms (Flores and Herrero 2010). One bacterium can also exist in either a unicellular or multicellular state, like the myxobacteria, which are single-celled but form social groups capable of complex multicellular behaviors and cellular differentiation (Velicer and Vos 2009; Chapter "Signaling in swarming and aggregating myxobacteria Dale Kaiser").

Why would a unicellular organism benefit from a multicellular lifestyle? In general, multicellularity provides a mean to build and scale up elaborate structures and to generate greater forces than single cells can. Multicellularity also induces cooperation between cells. Among the advantages that bacteria can obtain from a multicellular lifestyle are:

(i) avoidance of predation
(ii) improvement in efficiency of nutrient acquisition;
(iii) avoidance of non-cooperative individuals;
(iv) division of labor.

In the following sections of this chapter, we will give a few examples that illustrate some of the survival benefits that multicellularity confers to an otherwise unicellular organism. The building of architecturally complex assemblages, or biofilms, will be highlighted as an example of a ubiquitous mechanism that bacteria utilize to thrive by means of multicellularity.

Multicellularity in Cyanobacteria and Myxobacteria

The transition from unicellularity to multicellularity in bacteria is thought to have occurred independently multiple times in history (Grosberg and Strathmann 2007). By becoming multicellular, organisms opened the way for a plethora of remarkable innovations that would have been difficult, if not impossible, for single cells to accomplish. One of the immediate consequences of multicellularity for an organism is the increased capability to respond to always-changing environmental conditions. There are numerous examples that illustrate how these adaptations were successfully acquired in bacteria. In this section, we will give a glimpse into the division of labor of two organisms traditionally considered as examples of multicellularity: cyanobacteria and myxobacteria.

Division of Labor in Filamentous Cyanobacteria

Cyanobacteria are among the oldest known prokaryotic organisms and fossil records indicate that they may have already existed 3.5 billion years ago (Bya) (Schopf 2006), when the Earth was just 1 billion years old. Recent evidence suggests that the transition to multicellularity in cyanobacteria coincided with the onset of the "Great Oxidation Event" (GOE), around 2.5 Bya (Schirrmeister et al. 2013). This event increased the atmospheric oxygen levels, enabling the evolution of aerobic life on the planet.

Also known as blue-green algae, these organisms were classified into five subsections, with subsections I and II containing unicellular cyanobacteria and subsections III, IV and V consisting of filamentous cyanobacteria (Shapiro and Dworkin 1997). The growth as multicellular filamentous clumps or colonies confers cyanobacteria with a number of advantages, including protection against predation, increased ability to colonize substrates and also resistance to desiccation. Although the underlying mechanisms are not fully understood, cells within filamentous cyanobacteria can communicate with each other to allow redistribution of nutrients (Flores and Herrero 2010).An additional advantage of the multicellular lifestyle for filamentous cyanobacteria is the ability to translocate by gliding motility (they lack flagella) to regulate their exposure to oxygen and light (Tamulonis et al. 2011).

Cyanobacteria are photoautotrophs, able to perform photosynthesis (carbon fixation) in combination with diazotrophy (N_2 fixation). However, the nitrogenase, the enzyme needed for N_2 fixation, is irreversibly inactivated by the O_2 produced as a by-product of photosynthesis, making these two processes incompatible within the same cell. While to some extent performing photosynthesis in the daytime and nitrogen fixation at night can reduce this problem (Bergman et al. 1997), the metabolic incompatibility between photosynthesis and nitrogen fixation was solved in some filamentous cyanobacteria by division of labor: while most cells perform photosynthesis, a proportion of their vegetative cells differentiate into N_2-fixing heterocysts. The heterocysts are non-photosynthetic cells that provide fixed nitrogen to the neighboring cells when it becomes limiting in exchange for fixed carbon provided by the vegetative cells (Flores and Herrero 2010). Vegetative cells can respond to adverse environmental conditions like light limitation or phosphate deprivation by developing into a third cell-type, the akinetes. Akinetes are spore-like cells resistant to cold and desiccation, which will germinate when conditions are favorable again. Cyanobacteria can also differentiate into a fourth cell-type, the hormogonia, which are able to move using gliding motility or gas vacuole-mediated buoyancy. Hormogonia have a role as dispersal units and are released by fragmentation of the producing vegetative cell. After a period of time they return to sessility, growing as vegetative filaments with the potential to differentiate again, if needed (Flores and Herrero 2010; Meeks and Elhai 2002).

Cooperative Behavior of Myxobacteria

The myxobacteria, belonging to the delta-subdivision of the proteobacteria, provide another well-known example of bacterial multicellularity. In contrast to the cyanobacteria, myxobacteria do not grow as filaments but use a self-secreted extracellular matrix as a basis for their multicellular lifestyle (Velicer and Vos 2009). The best-studied species, *Myxococcus xanthus,* displays behaviors that involve cooperation among cells, highlighting the social nature of this organism; these include: fruiting body formation, swarming and predation (Velicer and Vos 2009).

In conditions of high cell density and nutrient depletion, *M. xanthus* forms fruiting body aggregates of about 10^5 cells. Cell differentiation inside fruiting bodies results in the formation of spores, which are resistant to heat, UV light and desiccation (Fremgen et al. 2010). In this process, *M. xanthus* relies both on gliding motility and cell-to-cell communication through extracellular chemical signals as well as through physical contact. Upon starvation, *M. xanthus* cells secrete a mixture of six amino acids, referred to as the A-signal. Cells in the population respond to the A-signal when it reaches a threshold concentration, known as the quorum (Kaiser 2003; Chapter "Signaling in swarming and aggregating myxobacteria Dale Kaiser"). Once the quorum is reached, the cells start to form aggregates that eventually become fruiting bodies. In addition to the A-signal another signaling molecule, a 17-kDa cell-surface-bound protein referred to as the C-signal, is involved in establishing the morphological shape and the location of the spores within *M. xanthus* fruiting bodies (Kaiser 2003). But multicellularity appears to come with a cost for the single-celled bacterium and can be lost if not needed. This phenomenon was studied in the laboratory using experimental evolution, under conditions in which multicellularity is not advantageous (e.g. liquid, shaken cultures). Interestingly, it was observed that defects in fruiting body formation, sporulation and motility emerged after only one thousand generations, making the authors conclude that these functions were all unimportant for fitness (Velicer et al. 1998).

Predation is another example of multicellular behavior that has evolved in myxobacteria. *M. xanthus* can feed on other microbial cells, hunting for prey using a strategy that has been compared to a bacterial wolfpack (Velicer and Vos 2009). This wolfpack swarms towards the prey organism using two complementary flagella-independent motility forms, namely A- and S-motility (Fremgen et al. 2010). The microbial prey cells are then killed by predation-associated molecules (e.g. hydrolytic enzymes) and the remains of the lysed cells benefit both the *M. xanthus* secreting cells as well as the non-secreting cells (Mendes-Soares and Velicer 2013).

Morphogenesis of a Biofilm: Multicellularity in Action

As discussed above, there are many examples that illustrate how bacteria embraced multicellularity as a general physiological trait. It is not our intention to provide a full review of every bacterial species for which multicellularity has been described.

However, it is worth mentioning that findings of the past few decades suggest that rather than being a curious phenomenon carried out by an exotic group of microorganisms, multicellularity is actually the rule and not the exception in the bacterial world (Shapiro 1998; Shapiro and Dworkin 1997).

It is well accepted that in natural settings bacteria are predominantly found adhered to surfaces, on which they develop into highly structured and architecturally complex communities by the production of an extracellular matrix (Aguilar et al. 2009; Branda et al. 2005; O'Toole et al. 2000). This is the fundamental definition of a biofilm and, not surprisingly, bacterial biofilms represent an excellent setting for the study of multicellularity.

Biofilm development has been extensively studied in both Gram-positive and -negative bacteria, with the last decades witnessing an explosion of knowledge regarding the underlying molecular mechanisms involved in this process. In the following section, we will summarize the findings of many labs working in the field in an attempt to create a picture of how biofilms develop, with the main focus on the results obtained for single-species biofilms.

Biofilm Morphotypes in B. subtilis and P. aeruginosa

Bacterial biofilms can develop into highly structured communities that show distinct morphologies. The morphology of the biofilm can vary greatly depending on the bacteria in study and on growth conditions (Branda et al. 2005; López et al. 2010). As an example of the structural features that can be found in a biofilm we will briefly describe the overall morphology of the biofilms formed by the Gram-positive *B. subtilis* and the Gram-negative *P. aeruginosa.*

When aliquots of the model Gram-positive *B. subtilis* are spotted on an agar surface containing the necessary nutrients required for biofilm development, colonies with a complex and elaborate architecture form (Branda et al. 2001). Under these conditions, *B. subtilis* develops into a biofilm with a number of characteristic morphological features (Fig. 1). One of these features is an asymmetric, radial pattern of wrinkles that was recently shown to be the result of the combination of localized cell death and the mechanical stiffness of the biofilm itself (Asally et al. 2012). Interestingly, localized or programmed cell death is a well-described phenomenon in developing multicellular eukaryotes. In addition to *B. subtilis,* programmed cell death has also been described as an intrinsic process during biofilm development in both *Escherichia coli* and *Staphylococcus aureus* (Kolodkin-Gal et al. 2009; Sadykov and Bayles 2012).

Another characteristic morphological feature of biofilms in *B. subtilis* is the presence of aerial projections at the edges of the growing colony (Fig. 1). In one of the first examples of the use of an "undomesticated" strain, *B. subtilis* NCIB3610 (a "wild-type" strain that has not been subjected to the extensive rounds of cultures and sub-cultures of a typical "domesticated", laboratory strain) was genetically engineered to express a fusion of the promoter of the late-sporulation gene *sspE* to the reporter *lacZ.* It was found that sporulation took place preferentially on the tips of

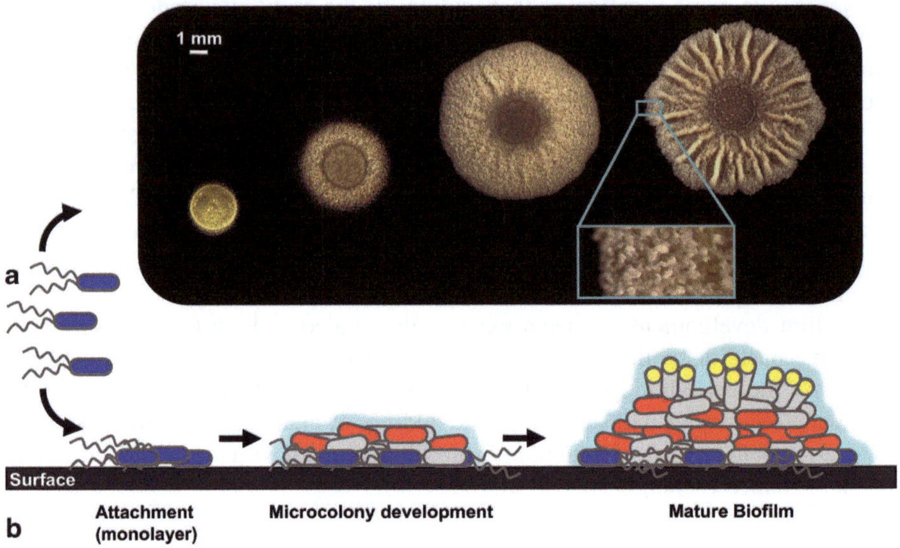

Fig. 1 Development of a *B. subtilis* biofilm. **a** Top view of a *B. subtilis* biofilm development over time. *B. subtilis* was spotted on an agar surface containing the necessary nutrients required for biofilm development. Pictures were taken at 12, 24, 48 and 72 h of biofilm development, respectively. After 72 h, a characteristic radial pattern of wrinkles is visible, as well as the aerial projections at the edge of the biofilm (magnified panel) (reprinted, with permission, from Vlamakis et al, 2008 [©Cold Spring Harbor Laboratory Press]). **b** Schematic representation of the biofilm shown in a, highlighting three cell-types that differentiate within the biofilm during development: swimmers (*blue*), matrix-producers (*red*) and sporulating cells (*yellow*). Cells are embedded in extracellular matrix, depicted as a *light-blue* zone surrounding the biofilm

these aerial projections (Branda et al. 2001). The aerial projections were termed *B. subtilis* fruiting body-like structures, due to their functional analogy to the sporulating fruiting bodies found in myxobacteria. This is a strong indication that in *B. subtilis* biofilms sporulation displays a defined spatiotemporal organization (see below). Remarkably, not just the aerial projections but all the structural complexity observed in NCIB3610 was absent in laboratory (domesticated) strains, suggesting that multicellularity and spatiotemporal organization can be lost in the process of strain domestication (Branda et al. 2001; McLoon et al. 2011). Strain domestication is likely the result of years of extensive manipulation of bacterial strains in the laboratory that, perhaps unknown to molecular geneticists, have resulted in the loss of multicellularity by selecting easily dispersible strains (Aguilar et al. 2007; McLoon et al. 2011).

The morphology and the development of *P. aeruginosa* biofilms have been primarily studied in flow-cell settings. Flow cells are small chambers with glass surfaces where submerged biofilms can develop under continuous flow of nutrients. The biofilms can be monitored by confocal scanning laser microscopy, allowing the capturing of images of biofilm development in real time (Christensen et al. 1999). Using this technique, the structured microcolonies formed by *P. aeruginosa* have been commonly referred to as "mushrooms". The initial steps of biofilm development involve the adhesion to the surface, where non-motile cells form the stalks of the mushroom. During subsequent biofilm maturation, motile sub-populations make the upper caps of the stalks, eventually completing the mushroom structure (Klausen et al. 2003).

Living Inside the Matrix: Adhesive Components of Biofilms

A key factor in maintaining multicellularity is the ability for individual cells to stick together. Bacteria can accomplish this in different and complementary ways. For example, they can adopt a filamentous mode of growth and/or use extracellular matrix material to adhere to each other. Among different bacterial species, and even in strains of the same species, the molecular mechanisms that contribute to the development of structurally complex biofilms can vary enormously. Despite this variation, several common features can be distinguished. One of the landmark features of biofilms is the presence of a self-produced extracellular matrix (composed by proteins, lipids, polysaccharides and DNA), which holds the cells together. Since the production of the extracellular matrix may be subject to a complex genetic regulation (Vlamakis et al. 2013), its composition may change depending on the environmental conditions in which the organism thrives or also on the developmental stage of the biofilm. The ability to build and maintain a biofilm is widespread in bacteria and in the following section we will summarize the contribution of some of the most important biomolecules present in the matrix of the biofilm contributing to the multicellularity of the bacterial population.

The extracellular DNA (eDNA) in the matrix has an important role as a cell-to-cell interconnecting compound, and its role in the establishment of multicellular communities has been documented for several bacterial species. For example, *P. aeruginosa* biofilm development was prevented by treatment with DNase I (Whitchurch et al. 2002), suggesting that eDNA has a role in the initial steps of biofilm formation. Further examples where eDNA has been demonstrated as an important matrix component include *Bacillus cereus* (Vilain et al. 2009), *Pseudomonas putida*, *Rhodococcus erythropolis*, *Variovorax paradoxus* (Steinberger and Holden 2005), *Haemophilus influenzae* (Jurcisek and Bakaletz 2007) and *S. aureus* (Izano et al. 2008).

While the mechanism by which the eDNA is released is still not fully understood, some evidence suggests that it may originate from processes that may or may not involve cell lysis. For example, eDNA is produced in a lysis-independent way in *B. subtilis* NCIB3610 even thought it does not seem to be required for biofilm formation (Zafra et al. 2012). In contrast, a recent study in *M. xanthus* showed that eDNA is very important for building and strengthening the biofilm and thus confers enhanced physical resistance to biological stresses (Hu et al. 2012). Interestingly, the eDNA participates in the *M. xanthus* biofilm structural development by directly interacting with other polymers present in the matrix, namely the exopolysaccharides (Hu et al. 2012). The exopolysaccharide component of the biofilm matrix has an important role for the biofilm community in addition to adhesion : in *B. subtilis*, it was proposed to generate osmotic pressure gradients that allow the cells to spread and find nutrients, making this process independent from motility (Seminara et al. 2012). In addition, the exopolysaccharides play a critical role in initiating and maintaining the structure of the biofilm as well as conferring protection to adverse environmental conditions (Aguilar et al. 2009).

Many bacteria can produce a number of polymeric macromolecules, however the contribution of different polymers to the biofilm is strain and condition dependent. For example, within the matrix of *B. subtilis* biofilms, at least two types of polymers have been identified as important in the process of biofilm development: poly-γ-DL-glutamic acid (γ-PGA) and exopolysaccharides (EPS). However, the undomesticated strain NCIB3610 produces primarily EPS (Branda et al. 2006; Branda et al. 2001) and the introduction of mutations in the genes for γ-PGA production did not affect the morphology of the biofilms (Branda et al. 2006). In contrast, γ-PGA enhanced the formation of surface-associated biofilms in the undomesticated strains RO-FF-1, B-1 and also in the domesticated strain JH642 (Morikawa et al. 2006; Stanley and Lazazzera 2005). In the case of *P. aeruginosa*, at least three distinct exopolysaccharides have been identified to be part of the extracellular matrix. Of these three, Pel and Psl are thought to be most important for biofilm structure under standard laboratory conditions. Although the third component, alginate, plays an important role in chronic lung infections of patients suffering from cystic fibrosis, it is only of very minor importance for biofilm formation on abiotic surfaces (Aguilar et al. 2009).

The integrity and robustness of the extracellular matrix of the biofilm is not only determined by its EPS composition or eDNA content. Proteinaceous adhesins in the matrix have also been shown to be of utmost importance for the overall structure of the biofilm. For example, the role of lectins in the structure and robustness of the biofilm has been studied in several bacterial species, including *Burkholderia cenocepacia* (Inhülsen et al. 2012), *P. aeruginosa* (Funken et al. 2012) and the plant-growth-promoting bacterium *Azospirilium brasilense* (Danhorn and Fuqua 2007). Lectins are sugar-binding proteins with a very specific recognition function that may bind to the EPS molecules of the matrix. Interestingly, lectins may also have additional binding sites for proteins, allowing the specific linkage of EPS molecules with proteins present at particular locations within the matrix, which may lead to the formation of an extracellular EPS-protein network (Neu and Lawrence 2009).

Other protein components have been implicated in the proper assembly of the matrix. For example, *B. subtilis* produces two extracellular proteins, BslA and TasA. Both these proteins are of crucial importance for the correct development of the biofilm. TasA forms long amyloid-like fibers that are anchored to the cell wall by the protein TapA (Romero et al. 2011, 2010) and in this way it is thought to contribute to the structural stability of the biofilm. BslA is an amphiphilic protein, important for surface hydrophobicity and also conferring resistance to antimicrobials. This protein works synergistically with TasA and the EPS component in the assembly of the biofilm matrix (Ostrowski et al. 2011). It was proposed that BslA may confer the biofilm with properties similar to the epithelia in higher organisms, giving protection from external stresses while providing an adequate environment for multicellular development (Kovács et al. 2012).

A family of large matrix proteins has also been identified as crucial in the development of multicellular communities in different organisms (Yousef and Espinosa-Urgel 2007; Reva and Tümmler 2008). These proteins are characterized by their negative net charge as well as by their enormous size, which in the case of LapA in *P. putida* can be over 8000 amino acids (Lasa and Penadés 2006). In addition, they have numerous repeats in their primary structure which, at the level of the gene, have been shown to be recombinogenic and may lead to the production of proteins of variable length within a biofilm population (Latasa et al. 2006). Members of this protein family have been identified in many Gram-negative and Gram-positive bacterial species, including *S. aureus, Streptococcus pyogenes, Lactobacillus reuteri, Bordetella bronchiseptica, P. fluorescens, Salmonella enteriditis, E. coli, Vibrio parahaemolyticus, Enterococcus faecium* and *E. faecalis* (Lasa and Penadés 2006; Latasa et al. 2006).

In some strains of staphylococcal species, the large surface proteins (biofilm associated proteins, Bap proteins) promote cell-cell interaction while also contributing to the extracellular matrix integrity (Lasa and Penadés 2006). In the human pathogen *B. cenocepacia*, it was recently shown that BapA is essential for biofilm development and that the expression of the operon, encoding for the genes responsible for the biosynthesis and export of this large surface protein, is under the control of a quorum sensing system (Inhülsen et al. 2012).

The Role of Chemical Signals in Biofilm Development

The expression of the different components of the biofilm matrix represents an important energetic investment for the cells in the biofilm. Thus, it makes sense that the expression of at least some of the matrix components is tightly regulated. Cell-to-cell communication, commonly referred to as quorum sensing (QS), is one of the mechanisms of genetic regulation that has emerged to play an important role in the establishment and development of bacterial biofilms (Aguilar et al. 2009). QS systems in bacteria rely on the production, diffusion, detection and response to small signal molecules. Although various signal molecules have been identified, the two most thoroughly investigated classes are the *N*-acyl-homoserine lactones (AHLs), which are produced by a number of Gram-negative bacteria, and small peptides, which are utilized by many Gram-positive species (Ng and Bassler 2009).

Bacteria use QS to regulate gene expression according to the size of the population: when a certain critical bacterial population, the quorum, has been attained, cells induce the expression of a particular set of genes that otherwise are silent. Biofilms appear to be the ideal locations where cell-to-cell communication can take place, since the cells are tightly bound in dense multicellular assemblages. Moreover, the biofilm matrix itself may act as a diffusion barrier for signal molecules, making the intercellular communication process more efficient and localized.

As mentioned above, the chemical composition of the exopolysaccharides present in the biofilm matrix can vary within species. Despite this, in a large number of bacteria QS has been shown to regulate the production and/or degradation of exopolysaccharides during biofilm development. An example of such a regulation can be found in the Gram-negative, corn pathogen, *Pantoea stewartii*. This bacterium uses the EsaI/EsaR QS circuitry (EsaI is necessary for the biosynthesis of AHLs, while EsaR regulates transcription of genes when threshold AHLs levels are reached in the cell) to regulate the production of stewartan, an acidic polymer of glucose, galactose and glucuronic acid present in mature biofilms. The EsaR QS-regulator represses transcription of the stewartan biosynthetic genes, which allows bacteria to attach to abiotic surfaces (von Bodman et al. 1998). In addition, *P. stewartii* mutants defective in AHL signal biosynthesis only form a flat and unstructured biofilm. Precise timing of stewartan expression seems to be essential for biofilm development since *esaR* mutants that synthesize stewartan constitutively do not attach properly to surfaces and only form fragile biofilms (von Bodman et al. 1998; Koutsoudis et al. 2006).

QS was also implicated in the development of *P. aeruginosa* biofilms since mutants in Pel biosynthesis or in the *lasI/lasR* QS network were severely impaired in the development of pellicles (biofilms at the air-liquid interface) and were defective in biofilms formed on submerged abiotic surfaces (Sakuragi and Kolter 2007). On the other hand, while some eDNA seems to be generated at basal levels, mutants in QS genes (*lasI/rhlI*) contained considerably less eDNA in the biofilm matrix, suggesting that cells may use QS for the generation of eDNA (Barken et al. 2008).

In addition to having a role in the establishment of the biofilm, QS also can be used by bacteria to trigger biofilm dispersal. This would allow bacteria to re-enter the planktonic mode of growth, allowing for the colonization of substrates in the vicinity. A well-documented example of such QS-dependent dispersal is the Gram-negative plant pathogen *Xanthomonas campestris*. This organism uses a diffusible signal factor (DSF) to sense population density and then regulate formation and maturation of biofilms. *X. campestris* mutants defective in DSF synthesis produce low levels of xanthan, an exopolysaccharide necessary for biofilm development (Dow et al. 2003; Torres et al. 2007). Interestingly, the DSF signal is also involved in the secretion of an endo-β-1,4-mannanase at high cell densities, which would degrade xanthan or another component of the matrix, causing disintegration of the biofilm matrix and releasing of bacterial cells from the biofilm (Dow et al. 2003). Similarly, QS mutants of *Rhodobacter sphaeroides* and *Yersinia pseudotuberculosis* form aggregates that could be dispersed by the addition of cognate AHLs, suggesting that escape from clumping is under QS-control (Atkinson et al. 1999; Puskas et al. 1997).

Cell Differentiation Within the Biofilm

A biofilm is composed of a large number of bacteria, which often arise from a single cell and thus represents a clonal population. However, the constituent cells of the biofilm can be found in a number of different physiological states, creating microenvironments to which the cells will adapt by altering their patterns of gene expression (An and Parsek 2007; Stewart and Franklin 2008). For example, investigations of *S. aureus* and *S. epidermis* biofilms revealed that DNA and protein synthesis took place only in discrete zones, suggesting that about two-thirds of the biofilm was metabolically inactive, despite the fact that the majority of the constituent cells were viable (Rani et al. 2007). In another example, laser capture microdissection microscopy combined with multiplex quantitative real-time reverse transcriptase PCR (LCMM/PCRqRT-PCR) was used to analyze biofilms formed by *P. aeruginosa* (Lenz et al. 2008). In this study, it was found that the distribution of the housekeeping gene *acpP* and two QS-regulated genes, *phzA1* and *aprA,* was not uniform throughout the biofilm. Moreover, big differences in expression of these genes were detected in cells in relatively close physical proximity.

The constituent cells of the biofilm and their unique patterns of gene expression were also investigated in *B. subtilis*. As mentioned above, sporulation could be detected in discrete zones of the biofilm (Branda et al. 2001; Veening et al. 2006). In addition to sporulation, cells within the biofilm can differentiate into further cell types that include matrix producers, motile cells, surfactin producers, cannibals, competent cells and degradative enzyme producers (López and Kolter 2010). Using thin-sectioning coupled to fluorescence microscopy, the distribution of three of these cell types was followed throughout biofilm development at the single-cell level (Vlamakis et al. 2008). By using transcriptional fusions to fluorescent proteins specific for cells undergoing sporulation, matrix production and motility, it was shown that these cell-types organized as distinct sub-populations within the biofilm. Moreover, the distribution of these cell-types clearly occurred in a dynamic spatiotemporal fashion, with motile cells as the dominant sub-population in early biofilm formation, matrix producers increasing at later stages and sporulating cells eventually starting from the upper regions of the biofilm (Fig. 1) (Vlamakis et al. 2008).

Mutants in the matrix components TasA or EPS showed both a dramatic reduction in the number of sporulating cells as well as an increase in the expression of the genes involved in matrix production. The activity of the sensor kinase KinD was at least partially responsible for this effect (Aguilar et al. 2010). KinD seems to have both kinase and phosphatase activities and it would function as a phosphatase for the global regulator Spo0A until the extracellular matrix or some of its components is sensed, thus acting as a checkpoint for sporulation in biofilm development. Low phosphorylation levels of Spo0A increase matrix production while preventing the cells to proceed to sporulation (Fujita et al. 2005). Once the matrix is correctly assembled, KinD functions as a kinase, increasing the intracellular levels of the phosphorylated form of Spo0A and allowing biofilm cells to continue their development towards sporulation (Aguilar et al. 2010).

The sensor kinase KinC has also been implicated in the development of *B. subtilis* biofilms. *B. subtilis* is able to trigger the production of matrix via KinC using the self-generated lipopeptide surfactin (López and Kolter 2010). Interestingly, the subpopulation of cells that produce surfactin (surfactin-producers) is different from the one that respond to surfactin (matrix-producers), and they are able to co-exist within the biofilm. This type of signaling adds a new level of sophistication to QS in bacteria and has been referred to as paracrine signaling. This unidirectional signaling is thought to allow for the compartmentalization of cellular differentiation and also to control the timing of gene expression in *B. subtilis* biofilms (López and Kolter 2010; Shank and Kolter 2011).

Biofilms and Human Health

Biofilms have a profound impact in medicine, industry and agriculture. As an example, biofilms associated to indwelling devices represent about half of the nosocomial infection cases reported every year in hospitals from the United States (Darouiche 2004). Indeed, microbial biofilms have been observed in most, if not all medical implant or devices, including cardiac pacemakers, joint prostheses, urinary catheters, mammary implants and ventricular assist devices (Darouiche 2004; Rodrigues 2011). The health care cost involved in treating these infections can vary from low cost in the case of non-surgical indwelling devices, as urinary catheters, to a very high cost in the case surgical implants, which can also lead to secondary complications if bacteremia is present (Lynch and Robertson 2008). One of the primary reasons why biofilms have such a negative impact in the clinical setting is that constituent cells acquire high levels of resistance against antimicrobial agents and host defenses, compared to the planktonic counterparts of the same microorganism (Bryers 2008; Costerton et al. 1999). The molecular mechanisms underlying this increased resistance are likely varied and it may consist in a combination of various events, including: (i) poor penetration of the antimicrobial agent across the dense biofilm matrix, which in some cases can even be inactivated in the outer layers of the biofilm faster than it diffuses (like reactive oxidants from phagocytic cells); (ii) the presence of persisters cells, which are subpopulations of dormant variants of regular cells that form stochastically within the biofilm and are not sensitive to antimicrobial treatments (Costerton et al. 1999; Lewis 2010).

We need a deeper understanding of the molecular basis of biofilm formation and maintenance, together with better insights of persister cell physiology. The development of new materials and coatings that help reduce the adhesion of microorganisms to surfaces (Rodrigues 2011) and the identification of molecular targets that increase the sensitivity of persister cells to antimicrobials (Van Acker et al. 2013) are approaches that will likely improve our ability to fight biofilm-related infections.

Summary

1. We have moved away from the old concept of bacteria being simple, selfish organisms. Bacteria are capable of complex behaviors, and the ability to stick together, to communicate with each other and to divide labor among its constituent cells reflects their true multicellular nature.

2. The advantages of multicellularity have been exploited by filamentous cyanobacteria, which physically separate metabolically incompatible processes, and by myxobacteria, which choreograph the behavior of the population to allow for the formation of complex fruiting bodies or for the predation of other bacteria.
3. In the natural environment, bacteria predominantly exist in the form of multicellular, architecturally complex aggregates called biofilms. Many bacterial species have been investigated with respect to their abilities to form biofilms and these studies showed that the extracellular matrix is of central importance for the correct assembly and maintenance of the biofilm. Complex regulatory networks, including QS circuitries, coordinate the production of matrix components in a spatiotemporal manner and control cell differentiation within the biofilm.
4. The mechanisms described above are examples of the different strategies that bacteria utilize to maximize their survival and their use of resources by being multicellular. These mechanisms have probably evolved to facilitate the response to changing environmental conditions, ensuring that the appropriate response is launched in the right place and at the right moment.
5. While many more players in the multicellular lifestyle of bacteria remain to be discovered, it is exciting to imagine that some of these mechanisms may be used as targets to combat pathogens and to prevent or eradicate detrimental biofilms.

Acknowledgments The authors thank Hera Vlamakis for critical reading of the manuscript. Financial support from the Swiss National Fund (Project 31003A_122013 to LE) and EU (grant 265862/Paravac to CE) is gratefully acknowledged.

References

Aguilar C, Vlamakis H, Losick R, Kolter R (2007) Thinking about *Bacillus subtilis* as a multicellular organism. Curr Opin Microbiol 10:638–643. doi:10.1016/j.mib.2007.09.006

Aguilar C, Carlier A, Riedel K, Eberl L (2009) Cell-to-cell communication in biofilms of Gram-negative bacteria. In: Krämer R, Jung K (eds) Bacterial Signaling. Wiley-VCH, Weinheim, pp 23–40

Aguilar C, Vlamakis H, Guzman A et al (2010) KinD is a checkpoint protein linking spore formation to extracellular-matrix production in *Bacillus subtilis* biofilms. MBio 1:1–7. doi:10.1128/mBio.00035-10

An D, Parsek MR (2007) The promise and peril of transcriptional profiling in biofilm communities. Curr Opin Microbiol 10:292–296. doi:10.1016/j.mib.2007.05.011

Asally M, Kittisopikul M, Rué P et al (2012) Localized cell death focuses mechanical forces during 3D patterning in a biofilm. Proc Natl Acad Sci U S A 109:18891–18896. doi:10.1073/pnas.1212429109

Atkinson S, Throup JP, Stewart GS, Williams P (1999) A hierarchical quorum-sensing system in *Yersinia pseudotuberculosis* is involved in the regulation of motility and clumping. Mol Microbiol 33:1267–1277. doi:10510240

Barken KB, Pamp SJ, Yang L et al (2008) Roles of type IV pili, flagellum-mediated motility and extracellular DNA in the formation of mature multicellular structures in *Pseudomonas aeruginosa* biofilms. Environ Microbiol 10:2331–2343. doi:10.1111/j.1462-2920.2008.01658.x

Bergman B, Gallon JR, Rai AN, Stal LJ (1997) N_2 Fixation by non-heterocystous cyanobacteria. FEMS Microbiol Rev 19:139–185

Branda SS, González-Pastor JE, Ben-Yehuda S et al (2001) Fruiting body formation by *Bacillus subtilis*. Proc Natl Acad Sci U S A 98:11621–11626. doi:10.1073/pnas.191384198
Branda SS, Vik S, Friedman L, Kolter R (2005) Biofilms: the matrix revisited. Trends Microbiol 13:20–26. doi:S0966-842X(04)00260-4
Branda SS, Chu F, Kearns DB et al (2006) A major protein component of the *Bacillus subtilis* biofilm matrix. Mol Microbiol 59:1229–1238. doi:10.1111/j.1365-2958.2005.05020.x
Bryers JD (2008) Medical biofilms. Biotechnol Bioeng 100:1–18. doi:10.1002/bit.21838
Christensen BB, Sternberg C, Andersen JB et al (1999) Molecular tools for study of biofilm physiology. Methods Enzym 310:20–42
Costerton JW, Stewart PS, Greenberg EP (1999) Bacterial biofilms: a common cause of persistent infections. Science 284:1318–1322 (80-)
Danhorn T, Fuqua C (2007) Biofilm formation by plant-associated bacteria. Annu Rev Microbiol 61:401–422. doi:10.1146/annurev.micro.61.080706.093316
Darouiche RO (2004) Treatment of infections associated with surgical implants. N Engl J Med 350:1422–1429. doi:10.1056/NEJMra035415
Dow JM, Crossman L, Findlay K et al (2003) Biofilm dispersal in *Xanthomonas campestris* is controlled by cell-cell signaling and is required for full virulence to plants. Proc Natl Acad Sci U S A 100:10995–11000. doi:12960398
Flores E, Herrero A (2010) Compartmentalized function through cell differentiation in filamentous cyanobacteria. Nat Rev Microbiol 8:39–50. doi:10.1038/nrmicro2242
Fremgen SA, Burke NS, Hartzell PL (2010) Effects of site-directed mutagenesis of mglA on motility and swarming of *Myxococcus xanthus*. BMC Microbiol 10:295. doi:10.1186/1471-2180-10-295
Fujita M, Gonza E, Losick R (2005) High- and low-threshold genes in the Spo0A regulon of *Bacillus subtilis*. J Bacteriol 187:1357–1368. doi:10.1128/JB.187.4.1357
Funken H, Bartels K, Wilhelm S et al (2012) Specific association of lectin LecB with the surface of *Pseudomonas aeruginosa*: role of outer membrane protein OprF. PLoS One 7:e46857. doi:10.1371/journal.pone.0046857
Grosberg RK, Strathmann RR (2007) The evolution of multicellularity: a minor major transition? Annu Rev Ecol Evol Syst 38:621–654. doi:10.1146/annurev.ecolsys.36.102403.114735
Hu W, Li L, Sharma S et al (2012) DNA builds and strengthens the extracellular matrix in *Myxococcus xanthus* biofilms by interacting with exopolysaccharides. PLoS One 7:e51905. doi:10.1371/journal.pone.0051905
Inhülsen S, Aguilar C, Schmid N et al (2012) Identification of functions linking quorum sensing with biofilm formation in *Burkholderia cenocepacia* H111. Microbiologyopen 1:225–242. doi:10.1002/mbo3.24
Izano EA, Amarante MA, Kher WB, Kaplan JB (2008) Differential roles of poly-*N*-acetylglucosamine surface polysaccharide and extracellular DNA in *Staphylococcus aureus* and *Staphylococcus epidermidis* biofilms. Appl Environ Microbiol 74:470–476. doi:10.1128/AEM.02073-07
Jurcisek JA, Bakaletz LO (2007) Biofilms formed by nontypeable *Haemophilus influenzae* in vivo contain both double-stranded DNA and type IV pilin protein. J Bacteriol 189:3868–3875. doi:10.1128/JB.01935-06
Kaiser D (2003) Coupling cell movement to multicellular development in myxobacteria. Nat Rev Microbiol 1:45–54. doi:10.1038/nrmicro733
Klausen M, Aaes-Jørgensen A, Molin S, Tolker-Nielsen T (2003) Involvement of bacterial migration in the development of complex multicellular structures in *Pseudomonas aeruginosa* biofilms. Mol Microbiol 50:61–68. doi:10.1046/j.1365-2958.2003.03677.x
Kolodkin-Gal I, Verdiger R, Shlosberg-Fedida A, Engelberg-Kulka H (2009) A differential effect of *E. coli* toxin-antitoxin systems on cell death in liquid media and biofilm formation. PLoS One 4:e6785. doi:10.1371/journal.pone.0006785
Koutsoudis M, Tsaltas D, Minogue T, von Bodman SB (2006) Quorum-sensing regulation governs bacterial adhesion, biofilm development, and host colonization in *Pantoea stewartii* subspecies *stewartii*. Proc Natl Acad Sci U S A 103:5983–5988. doi:10.1073/pnas.0509860103

Kovács AT, van Gestel J, Kuipers OP (2012) The protective layer of biofilm: a repellent function for a new class of amphiphilic proteins. Mol Microbiol 85:8–11. doi:10.1111/j.1365-2958.2012.08101.x

Lasa I, Penadés J (2006) Bap: a family of surface proteins involved in biofilm formation. Res Microbiol 157:99–107. doi:10.1016/j.resmic.2005.11.003

Latasa C, Solano C, Penadés J, Lasa I (2006) Biofilm-associated proteins. C R Biol 329:849–857. doi:10.1016/j.crvi.2006.07.008

Lenz AP, Williamson KS, Pitts B et al (2008) Localized gene expression in *Pseudomonas aeruginosa* biofilms. Appl Environ Microbiol 74:4463–4471. doi:10.1128/AEM.00710-08

Lewis K (2010) Persister cells. Annu Rev Microbiol 64:357–372. doi:10.1146/annurev.micro.112408.134306

López D, Kolter R (2010) Extracellular signals that define distinct and coexisting cell fates in *Bacillus subtilis*. FEMS Microbiol Rev 34:134–149. doi:10.1111/j.1574-6976.2009.00199.x

López D, Vlamakis H, Kolter R (2010) Biofilms. Cold Spring Harb Perspect Biol 2:a000398. doi:10.1101/cshperspect.a000398

Lynch A, Robertson G (2008) Bacterial and fungal biofilm infections. Annu Rev Med 59:415–428. doi:10.1146/annurev.med.59.110106.132000

McLoon AL, Guttenplan SB, Kearns DB et al (2011) Tracing the domestication of a biofilm-forming bacterium. J Bacteriol 193:2027–2034. doi:10.1128/JB.01542-10

Meeks JC, Elhai J (2002) Regulation of cellular differentiation in filamentous cyanobacteria in free-living and plant-associated symbiotic growth states. Microbiol Mol Bio Rev 66:94–121. doi:10.1128/MMBR.66.1.94

Mendes-Soares H, Velicer GJ (2013) Decomposing predation: testing for parameters that correlate with predatory performance by a social bacterium. Microb Ecol 65:415–423. doi:10.1007/s00248-012-0135-6

Morikawa M, Kagihiro S, Haruki M et al (2006) Biofilm formation by a *Bacillus subtilis* strain that produces γ-polyglutamate. Microbiology 152:2801–2807. doi:10.1099/mic.0.29060-0

Neu TR, Lawrence JR (2009) Extracellular polymeric substances in microbial biofilms. In: Moran AP, Holst O, Brennan PJ, von Itzstein M (eds) Microbial glycobiology. Elsevier, London, pp 735–758

Ng W, Bassler B (2009) Bacterial quorum-sensing network architectures. Annu Rev Genet 43:197–222. doi:10.1146/annurev-genet-102108-134304

O'Toole G, Kaplan HB, Kolter R (2000) Biofilm formation as microbial development. Annu Rev Microbiol 54:49–79

Ostrowski A, Mehert A, Prescott A et al (2011) YuaB functions synergistically with the exopolysaccharide and TasA amyloid fibers to allow biofilm formation by *Bacillus subtilis*. J Bacteriol 193:4821–4831. doi:10.1128/JB.00223-11

Puskas A, Greenberg EP, Kaplan S, Schaefer AL (1997) A quorum-sensing system in the free-living photosynthetic bacterium *Rhodobacter sphaeroides*. J Bacteriol 179:7530–7537. doi:9393720

Rani SA, Pitts B, Beyenal H et al (2007) Spatial patterns of DNA replication, protein synthesis, and oxygen concentration within bacterial biofilms reveal diverse physiological states. J Bacteriol 189:4223–4233. doi:10.1128/JB.00107-07

Reva O, Tümmler B (2008) Think big-giant genes in bacteria. Environ Microbiol 10:768–777. doi:10.1111/j.1462-2920.2007.01500.x

Rodrigues LR (2011) Inhibition of Bacterial Adhesion on Medical Devices. Adv Exp Med Biol 715:351–367. doi:10.1007/978-94-007-0940-9

Romero D, Aguilar C, Losick R, Kolter R (2010) Amyloid fibers provide structural integrity to *Bacillus subtilis* biofilms. Proc Natl Acad Sci U S A 107:2230–2234. doi:10.1073/pnas.0910560107

Romero D, Vlamakis H, Losick R, Kolter R (2011) An accessory protein required for anchoring and assembly of amyloid fibres in *B. subtilis* biofilms. Mol Microbiol 80:1155–1168. doi:10.1111/j.1365-2958.2011.07653.x

Sadykov MR, Bayles KW (2012) The control of death and lysis in staphylococcal biofilms: a coordination of physiological signals. Curr Opin Microbiol 15:211–215. doi:10.1016/j.mib.2011.12.010

Sakuragi Y, Kolter R (2007) Quorum-sensing regulation of the biofilm matrix genes (pel) of *Pseudomonas aeruginosa*. J Bacteriol 189:5383–5386. doi:10.1128/JB.00137-07

Schirrmeister BE, de Vos JM, Antonelli A, Bagheri HC (2013) Evolution of multicellularity coincided with increased diversification of cyanobacteria and the Great Oxidation Event. Proc Natl Acad Sci U S A. doi:10.1073/pnas.1209927110

Schopf JW (2006) Fossil evidence of Archaean life. Philos Trans R Soc Lond B Biol Sci 361:869–885. doi:10.1098/rstb.2006.1834

Seminara A, Angelini TE, Wilking JN et al (2012) Osmotic spreading of *Bacillus subtilis* biofilms driven by an extracellular matrix. Proc Natl Acad Sci U S A 109:1116–1121. doi:10.1073/pnas.1109261108

Shank EA, Kolter R (2011) Extracellular signaling and multicellularity in *Bacillus subtilis*. Curr Opin Microbiol 14:741–747. doi:10.1016/j.mib.2011.09.016

Shapiro JA (1988) Bacteria as multicellular organisms. Sci Am 258:82–89

Shapiro JA (1998) Thinking about bacterial populations as multicellular organisms. Annu Rev Microbiol 52:81–104. doi:10.1146/annurev.micro.52.1.81

Shapiro JA, Dworkin M (1997) Bacteria as multicellular organisms. Oxford University Press, New York

Stanley NR, Lazazzera BA (2005) Defining the genetic differences between wild and domestic strains of *Bacillus subtilis* that affect poly-γ-dl-glutamic acid production and biofilm formation. Mol Microbiol 57:1143–1158. doi:10.1111/j.1365-2958.2005.04746.x

Steinberger RE, Holden PA (2005) Extracellular DNA in single- and multiple-species unsaturated biofilms. Appl Environ Microbiol 71:5404–5410. doi:10.1128/AEM.71.9.5404

Stewart PS, Franklin MJ (2008) Physiological heterogeneity in biofilms. Nat Rev Microbiol 6:199–210. doi:10.1038/nrmicro1838

Tamulonis C, Postma M, Kaandorp J (2011) Modeling filamentous cyanobacteria reveals the advantages of long and fast trichomes for optimizing light exposure. PLoS One 6:e22084. doi:10.1371/journal.pone.0022084

Torres P, Malamud F, Rigano L et al (2007) Controlled synthesis of the DSF cell-cell signal is required for biofilm formation and virulence in *Xanthomonas campestris*. Environ Microbiol 9:2101–2109. doi:EMI1332

Van Acker H, Sass A, Bazzini S et al (2013) Biofilm-grown *Burkholderia cepacia* complex cells survive antibiotic treatment by avoiding production of reactive oxygen species. PLoS One 8:e58943. doi:10.1371/journal.pone.0058943

Veening J, Kuipers OP, Brul S et al (2006) Effects of phosphorelay perturbations on architecture, sporulation, and spore resistance in biofilms of *Bacillus subtilis*. J Bacteriol 188:3099–3109. doi:10.1128/JB.188.8.3099

Velicer GJ, Vos M (2009) Sociobiology of the myxobacteria. Annu Rev Microbiol 63:599–623. doi:10.1146/annurev.micro.091208.073158

Velicer GJ, Kroos L, Lenski RE (1998) Loss of social behaviors by *Myxococcus xanthus* during evolution in an unstructured habitat. Proc Natl Acad Sci U S A 95:12376–12380

Vilain S, Pretorius JM, Theron J, Brözel VS (2009) DNA as an adhesin: *Bacillus cereus* requires extracellular DNA to form biofilms. Appl Environ Microbiol 75:2861–2868. doi:10.1128/AEM.01317-08

Vlamakis H, Aguilar C, Losick R, Kolter R (2008) Control of cell fate by the formation of an architecturally complex bacterial community. Genes Dev 22:945–953. doi:10.1101/gad.1645008

Vlamakis H, Chai Y, Beauregard P et al (2013) Sticking together: building a biofilm the *Bacillus subtilis* way. Nat Rev Microbiol 11:157–168. doi:10.1038/nrmicro2960

Von Bodman SB Majerczak DR Coplin DL (1998) A negative regulator mediates quorum-sensing control of exopolysaccharide production in *Pantoea stewartii* subsp. *stewartii*. Proc Natl Acad Sci U S A 95:7687–7692

Whitchurch CB, Tolker-Nielsen T, Ragas PC, Mattick JS (2002) Extracellular DNA required for bacterial biofilm formation. Science 295:1487. doi:11859186 (80-)

Yousef F, Espinosa-Urgel M (2007) In silico analysis of large microbial surface proteins. Res Microbiol 158:545–550. doi:10.1016/j.resmic.2007.04.006

Zafra O, Lamprecht-Grandío M, González de Figueras C, González-Pastor JE (2012) Extracellular DNA release by undomesticated *Bacillus subtilis* is regulated by early competence. PLoS One 7:e48716. doi:10.1371/journal.pone.0048716

Part II
Model-Systems

Choanoflagellates: Perspective on the Origin of Animal Multicellularity

Stephen R. Fairclough

Abstract For more than a century and half choanoflagellates, the closest living relatives of animals, have fascinated evolutionary biologists. By characterizing the similarities and differences between choanoflagellates and animals, biologists have gained perspective on the biology of their last common ancestor, the "Urchoanimal", as well as the evolutionary foundations of multicellularity and the origin of animals. The best-studied colonial choanoflagellate, *Salpingoeca rosetta*, forms colonies by cell division and not by cell aggregation. The observation that cytoplasmic bridges connect cells in *S. rosetta* colonies and other colonial choanoflagellates, as well as cells in sponges, suggests that this mechanism of colony formation may be ancestral within the choanoflagellate lineage and may have been present in the Urchoanimal as well. The comparison of choanoflagellate gene content and gene function with animal gene content and gene function has revealed that many of the basic mechanisms of cell adhesion, signaling, and differentiation that were previously thought to be unique to animals are also present in choanoflagellates, indicating that these genes were present prior to the evolution of animals. These insights refine our understanding of genes that emerged on the stem lineage leading to the last common ancestor of all animals, the "Urmetazoan". Taken together the data from choanoflagellates have provided deep insights into the biology of the Urchoanimal and the evolution of animal multicellularity.

Keywords Choanoflagellate · Cadherin · Development · Evolution · Hedgling · *Monosiga brevicollis* · Multicellularity · RTK · *Salpingoeca rosetta*

Introduction

Reconstructing the Origin of Animals

Reconstructing the evolutionary events that led to the origin of animals requires an understanding of the biology of ancient animals and their ancestors (King 2004). However, the fossil record, which often provides insight into the biology of more

S. R. Fairclough (✉)
Department of Energy, Joint Genome Institute, Walnut Creek, CA, USA
e-mail: srfairclough@gmail.com

I. Ruiz-Trillo, A. M. Nedelcu (eds.), *Evolutionary Transitions to Multicellular Life*,
Advances in Marine Genomics 2, DOI 10.1007/978-94-017-9642-2_5

recent ancestors, has remained largely silent with regard to these organisms, presumably because of their antiquity, small size, and composition (Chapters "Fossil and transcriptomic perspectives on the origins and success of metazoan multicellularity" and "Origin of Metazoan Developmental Toolkits and their Expression in the Fossil Record"). A complementary approach to understand ancient biology is to compare the biology of extant organisms and infer which shared characteristics were present in their last common ancestor by parsimony. Choanoflagellates, the closest living relatives of animals (Fig. 1a), have fascinated evolutionary biologists because of their morphological and behavioral similarities to sponge feeding cells, choanocytes (Saville-Kent 1880). By identifying the similarities and differences between choanoflagellates and animals, biologists have gained understanding of the biology of their last common ancestor, the Urchoanimal (Saville-Kent 1880; King 2004). The application of molecular biology and genomics to the study of choanoflagellates has revealed that many of the genes involved in the basic mechanisms of cell adhesion, signaling, and differentiation that were previously thought to be unique to animals are also present in choanoflagellates. Such findings indicate that these genes were present in the Urchoanimal (King et al. 2003; King et al. 2008; Fairclough et al. 2013). These observations also refine our understanding of genes that emerged on the stem lineage leading to the last common ancestor of all animals, the Urmetazoan. By continuing to investigate the biology and gene content of choanoflagellates and their relatives we can reconstruct the foundational events that preceded the origin of animals and their diversification into myriad modern forms (also see Chapters "A comparative genomics perspective on the origin of multicellularity and early animal evolution" and "Transcription factors and the origin of animal multicellularity").

Choanoflagellate Biology

Choanoflagellates are free-living heterotrophic eukaryotes that occupy both pelagic and benthic zones in marine, brackish, and freshwater environments around the globe. The life cycles of all choanoflagellates contain a single-celled phase and many species are also capable of forming multicellular colonies of morphologically similar cells (Saville-Kent 1880; Leadbeater 1983a; Leadbeater and Thomsen 2000; Dayel et al. 2011). Depending on the species, solitary and colonial forms are variously capable of free-swimming in the water column or adhering to substrates directly or through either the periplast or a thin pedicel (Saville-Kent 1880; Leadbeater 1983a; Dayel et al. 2011). Choanoflagellates grow vegetatively, with many species undergoing longitudinal fission (Karpov and Leadbeater 1998). Under nutrient limiting conditions, *S. rosetta* cells are capable of producing morphologically distinct haploid cells that can fuse to form diploid cells (Levin and King 2013).

As their name implies, choanoflagellates have a distinct morphology typified by an ovoid or spherical cell body that is 3–10 μm in diameter and an apical flagellum surrounded by a collar of 30–40 microvilli (Fig. 1b) (Leadbeater and Thomsen 2000; Leadbeater 2008; Dayel et al. 2011). Movement of the flagellum creates water

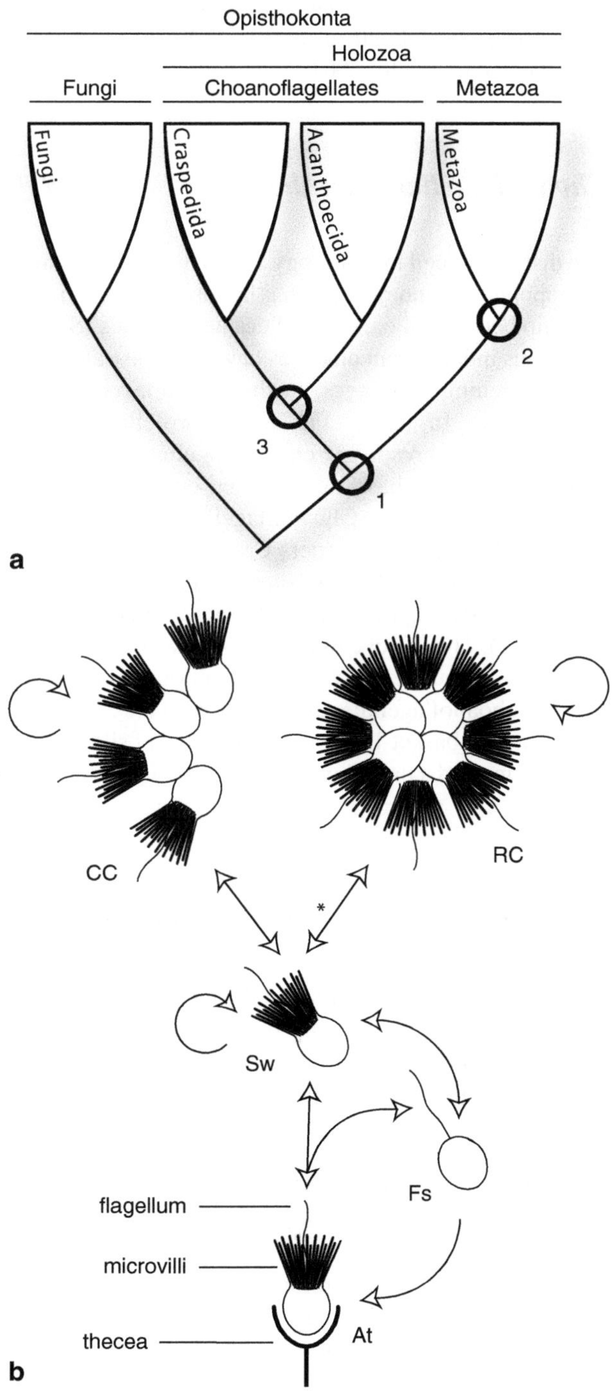

Fig. 1 Choanoflagellates are the closest known living relatives of animals. **a** Reconstructing the biology of last common ancestor of animals and choanoflagellates, i.e. the "Urchoanimal" (1), is critical for understanding the evolutionary innovations that occurred on the stem lineages leading to the last common ancestor of animals (2; the "Urmetazoan") and the last common ancestor of choanoflagellates (3). **b** Choanoflagellate cell morphology is typified by an ovoid or spherical cell body that is 3–10 μm in diameter and an apical flagellum surrounded by a collar of 30–40 microvilli. *S. rosetta* is a colony-forming member of the Craspedida. *S. rosetta* can differentiate into at least five morphologically and behaviorally distinct cell types. Solitary "thecate" cells attached to a substrate (*Th*) can produce solitary swimming (*Sw*) cells or fast swimming solitary (*FS*) cells, either through cell division or theca abandonment. Solitary swimming cells can divide completely to produce solitary daughter cells or remain attached after undergoing incomplete cytokinesis to produce either chain colonies (*CC*), or rosette colonies (*RC*) in the presence of the bacterium *Algoriphagus machipongonensis* (*). (Adapted from Fairclough et al. 2013)

currents that can propel free-swimming choanoflagellates through the water column or trap bacteria and detritus against the collar of microvilli where they are engulfed (Leadbeater and Kelly 2001; Dayel et al. 2011).

Choanoflagellates are the Closest Living Relatives of Animals

The distinctive cell morphology and method of feeding of choanoflagellates is shared with sponge choanocytes and has historically been used as evidence for the close relationship between choanoflagellates and animals (Saville-Kent 1880). More recently, numerous independent analyses have demonstrated that animals are a monophyletic group containing sponges, with choanoflagellates as their closest known living relatives (Fig. 1a) (Lang et al. 2002; Torruella et al. 2012).

In addition to resolving choanoflagellates as the sister group to animals, molecular phylogenetics has also informed our understanding of relationships within the choanoflagellate clade. Choanoflagellates form two monophyletic groups that correspond to the taxonomic orders Craspedida and Acanthoecida (Nitsche et al. 2011). The Craspedida are distinguished by an extracellular investment that is entirely organic and does not project above the anterior end of the extended feeding cell (Nitsche et al. 2011). The life cycles of Craspedida generally contain a vegetative stage that is sedentary and stalked and a solitary or colonial motile stage. All documented colony forming choanoflagellates are members of the Craspedida. The Acanthoecida are distinguished by basket-like lorica composed of siliceous costae comprising rod-shaped costal strips and a partial or entire organic investment on inner surface (Nitsche et al. 2011).

Genome Structure and Gene Annotation

The sequencing of choanoflagellate genomes has provided new perspectives on choanoflagellate biology and overturned our thinking on the evolution of several important animal genes by revealing that numerous signaling and adhesion genes critical to animal development are present in choanoflagellate genomes, and thus predate the Urchoanimal ancestor (King et al. 2003, 2008; Fairclough et al. 2013). To date, two choanoflagellate genomes have been sequenced: those of the colonial *S. rosetta* and the solitary *M. brevicollis*. The 55.4 Mb *S. rosetta* genome was sequenced to 33x average coverage assembled into 154 scaffolds with 50 % of the scaffolds longer than 1.52 Mb and identified a minimum of 31 chromosomes containing 11,629 genes (Fairclough et al. 2013). The 41.6 Mb *M. brevicollis* genome was sequenced to 8x average coverage and assembled into 218 scaffolds and identified a minimum of 19 chromosomes containing approximately 9,200 genes (King et al. 2008).

Comparisons of the *S. rosetta* and *M. brevicollis* genomes with those of diverse animals and eukaryotes reveal the minimal gene content of long-extinct ancestors and the patterns of gene gain and loss that accompanied their evolution (Fairclough et al. 2013). Although the Urmetazoan and the Urchoanimal genomes each contained at least 10,000 genes, ~10 % of genes in the Urchoanimal were replaced with novel genes along the metazoan stem lineage. In contrast, the evolution of choanoflagellates from the Urchoanimal was characterized by extensive gene loss (~40 % of the Urchoanimal genome) that was only weakly counteracted by gene gain. In addition, the *S. rosetta* and *M. brevicollis* lineages have each experienced massive gene gain, with 36 % and 33 %, respectively, of each species' gene content being unique (Fairclough et al. 2013).

Choanoflagellate Multicellularity

In its most basic form, multicellularity can be defined as multiple cells physically attached to each other. The existence of colonial choanoflagellates has made choanoflagellates an organism of choice for testing hypotheses about multicellularity in general and the origin of animals in particular (King 2004; Carr et al. 2008; King et al. 2009). The origin of multicellularity within choanoflagellates is an unresolved question. Current data from phylogenetics and the reconstruction of ancestral character states within the choanoflagellate group indicates that Craspedida contain species that form colonies, but there is no documentation of multicellular forms present in Acanthoecida (Fig. 1a) (Nitsche et al. 2011). This distribution of multicellularity indicates that colony formation likely evolved after the diversification of Craspedida and Acanthoecida. However, it is also possible that multicellularity evolved multiple times independently, or that the last common ancestor of animals and choanoflagellates was capable of forming multicelled colonies (Carr et al. 2008).

Salpingoeca rosetta, the most thoroughly studied colony forming choanoflagellate, has a life history containing at least five morphologically and behaviorally differentiated cell types (Fig. 1b) (Dayel et al. 2011). Its life history includes two colonial forms, rosette and linear, as well as three solitary cell types: slow swimming, fast swimming, and substrate attached (also known as 'thecate') (Dayel et al. 2011). The slow swimming cell plays a central role in the *S. rosetta* life history because it is competent to differentiate into each of the other identified cell types. In *S. rosetta* the development of colonies from solitary cells occurs by cell division (Fig. 2), with sister cells remaining stably attached by cytoplasmic bridges and extracellular material (Fig. 3) (Fairclough et al. 2010; Dayel et al. 2011). Although the process of colony formation has not been documented in other species, cytoplasmic bridges have been observed, suggesting a similar mechanism (Hibberd 1975; Karpov and Coupe 1998). *S. rosetta* cell types have distinct transcriptional profiles, suggesting that colony formation is a regulated developmental process (Fairclough et al. 2013). In mature colonies the transcriptome is dramatically enriched for genes unique to *S. rosetta*, with relatively little contribution from genes uniquely shared with a solitary

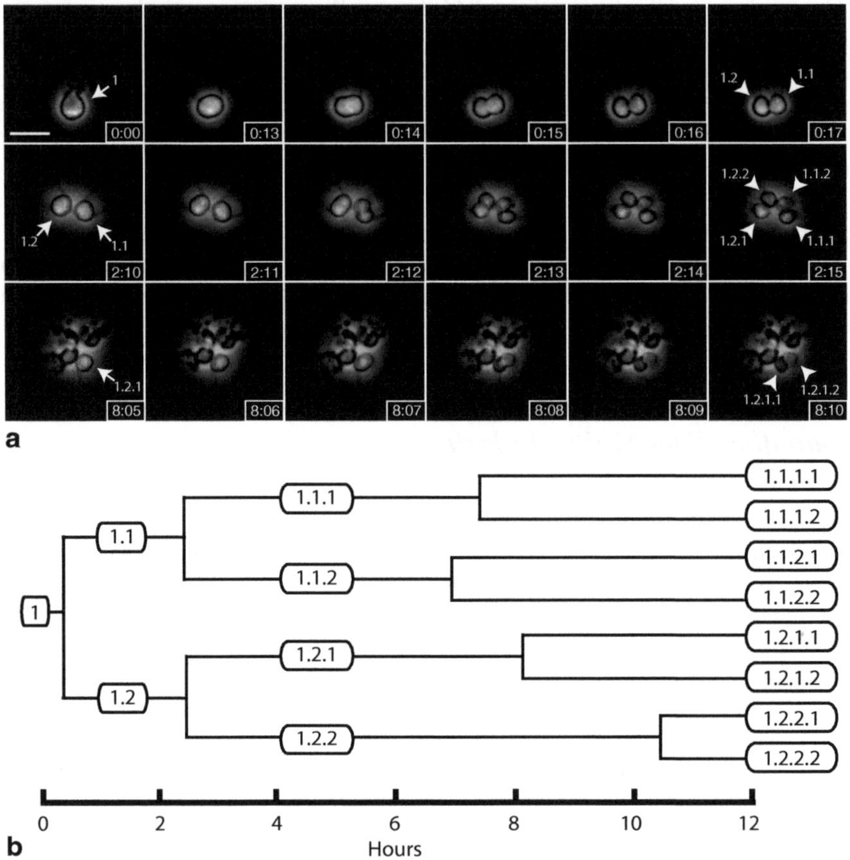

Fig. 2 *Salpingoeca rosetta* colonies develop through cell division, not aggregation. **a** Still images from a time-lapse movie show *S. rosetta* cells dividing (*arrows*) during colony development. In each case, the sister cells (*arrowheads*) remain attached. Extra-colonial cells were never observed to join a colony through aggregation. Scale bar represents 10 μm. The founder cell and its progeny were marked 1–1.2.1.2 to generate the cell pedigree in panel b. Time since start of movie (hours:minutes) is indicated in *lower right* of each panel. **b** The mapping of cell pedigree as a function of time, based on the time-lapse movie in panel **a**, shows that cells divide asynchronously during colony formation. (Republished from Fairclough et al. 2010)

choanoflagellate, *M. brevicollis,* indicating that major aspects of *S. rosetta* colony biology evolved after the divergence from *M. brevicollis* or were lost from *M. brevicollis*. Intriguingly, developing colonies also upregulate a disproportionate number of genes shared by *S. rosetta* and animals. Therefore, it is possible that genes that regulate colony development were also instrumental in the origin of animals (Fairclough et al. 2013).

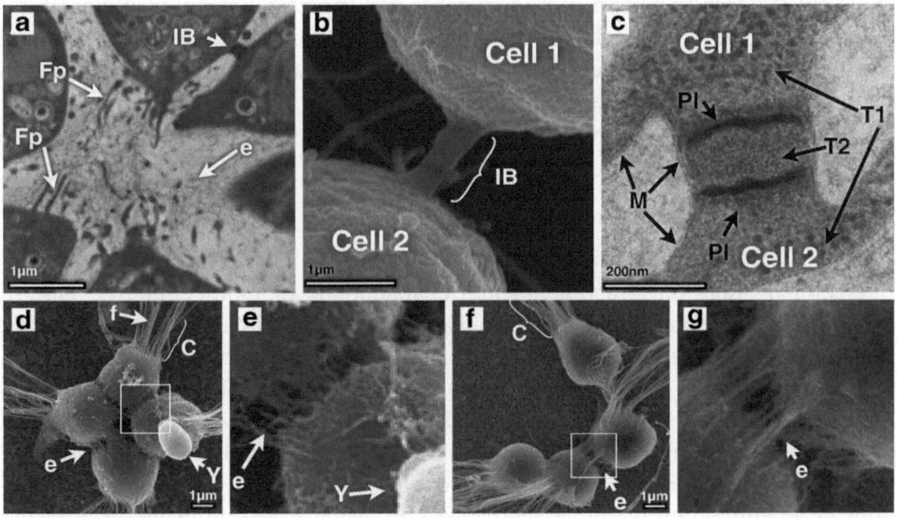

Fig. 3 Intercellular attachment in *S. rosetta* colonies. Cells within rosette colonies are attached to one another with ECM, filopodia, and intercellular bridges. **a** TEM of a thin section through a rosette shows filopodia extending into the central ECM and an intercellular bridge connecting neighboring cells. **b** SEM of an intercellular bridge. **c** TEM of a thin section through an intercellular bridge shows two electron-dense plates trisecting the bridge. The texture of the cytoplasm within cells (T1) differs from that within the bridge (T2). **d, e** SEM of a rosette colony shows a shared, filamentous ECM linking *S. rosetta* cell bodies. ECM is absent from co-cultured yeast cells (included as a negative control). **f, g** Cells in chain colonies share ECM filaments (e), but lack filopodia. Key: *f* flagellum, *C* collar, *IB* intercellular bridge, *Fp* filopodia, *E* ECM, *Y* yeast, *Pl* bridge plate, *M* cell membrane. *Scale bars* as marked. (Republished from Dayel et al. 2011)

Environmental Influence on Multicellular Development

The environment is a powerful contributor to organismal development. The recent characterization of the human microbiome has highlighted the significant role that the microbial environment plays in the biology of humans, but this trend has been recognized across other organisms for several decades (McFall-Ngai et al. 2013). In *S. rosetta*, rosette colony formation is triggered by the presence of the prey bacterium *Algoriphagus machipongonensis* (Fairclough et al. 2010; Dayel et al. 2011; Alegado et al. 2012), a member of the Bacteroidetes phylum that includes species known to influence animal development (Alegado et al. 2011). However, *Algoriphagus machipongonensis* appears to trigger *S. rosetta* colony development through a sulfonolipid, RIF-1, that is not found in animals (Alegado et al. 2012). Furthermore, neither the application of *A. machipongonensis* nor RIF-1 to other choanoflagellates, has elicited a colony response, suggesting that this sulfonolipid triggered colony response is unique to *S. rosetta*. Different *S. rosetta* cell types that were fed different environmental bacteria have distinct transcriptional profiles, but the transduction and downstream signaling of these interactions remains to be characterized (Fairclough et al. 2013). Despite its apparent uniqueness, the triggering of multicellular development by bacteria in the closest living relatives of animals highlights the potential role the bacterial environment may have played in the evolution of animals and in the evolution of multicellularity more broadly.

Multicellular Cell Adhesion

Intercellular adhesion is a fundamental characteristic of multicellularity. One of the most iconic examples of intercellular adhesion in animals is the epithelium. Eumetazoan epithelia are replete with special cell-cell junctions (including adherens, demosomal, gap, and tight/septate junctions) and cell-substrate junctions regulated by hemidesmosomes and focal adhesions that attach to a basal lamina (Tyler 2003). Although there are pictures and drawings of hollow choanoflagellate colonies (Saville-Kent 1880), there is no documentation of any structure that functions as an epithelium by either segregating an inside from an outside or top from bottom. Despite the absence of epithelia tissue, the sequencing of choanoflagellate genomes and transcriptomes indicates that genes associated with epithelial structures and functions in Bilateria arose prior to the evolution of animals (Abedin and King 2008; Sebé-Pedrós et al. 2010; Srivastava et al. 2010; Chapter "A comparative genomics perspective on the origin of multicellularity and early animal evolution").

In choanoflagellates, intercellular adhesion is mediated by three physical structures: cytoplasmic bridges, protein mediated cell-cell adhesion, and extracellular material (Fig. 3). In *S. rosetta* colonies, the initial mechanism of intercellular adhesion is the cytoplasmic bridge that is thought to arise from incomplete cytokinesis (Figs. 3b, 3c). Although there is membrane continuity between the cells, many of the bridges contain perpendicular electron dense bands that appear to block bulk cytoplasmic continuity and potentially block all molecular transfer (Fig. 3c) (Dayel et al. 2011). The process by which the cytoplasmic bridges form remains to be fully characterized, but gene expression data show that septins, proteins known for regulating cytokinesis and stabilizing cytoplasmic bridges, are upregulated in colonial cells (Fairclough et al. 2013).

In addition to cellular continuity in multicellular choanoflagellates, there also appears to be protein mediated cell adhesion in which cells are attached to each other via the microvilli (Leadbeater 1983b). Although the proteins mediating this cell adhesion have not been characterized, a number of genes associated with epithelial structures and functions in Bilateria are present in choanoflagellates. One such family of cell adhesion genes thought to have played important roles in early animal evolution also found in choanoflagellates is the cadherins. The *M. brevicollis* genome encodes at least 23 cadherins (Abedin and King 2008) and the *S. rosetta* genome encodes 29 cadherin, numbers that are comparable to those in the genomes of many animals (including *D. melanogaster* and *C. intestinalis*) (King et al. 2008; Abedin and King 2008; Nichols et al. 2012). Despite the comparable numbers of cadherins, relationships between metazoan and choanoflagellate cadherins indicate that these genes have either largely expanded independently of each other or dramatically diverged in the choanoflagellate and animal lineages. At least three modern cadherin families, lefftyrins, coherins, and hedglings, were present in the Urchoanimal. However, these three families are absent from Bilateria (Nichols et al. 2012) making the assignment of function by homology more tenuous.

Animal cadherins are best known for their roles in cell adhesion and intercellular signaling, and they can also act as docking sites for pathogenic bacteria (Abedin and King 2010). By extension from these known animal functions, choanoflagellate cadherins may facilitate cell adhesion in colonies, transduce signals, function in prey capture, or may act in choanoflagellate-specific processes, such as attachment to theca or microvilli stabilization. Investigation of gene expression offers insight into potential gene function. Of the 29 cadherins present in the *S. rosetta* genome, eleven are upregulated in substrate-attached cells where they may help regulate attached cell differentiation, potentially through functions related to signaling or environmental substrate attachment. A different set of ten cadherins appears to be specifically upregulated in colonies. Many of these colony-specific cadherins are arguably too short to be mediating direct adhesion between neighboring cells (from 565–8158 a.a.). Instead, they may act either as signaling molecules or by interacting with the loose extracellular material surrounding cells in colonies (Dayel et al. 2011). Notably, there is no correlation between the expression of specific cadherins and the species of bacterial prey fed to a given samples, suggesting either that cadherins are not specific for particular species of bacterial prey or that cadherins with functions related to prey capture are not regulated at the transcriptional level.

Cell adhesion and tissue integrity in Eumetazoa are also mediated by interactions between epithelia and the basal lamina, a layer of extracellular matrix comprised of collagen, laminin, fibronectin, proteoglycans and polysaccharides, that anchors epithelial cells and helps maintain cell polarity (Nichols et al. 2009; Abedin and King 2010). As with cell-cell adhesion, many of the molecules required for cell-substrate attachment emerged prior to or early in animal evolution (Hynes 2012). Integrins, which connect cells to the extracellular material, as well as the extracellular material proteins that comprise the basal lamina, have been detected in all major animal phyla, including some sponges, despite the fact that most Porifera lack an identifiable basal lamina (Leys et al. 2009; Abedin and King 2010). Outside of animals, integrin subunits and their predicted intracellular binding partners are found in the genome of a unicellular relative of animals, *C. owczarzaki*, and an integrin alpha domain is present in choanoflagellates (King et al. 2008; Sebé-Pedrós et al. 2010). As with many of the "epithelial" genes, the early emergence of integrins and extracellular material components indicates that these genes evolved prior to their function in the multicellular context of animals, suggesting that they had functions in the unicellular and colonial progenitors of animals. It is possible that modern-day epithelial polarity genes in animals had ancestral functions in the establishment of cell polarity. Alternatively, extracellular material and adhesion genes may have mediated the interactions of cells with inert environmental substrates through the deposition of extracellular material, such as the choanoflagellate theca, but these hypothesis and others await experimentation.

Developmental Signaling: Organizing in Space and Time

Development, the coordination of multicellular growth, is often regulated by intercellular signaling. In animals, an ever-growing body of data from the study of the evolution of development demonstrates that a core set of signaling pathways has had a prominent role in morphological evolution. Seven intercellular signaling pathways are traditionally considered critical for and unique to animal development: nuclear hormone receptors, WNT, TGF-b, Jak/STAT, Notch/Delta, Hedgehog, and receptor tyrosine kinases (RTK). Conserved components of all of these pathways are expressed in sponges, indicating they were present in the last common ancestor of animals (Nichols et al. 2006; Bridgham et al. 2010; Srivastava et al. 2010; Chapter "A comparative genomics perspective on the origin of multicellularity and early animal evolution"). In contrast, there is little evidence from the gene content of the choanoflagellates sequenced to date for the complete multi-component developmental signaling pathways typically thought to function in animals (King et al. 2008; Fairclough et al. 2013). While this could be due to our inability to detect homologs in choanoflagellates because of the deep divergence time, it seems likely that many of the components are simply not present in choanoflagellates. Although complete pathways are not present, some signaling pathway components and signaling domains from animals are present in choanoflagellates, including RTKs; NL domains; the Hedgehog signal domain, and the Hint domain (which together make up animal hedgehog proteins); the hedgehog receptor, Patched; Dispatched; and STAT (King et al. 2003, 2008; Manning et al. 2008; Fairclough et al. 2013). The presence of these components in choanoflagellates suggests that rudiments of some animal developmental signaling pathways were present in the last common ancestor of choanoflagellates and animals. Interestingly, the protein domains diagnostic of proteins in these pathways are frequently found in novel configurations compared to those in Bilateria, indicating that animal developmental signaling pathways were assembled (at least in part) by domain shuffling of preexisting domains. Alternatively, the fully assembled signaling pathways may have predated the origin of choanoflagellates, but their component proteins were shuffled within the choanoflagellate lineage (King et al. 2008; Gazave et al. 2009).

The evolutionary history of the Hedgehog signal domain is an excellent example of the importance of domain shuffling in the metazoan stem lineage (Fig. 4). The Hedgehog signal domain is best known as a secreted ligand in the Hedgehog signaling pathway that regulates developmental patterning in bilaterians. In this context the Hedgehog ligand is composed of an N-terminal signaling domain that is autocatalytically cleaved from (and by) the C-terminal HINT domain (Lee et al. 1994). The genome sequencing of *M. brevicollis*, *S. rosetta*, and the sponge, *A. queenslandica,* revealed new molecular contexts for this domain on the extracellular N-terminus of a cadherin molecule, dubbed Hedgling (Fig. 3, see also Chapter "Developmental signalling and emergence of animal multicellularity")(Adamska et al. 2007; King et al. 2008; Fairclough et al. 2013).

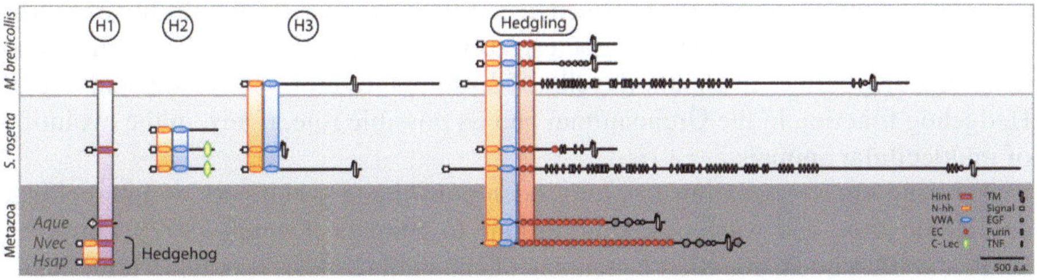

Fig. 4 The diversity of choanoflagellate Hedgehog signal domain proteins. The archetypal Hedgehog protein represented by proteins from Nematostella vectensis (*Nvec*) and Homo sapiens (*Hsap*) contains both an N-terminal Hedgehog signal domain and a Hint domain (H1). The choanoflagellate proteome contains proteins that contain Hint domains and a number of Hedgehog signal domains in diverse and novel protein architectures. Both *M. brevicollis* and *S. rosetta* contain Hedgehog proteins associated with transmembrane domains (H3), a subset of which also contain cadherin domains (Hedglings). *S. rosetta* also contains three short peptides that contain a signal sequence and Hedgehog signal domain (H2), two of which also contain C-lectin domains. (Adapted from King et al. 2008)

These genomes also encode HINT domains in other proteins, indicating that these domains evolved independently. The *S. rosetta* genome also contains several additional Hedgehog signal domain containing proteins (Nichols et al. 2012) that lack either cadherin or HINT domains, but contain additional protein domains that expand their potential functions and complicate efforts to reconstruct the evolutionary history of the Hedgehog domain (Fig. 4) (Adamska et al. 2007; Nichols et al. 2012; Fairclough et al. 2013). The absence of these short Hedgehog signal domain proteins in other lineages suggest that these are independent expansions in *S. rosetta.* Although Hedgling and Hedgehog genes are absent from the *Trichoplax* genome, Hedgling, HINT domains, and the classically described Hedgehog protein are encoded by the genome of the cnidarian *Nematostella vectensis*, suggesting that the Hedgehog signal domain and the HINT domain were brought together by domain shuffling and that Hedgling was lost prior to the evolution of bilaterians (Adamska et al. 2007). The choanoflagellate genomes also contain the canonical Hedgehog receptor Patched. If these Hedgehog signal proteins interact with homologs of the Hedgehog receptor, Patched, we may infer that this interaction was present in the Urchoanimal and potentiated the origin of the Hedgehog signaling pathway.

The sequencing of the *S. rosetta* transcriptome provided the first comprehensive look at gene expression in a choanoflagellate and facilitated insight into the functions of important developmental genes, such as the *Hedgehog/Hedgling* family, in choanoflagellates (Fairclough et al. 2013). Interestingly, the genes containing predicted membrane-bound Hedgehog signals (including Hedgling) are upregulated in substrate-attached cells suggesting they may be mediating substrate attachment. In contrast, the three *S. rosetta* Hedgehog domain-containing genes that lack a predicted transmembrane domain and contain predicted signal peptides, suggesting that they may act as secreted ligands, are robustly upregulated in colonial cells. In *S. rosetta,* the *Patched* homologs also show differential expression, with some upregulated in colonies and others upreglated in attached cells leaving open the possibility of ancestral interaction. The functional characterization of the Hedgehog signal domain and its interacting partner in choanoflagellates promises to inform our understanding of Hedgehog function in the Urchoanimal and its possible role, if any, in the evolution of multicellular animals.

Tyrosine kinase signaling has long been considered a hallmark of intercellular communication, and was considered unique to animals (Darnell 1997). However, ESTs and subsequent genome sequencing of choanoflagellates revealed a remarkable number and diversity of tyrosine kinase signaling molecules (King et al. 2003). The *S. rosetta* genome, like the genome of *M. brevicollis*, contains a diverse and abundant repertoire of tyrosine kinases (King et al. 2003, 2008; Suga et al. 2008, 2012; Li et al. 2008; Manning et al. 2008; Pincus et al. 2008; Young et al. 2011), although only a handful are orthologous with animal tyrosine kinases. Among those shared with animals are Eph tyrosine kinases, which are important mediators of cell migration and were previously thought to be unique to metazoans. Choanoflagellates also possess tyrosine kinases with novel combinations of signaling protein domains, including combinations of signaling domains that do not occur in metazoans, thus potentially illuminating the early evolution of pTyr signaling.

Two choanoflagellate Src kinases, MbSrc1 and MbSrc4, have received experimental attention that offers insight into their function in the Urchoanimal. In animals, *Src* is a proto-oncogene whose functions included the regulation of cell adhesion, motility, shape, and differentiation. Li and colleagues cloned and purified the non-receptor tyrosine kinase MbSrc1 from *M. brevicollis* and found that the individual Src homology 3 (SH3), SH2, and catalytic domains have similar functions to their mammalian counterparts (Li et al. 2008). However, in contrast to mammalian c-Src, the SH2 and catalytic domains of MbSrc1 do not appear to be functionally coupled, suggesting that Src autoinhibition likely evolved more recently within the metazoan lineage.

Li and colleagues also cloned MbSrc4, which contains a lipid-binding C2 domain in the N-terminus of a protein with SH3-SH2-kinase domains and found that the enzyme is highly active as a tyrosine kinase and that the C2, SH3, and SH2 domains function to localize the kinase in a manner similar to mammalian Src-like kinases. The membrane-binding activity of the C2 domain functions similarly to the myristoylation signal of c-Src, suggesting that, like c-Src, MbSrc 4 interacts with membranes, but using mechanisms that arose through convergent evolution. When expressed in mammalian cells, full-length MbSrc4 displays low activity toward mammalian proteins, and it cannot functionally substitute for mammalian c-Src in a reporter gene assay. Removal of the MbSrc4 C2 domain leads to increased phosphorylation of cellular proteins. This suggests that in contrast to the related *M. brevicollis* Src-like kinase MbSrc1, MbSrc4 is not targeted properly to mammalian Src substrates, suggesting that the C2 domain plays a specific role in *M. brevicollis* signaling (Li et al. 2008). Functional studies such as these have the power to illuminate evolutionary processes, including domain shuffling, that contributed to the origin of the animals.

Transcriptional Insights into S. rosetta Cell Differentiation

One advantage of multicellularity is the ability to subdivide function between cells, which manifests as cellular differentiation. New mechanisms for regulating gene expression are the basis for cellular differentiation and are an important source of developmental novelty. Since multicellular organisms evolve from unicellular predecessors, mechanisms for maintaining spatial differentiation can be coopted from mechanisms of temporal or environmental regulation that exist in the unicellular ancestor. The phylogenetic relationship of choanoflagellates to animals makes the understanding of choanoflagellates gene regulation key to understanding the early evolution of animal gene regulation. Furthermore, the *S. rosetta* life history features five described cell types including both unicellular and multicellular stages making it an ideal organism for studying the role of gene regulation in the transition between unicellular and multicellular lifestyles in a developmental context. The sequencing of the *S. rosetta* transcriptome provided the first comprehensive look at gene expression in a choanoflagellate and facilitated insight into the regulation of gene expression (Fairclough et al. 2013). The transcriptional profiles revealed that morphologically and behaviorally different cells have distinct patterns of gene expression. Genes, including those containing signaling domains such as the hedgehog signal domain and adhesion domains such as cadherins, display distinct patterns of expression consistent with cell type specific function. The *cis*-regulatory elements of choanoflagellates, such as promoters, enhancers and other regulatory features, have yet to be explored. However, the regulatory networks of animals are critical to their development, so investigations of gene regulation in early branching animals and choanoflagellates may have important implications for reconstructing animal origins and for determining the core gene networks around which animal development evolves. The sequencing of choanoflagellate genomes has revealed their gene content (including transcription factor content) while transcriptome sequencing has provided information about gene expression and location and timing (Fairclough et al. 2013). Finally, tools for manipulating gene function are currently being developed for choanoflagellates. These, coupled with *in vitro* biochemical approaches (Li et al. 2008; Fairclough et al. 2013), open the door to future efforts to explore the regulatory landscape of multicellularity in both a developmental and evolutionary context.

Transcription Factors

The cellular differentiation observed in the most familiar forms of multicellularity is a manifestation of transcriptional regulation mediated by transcription factors. Thus to understand the origin of animal development it is necessary to understand the evolutionary history of transcription factors and their function in choanoflagellates. Analysis of opisthokont genomes indicates that many transcription factors families involved in animal development such as HMG box, homeodomain (both

TALE and non-TALE), bHLH, bZIP, or Mef2-like have deep eukaryotic roots (Degnan et al. 2009; Sebé-Pedrós et al. 2011; Chapter "Transcription factors and the origin of animal multicellularity"). Some of these families such bzip and bhlh underwent diversification prior to the divergence of opisthokonts while other, such as Sox and homeodomains, have diversified within animals. The choanoflagellate lineage appears to have lost several families of transcription factors including Churchill, T-box, grainy-head like, and Runx. (Sebé-Pedrós et al. 2011; Fairclough et al. 2013). These analyses also revealed that ETS, Smad, and NRs as well as specific classes of homeobox, bZIPs, bHLH, and HMG box are animal specific under the current taxon sampling. The sequencing of additional opisthokont genomes may alter the patterns of transcription factor gain and loss ((Sebé-Pedrós et al. 2011); for a more detail discussion see Chapter "Transcription factors and the origin of animal multicellularity").

Understanding the function of transcription factors in choanoflagellates may provide critical understanding of their ancestral function. Myc is a critical regulator of cell growth, proliferation, and death previously thought to be unique to metazoans. The sequencing of choanoflagellate genomes has revealed homologs of Myc and its binding partner, Max. In *M. brevicollis*, MbMyc protein is observable in flagellar and nuclear regions, consistent with a role in transcription (Young et al. 2011). The interaction of Myc with Max also appears to have arisen prior to the divergence of the choanoflagellate and metazoan lineages. *M. brevicollis* homologs of Myc and Max recognize both canonical and noncanonical E-boxes, the DNA-binding sites through which metazoan Myc proteins act. Moreover, as with metazoan Max proteins, MbMax can form homodimers that bind to E-boxes. However, cross-species dimerization between Mb and human Myc and Max proteins was not observed, suggesting that the binding interface has diverged (Young et al. 2011).

miRNAs Appear Absent from Choanoflagellates

Both miRNAs and piRNAs are inferred to have been present in stem Bilateria and deep sequencing of small RNAs from early branching animals has indicated that miRNAs and piRNAs have been present since the evolution of animals (Grimson et al. 2008; Sperling et al. 2009). Because canonical miRNAs and piRNAs have not been detected in choanoflagellates (Grimson et al. 2008), their evolution may have helped to usher in the era of multicellular animals. Animal miRNA evolution seems to have been very dynamic: pre-miRNAs of Porifera, Cnidaria and Bilateria have evolved distinct lengths and all miRNAs have been lost in *Trichoplax* (Grimson et al. 2008; Sperling et al. 2009). In addition, none of the identified miRNAs have recognizable conservation between Porifera, Cnidaria and Bilateria and only a single *Nematostella vectensis* miRNA has recognizable homology to a bilaterian miRNA. This could be either because it is, in fact, the only homolog of extant bilaterian miRNAs or because deep divergence of these organisms masks the common ancestry of their miRNAs. These observations of miRNAs mirror reports of miRNA–target interactions within the nematode, fly and vertebrate lineages, where very few appear to be conserved throughout all three lineages (Grimson et al. 2008). Although the mechanism of miRNA action was present in the Urmetazoan, the influence of miRNAs on early animal evolution is difficult to decipher due to the absence of clear homology.

Conclusions

The overall picture emerging from the study of opisthokonts and early branching animals is consistent with the view of evolution as a tinker (Jacob 1977). Choanoflagellate multicellularity provides a unique perspective on the evolution of animal multicellularity, and the discovery that choanoflagellates can form colonies of clonal cells through cell division in a developmentally controlled way increases their potential relevance to understanding this significant evolutionary event. The observation that bacteria trigger colony development highlights the profound influence that the microbial environment can have on development and suggests that the microbial environment may have played a significant role in the evolution of animal multicellularity. The sequencing of choanoflagellate genomes has overturned our thinking on the evolution of several important animal gene families including those integral to animal development such as cadherins, Hedgehog, and receptor tyrosine kinases. The traces of many of the mechanisms used by animals for cell adhesion, signaling and differentiation in unicellular and multicellular organisms suggests that many of these domains and genes were functioning in unicellular contexts and have been co-opted to new functions in animals. The transcriptional profiling of the *S. rosetta* genome has provided a comprehensive look at the utilization of its gene content over the course of its life history and facilitated the testing of hypotheses about the functions of animal genes in choanoflagellates. Mapping of gene expression to cell biology revealed that choanoflagellate genes are deployed both in novel contexts and in conditions consistent with our understanding of their function in animals. By developing new experimental techniques to examine choanoflagellate gene content and function, we can reconstruct the evolution of animal genomes and determine how the evolution of gene functions contributed to the origin of animal multicellularity.

Summary

1. Choanoflagellates are the closest known living relatives of the Metazoa.
2. Choanoflagellates are divided into two clades: the Craspedida—which contain all known colonial choanoflagellates, and the Acanthoecida—which have loricates.
3. *Salpingoeca rosetta* colonies develop through cell division, not aggregation.
4. The diversity of choanoflagellates contain animal signaling and adhesion domains, such as the Hedgehog signal and cadherin domains, in conserved and novel contexts.

References

Abedin M, King N (2008) The premetazoan ancestry of cadherins. Science 319:946–948. doi:10.1126/science.1151084

Abedin M, King N (2010) Diverse evolutionary paths to cell adhesion. Trends Cell Biol 20:734–742. doi:10.1016/j.tcb.2010.08.002

Adamska M, Matus DQ, Adamski M et al (2007) The evolutionary origin of hedgehog proteins. Curr Biol 17:R836–7. doi:10.1016/j.cub.2007.08.010

Alegado RA, Ferriera S, Nusbaum C et al (2011) Complete genome sequence of *Algoriphagus sp.* PR1, bacterial prey of a colony-forming choanoflagellate. J Bacteriol 193:1485–1486. doi:10.1128/JB.01421-10

Alegado RA, Brown LW, Cao S et al (2012) A bacterial sulfonolipid triggers multicellular development in the closest living relatives of animals. Elife 1:e00013. doi:10.7554/eLife.00013

Bridgham JT, Eick GN, Larroux C et al (2010) Protein evolution by molecular tinkering: diversification of the nuclear receptor superfamily from a ligand-dependent ancestor. PLoS Biol. doi:10.1371/journal.pbio.1000497

Carr M, Leadbeater BSC, Hassan R et al (2008) Molecular phylogeny of choanoflagellates, the sister group to Metazoa. Proc Natl Acad Sci USA 105:16641–16646. doi:10.1073/pnas.0801667105

Darnell JE (1997) Phosphotyrosine signaling and the single cell:metazoan boundary. Proc Natl Acad Sci U S A 94(22):11767–11769

Dayel MJ, Alegado RA, Fairclough SR et al (2011) Cell differentiation and morphogenesis in the colony-forming choanoflagellate *Salpingoeca rosetta.* Dev Biol 357:73–82. doi:10.1016/j.ydbio.2011.06.003

Degnan BM, Vervoort M, Larroux C, Richards GS (2009) Early evolution of metazoan transcription factors. Curr Opin Genet Dev 19:591–599. doi:10.1016/j.gde.2009.09.008

Fairclough SR, Dayel MJ, King N (2010) Multicellular development in a choanoflagellate. Curr Biol 20:R875–876. doi:10.1016/j.cub.2010.09.014

Fairclough SR, Chen Z, Kramer E et al (2013) Premetazoan genome evolution and the regulation of cell differentiation in the choanoflagellate *Salpingoeca rosetta.* Genome Biol 14:R15. doi:10.1186/gb-2013-14-2-r15

Gazave E, Lapébie P, Richards GS et al (2009) Origin and evolution of the Notch signaling pathway: an overview from eukaryotic genomes. BMC Evol Biol 9:249. doi:10.1186/1471-2148-9-249

Grimson A, Srivastava M, Fahey B et al (2008) Early origins and evolution of microRNAs and Piwi-interacting RNAs in animals. Nature 455:1193–1197. doi:10.1038/nature07415

Hibberd DJ (1975) Observations on the ultrastructure of the choanoflagellate *Codosiga botrytis* (Ehr.) Saville-Kent with special reference to the flagellar apparatus. J Cell Sci 17:191–219

Hynes RO (2012) The evolution of metazoan extracellular matrix. J Cell Biol 196:671–679. doi:10.1083/jcb.201109041

Jacob F (1977) Evolution and tinkering. Science 196:1161–1166

Karpov SA, Coupe SJ (1998) A revision of choanoflagellate genera *Kentosiga schiller*, 1953 and *Desmarella kent*, 1880. Acta protozool 37:23–27

Karpov SA, Leadbeater BS (1998) Cytoskeleton structure and composition in Choanoflagellates. J Eukaryot Microbiol 45:361–367. doi:10.1111/j.1550-7408.1998.tb04550.x

King N (2004) The unicellular ancestry of animal development. Dev Cell 7:313–325. doi:10.1016/j.devcel.2004.08.010

King N, Hittinger CT, Carroll SB (2003) Evolution of key cell signaling and adhesion protein families predates animal origins. Science 301:361–363. doi:10.1126/science.1083853

King N, Westbrook MJ, Young SL et al (2008) The genome of the choanoflagellate *Monosiga brevicollis* and the origin of metazoans. Nature 451:783–788. doi:10.1038/nature06617

King N, Young SL, Abedin M et al (2009) The choanoflagellates: heterotrophic nanoflagellates and sister group of the metazoa. Cold Spring Harb Protoc 2009:pdb.emo116. doi:10.1101/pdb.emo116

Lang BF, O'kelly C, Nerad T et al (2002) The closest unicellular relatives of animals. Curr Biol. doi:10.1016/S0960-9822(02)01187-9

Leadbeater BS (1983a) Life-history and ultrastructure of a new marine species of *Proterospongia* (Choanoflagellida). J Mar Biol Assoc UK 63:135–160. doi:10.1017/S0025315400049857

Leadbeater BSC (1983b) Distribution and chemistry of microfilaments in choanoflagellates, with special reference to the collar and other tentacle systems. Protistologica 19:157–166

Leadbeater B (2008) Choanoflagellate evolution: the morphological perspective. Protistology 5:256–267

Leadbeater BS, Kelly M (2001) Evolution of animals-choanoflagellates and sponges. Water Atmos 9:9–11

Leadbeater BS, Thomsen H (2000) Illustrated Guide to the Protozoa, 2nd ed. Wiley-Blackwell, pp 14–38

Lee JJ, Ekker SC, Kessler von DP et al (1994) Autoproteolysis in hedgehog protein biogenesis. Science 266:1528–1537

Levin TC, King N (2013) Evidence for sex and recombination in the Choanoflagellate *Salpingoeca rosetta*. Curr Biol 23:2176–2180. doi:10.1016/j.cub.2013.08.061

Leys SP, Nichols SA, Adams EDM (2009) Epithelia and integration in sponges. Integr Comp Biol 49:167–177. doi:10.1093/icb/icp038

Li W, Young SL, King N, Miller WT (2008) Signaling properties of a non-metazoan Src kinase and the evolutionary history of Src negative regulation. J Biol Chem 283:15491–15501. doi:10.1074/jbc.M800002200

Manning G, Young SL, Miller WT, Zhai Y (2008) The protist, *Monosiga brevicollis*, has a tyrosine kinase signaling network more elaborate and diverse than found in any known metazoan. Proc Natl Acad Sci USA 105:9674–9679. doi:10.1073/pnas.0801314105

McFall-Ngai M, Hadfield MG, Bosch TCG et al (2013) Animals in a bacterial world, a new imperative for the life sciences. Proc Natl Acad Sci USA 110:3229–3236. doi:10.1073/pnas.1218525110

Nichols SA, Dirks W, Pearse JS, King N (2006) Early evolution of animal cell signaling and adhesion genes. Proc Natl Acad Sci USA 103:12451–12456. doi:10.1073/pnas.0604065103

Nichols SA, Dayel MJ, King N (2009) Animal evolution: genomes, fossils and trees. Oxford University Press, pp 24–32

Nichols SA, Roberts BW, Richter DJ et al (2012) Origin of metazoan cadherin diversity and the antiquity of the classical cadherin/β-catenin complex. Proc Natl Acad Sci USA 109:13046–13051. doi:10.1073/pnas.1120685109

Nitsche F, Carr M, Aarndt H, Leadbeater BS (2011) Higher level taxonomy and molecular phylogenetics of the choanoflagellatea. J Eukaryot Microbiol 58:452–462. doi:10.1111/j.1550-7408.2011.00572.x

Pincus D, Letunic I, Bork P, Lim WA (2008) Evolution of the phospho-tyrosine signaling machinery in premetazoan lineages. Proc Natl Acad Sci USA 105:9680–9684. doi:10.1073/pnas.0803161105

Saville-Kent W (1880) A manual of the Infusoria: including a description of all known flagellate, ciliate, and tentaculiferous protozoa, British and foreign, and an account of the organization and the affinities of the sponges. London: D. Bogue

Sebé-Pedrós A, de Mendoza A, Lang BF et al (2011) Unexpected repertoire of metazoan transcription factors in the unicellular holozoan *Capsaspora owczarzaki*. Mol Biol Evol 28:1241–1254. doi:10.1093/molbev/msq309

Sebé-Pedrós A, Roger AJ, Lang FB et al (2010) Ancient origin of the integrin-mediated adhesion and signaling machinery. Proc Natl Acad Sci USA 107:10142–10147. doi:10.1073/pnas.1002257107

Sperling EA, Peterson KJ, Pisani D (2009) Phylogenetic-signal dissection of nuclear housekeeping genes supports the paraphyly of sponges and the monophyly of Eumetazoa. Mol Biol Evol 26:2261–2274. doi:10.1093/molbev/msp148

Srivastava M, Simakov O, Chapman J, et al. (2010) The *Amphimedon queenslandica* genome and the evolution of animal complexity. Nature 466:720–726. doi:10.1038/nature09201
Suga H, Sasaki G, Kuma K-I et al (2008) Ancient divergence of animal protein tyrosine kinase genes demonstrated by a gene family tree including choanoflagellate genes. FEBS Lett 582:815–818. doi:10.1016/j.febslet.2008.02.002
Suga H, Dacre M, de Mendoza A, et al. (2012) Genomic survey of premetazoans shows deep conservation of cytoplasmic tyrosine kinases and multiple radiations of receptor tyrosine kinases. Sci Signal 5:ra35. doi:10.1126/scisignal.2002733
Torruella G, Derelle R, Paps J et al (2012) Phylogenetic relationships within the opisthokonta based on phylogenomic analyses of conserved single-copy protein domains. Mol Biol Evol 29:531–544. doi:10.1093/molbev/msr185
Tyler S (2003) Epithelium–the primary building block for metazoan complexity. Integr Comp Biol 43:55–63. doi:10.1093/icb/43.1.55
Young SL, Diolaiti D, Conacci-Sorrell M et al (2011) Premetazoan ancestry of the Myc-Max network. Mol Biol Evol 28:2961–2971. doi:10.1093/molbev/msr132

Filastereans and Ichthyosporeans: Models to Understand the Origin of Metazoan Multicellularity

Hiroshi Suga and Iñaki Ruiz-Trillo

Abstract The origin of animals or metazoans from their unicellular ancestors is one of the most important evolutionary transitions in the history of life. To decipher the molecular mechanisms involved in this transition, it is crucial to understand both the early evolution of animals and their unicellular prehistory. Recent phylogenomic analyses have shown that there are at least three distinct unicellular or colonial lineages closely related to metazoans: choanoflagellates, ichthyosporeans and filastereans. However, until recently, choanoflagellates had been the only lineage for which an entire genome sequence was available. Moreover, the lack of transgenesis tools in any of these unicellular lineages had precluded the possibility of performing functional analyses. To better understand the unicellular prehistory of animals, we have recently obtained the genome sequences of both filastereans and ichthyosporeans. Analyses of their genomes identified many important genes for metazoan multicellularity and development, some of which are absent from the choanoflagellate genomes and thus were thought to be metazoan-specific. We have also established methods for transgenesis and gene silencing in ichthyosporeans. The combination of genomic information and molecular tools in filastereans and ichthyosporeans facilitate efficient functional analyses to understand how the key genes in the evolution of multicellularity were co-opted during the unicellular-to-multicellular transition that gave rise to metazoans. We propose that filastereans and ichthyosporeans are ideal model organisms for investigating the origin of metazoan multicellularity.

Keywords Filastereans · Ichthyosporeans · Evolution of multicellularity · Genome sequencing · Transgenesis

H. Suga (✉)
Faculty of Life and Environmental Sciences, Prefectural University of Hiroshima, Nanatsuka 562, Shobara, Hiroshima 727-0023, Japan

I. Ruiz-Trillo
Institut de Biologia Evolutiva (CSIC–Universitat Pompeu Fabra), Passeig Marítim de la Barceloneta 37–49, 08003 Barcelona, Spain
e-mail: inaki.ruiz@ibe.upf-csic.es

Departament de Genètica, Universitat de Barcelona, Barcelona, Spain

Institució Catalana de Recerca i Estudis Avançats (ICREA), Barcelona, Spain

I. Ruiz-Trillo, A. M. Nedelcu (eds.), *Evolutionary Transitions to Multicellular Life*, Advances in Marine Genomics 2, DOI 10.1007/978-94-017-9642-2_6

Introduction

The emergence of multicellular organisms from their unicellular ancestors is one of the major evolutionary transitions in the history of life. Phylogenetic analyses have shown that multicellularity was independently acquired several times in the course of eukaryote evolution, such as in animals, charophyte algae (and their descendants, land plants), fungi, slime molds, red algae and brown algae (Bonner 1998; King 2004; Grosberg and Strathmann 2007; Ruiz-Trillo et al. 2007; Rokas 2008a, b; Knoll 2011; see also Chapter "Timing the origins of multicellular eukaryotes through phylogenomics and relaxed molecular clock analyses"). Although the evolution of animals or metazoans from their unicellular ancestors is one of the most important evolutionary transitions in the history of life, little is known about the molecular mechanisms involved in the origin of multicellularity (also discussed in Chapter "Choanoflagellates: Perspective on the origin of animal multicellularity", "A comparative genomics perspective on the origin of multicellularity and early animal evolution", "Transcription factors and the origin of animal multicellularity" and "Developmental signalling and emergence of animal multicellularity").

Until recently, the origin of metazoans could only be studied by comparing the genomes of metazoans with those of other distantly-related eukaryotes, such as fungi, plants, and amoebozoans. Consequently, many of the molecules involved in the multicellular development of animals were considered to be unique to metazoans because they were not present in any of the other eukaryotes analyzed. This metazoan-specific repertoire included genes involved in cell adhesion, cell differentiation, and cell-cell communication. They were thus considered animal innovations and critical for the evolution of metazoan multicellularity (Rokas 2008b). In these studies, however, key eukaryote lineages that are more closely-related to metazoans than to fungi and plants were missing. Here, we first describe how these newly recognized eukaryote lineages were discovered, then review current knowledge of these organisms and their available genome data, and finally introduce our recent innovations on molecular tools in ichthyosporeans and discuss their potential as model organisms.

Exploring the Closest Unicellular and Colonial Relatives of Metazoans

The relationship between animals and unicellular/colonial choanoflagellates was first proposed by James-Clark in 1866 (James-Clark 1866) on the basis of the morphological resemblance between choanoflagellates and a particular cell type in sponges, the choanocyte (King 2004). No major contribution had been made until Cavalier-Smith proposed that animals, choanoflagellates and fungi are allied taxa with two synapomorphies: the distinctive combination of flattened but not discoidal mitochondrial cristae, and the single posterior flagellum in motile flagellated cells. The clade comprising these three lineages was thus named Opisthokonta (posterior flagellum) (Cavalier-Smith 1987). The first ribosomal DNA (rDNA) phylogeny to support the close relationship between choanoflagellates and metazoans was published in 1993 (Wainright et al. 1993), which was further confirmed by another phylogenetic analysis on mitochondrial genes in 2002 (Lang et al. 2002).

In 1995, the enigmatic rosette agent of Chinook salmon, initially described in 1986 (Elston et al. 1986; Harrell et al. 1986), was suggested to be another close unicellular relative of metazoans and choanoflagellates (Kerk et al. 1995). A year later, an rDNA phylogeny with additional data demonstrated the existence of a clade of protists near fungi and metazoan, all of which were parasites of aquatic animals (Ragan et al. 1996). This clade was initially named DRIP after the included species Dermocystidium, rosette agent, Ichthyophonus, and Psorospermium, and then renamed as Ichthyosporea (Cavalier-Smith 1998) or Mesomycetozoea (Mendoza et al. 2002). Further, more recent molecular phylogenies using protein coding genes showed that ichthyosporeans are indeed more closely related to metazoans and choanoflagellates than to fungi (Ruiz-Trillo et al. 2004, Ruiz-Trillo et al. 2006; Steenkamp et al. 2006), and the number of known ichthyosporeans has been growing ever since (Jøstensen et al. 2002; Arkush et al. 2003; Pekkarinen et al. 2003; Marshall et al. 2008; Marshall and Berbee 2010). Most ichthyosporeans appear to be symbionts, isolated from the digestive organs of various marine vertebrates and invertebrates, although a recent analysis of environmental rDNA clone libraries suggests the presence of free living species or a free-living stage (Del Campo and Ruiz-Trillo 2013).

Molecular phylogenies have shown that there are three additional unicellular lineages closely related to metazoans, choanoflagellates, and ichthyosporeans. These include the symbiotic amoeboid *Capsaspora owczarzaki*, and the heterotrophic marine amoeboids *Ministeria vibrans* and *Corallochytrium limacisporum* (Cavalier-Smith and Allsopp 1996; Cavalier-Smith and Chao 2003; Medina et al. 2003; Ruiz-Trillo et al. 2004; Ruiz-Trillo et al. 2006; Steenkamp et al. 2006). A multi-gene tree indicated that *C. owczarzaki* and *M. vibrans* are related lineages that form a clade known as Filasterea (Shalchian-Tabrizi et al. 2008). Recent multi-gene and phylogenomic analyses positioned the Filasterea as the closest known sister group to a clade comprising the Metazoa and the Choanoflagellata, and the Ichthyosporea as likely sister lineage of these three clades (Fig. 1) (Shalchian-Tabrizi et al. 2008; Torruella et al. 2012; Paps et al. 2013). The position of *Corallochytrium limacisporum* remains unclear but appears closely related to ichthyosporeans or filastereans (Paps et al. 2013).

Biology of Filastereans

The filasterean *C. owczarzaki* was isolated from the hemolymph of the pulmonate snail *Biomphalaria glabrata*, the intermediate host of the blood fluke *Schistosoma mansoni* (Stibbs et al. 1979). In the axenic culture, *C. owczarzaki* proliferates through

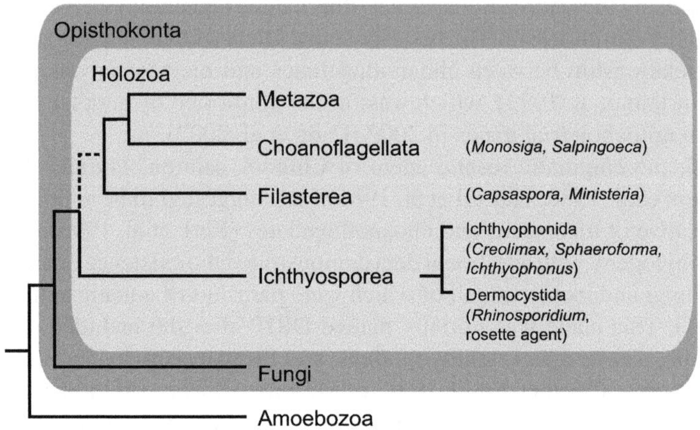

Fig. 1 Schematic drawing of the Opisthokonta phylogenetic tree. A consensus tree of the opisthokonts on the basis of two phylogenomics/multigene studies is shown (Torruella et al. 2012; Paps et al. 2013). The sister group position of the Ichthyosporea with respect to the Metazoa + Choanoflagellata + Filasterea is shown by a *dotted line* because it has not yet received full support. Some genera mentioned in the text are shown in parentheses

simple cell division (Fig. 2). When the culture matures, it encysts and drifts out, retracting the pseudopodia (Sebé-Pedrós et al. 2013). In some culturing conditions, *C. owczarzaki* cells aggregate and form a colony-like structure (Sebé-Pedrós et al. 2013). Sexual reproduction has not been reported, although the related genes are abundant in its genome (Suga et al. 2013). The lifecycle of another genus of filasterean, *Ministeria*, is largely unknown. The only culture available for this lineage is that of *M. vibrans*, a free-living marine amoeba that eats bacteria, similar to choanoflagellates (Tong 1997).

Biology of Ichthyosporeans

Due to their impact on human life, ichthyosporeans have been studied for many years. For example, the first study of *Rhinosporidium seeberi* dates back well over a century (Seeber 1900). *R. seeberi* is the etiologic agent of rhinosporidiosis, a tropical disease of the mucous membranes of various tissues. *Ichthyophonus hoferi*, a fish parasite, is another well-known species due to its impact on fisheries (McVigar 1982). While mainly dwelling in the internal organs of metazoans, ichthyosporeans do not always harm their hosts. The presence of some free-living species has also been suggested (van Hannen et al. 1999; Mendoza et al. 2002; Del Campo and Ruiz-Trillo 2013).

Ichthyosporeans are divided into two groups: the orders Dermocystida and Ichthyophonida. Their propagating strategies are similar, releasing numerous daughter cells from a mother cell after growing in size. The difference is in the form of the released cells, which are considered to be the infecting units (Mendoza et al. 2002);

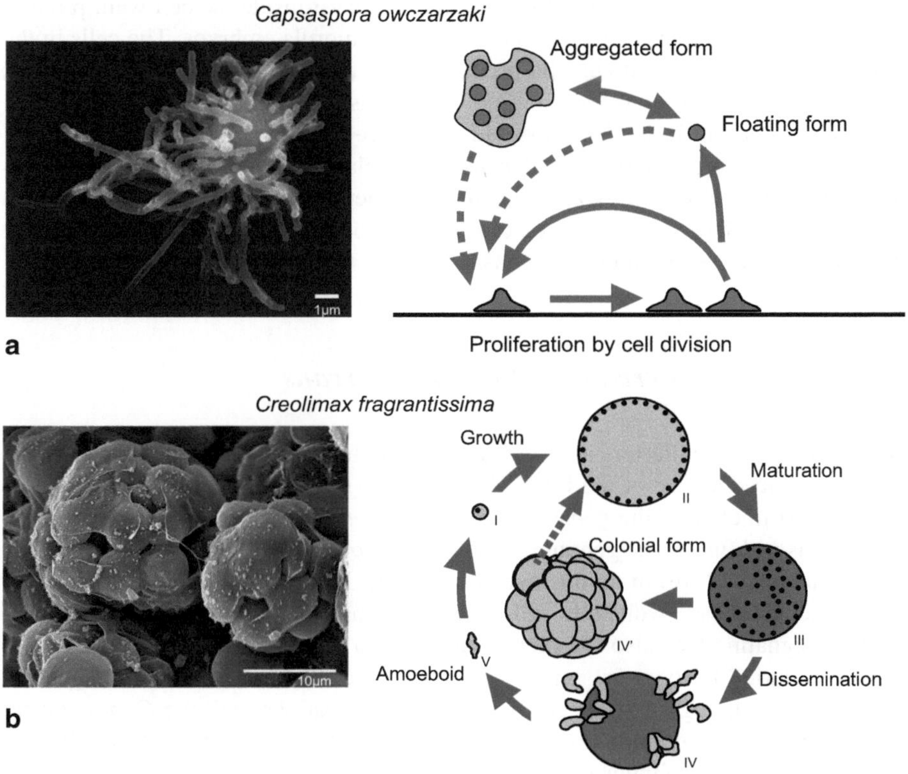

Fig. 2 *Capsaspora* and *Creolimax*. **a** the filasterean *Capsaspora owczarzaki*. A scanning electron microscope (*SEM*) image is shown on the left, and a schematic diagram of its lifecycle on the right. **b** the ichthyosporean *Creolimax fragrantissima*. A SEM image of colonial-stage cells is shown on the left, and its lifecycle is depicted on the right. *Dotted arrows* indicate the transfer to fresh medium. Photos courtesy of Arnau Sebé-Pedrós and available under CC BY at wikimedia

generally, dermocystids produce uniflagellate zoospores, while ichthyophonids release motile amoeboid cells.

The biology of two ichthyophonids (*Creolimax fragrantissima* and *Sphaeroforma arctica*) have been studied in culture. The lifecycle of *C. fragrantissima* is divided into growth, maturation, dissemination, and amoeboid stages (Fig. 2). In the growth stage, a uninucleate cell of 6–8 μm in diameter grows up to ~70 μm, allowing many rounds of nuclear division without a cell division (Marshall et al. 2008). In fresh medium, it grows for approximately two days and then begins maturation. Compared to the growth process, maturation is much more rapid, taking only 2–3 h, although the cell morphology changes dynamically; first, a rapid rearrangement of cytoplasm and nuclei occurs, then the syncytium cellularizes (cell membrane is formed), and finally amoebae are produced within the cell wall. The amoebae then exit the mother cell through tears in the cell wall (see movie in (Suga and Ruiz-Trillo 2013)). After crawling on substrate, they settle, encyst, and the cycle starts again. When the culture matures, the mother cell allows the amoebae to encyst inside the cell wall, producing colonies that superficially resemble metazoan morula embryos. The cells in the colony can start the usual amoeba-production in fresh medium. *S. arctica*, another ichthyophonid species (Jøstensen et al. 2002), has a similar lifecycle to that of *C. fragrantissima*, but lacks the amoeboid stage (see http://youtu.be/NwQk6xsqLJo). It is however worth mentioning that its close relative *Sphaeroforma tapetis*, formerly known as *Pseudoperkinsus tapetis*, has brief amoeboid or plasmodial stages (Marshall and Berbee 2013). To date, sexual reproduction has not been reported or induced in cultures of ichthyosporeans.

Genomes of Filastereans and Ichthyosporeans

The genomes of the filasterean *C. owczarzaki* and the ichthyosporean *S. arctica* have been sequenced as part of the UNICORN project (Ruiz-Trillo et al. 2007), the goal of which is to explore the genes involved in the origins of multicellularity in both metazoans and fungi. The genome sequence of *C. owczarzaki* is made up of 84 scaffolds that span 28 Mb in total (Suga et al. 2013). Importantly, the 8,657 predicted genes include those involved in cell-adhesion, transcriptional regulation, and intracellular signaling (see also Chapter "Transcription factors and the origin of animal multicellularity"). However, some of these genes are not present in choanoflagellates and were thus thought to be metazoan-specific innovations until the *C. owczarzaki* genome became available. Therefore, the genome of *C. owczarzaki* shows that the unicellular ancestor of metazoans already had a complex repertoire of genes involved in multicellular functions, suggesting that both the adoption of existing genes for new functions (gene co-option) and the generation of new genes were responsible for the origin of metazoan multicellularity (Suga et al. 2013).

Sequencing of the *S. arctica* genome is still in progress within UNICORN, with approximately 20 % of its genome estimated to be repeat sequences, making the assembly difficult. The genome of the ichthyosporean *C. fragrantissima* is also being sequenced. While closely-related to *S. arctica* (Paps et al. 2013), *C. fragrantissima*'s genome is much more compact and less repetitive. It is noteworthy that preliminary analyses of ichthyosporean genome data reveal that these organisms also have genes that are important for metazoan multicellularity (Sebé-Pedrós and Ruiz-Trillo 2010; Suga et al. 2014).

C. owczarzaki and most of the ichthyosporeans show symbiotic lifestyles. In general, a tight symbiotic connection to the host strongly affects the genome organization of symbionts, leading to gene loss, reduced genome size, and decreased GC content (Moya et al. 2008; McCutcheon and Moran 2012). However, the *C. owczarzaki* genome does not appear to show such characteristics, except for genome size (28 Mb), which is considerably smaller than that of the free-living choanoflagellate *M. brevicollis* (42 Mb) (King et al. 2008; Suga et al. 2013). Moreover, no trace of lateral gene transfer from the host snail has been detected (Suga et al. 2013). This casts doubt on whether *C. owczarzaki* is an obligatory symbiont. Similar to

C. owczarzaki, the genomes of ichthyosporeans do not appear to be significantly influenced by symbiosis.

Molecular Manipulation Tools

The genome sequences of unicellular relatives (choanoflagellates, filastereans, and ichthyosporeans) of animals are revealing that the unicellular ancestor of metazoans already had a rich repertoire of genes involved in the development and maintenance of the multicellular bodies of animals (King et al. 2008; Manning et al. 2008; Sebé-Pedrós et al. 2010; Sebé-Pedrós et al. 2011; Sebé-Pedrós et al. 2012; Suga et al. 2012; Suga et al. 2013; Suga et al. 2014). These genes may seem unnecessary for protists, who are, in principle, living independently of each other. It is therefore likely that these genes were co-opted for multicellularity-related functions when the first animal evolved. To understand how they were co-opted, the ancestral functions of these proteins must be clarified. One good example is a study of choanoflagellate cadherins (Abedin and King 2008) in which the authors used antibodies to show that cadherins were localized at the feeding collar, suggesting a role in the recognition and capture of bacterial prey. However, for a deeper understanding of their in vivo functions, gene-level manipulations such as transgenesis and gene silencing are essential. Unfortunately, such tools had not been available for any of the closest unicellular relatives of metazoans.

To overcome this limitation, we are developing transgenesis tools in both ichthyosporeans and filastereans, and have recently succeeded in transgenesis of *C. fragrantissima* cells by electroporation (Suga and Ruiz-Trillo 2013). Instead of a standard electroporator with a cuvette-type chamber, we used a recently developed electroporator with a long thin chamber and a wire-type electrode, which are considered to increase the survival rate of transformants (Kim et al. 2008). Transformation efficiency is still around 7 % at most (approximated by the number of transformants per survivor), and no stable integration of electroporated constructs into the genomic DNA has been observed so far. However, the relatively short ($\sim$ 2 days) turnover of *C. fragrantissima* generation allows the expressed proteins, e.g. fluorescent proteins, to survive during the whole lifecycle. To test the possibility of gene silencing, we have also examined the effect of synthetic small interfering (si) RNA and Morpholino, which can easily be brought into the cell using the same technique, and have successfully silenced protein expression (Suga and Ruiz-Trillo 2013). The successful delivery of Morpholino does not require labeling with a weakly-charged fluorescein tag, which has been considered to help the delivery by electroporation (Kos et al. 2003).

Our group has also obtained positive results for transforming *C. owczarzaki* (in preparation), although considerable improvement of the methodology is still required to increase the transformation efficiency.

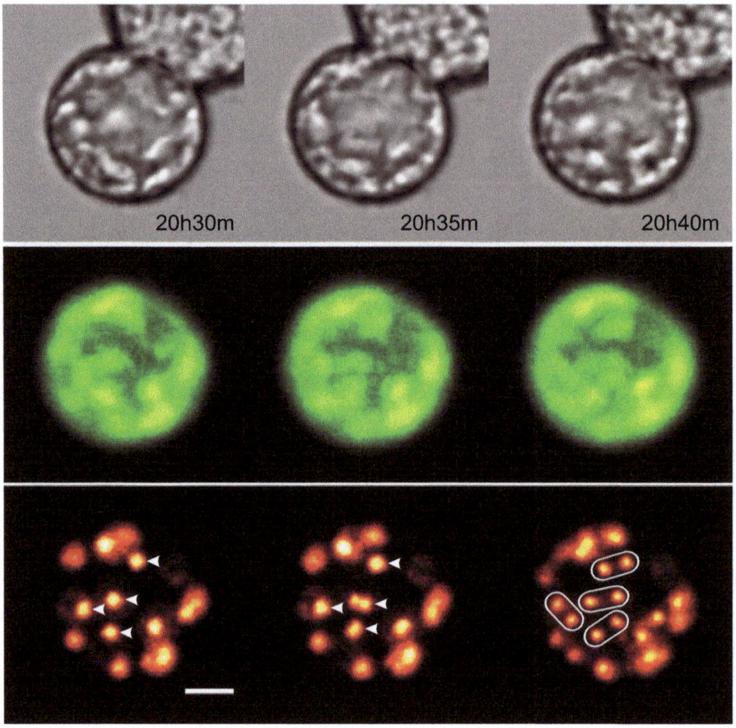

Fig. 3 Synchronized nuclear divisions in a *Creolimax* syncytial growth-stage cell. Stills from a time-lapse movie (the full movie available in (Suga and Ruiz-Trillo 2013)). Bright field images (*top*) with the elapsed time since the start of the movie, which is approximately 12 h after the electroporation, green fluorescence in cytoplasm (*middle*), and red fluorescence in nuclei (*bottom*) are shown. Four examples of synchronized nuclear divisions are highlighted by *arrowheads* and *ovals*. Scale bar, 10 μm

Colony Formation of Protists and the Evolution of Multicellularity

Using the newly developed transgenesis tools, we have obtained some insights into the colony formation of ichthyosporeans (Suga and Ruiz-Trillo 2013). Specifically, we transformed *C. fragrantissima* cells using a construct expressing histone H2B tagged with a fluorescent protein. The expressed protein was successfully incorporated into the chromosomes and we could visualize the movement of nuclei in a live cell (Fig. 3). The transformants demonstrated that during the growth stage *C. fragrantissima* cells undergo nuclear divisions without cell division, thus creating a syncytium. In addition, nuclear divisions are strictly synchronized throughout the entire growth stage. Interestingly, the *C. fragrantissima* colonies, which resemble metazoan morula embryos, were formed through cellularization of the syncytium.

These observations provide a new view on the origin of metazoan multicellularity. Traditionally, three different scenarios have been hypothesized as possible mechanisms for the initial multicellular development of metazoans: serial incomplete cell division, cell aggregation, and cellularization of a syncytium (Willmer 1990; Grosberg and Strathmann 2007; Fairclough et al. 2010). Of these, incomplete cell division has received much support, especially following the observation that the choanoflagellate *Salpingoeca rosetta* (See Chapter "Choanoflagellates: Perspective on the origin of animal multicellularity") (Fairclough et al. 2010) forms colonies only by cell division. This study showed that the synchrony of cell division is gradually lost as the *S. rosetta* colony grows (Fairclough et al. 2010), as seen in many metazoan embryos. The second hypothesis is less favored, since metazoans do not generally allow genetic heterogeneity (however the filasterean *C. owczarzaki* colony appears to be formed only by aggregation (Sebé-Pedrós et al. 2013)). The third hypothesis, i.e. cellularization of syncytium, has been supported by observations on the development of some slime molds and fungi, where a multi-cell-like structure is formed after development of the syncytium (Willmer 1990; Bonner 1998; Grosberg and Strathmann 2007). In metazoans, this strategy is typically seen in insect embryogenesis. Although seen in various eukaryotic lineages, this third mechanism does not appear to have been enthusiastically endorsed for the multicellularity evolution. In our study, however, the colonies of ichthyosporeans are shown to be formed by the cellularization of syncytia. We thus consider that this mechanism represents an important strategy for developing multicell-like structures in animals and their relatives, in addition to the incomplete cell division and cell aggregation. However, it is still possible that the morphological resemblance between metazoan embryogenesis and colony-formation of extant pre-metazoans is only superficial and nothing to do with their evolutionary origins. Molecular-level investigation into the mechanisms of colony formation by these organisms is required to obtain deeper insights.

Future Prospects

Comparative genomics between metazoans and their most closely related unicellular lineages has provided important insights, showing that the unicellular ancestor of metazoans already had a complex gene repertoire for cell adhesion, cell communication and cell differentiation. This indicates that co-option of ancestral genes into new functions played an important role in animal origins. Then, what was the ancestral function of those proteins, and how were they co-opted into new functions within the complex multicellular framework? To answer these questions, comparative genomics or simple expression studies are not enough; gene-level manipulation performed directly in those animal relatives is indispensable.

In this regard, the establishment of transgenesis and gene silencing techniques on the metazoan-relatives will drastically change our approach to the origin of animal multicellularity. Using our model organisms, one could unravel the function of these proteins in the context of a unicellular lifestyle, and better understand the process of gene co-option and ultimately the evolution of the first animal.

Summary

- Ichthyosporeans and filastereans are recently (re)discovered clades. They are the closest metazoan-relatives besides choanoflagellates.
- The lifecycles of ichthyosporeans and filastereans have been described under laboratory conditions. Both form characteristic colonies or aggregations.
- The genomes of one filasterean, *Capsaspora owczarzaki*, and two ichthyosporeans, *Sphaeroforma arctica* and *Creolimax fragrantissima* are either completely sequenced or in the process of sequencing. Analysis of these genomes has greatly clarified evolutionary history from unicellular protists to multicellular metazoans.
- To directly analyze the functions of genes that appear to have been co-opted during the transition from unicellular to multicellular systems, we have developed a transgenesis technique in the ichthyosporean *Creolimax fragrantissima*.
- In addition to the widely-accepted hypothesis of incomplete serial cell division scenario, cellularization of a syncytium could also be considered as a possible mechanism for the initial development of metazoan multicellularity. Molecular-level insights should provide more concrete clues.

Acknowledgments We thank the UNICORN project for the genome sequencing. Figures reprinted from Developmental Biology, vol 377, Issue 1, H. Suga and I Ruiz-Trillo, Development of ichthyosporeans sheds light on the origin of metazoan multicellularity, ©Elsevier Inc. (2013), with permission from Elsevier. This work is supported by the Marie Curie Intra-European Fellowship (H. S.), a European Research Council Starting Grant (ERC-2007-StG- 206883), and a grant (BFU2011–23434) from the Ministerio de Economía y Competitividad (MINECO) to I. R.-T.

References

Abedin M, King N (2008) The premetazoan ancestry of cadherins. Science 319:946–948

Arkush KD, Mendoza L, Adkison MA et al (2003) Observations on the life stages of *Sphaerothecum destruens* n. g., n. sp., a mesomycetozoean fish pathogen formerly referred to as the rosette agent. J Eukaryot Microbiol 50:430–438

Bonner J (1998) The origins of multicellularity. Integr Biol 1:27–36

Cavalier-Smith T (1987) The origin of fungi and pseudofungi. In: Rayner ADM, Brasier CM, Moore D (eds) Evolutionary biology of fungi. Cambridge University Press, Cambridge, pp 339–353

Cavalier-Smith T (1998) A revised six-kingdom system of life. Biol Rev 73:203–266

Cavalier-Smith T, Allsopp MTEP (1996) *Corallochytrium*, an enigmatic non-flagellate protozoan related to choanoflagel-lates. Eur J Protistol 32:306–310

Cavalier-Smith T, Chao EE (2003) Phylogeny of choanozoa, apusozoa, and other protozoa and early eukaryote megaevolution. J Mol Evol 56:540–563

Del Campo J, Ruiz-Trillo I (2013) Environmental survey meta-analysis reveals hidden diversity among unicellular opisthokonts. Mol Biol Evol 30:802–805

Elston RA, Harrell LW, Wilkinson MT (1986) Isolation and in vitro characteristics of chinook salmon (*Oncorhynchus tshawytscha*) rosette agent. Aquaculture 56:1–21

Fairclough SR, Dayel MJ, King N (2010) Multicellular development in a choanoflagellate. Curr Biol 20:R875–R876

Grosberg EK, Strathmann RR (2007) The evolution of multicellularity: a minor major transition? Annu Rev Ecol Evol Syst 38:621–654

Harrell LW, Elston RA, Scott TM et al (1986) A significant new systemic disease of net-pen reared chinook salmon (*Oncorhynchus tshawytscha*) brood stock. Aquaculture 55:249–262

James-Clark H (1866) Note on the infusoria flagellata and the spongiae ciliatae. Am J Sci 1:113–114

Jøstensen J-P, Sperstad S, Johansen S et al (2002) Molecular-phylogenetic, structural and biochemical features of a cold-adapted, marine ichthyosporean near animal-fungal divergence, described from *in vitro* cultures. Eur J Protistol 38:93–104

Kerk D, Gee A, Standish M et al (1995) The rosette agent of chinook salmon (*Oncorhynchus tshawytscha*) is closely related to choanoflagellates, as determined by the phylogenetic analyses of its small ribosomal subunit RNA. Mar Biol 122:187–192

Kim JA, Cho K, Shin MS et al (2008) A novel electroporation method using a capillary and wire-type electrode. Biosens Bioelectron 23:1353–1360

King N (2004) The unicellular ancestry of animal development. Dev. Cell 7:313–325

King N, Westbrook MJ, Young SL et al (2008) The genome of the choanoflagellate *Monosiga brevicollis* and the origin of metazoans. Nature 451:783–788

Knoll AH (2011) The multiple origins of complex multicellularity. Annu Rev Earth Planet Sci 39:217–239

Kos R, Tucker RP, Hall R et al (2003) Methods for introducing morpholinos into the chicken embryo. Dev Dyn 226:470–477

Lang BF, O'Kelly C, Nerad T et al (2002) The closest unicellular relatives of animals. Curr Biol 12:1773–1778

Manning G, Young SL, Miller WT et al (2008) The protist, *Monosiga brevicollis*, has a tyrosine kinase signaling network more elaborate and diverse than found in any known metazoan. Proc Natl Acad Sci U S A 105:9674–9679

Marshall WL, Berbee ML (2010) Facing unknowns: living cultures (*Pirum gemmata* gen. nov., sp. nov., and *Abeoforma whisleri*, gen. nov., sp. nov.) from invertebrate digestive tracts represent an undescribed clade within the unicellular Opisthokont lineage ichthyosporea (Mesomycetozoea). Protist 162:33–57

Marshall WL, Berbee ML (2013) Comparative morphology and genealogical delimitation of cryptic species of sympatric isolates of *Sphaeroforma* (Ichthyosporea, Opisthokonta). Protist 164:287–311

Marshall WL, Celio G, McLaughlin DJ et al (2008) Multiple isolations of a culturable, motile Ichthyosporean (Mesomycetozoa, Opisthokonta), *Creolimax fragrantissima* n. gen., n. sp., from marine invertebrate digestive tracts. Protist 159:415–433

McCutcheon JP, Moran NA (2012) Extreme genome reduction in symbiotic bacteria. Nat Rev Microbiol 10:13–26

McVigar AH (1982) *Ichthyophonus* infections of fish. In: Roberts RJ (eds) Microbial diseases of fish. Academic Press, London, p 243–269

Medina M, Collins AG, Taylor JW et al (2003) Phylogeny of opisthokonta and the evolution of multicellularity and complexity in fungi and metazoa. Int J Astrobiol 2:203–211

Mendoza L, Taylor JW, Ajello L (2002) The class Mesomycetozoea: a heterogeneous group of microorganisms at the animal-fungal boundary. Annu Rev Microbiol 56:315–344

Moya A, Pereto J, Gil R et al (2008) Learning how to live together: genomic insights into prokaryote-animal symbioses. Nat Rev Genet 9:218–229

Paps J, Medina-Chac6n LA, Marshall WL et al (2013) Molecular phylogeny of unikonts: new insights into the position of apusomonads and ancyromonads and the internal relationships of opisthokonts. Protist 164:2–12

Pekkarinen M, Lom J, Murphy CA et al (2003) Phylogenetic position and ultrastructure of two dermocystidium species (Ichthyosporea) from the common perch (*Perca fluviatilis*). Acta Protozool 42:287–307

Ragan MA, Goggin CL, Cawthorn RJ et al (1996) A novel clade of protistan parasites near the animal-fungal divergence. Proc Natl Acad Sci U S A 93:11907–11912

Rokas A (2008a) The molecular origins of multicellular transitions. Curr Opin Genet Dev 18: 472–478

Rokas A (2008b) The origins of multicellularity and the early history of the genetic toolkit for animal development. Annu Rev Genet 42:235–251

Ruiz-Trillo I, Inagaki Y, Davis LA et al (2004) *Capsaspora owczarzaki* is an independent opisthokont lineage. Curr Biol 14:R946–R947

Ruiz-Trillo I, Lane CE, Archibald JM et al (2006) Insights into the evolutionary origin and genome architecture of the unicellular opisthokonts *Capsaspora owczarzaki* and *Sphaeroforma arctica*. J Eukaryot Microbiol 53:379–384

Ruiz-Trillo I, Burger G, Holland PW et al (2007) The origins of multicellularity: a multi-taxon genome initiative. Trends Genet 23:113–118

Sebé-Pedrós A, Ruiz-Trillo I (2010) Integrin-mediated adhesion complex: cooption of signaling systems at the dawn of Metazoa. Commun Integr Biol 3:475–477

Sebé-Pedrós A, Roger AJ, Lang FB et al (2010) Ancient origin of the integrin-mediated adhesion and signaling machinery. Proc Natl Acad Sci U S A 107:10142–10147

Sebé-Pedrós A, de Mendoza A, Lang BF et al (2011) Unexpected repertoire of metazoan transcription factors in the unicellular holozoan *Capsaspora owczarzaki*. Mol Biol Evol 28:1241–1254

Sebé-Pedrós A, Zheng Y, Ruiz-Trillo I et al (2012) Premetazoan origin of the Hippo signaling pathway. Cell Rep 1:13–20

Sebé-Pedrós A, Irimia M, Del Campo J et al (2013) Regulated aggregative multicellularity in a close unicellular relative of metazoa. eLife 2:e01287

Seeber GR (1900) Un nuevo sporozuario parasito del hombre: dos casos encontrados en polypos nasals. Dissertation, Universidad Nacional de Buenos Aires

Shalchian-Tabrizi K, Minge MA, Espelund M et al (2008) Multigene phylogeny of choanozoa and the origin of animals. PLoS One 3:e2098

Steenkamp ET, Wright J, Baldauf SL (2006) The protistan origins of animals and fungi. Mol Biol Evol 23:93–106

Stibbs HH, Owczarzak A, Bayne CJ et al (1979) Schistosome sporocyst-killing amoebae isolated from *Biomphalaria glabrata*. J Invertebr Pathol 33:159–170

Suga H, Ruiz-Trillo I (2013) Development of ichthyosporeans sheds light on the origin of metazoan multicellularity. Dev Biol 377:284–292

Suga H, Dacre M, de Mendoza A et al (2012) Genomic survey of pre-metazoans shows deep conservation of cytoplasmic tyrosine kinases and multiple radiations of receptor tyrosine kinases. Sci Signal 5:ra35

Suga H, Chen Z, De Mendoza A et al (2013) The genome of *Capsaspora* reveals a complex unicellular prehistory of animals. Nat Commun 4:2325

Suga H, Torruella G, Burger G et al (2014) Earliest holozoan expansion of phosphotyrosine signaling. Mol Biol Evol 31:517–528

Tong SM (1997) Heterotrophic flagellates and other protists from Southampton Water, U.K. Ophelia 47:71–131

Torruella G, Derelle R, Paps J et al (2012) Phylogenetic relationships within the Opisthokonta based on phylogenomic analyses of conserved single copy protein domains. Mol Biol Evol 20:531–544

van Hannen EJ, Mooij W, van Agterveld MP et al (1999) Detritus-dependent development of the microbial community in an experimental system: qualitative analysis by denaturing gradient gel electrophoresis. Appl Environ Microbiol 65:2478–2484

Wainright PO, Hinkle G, Sogin ML et al (1993) Monophyletic origins of the metazoa: an evolutionary link with fungi. Science 260:340–342

Willmer P (1990) The origin of the Metazoa. In: Invertebrate relationships: pattern in animal evolution. Cambridge University Press, Cambridge, pp 163–198

Volvocine Algae: From Simple to Complex Multicellularity

Matthew D. Herron and Aurora M. Nedelcu

Abstract The evolution of multicellularity provided new ways for biological systems to increase in complexity. However, although high levels of complexity have indeed been attained in several multicellular lineages, natural selection does not necessarily favor complex biological systems. Why and how, then, has complexity increased in some lineages? We argue that the volvocine green algae (*Volvox* and its relatives) are a uniquely valuable model system for understanding the evolution of multicellular complexity. Using a general framework for the evolution of complexity, we discuss the various levels of morphological and developmental complexity achieved in this group, and consider both the why and the how underlying the changes in complexity levels that took place in this group.

Keywords Cell differentiation · *Chlamydomonas* · Chlorophyta · Complexity · Genetics · Multicellularity · Natural selection · *Volvox*

The Issue of Complexity

It is absurd to talk of one animal being higher than another.—C. R. Darwin (1837, p. 74)

Darwin famously reminded himself against using such value-laden terms as 'higher' and 'lower', but their use has continued (including in Darwin's own writing; see Richards 1988) right up to the present. Ideas of progress have been present in the literature of evolutionary theory as long as there has been such a literature. These ideas have changed over time and differed among authors, but generally include the notion of improvement over time as either a passive or an actively (selectively) driven trend. "Improvement", of course, implies some standard against which organisms are measured, and it is in the choice of standard that subjectivity is introduced (Ayala

M. D. Herron (✉)
Division of Biological Sciences, University of Montana, Missoula, MT, USA
e-mail: xprinceps@gmail.com

A. M. Nedelcu
Department of Biology, University of New Brunswick, Fredericton, NB, Canada
e-mail: anedelcu@unb.ca

I. Ruiz-Trillo, A. M. Nedelcu (eds.), *Evolutionary Transitions to Multicellular Life,* Advances in Marine Genomics 2, DOI 10.1007/978-94-017-9642-2_7

1988). Large, complex, intelligent organisms are considered 'advanced' or 'higher' and therefore are thought to represent an improvement over the 'primitive' or 'lower' organisms, which lack these presumed benefits.

These ideas are the intellectual descendants of the great chain of being (Lovejoy 1936), which was in turn based on the self-evident truth that humans are the pinnacle of evolution (Hull 1988). This framework, not surprisingly, leads to the circular conclusion that humans are the most advanced species, and the criterion for judging other species as advanced or primitive effectively reduces to degree of similarity to humans.

In spite of Darwin's well-known doubts, and those of numerous subsequent authors, assumptions of progress in evolution persist. Extant organisms are referred to as more or less 'primitive' or 'advanced', or even as basal to other extant organisms (Krell and Cranston 2004). The misconceptions that lead to these statements are not always merely semantic, and they can have substantial misleading influence on the interpretation of historical evidence, typically when the ancestors of a species-rich clade are assumed to have been similar to extant members of a species-poor sister clade (for example, that early mammals must have been monotreme-like). The influence of this misconception on ideas of ancestral mammals and angiosperms was recently reviewed by Crisp and Cook (2005).

Recent years have seen efforts to recast discussions of progress in terms of explicit, objective, *a priori* criteria. An example of this trend is the shift in focus from progress, which requires value judgments (Ayala 1988), to increases in complexity, which are in some sense quantifiable. This leaves, of course, the problem of how complexity is to be defined and measured. As McShea (1996) pointed out, investigations of trends in complexity can only provide nontrivial answers when the criteria for complexity are set in advance. To get meaningful answers to questions about how and why, and even if, complexity has increased through time in a given clade, we must estimate complexity based on criteria that are defined independently of the members of the clade (McShea 1996).

In this chapter, we first discuss several aspects relevant to defining and measuring complexity in biological systems as well as to understanding how and why complexity increased in some (but not all) lineages. Then we focus on a particular type of complexity—that is, multicellular complexity; we provide an overview of multicellular systems and discuss the proximate and ultimate causes for their increase in complexity. Lastly, we focus on a specific multicellular lineage—the volvocine algae—and argue that this group is a uniquely valuable model system for understanding the evolution of multicellular complexity. Using the general framework for the evolution of complexity introduced earlier, we review the various levels of morphological and developmental complexity achieved in this group, and consider both the why and the how underlying the changes in complexity levels that took place in this group.

What is Complexity?

Complexity itself is a complex concept, which has proven rather difficult to define (see also Chap. 15). Dozens of definitions have been offered, all with specific shortcomings. Going back to the origin of the word, '*complexus*' in Latin means 'entwined' or 'twisted together.' The term implies the existence of two or more components that are difficult to separate. In other words, complexity requires two attributes: the existence of (i) distinct parts and (ii) connections. Heylighen (1999) defines 'distinction' and 'connection' as the two aspects that characterize complexity in any system. '*Distinction*' corresponds to variety, heterogeneity, and different behaviors; '*connection*' corresponds to constraint and dependency. In this framework, complexity increases when the variety (distinction) and dependency (connection) of parts increase in at least one of the following dimensions: space, time, spatial scale, or temporal scale (Heylighen 1999). The process of increase in variety corresponds to *differentiation*; the process of increase in the number or strength of connections defines *integration*. The complexity produced by differentiation and integration in the spatial dimension can be called '*structural*', in the temporal dimension '*functional*', in the spatial scale dimension '*structural hierarchical*', and in the temporal scale dimension '*functional hierarchical*'. The scale dimension reflects in the number of hierarchical levels at which structure or function can be detected. For example, a multicellular organism has one more spatial hierarchical level than a unicellular individual—that is, cells within the organism. Similarly, multicellular development adds a level of functional hierarchy not found in unicells, namely interactions among cells, such as during embryonic development.

Similarly, McShea (1996) defines four facets of complexity, which have the potential to vary at least somewhat independently. Complexity is viewed in terms of objects or processes, either of which can be hierarchical or nonhierarchical. In terms of objects, hierarchical complexity refers to the number of levels of nestedness (parts within wholes; e.g., organelles within cells, cells within tissues, tissues within organs, organs within organisms), while nonhierarchical complexity refers to the number of different part types at a given level of nestedness (e.g., the number of cell types within a multicellular organism). Process complexity refers to causal relationships, which can be hierarchical (the number of levels of a causal hierarchy) or nonhierarchical (the number of independent interactions among parts at a given level of the hierarchy). In biological terms, McShea uses object complexity to address morphological questions and process complexity to address questions related to development.

Criteria

Complexity is not only a concept that is hard to define, but also a trait that is hard to measure. Several criteria have been used to compare biological systems in terms of complexity levels (see Szathmáry et al. 2001 for discussion and references). Complexity has been evaluated in terms of the number of cell types, the amount of DNA content, the number of genes, the number of transcription factors, and the number of transcriptome states. All of these criteria have proven to be incommensurable across the entire range of biological systems. For instance, although vertebrates have a higher number of genes than invertebrates, within invertebrates the simple worm *Caenorhabditis elegans* has more genes than the more complex (in terms of number of cell types) fly *Drosophila melanogaster* (Carroll 2001). More recently, the number of transcription factors together with the number of genes they regulate (i.e., connectivity) has been proposed as a better indicator of complexity, but such estimates are harder to achieve (Szathmáry et al. 2001).

How Biological Systems Increase in Complexity (Proximate Causes)

Several processes and mechanisms can be identified as responsible for the observed increases in complexity in some biological systems. Major increases in hierarchical complexity have been achieved during transitions in individuality—through symbiosis (e.g., during the evolution of the eukaryotic cell) and cooperation and division of labor (such as during the evolution of multicellularity and eusociality; see Chap. 9). More subtle and gradual increases in complexity levels throughout evolution involved gene duplication followed by functional diversification, domain shuffling, alternative splicing, and changes in gene regulation— in other words changes in genome complexity.

As in this chapter we are concerned with the evolution of multicellularity, here we are only focusing on some aspects relevant to the proximate causes responsible for the increases in morphological complexity that took place in multicellular lineages. As mentioned earlier, there is no perfect/universal direct correlation between the number of total genes in a genome and morphological complexity (i.e., number of cell types); thus, changes in the number of genes are not likely to be fully responsible for the observed increases in complexity. Nevertheless, it has been suggested that an increase in the number of specific genes (e.g., genes involved in particular developmental functions) might be relevant to achieving increased morphological complexity (Carroll 2001). Among these, transcription factors (TFs) are key regulators of cell differentiation (by affecting cell-specific expression of genes), and as such they are likely involved in changes in complexity levels. Yet, although an expansion in the number of TFs offers the potential for an increase in complexity, it is not necessary for the evolution of increased morphological complexity (Carroll 2001). In fact, the number of TF genes in a genome is rather small; also, a small number of TFs can be responsible for large differences in gene expression patterns among cell types (see Carroll 2001, de Mendoza et al. 2013, Chaps. 15 and 18 for further discussion and references).

Since differences in morphological and developmental complexity cannot be solely attributed to differences in gene content, increases in complexity are likely due to changes in regulatory elements that act in *cis* to control gene expression. The expansion of regulatory elements in a genome can be interpreted as an increase in genomic complexity in all four senses described above—in the number of different parts (i.e., regulatory elements) in a regulatory system (*structural complexity*), in the number of different interactions of these parts (*functional complexity*), in the number of levels in developmental hierarchies (*functional hierarchical*), and in the number of parts and interactions at a given spatial scale (*structural hierarchical*) (Carroll 2001).

Why Biological Systems Increase in Complexity (Ultimate Causes)

There is no *a priori* reason that more complex systems would be preferred by natural selection. Evolution selects for increases in fitness, but high fitness can be achieved both by very complex (e.g., animals) and very simple systems (e.g., bacteria). So, why did complexity increase in some lineages? Below, we describe a general framework based on cybernetics principles (proposed by Heylighen 1999), which emphasizes the role of the environment in driving the evolution of biological systems.

High fitness can be achieved if a system is very stable and/or if it is likely that many copies of that system will be produced (Heylighen 1999). Thus, a system will be selected if: (1) its parts 'fit together', i.e., form an intrinsically stable whole ('intrinsic' fitness), and (2) the whole 'fits' its environment, i.e., it can resist external perturbations and profit from external resources to reproduce ('extrinsic' fitness). Variation will result in differentiation, and selection of fit relationships will simultaneously result in integration by adding or strengthening connections between parts; the end result will be an increase in *structural complexity*.

As the environment changes, the system needs to maintain an invariant configuration in spite of variable disturbances; that is, *homeostasis*. In cybernetics terms (Ashby 1956), homeostasis can be achieved by control, i.e., "the compensation of external perturbations by the appropriate counter-actions so that a desired goal is reached or maintained" (Heylighen 1999). Ashby's (1956, 1958) Law of Requisite Variety states that "in order to achieve control (and maintain homeostasis), the variety of actions a control system is able to execute must be at least as great as the variety of environmental perturbations that need to be compensated." The larger the variety of actions performed by the system, the larger the range of disturbances that can be counteracted and the set of environmental situations in which the system can survive. All other things being equal, greater control variety implies greater fitness. A larger repertoire of possible actions allows the system to survive in a larger variety of situations. As evolution through natural selection would tend to increase control, internal variety will also increase. This can be interpreted as a functional differentiation, which will result in an increase in *functional complexity* (i.e., the emergence of more diverse activities or functions).

The variety of an evolving system will slowly increase towards, but will never actually match, the limitless variety of the environment. Depending on the variety of perturbations in its environment, the evolving system will reach a trade-off level where further increase in complexity will be costly; for instance, for viruses this trade-off level is characterized by a low functional variety (Heylighen 1999). However, as the environment of a system consists itself of evolving systems, the increase in variety in one system generates a stronger need for variety increase in the other since it will now need to control a more complex environment (cf. Waddington 1969). This self-reinforcing interaction is an illustration of the Red Queen Principle (Van Valen 1973), which states that a system must change continuously in order to merely maintain its fitness relative to the systems it co-evolves with. The end result is that many evolutionary systems that are in direct interaction with each other will tend to grow more complex as they need to control more complex environments (Heylighen 1999). Nevertheless, not all evolutionary systems will increase in complexity; those that have attained a good trade-off point and are not challenged by an environment putting more complex demands on them will maintain their current level of complexity. A shift to a less variable environment, as often accompanies a parasitic or endosymbiotic lifestyle, can even lead to a reduction in complexity.

To sum up, although fitness is relative to the environment, it has two components that can increase in an absolute sense, (i) intrinsic fitness (stability) and (ii) extrinsic fitness (control). Selection for increased stability and control, when unopposed by trade-offs, will thus tend to be accompanied by respective increases in structural complexity (number and strength of linkages between components) and functional complexity (the number of environmental perturbations that can be counteracted) (for more discussion, see Heylighen 1999).

Multicellular Complexity

Overview

Here, we define multicellularity as a category of phenotypes that are based on more than one cell. Such phenotypes can be stable and represent the longest part of a life-cycle or be transient (induced in response to external stimuli) and represent a small (or facultative) portion of a life cycle—as in, for instance, myxobacteria (Chap. 22), cellular slime molds (Chap. 21) and some choanoflagellates (e.g., Bonner 2003; Velicer and Vos 2009; Dayel et al. 2011). Multicellular phenotypes can consist of cells that are identical in terms of differentiation potential (here, referred to as simple multicellularity since the parts are identical; i.e., low structural/functional complexity) or a mixture of 2 or more cell types with distinct differentiation potentials (complex multicellularity; high structural and functional complexity); note that this distinction differs from that of Knoll (2011) (also Wolpert and Szathmáry 2002; Schaap et al. 2006; Butterfield 2009) who restricts the term "complex multicellularity" to multicellular organisms displaying intercellular communication and differentiated tissues.

In most lineages, multicellularity develops from a single cell (spore or zygote) whose mitotic products fail to separate (clonal/unitary development). However, multicellular forms that develop via the aggregation of single-celled individuals (aggregative development) are also known (e.g., myxobacteria, Filasterea and cellular slime molds; see Chaps. 6, 21 and 22). These two developmental pathways result in multicellular phenotypes that differ with respect to the degree of relatedness among cells and the level of complexity they achieved; lower cell relatedness and lower complexity levels characterize lineages in which multicellularity involves aggregation (for a detailed treatment of the issues associated with the two types of development see Grosberg and Strathmann 2007). In this chapter, we are mainly concerned with the evolution of complex multicellularity in lineages with clonal development.

Although multicellularity has evolved independently in at least 25 separate lineages from all three domains of life—Archaea, Eubacteria and Eukaryota (see Chap. 1, King 2004, and Grosberg and Strathmann 2007 for examples and references), multicellular forms with differentiated cell types are only known in a handful of groups (e.g., cyanobacteria and myxobacteria; ciliates; cellular slime molds; red, green and brown algae; land plants; fungi; animals). How and why complex multicellularity evolved, and why some multicellular lineages increased in complexity more than others, are still challenging questions.

Proximate Causes

The transition to multicellularity requires a series of specific mechanisms to ensure (i) the physical unity/stability of the multicellular individual (though generally referred to as adhesion, such mechanisms are rather different among multicellular lineages; see Abedin and King 2010 for a discussion), (ii) communication and recognition among cells (to ensure functional unity/stability), and (iii) regulation of cell growth, proliferation and differentiation (to ensure reproductive unity/stability). Current data (see below) indicate that components of many of these mechanisms were already present in the unicellular ancestors of multicellular lineages.

Indeed, the evolution of simple multicellularity appears to mainly have involved the co-option of existing mechanisms rather than the invention (de novo) of multicellular-specific genes and pathways. For instance, genes that code for proteins associated with adhesion (e.g., integrins, cadherins), cell signaling and cell-cell communication (e.g., tyrosine kinases) predate the evolution of Metazoa (e.g., King et al. 2003, 2008; Abedin and King 2010; Sebé-pedrós et al. 2010, Suga et al. 2013; Chaps. 5, 14, 20). Similarly, in volvocine algae, genes coding for components of the extracellular matrix (which ensures the physical unity and structural stability of the group) have evolved from genes already present in their unicellular ancestors (Prochnik et al. 2010).

Multicellular development and cell differentiation pathways (ensuring functional and reproductive unity and resulting in an increase in structural and functional complexity) have also evolved from pathways present in unicellular lineages—as in, for instance, the cellular slime mold *Dictyostelium discoideum* (Schaap 2011; Chap. 21) and the green alga *Volvox carteri* (Nedelcu 2009b). Likewise, genes involved in moss development have been found in the closest unicellular relatives of land plants (Nedelcu et al. 2006). Lastly, programmed cell death—thought to be an important developmental mechanism in multicellular lineages—is widespread in the unicellular world (e.g., Nedelcu et al. 2011); and many programmed cell death genes have been found in the genomes of single-celled species (e.g., Nedelcu 2009a). Nevertheless, some gene families coding for proteins involved in multicellular development have evolved specifically during the evolution of multicellularity (e.g., some transcription factor familes in metazoans; Degnan et al. 2009; also see Sebé-Pedrós et al. 2011, Suga et al. 2013, Chap. 18). Relative to simple multicellularity, the evolution of complex multicellularity typically entailed an increase in the number of genes involved in signal transduction and transcriptional regulation (through duplication followed by diversification), the evolution of new protein domains and/or the shuffling of preexisting domains, and tinkering with the basic genetic toolkit via the modification of patterns of gene expression (through the evolution of new *cis*-regulatory elements) (see Rokas 2008 for further discussion and examples).

Ultimate Causes

Various benefits have been put forward to explain why complex multicellularity evolved. Discussions are mainly centered around the evolution of cell differentiation (used here to refer to spatial cell differentiation resulting in specialized cell types and an increase in complexity), especially in the context of the evolution of specialized reproductive (germ) and somatic cells. These include conflict mediation (i.e., by restricting access to the germ line), division of labor, or overcoming life history trade-offs associated with reproducing a large body (e.g., Buss 1987; Maynard Smith and Szathmáry 1997; Michod 2006; Michod et al. 2006). In addition to explanations involving selective forces, non-adaptive scenarios invoking thermodynamic laws (Otsuka 2008), genetic drift (Lynch and Conery 2003), or passive outcomes of local environmental effects (Schlichting 2003) have also been proposed to explain the increase in complexity during the evolution of multicellularity.

Volvocine Algae as a Model System

Overview

The volvocine algae are a group of green algae (in the Chlorophyceae) comprising both single-celled species, such as *Chlamydomonas*, and multicellular species with various numbers and types of cells (Table 1 and Fig. 1). Volvocine algae have fascinated biologists ever since Antonie van Leeuwenhoek first saw *Volvox* "drive and move in the water" (van Leeuwenhoek 1700, p. 511). In his *Systema Naturae*, Linnaeus, impressed with their ability to roll around without limbs ("*Volvendo*

Table 1 Taxonomy of colonial volvocine algae. Numbers of species are approximations, as the validity of many described species is questionable. Numbers of cells are restricted to powers of 2 (with the possible exception of *Volvox*), so, for example, "8–32" should be understood as 8, 16, or 32 cells

Family	Genus	# of species	# of cells	Differentiated cells	% somatic	Colony morphology
Tetrabaenaceae	*Basichlamys*	1	4	No	0	Cluster
	Tetrabaena	1	4	No	0	Cluster
Goniaceae	*Astrephomene*	2	32–128	Somatic	6–12	Spheroid
	Gonium	10	8–32	No	0	Flat/bowl
Volvocaceae	*Colemanosphaera*	2	16–32	No	0	Spheroid
	Eudorina	8	16–32	No	0	Spheroid
	Pandorina	6	16–32	No	0	Spheroid
	Platydorina	1	16–32	No	0	Flattened[a]
	Pleodorina	6	32–128	Somatic	12–50	Spheroid
	Volvox	25	500–50,000	Somatic & germ	98 +	Spheroid
	Volvulina	4	8–16	No	0	Spheroid
	Yamagishiella	1	16–32	No	0	Spheroid

[a] *Platydorina* develops as a spheroid, undergoing complete inversion, but is secondarily flattened

seque rotando celeriter movens absque artubus!"), gave van Leeuwenhoek's "great round particles" the formal name *Volvox* ("to roll") (Linnaeus 1758, p. 821). An additional 11 genera have been described since: *Gonium* (Müller 1773), *Pandorina* (Bory de Saint-Vincent 1824), *Eudorina* (Ehrenberg 1832), *Pleodorina* (Shaw 1894), *Platydorina* (Kofoid 1899), *Volvulina* (Playfair 1915), *Astrephomene* (Pocock 1954), *Yamagishiella* (Nozaki and Kuroiwa 1992), *Colemanosphaera* (Nozaki et al. 2014), *Basichlamys* and *Tetrabaena* (Nozaki and Itoh 1994; Nozaki et al. 1996). Genus-level taxonomy within the Volvocales is badly in need of revision, as most nominal genera are polyphyletic. It is hard to give an exact number of volvocine species, since many described species are almost certainly synonymous, but by the end of the twentieth century the number of valid described species was probably on the order of 50. In the twenty-first century, new volvocine species are being described at a rate approaching one per year, nearly all by Hisayoshi Nozaki and colleagues (e.g., Nozaki and Krienitz 2001; Nozaki et al. 2006, 2014; Hayama et al. 2010; Nozaki and Coleman 2011; Isaka et al. 2012).

Collectively, the volvocine algae have a cosmopolitan distribution, although the known ranges of particular species can be anywhere from a single pond to multiple continents. The most common habitat is warm, eutrophic freshwater ponds and pools (Kirk 1998), but some species are also found in oligotrophic lakes (Coleman 2001), rivers (Kofoid 1899, 1900; Znachor and Jezberová 2005), rice paddies (Nozaki 1983), damp soils (Bold 1949), snow, and ice (Nozaki and Ohtani 1992; Hoham et al. 2002). Most species are obligate photoautotrophs, but a few, notably *Astrephomene*

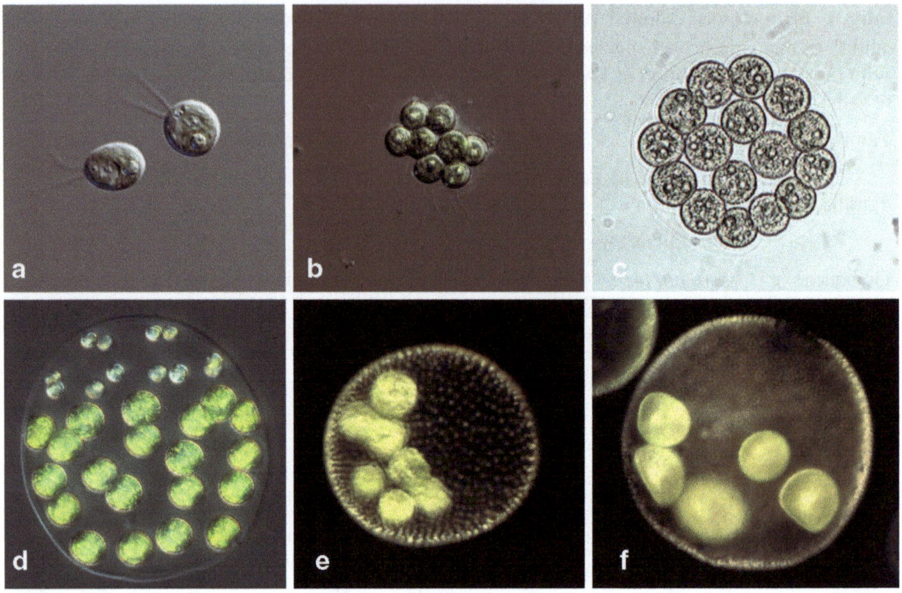

Fig. 1 Representative volvocine algae. **a** *Chlamydomonas reinhardtii*, a unicell. **b** *Gonium pectorale*, a flat plate of 8–32 cells (A and B by D. Shelton). **c** *Eudorina elegans*, a spheroid with up to 32 undifferentiated cells. **d** *Pleodorina starrii*, a partially differentiated spheroid with up to 64 cells. The small cells near the anterior (top) are somatic cells specialized for motility; the larger cells perform both reproductive and motility functions. **e** *Volvox carteri*, a spheroid with ~2000 small somatic cells and a few much larger reproductive cells. **f** *V. barberi*, a spheroid with ~30,000 somatic cells and a few reproductive cells

spp. and some species of *Chlamydomonas*, are mixotrophs that can consume acetate (Pringsheim and Wiessner 1960; Brooks 1972).

The life histories of all known volvocine algae are facultatively sexual, with asexual reproduction occurring in the haploid phase. Like most multicellular organisms, multicellular volvocine algae develop clonally through mitosis from a single cell. However, cell division in asexual development takes an unusual form, called palintomy, in which cells grow to many times their original size, and then undergo several rounds of division without intervening growth (Coleman 1979; Sleigh 1989). In single-celled species (e.g., *Chlamydomonas* spp., *Vitreochlamys* spp.), palintomy is followed by the release of up to 32 daughter cells. In multicellular species, the mitotic products of a given reproductive cell form a daughter colony, which is subsequently released; this process is known as autocolony formation.

The sexual phase of the life cycle is triggered by environmental conditions (nitrogen-deprivation in *Chlamydomonas* and some colonial volvocine algae, or heat-stress in *V. carteri*) or, in some cases, by signaling molecules released by "spontaneously" developed sexual colonies (Kirk and Kirk 1986). In anisogamous and oogamous species, asexual colonies are indistinguishable; the differences between males and females only become apparent upon entry into the sexual phase. Volvocine sexual reproduction spans a range from isogamy (equally-sized flagellated gametes) to anisogamy (flagellated gametes of unequal size) to oogamy (one gamete is large and non-motile), with larger species tending toward more unequal gametes (Nozaki 1996). Both heterothallic (genetically determined mating types or sexes) and homothallic (both mating types or sexes within an isogenic strain) species exist, and, in homothallic species, individual colonies may produce both types of gametes (monoecy) or there may be separate male and female colonies (dioecy). Although here we are not concerned with aspects related to sexual development, this group is also an ideal system to investigate the evolution of sexes and sexual reproduction (e.g., Nozaki 1996, 2014; Hiraide et al. 2013).

The volvocine algae are an ideal model system for understanding the evolution of multicellular complexity. The origin of multicellularity in this group was probably around 200 million years ago (MYA), much more recent than those of complex multicellular taxa such as animals, land plants, fungi, and red algae (Herron et al. 2009). Furthermore, the years have been kind to this group, as many species with intermediate degrees of complexity survive to this day. For example, the basic body plans of *Astrephomene*, *Gonium*, and *Yamagishiella* appear to be unchanged from their origins ~ 150 MYA (Herron et al. 2009). The two species of volvocine algae that have been intensively studied are at the two extremes of the complexity range in this group: *Chlamydomonas reinhardtii*—a unicellular relative of the colonial volvocine algae (Harris 2001, 2009), and *Volvox carteri* forma *nagariensis*—a multicellular species with ~ 2000 cells and a complete germ-soma division of labor (Kirk 1998, 2005; Fig. 1). Both of these species have sequenced genomes (Merchant et al. 2007; Prochnik et al. 2010), and work is progressing on several other volvocine species (Umen and Olson 2012).

Here, we advocate the development of the entire volvocine clade, that is, the families Volvocaceae, Goniaceae, and Tetrabaenaceae along with closely related unicellular algae, as a model system with which to study the evolution of complexity. Doing so leverages what, in our opinion, is the most attractive feature of this group, the existence of extant species with nearly every conceivable degree of complexity from single cells to differentiated multicellular organisms. Taking this broad view encourages comparative analyses, and this approach has already been successful in addressing questions related to the evolution of cooperation, multicellularity, cellular differentiation, morphology and development, and anisogamy, as well as questions about biomechanics and hydrodynamics. Furthermore, many important traits have evolved convergently, allowing questions about how similar the genetic and developmental mechanisms underlying these traits in different lineages are. In addition, some traits vary within as well as among species, allowing studies that bridge micro- and macro-evolution. Lastly, experimental evolution studies have demonstrated that simple multicellularity can be easily evolved in the lab from various unicellular species, including *C. reinhardtii* (Boraas et al. 1998; Ratcliff et al. 2012, 2013), and volvocine algae can also be used to experimentally evolve cell differentiation. Such studies will allow for experimentally addressing a variety of issues related to major evolutionary transitions, such as biological scaling and multilevel selection. The results of these experiments can be interpreted comparatively in the context of the extensive extant diversity of volvocine life histories.

What do the Volvocine Algae tell us About the Evolution of Complexity?

The evolution of the volvocine algae has often been viewed in the framework of the 'volvocine lineage hypothesis' (Lang 1963; Van de Berg and Starr 1971; Pickett-Heaps 1975)—the idea that the group members represent a progressive increase in size and complexity from unicellular *Chlamydomonas* to multicellular *Volvox* and that the phylogeny of the group reflects this progression. Within the colonial species, *Gonium* was considered the most 'primitive' (Pickett-Heaps 1975), and *Volvox* was a 'culminating member' of the volvocine lineage (Nozaki and Itoh 1994) and '... the ultimate expression of colonial development' (Pickett-Heaps 1975). Within the genus *Volvox*, *V. powersii* and *V. gigas* were viewed as the most 'primitive', and either *V. carteri* and *V. obversus* or the members of the section *Volvox* (e.g., *V. barberi, V. globator*) were thought to be the most 'advanced' (Desnitski 1995). Nevertheless, we now know that the volvocine lineage hypothesis is an oversimplification of volvocine phylogeny; in fact, complex multicellularity evolved independently in several '*Volvox*' lineages (Herron and Michod 2008; Herron et al. 2010; Fig. 2).

Below, we discuss morphological and developmental complexity in this group, using Heylighen's and McShea's frameworks. In doing so, we address both why and how complexity increased in this group (in several lineages).

Morphological/Structural Complexity

Regardless of developmental type, the cells in a multicellular organism must have some way of adhering to each other; in Heylighen's (1999) framework, the parts/cells have to fit together to form an intrinsically stable whole. Various multicellular groups achieve this in different ways (Abedin and King 2010). In the volvocine algae, cells are held together by an extracellular matrix (ECM) that is homologous to the cell wall in unicellular relatives (Kirk et al. 1986). ECM ensures both physical unity and structural stability ('intrinsic' fitness). Compared to unicellular algae, whose daughter cells are free to realize independent fates, the fates of daughter cells in multicellular algae are inextricably bound. The daughters of a given reproductive cell share a physical location and a set of internal and external conditions. As parts of the newly formed multicellular group, the daughter cells contribute to the increase in the *structural hierarchical complexity* of the system.

Interestingly, transitions between compact colonies such as *Pandorina*, which contain very little ECM, and larger colonies such as *Eudorina* (Fig. 1), with large volumes of ECM, have occurred multiple times in both directions (Herron and Michod 2008; Fig. 2). The volume of ECM scales allometrically with colony size,

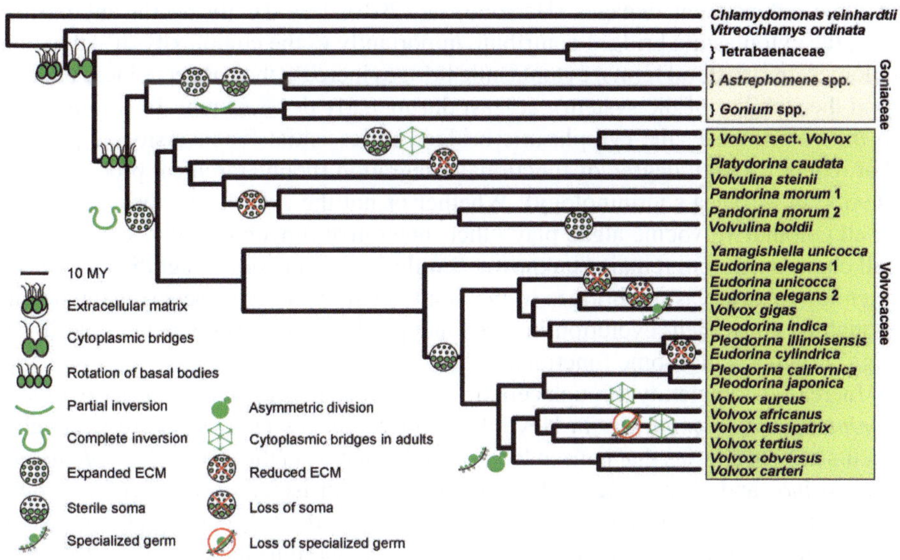

Fig. 2 Reconstructed changes in ancestral character states. Developmental changes shown are those supported by hypothesis tests in Herron and Michod 2008, Herron et al. 2009, 2010. Note that some nominal genera and species are polyphyletic and thus appear in more than one place in the tree

with the largest colonies (*Volvox* spp.) consisting of > 99 % ECM. This allometric increase suggests that the ECM might serve additional roles or have additional selective benefits. Indeed, in the larger species, the ECM is a complex organ with a great deal of internal structure that varies among species (Kirk et al. 1986; Nozaki and Kuroiwa 1992). Functionally, the ECM plays important roles in sex induction (Starr 1970; Gilles et al. 1983), inversion (discussed below; Ueki and Nishii 2009), and possibly nutrient storage (Koufopanou and Bell 1993; also see Chap. 11). In the *V. carteri* lineage, expansion of the ECM was accompanied by expansions of two of the gene families involved in its construction (Prochnik et al. 2010; Umen and Olson 2012). Thus, although the ECM evolved from a pre-existing structure (the cell wall) and a pre-existing set of genes, its expansion and differentiation have contributed to both *structural and functional complexity*.

In all members of the Goniaceae and Volvocaceae, during early embryonic development cells are connected through cytoplasmic bridges resulting from incomplete cytokinesis (Gerisch 1959; Bisalputra and Stein 1966; Gottlieb and Goldstein 1977; Marchant 1977; Fulton 1978; Green et al. 1981; Iida et al. 2013). These bridges ensure the stability of the early embryo. In most species, the bridges break down later in development, leaving cells physically unconnected; nevertheless, the stability of the system (and its 'intrinsic' fitness) is maintained by the ECM. In some species of *Volvox*, though, cytoplasmic bridges are retained in adult colonies, a paedomorphic trait (juvenile trait retained in adults) that has apparently arisen convergently in three independent lineages (Herron et al. 2010; Fig. 2). By nearly any definition, the cytoplasmic bridges contribute importantly to the complexity of volvocine colonies. In structural terms, cytoplasmic bridges increase the connectedness of cells (and the stability of the system), at least during early development. In *Gonium* and in the Volvocaceae, the cytoplasmic bridges are essential for inversion, a process that requires a high degree of functional integration (nonhierarchical process complexity, in McShea's terminology). Whether or not the cytoplasmic bridges in any or all colonial volvocine algae play other integrative functional roles, e.g., in cell-cell communication, remains unknown. Similarly, the functional significance of the retention of cytoplasmic bridges in the adults of some *Volvox* species is unknown, though its convergent evolution in three independent lineages (Herron et al. 2010; Fig. 2) suggests that some functional role exists.

Increased complexity through cell differentiation (*structural complexity* or *object nonhierarchical complexity* in Heylighen's or McShea's fameworks, respectively) occurs only in species that have at least 32 cells; that is, in the genera *Astrephomene*, *Pleodorina*, and *Volvox* (Table 1). This trend is consistent with the general view that increase in organismal size through increase in the number of cells increases the potential for increase in diversity of cell types (and thus increase in complexity) (Carroll 2001). *Astrephomene* and *Pleodorina* species possess only one specialized cell type. In these species, initially, all cells look alike (identical parts); later, most cells lose flagella and become reproductive, while several cells (at the posterior and anterior pole in *Astrephomene* and *Pleodorina,* respectively) remain flagellated and act as terminally differentiated, non-replicative somatic cells (Table 1). Two specialized cell types—terminally differentiated somatic and germ cells (gonidia) lacking flagella—are only present in *Volvox* species, which are also the largest in terms of number of total cells (Table 1).

Developmental/Functional Complexity

In addition to the broad range of morphological complexities, the volvocine algae also exhibit varying degrees of functional complexity, which are apparent in their developmental processes, including organismal polarity, inversion, and cell differentiation.

In the Goniaceae and Volvocaceae, organismal polarity is established through rotation of the basal bodies, which are attached to the flagella. In single-celled volvocine algae such as *Chlamydomonas*, the basal bodies are arranged in such a way that the two flagella beat in opposite directions (Kirk 2005). As a result, the cell swims in a "breast stroke," and the cell has an anterior-posterior polarity defined by the direction of swimming (Kirk 2005). This arrangement is retained in the four-celled species, *Tetrabaena socialis* and *Basichlamys sacculiferum*, which are the only known members of the family Tetrabaenaceae (Stein 1959; Nozaki and Itoh 1994). In *Gonium*, however, the basal bodies of the peripheral cells are rotated such that the flagella beat in the same direction, toward the periphery of the colony (Greuel and Floyd 1985; Kirk 2005). The four central cells retain the ancestral orientation, giving *Gonium* colonies a center-to-edge polarity at the colony level (Kirk 2005). The resulting difference between cells in the center and those on the periphery constitute an increase (relative to colonies without basal body rotation) in structural and functional differentiation. In spheroidal colonies, rotation of the basal bodies results in all flagella beating in roughly the same direction, establishing an anterior-posterior polarity, as in *Chlamydomonas*, based on the direction of swimming (Kirk 2005).

In *Gonium* and in the Volvocaceae, embryos at the end of cell division find themselves in an awkward configuration. Embryos at this stage are shaped as shallow bowls or spheres, but in either case, the flagella are on the wrong (concave or interior) surface for locomotion (Kirk 2005). Through a process of partial (*Gonium*) or complete inversion (Volvocaceae), the embryos change their topology so that the flagella end up on the convex (*Gonium*) or exterior (Volvocaceae) surface (Stein 1965; Fulton 1978; Kirk 2005). Inversion requires a high degree of functional integration among cells, as it is the movement of individual cells relative to the their cytoplasmic bridges that generates the emergent phenomenon of coordinated collective movement (*functional/process hierarchical complexity*) (Green et al. 1981).

Cellular differentiation occurs in three volvocine genera: *Astrephomene*, *Pleodorina*, and *Volvox*. In *Pleodorina* and *Astrephomene* embryos, all cells start at the same size, but a subset near the anterior (*Pleodorina*) or posterior (*Astrephomene*) either fail to grow or grow at a slower rate, resulting in adult colonies with two cell sizes. The larger cells behave as cells in undifferentiated species, initially providing flagellar motility but then eventually reproducing and losing the flagella. The smaller cells, though, perform vegetative functions only and never reproduce. The evolution of cellular specialization in these genera constitutes an increase not only in variety/differentiation (*structural complexity*), but also in functional integration (*functional/process non-hierarchical complexity)*, as cells become completely dependent on each other for the basic life-history functions of survival and reproduction.

In some *Volvox* species, as in *Pleodorina* and *Astrephomene*, at the end of embryogenesis all cells are similar in size. Some cells then grow slightly or not at all and differentiate as somatic cells, while other cells lose their flagella and grow to thousands of times their original size before producing a new generation. In a few closely related species of *Volvox*, though, during embryogenesis some cells undergo several rounds of asymmetric divisions, resulting in an embryo with two cell sizes (Fig. 2). The smaller of these cells will differentiate as somatic cells, while the larger cells become specialized reproductive cells (gonidia), with no flagella and no contribution to the motility of the individual. The evolution of asymmetric divisions (which take place only in half of the embryo, by an unknown mechanism) contributed to increased functional/process complexity in these lineages.

Both somatic cells and gonidia became specialized cells by losing functions that were present in the ancestral, generalist cells: somatic cells have lost reproductive functions, while reproductive cells have lost motility/survival functions. This is consistent with the pattern described by McShea (2002) that lower-level entities that have combined to form a higher-level entity tend to undergo a reduction in their own complexity.

Genomic Complexity vs. Morphological and Developmental Complexity

Despite the fact that *V. carteri* is morphologically and developmentally more complex than *C. reinhardtii,* a comparison between these two volvocine genomes did not reveal a significant difference in the total number of predicted genes (14,520 in *V. carteri vs.* 14,516 in *C. reinhardtii*) (Prochnik et al. 2010). Nevertheless, the number of genes involved in specific multicellular structures and developmental functions appears to be higher in *V. carteri* relative to *C. reinhardtii.* Specifically, *V. carteri* possesses a higher number of genes involved in ECM and cell-cycle regulation; furthermore, the *V. carteri* genome is enriched in volvocine-specific genes of unknown function, some of which might be involved in complexity-related traits specific to this group (Prochnik et al. 2010). Also, the TAZ family of transcription factors is more represented in *V. carteri* relative to *C. reinhardtii* (our unpublished data).

Interestingly, orthologs of *regA*—the gene responsible for the differentiation of somatic cells in *V. carteri*—have been recently found in several distantly related *Volvox* species (e.g., *V. gigas*), suggesting that this gene was already present in a volvocine ancestor without specialized somatic cells (Hanschen et al. 2014; Fig. 2). This scenario implies that the evolution of somatic cell differentiation and the increase in complexity observed in the lineage leading to *V. carteri* involved changes in the regulatory elements of *regA* (whether *regA* has a role in somatic cell differentiation in the other *Volvox* species is not yet known). Furthermore, *regA* appears to have evolved from a *regA*-like gene already present in the unicellular ancestor of volvocine algae; its co-option might have involved changing its expression pattern from a temporal context (in response to environment) into a spatial (developmental) context (Nedelcu and Michod 2006; Nedelcu 2009b).

Similarly, two other genes involved in complexity-related traits in *V. carteri* also have orthologs in the unicellular *C. reinhardtii*. These are the *glsA* gene involved in the asymmetric divisions responsible for setting aside the large cells that will develop into gonidia (Kirk et al. 1991); and the *invA* gene involved in the process of embryonic inversion (Nishii et al. 2003). In both cases, the *C. reinhardtii* ortholog can rescue a *V. carteri* mutant (Kirk et al. 1986; Kirk 2005), indicating that the difference in function between the two orthologs does not involve changes at the protein level.

Altogether, the available genomic information supports the idea that overall gene content is not a good indicator of organismal complexity, and points towards an increase in genome complexity through gene duplication and co-option via changes in regulatory elements as being mainly responsible for the observed increase in morphological and developmental complexity in this group. These findings are consistent with the general trends identified during the evolution of morphological complexity (discussed earlier).

The sequencing of additional volvocine genomes from lineages with different grades of morphological and developmental complexity (which is underway in several labs) will make it possible to further investigate the relationship between genomic complexity and organismal complexity among closely related species, thus avoiding some of the confounding factors associated with comparing species that have diverged a long time ago. The significance and relative contribution of mutations in coding regions vs. changes in gene regulation to the genetic basis for the evolution of new morphological traits is currently an issue of heated debate (Hoekstra and Coyne 2007; Carroll 2008). Due to their relatively low but variable levels of complexity, simpler underlying genetics and recent evolutionary history, volvocine algae have the potential to provide significant insights into this debate.

Independent Increases in Complexity

Interestingly, multicellular volvocine algae are monophyletic, suggesting that simple multicellularity evolved only once in this group (Nozaki 2003). However, complex (differentiated) multicellularity appears to have evolved independently several times. For instance, sterile somatic cells, specialized germ cells, and retention of cytoplasmic bridges in adult spheroids each occurred independently, multiple times, within the volvocine algae (Herron and Michod 2008; Herron et al. 2010; Fig. 2). Nevertheless, despite the apparent ease of evolving complexity-related traits, the number of cell types remained low in all volvocine species. Furthermore, it appears that in some cases, morphologically-complex lineages evolved towards simplification. For instance, both forms of cellular differentiation have apparently been lost in some lineages within the Volvocaceae (Herron and Michod 2008; Fig. 2). These observations raise a number of questions. What were the factors that contributed to the independent increases and decreases in complexity in this group? Why did not all volvocine species reach the complexity levels attained by some *Volvox* species? And why did none of the *Volvox* lineages reach even higher complexity levels?

Following his analysis of morphological complexity, Carroll (2001) concluded that "the observed limits of form seem to be due to a combination of both chance and necessity, a product of historical contingency and imposed by external agents (for example, selection) and internal rules (for example, constraints)". He further argues that selection cannot be the whole story, and that the internally imposed constraints also shape the range of possible morphologies and can themselves evolve. Volvocine algae exemplify these statements very well; the evolution of complexity in this group is likely a reflection of (i) historical contingencies associated with the specific cellular and genetic background of the *Chlamydomonas*-like ancestor, (ii) specific developmental constraints, and (iii) diverse selective pressures.

For instance, the evolution of multicellularity in the volvocine algae is thought to have been facilitated by the specific type of cell division (palintomy) that was inherited from the *Chlamydomonas*-like ancestor (Kirk 1998). In addition, the mechanistic basis for the evolution of somatic cell differentiation (at least in *V. carteri*) can also be traced back to the ability of single-celled volvocine ancestors to temporarily repress their reproduction to increase survival, as part of their general photo-acclimation response to limiting or stressful conditions (Nedelcu and Michod 2006; Nedelcu 2009b). However, the very same factors that allowed some lineages in this group to achieve high levels of complexity may have affected how complex they could become. For instance, although palintomy has been replaced by binary fission in some *Volvox* lineages (Herron et al. 2010), other *Volvox* species have retained palintomy, which has limited their potential to evolve increased numbers of cells (to produce *n* cells, gonidia need to grow 2^n-fold in volume). More importantly, the fact that the evolution of somatic cells involved the permanent suppression of cell division (instead of its temporal and/or spatial regulation) has likely limited its potential to evolve new cell types and thus affected the evolvability of this lineage (Nedelcu and Michod 2004).

The independent increases and decreases in size and complexity observed in this group suggest that strong selective pressures to increase or decrease complexity levels did (do) exist and that such pressures differ among the environments or the ecological niches these algae inhabit. These pressures are thought to include predator avoidance, motility, and nutrient availability (see Chap. 11), but a detailed analysis looking for a correlation between specific environmental factors, organismal size and complexity levels displayed by the volvocine algae inhabiting a specific environment/ecological niche has not been performed. Based on Ashby's Law of Requisite Variety (discussed above) we predict that in lineages that evolved large colony sizes (in response to selection for large body size) increased levels of complexity correlate to increased variability in their environment/niche. In other words, we suggest that more variable environments selected for more complex forms. This relationship between complexity increase and environmental variability might also be used to address why complex volvocine algae evolved from a *Chlamydomonas reinhardtii*-like ancestor; or why complex multicellularity (with specialized somatic and reproductive cells) did not evolve from other unicellular volvocalean ancestors. Specifically, like the extant *C. reinhardti*, it is possible that such an ancestor was adapted to variable environments and already possessed mechanistic and genetic factors that allowed control over the environment. Lineages that evolved multicellularity in response to selection for increased size have co-opted these mechanisms and use them to control variable environments in new ways. That not all lineages reached the complexity levels achieved by some *Volvox* species might reflect the fact that some species attained a good trade-off point and are not challenged anymore by their environment.

Concluding Remarks

Volvocine algae have been studied for a long time and used to address various questions from very diverse fields. Here, we argue that because complexity levels have increased or decreased independently in several lineages, this group is an ideal model-system to investigate the evolution of complexity. The independent acquisitions and losses of traits associated with complexity in this group represent an unprecedented opportunity to (i) explore the genetic basis and the selective pressures responsible for such changes in complexity levels, (ii) distinguish between the mechanisms that have been proposed to explain increases in complexity in biological systems—passive (random) vs. active/"driven" (non-random) processes, and external (affected by selection, ecology, environment) vs. internal (under genetic, developmental, biomechanical control) (McShea 1994; Carroll 2001), and (iii) identify general principles underlying the evolution of complexity.

Summary

1. Complexity is a complex concept, which has proven rather difficult to define and measure.
2. There is no a priori reason that more complicated systems would be preferred by natural selection. Evolution selects for increases in fitness, but high fitness can be achieved both by very complex and very simple systems.
3. Selection for increased stability and control over the environment will tend to be accompanied by respective increases in structural/morphological complexity and functional/developmental complexity.
4. How and why complex multicellularity evolved, and why some multicellular lineages increased in complexity more than others, are still challenging questions.
5. The volvocine algae—comprising species with nearly every conceivable degree of complexity from single cells to differentiated multicellular organisms—are an ideal model system for understanding the evolution of multicellular complexity.
6. The independent acquisitions and losses of traits associated with complexity in the volvocine group represent an unprecedented opportunity to (i) explore the genetic basis and the selective pressures responsible for such changes in complexity levels, (ii) distinguish between the mechanisms that have been proposed to explain increases in complexity in biological systems, and (iii) identify general principles underlying the evolution of complexity.

Acknowledgements We thank Erik Hanschen and Deborah Shelton for comments on the manuscript; we also thank Deborah Shelton for providing pictures of volvocine algae. We gratefully acknowledge support from a NASA Astrobiology Institute postdoctoral fellowship and from the John Templeton Foundation (MDH) and NSERC (AMN).

References

Abedin M, King N (2010) Diverse evolutionary paths to cell adhesion. Trends Cell Biol 20:742–734

Ashby WR (1956) Introduction to cybernetics. Methuen, London

Ashby WR (1958) Requisite variety and its implications for the control of complex systems. Cybernetica 1:83–99

Ayala FJ (1988) Can "progress" be defined as a biological concept? In: Nitecki MH (ed) Evolutionary progress. The University of Chicago Press, Chicago, pp 75–96

Bisalputra T, Stein R (1966) The development of cytoplasmic bridges in *Volvox aureus*. Can J Bot 44:1697–1702

Bold HC (1949) The morphology of *Chlamydomonas chlamydogama*, sp. nov. Bull Torrey Bot Club 76:101–108

Bonner JT (2003) On the origin of differentiation. J Biosci 28:523–528

Boraas ME, Seale DB, Boxhorn JE (1998) Phagotrophy by a flagellate selects for colonial prey: a possible origin of multicellularity. Evol Ecol 12:153–164

Bory de Saint-Vincent JBGM (1824) Pandorina. In: Lamouroux JV, Bory de Saint-Vincent JBGM, Deslongschamps E (eds) Encyclopédie méthodique ou par ordre de matières. Histoire naturelle des zoophytes, ou animaux rayonnés, faisant suite Ã l'histoire naturelle des vers de Bruguière. Mme veuve Agasse, Paris, p 600

Brooks AE (1972) The physiology of *Astrephomene gubernaculifera*. J Eukaryot Microbiol 19: 195–199
Buss LW (1987) The evolution of individuality. Princeton University Press, Princeton
Butterfield NJ (2009) Modes of pre-Ediacaran multicellularity. Precambrian Res 173:201–211
Carroll SB (2001) Chance and necessity: morphological complexity and diversity. Nature 409: 1102–1109
Carroll SB (2008) Evo-devo and an expanding evolutionary synthesis: a genetic theory of morphological evolution. Cell 134:25–36
Coleman AW (1979) Sexuality in colonial green flagellates. In: Levandowsky M, Hutner SH (eds) Biochemistry and physiology of protozoa. Academic Press, New York, pp 307–340
Coleman AW (2001) Biogeography and speciation in the *Pandorina/Volvulina* (Chlorophyta) superclade. J Phycol 37:836–851
Crisp MD, Cook LG (2005) Do early branching lineages signify ancestral traits? Trends Ecol Evol 20:122–128
Darwin CR (1837) Notebook B: [Transmutation of species (1837–1838)]. CUL–DAR121. Darwin Online, http://darwin-online.org.uk/.
Dayel MJ, Alegado RA, Fairclough SR, Levin TC, Nichols SA, McDonald K, King N. 2011. Cell differentiation and morphogenesis in the colony-forming choanoflagellate *Salpingoeca rosetta*. Dev Biol 357:73–82 (Elsevier Inc)
de Mendoza A Sebé-Pedrós A Šestak MS Matejčić M Torruella G Domazet-Lošo T Ruiz-Trillo I (2013) Transcription factor evolution in eukaryotes and the assembly of the regulatory toolkit in multicellular lineages. Proc Natl Acad Sci U S A 110:E4858–E4866
Degnan BM, Vervoort M, Larroux C, Richards GS (2009) Early evolution of metazoan transcription factors. Curr Opin Genet Dev 19:591–599
Desnitski AG (1995) A review on the evolution of development in *Volvox*—morphological and physiological aspects. Eur J Protistol 31:241–247
Ehrenberg CG (1832) Über die Entwickelung und Lebensdauer der Infusionsthiere; nebst ferneren Beiträgen zu einer Vergleichung ihrer organischen Systeme. Abh K Abhandlungen der Königlichen Akad Wissenschaften zu Berlin. Phys Kl 1831:1–154
Fulton AB (1978) Colonial development in *Pandorina morum*: II. Colony morphogenesis and formation of the extracellular matrix. Dev Biol 251:236–251
Gerisch G (1959) Die Zelldifferenzierung bei *Pleodorina californica* Shaw und die Organisation der Phytomonadinenkolonien. Arch Protistenkd 104:292–358
Gilles R, Gilles C, Jaenicke L (1983) Sexual differentiation of the green alga *Volvox carteri*. Naturwissenschaften 70:571–572
Gottlieb B, Goldstein ME (1977) Colony develompent in *Eudorina elegans* (Chlorophyta, Volvocales). J Phycol 13:358–364
Green KJ, Viamontes GI, Kirk DL (1981) Mechanism of formation, ultrastructure and function of the cytoplasmic bridge system during morphogenesis in *Volvox*. J Cell Biol 91:756–769
Greuel BT, Floyd GL (1985) Development of the flagellar apparatus and flagellar orientation in the colonial green alga *Gonium pectorale* (Volvocales). J Phycol 21:358–371
Grosberg RK, Strathmann RR (2007) The evolution of multicellularity: a minor major transition? Annu Rev Ecol Evol Syst 38:621–654
Hanschen ER, Ferris PJ, Michod RE (2014) Early evolution of the genetic basis for soma in the Volvocaceae. Evolution 68:2014–2025
Harris EH (2001) *Chlamydomonas* as a model organism. Annu Rev Plant Physiol Plant Mol Biol 52:363–406
Harris EH (2009) The Chlamydomonas sourcebook, second edition. Academic Press, Oxford
Hayama M, Nakada T, Hamaji T, Nozaki H (2010) Morphology, molecular phylogeny and taxonomy of *Gonium maiaprilis* sp. nov. (Goniaceae, Chlorophyta) from Japan. Phycologia 49:221–234
Herron MD, Michod RE (2008) Evolution of complexity in the volvocine algae: transitions in individuality through Darwin's eye. Evolution 62:436–451
Herron MD, Hackett JD, Aylward FO, Michod RE (2009) Triassic origin and early radiation of multicellular volvocine algae. Proc Natl Acad Sci U S A 106:3254–3258

Herron MD, Desnitskiy AG, Michod RE (2010) Evolution of developmental programs in *Volvox* (Chlorophyta). J Phycol 46:316–324

Heylighen F (1999) Growth of complexity. In: Heylighen F, Bollen J, Riegler A (eds) The evolution of complexity. Kluwer Academic, Dordrecht, pp 17–44

Hiraide R, Kawai-Toyooka H, Hamaji T, Matsuzaki R, Kawafune K, Abe J, Sekimoto H, Umen J, Nozaki H (2013) The evolution of male-female sexual dimorphism predates the gender-based divergence of the mating locus gene MAT3/RB. Mol Biol Evol 30:1038–1040

Hoekstra HE, Coyne JA (2007) The locus of evolution: evo devo and the genetics of adaptation. Evolution 61:995–1016

Hoham RW, Bonome TA, Martin CW, Leebens-mack JH (2002) A combined 18S rDNA and rbcL phylogenetic analysis of *Chloromonas* and *Chlamydomonas* (Chlorophyceae, Volvocales) emphasizing snow and other cold-termperature habitats. J Phycol 38:1051–1064

Hull DL (1988) Progress in ideas of progress. In: Nitecki MH (ed) Evolutionary progress. The University of Chicago Press, Chicago, pp 27–48

Iida H, Ota S, Inouye I (2013) Cleavage, incomplete inversion, and cytoplasmic bridges in *Gonium pectorale* (Volvocales, Chlorophyta). J Plant Res 126:699–707

Isaka N, Kawai-Toyooka H, Matsuzaki R, Nakada T, Nozaki H (2012) Description of two new monoecious species of *Volvox* sect. *Volvox* (Volvocaceae, Chlorophyceae), based on comparative morphology and molecular phylogeny of cultured material. J Phycol 48:759–767

King N (2004) The unicellular ancestry of animal development. Dev Cell 7:313–325

King N, Hittinger CT, Carroll SB (2003) Evolution of key cell signaling and adhesion protein families predates animal origins. Science 301:361–363

King N., Westbrook MJ, Young SL, Kuo A, Abedin M, Chapman J, Fairclough S, Hellsten U, Isogai Y, Letunic I, Marr M, Pincus D, Putnam N, Rokas A, Wright KJ, Zuzow R, Dirks W, Good M, Goodstein D, Lemons D, Li W, Lyons JB, Morris A, Nichols S, Richter DJ, Salamov A, JGI Sequencing, Bork P, Lim WA, Manning G, Miller WT, McGinnis W, Shapiro H, Tjian R, Grigoriev IV, Rokhsar D (2008) The genome of the choanoflagellate *Monosiga brevicollis* and the origin of metazoans. Nature 451:783–788

Kirk DL (1998) Volvox: molecular-genetic origins of multicellularity. Cambridge University Press, Cambridge

Kirk DL (2005) A twelve-step program for evolving multicellularity and a division of labor. Bioessays 27:299–310

Kirk DL, Kirk MM (1986) Heat shock elicits production of sexual inducer in *Volvox*. Science 231:51–54

Kirk DL, Birchem R, King N (1986) The extracellular matrix of *Volvox*: a comparative study and proposed system of nomenclature. J Cell Sci 80:207–231

Kirk DL, Kaufman MR, Keeling RM, Stamer KA (1991) Genetic and cytological control of the asymmetric divisions that pattern the *Volvox* embryo. Dev Suppl 1:67–8

Knoll AH (2011) The multiple origins of complex multicellularity. Annu Rev Earth Planet Sci 39:217–239

Kofoid CA (1899) Plankton studies. III. On *Platydorina*, a new genus of the family Volvocidae, from the plankton of the Illinois River. Bull Illinois State Lab Nat Hist 5:419–440

Kofoid CA (1900) Plankton Studies. II. On *Pleodorina illinoisensis*, a new species from the plankton of the Illinois River. Ann Mag Nat Hist 6:139–156

Koufopanou V, Bell G 1993. Soma and germ: an experimental approach using *Volvox*. Proc R Soc Lond B 254:107–113

Krell F-T, Cranston PS (2004) Which side of the tree is more basal? Syst Entomol 29:279–281

Lang NJ (1963) Electron microscopy of the Volvocaceae and Astrephomenaceae. Am J Bot 50: 280–300

Linnaeus C 1758. Systema naturae per regna tria naturae, secundum classes, ordines, genera, species, cum characteribus, differentiis, synonymis, locis. Tomus I. Editio decima, reformata. Editio decima revisa, vol 1. Holmiae, Stockholm

Lovejoy AO (1936) The great chain of being. Harvard University Press, Cambridge

Lynch M, Conery JS (2003) The origins of genome complexity. Science 302:1401–1404
Marchant HJ (1977) Colony formation and inversion in the green alga *Eudorina elegans*. Protoplasma 93:325–339
Maynard Smith J, Szathmáry E (1997) The major transitions in evolution. Oxford University Press, New York
McShea DW (1994) Mechanisms of large-scale evolutionary trends. Evolution 48:1747–1763
McShea DW (1996) Perspective: metazoan complexity and evolution: is there a trend? Evolution 50:477–492
McShea DW (2002) A complexity drain on cells in the evolution of multicellularity. Evolution 56:441–452
Merchant SS, Prochnik SE, Vallon O, Harris EH, Karpowicz SJ, Witman GB, Terry A, Salamov A, Fritz-Laylin LK, Maréchal-Drouard L, Marshall WF, Qu L-H, Nelson DR, Sanderfoot AA, Spalding MH, Kapitonov VV, Ren Q, Ferris PJ, Lindquist E, Shapiro H, Lucas SM, Grimwood J, Schmutz J, Cardol P, Cerutti H, Chanfreau G, Chen C-L, Cognat V, Croft MT, Dent R, Dutcher S, Fernández E, Fukuzawa H, González-Ballester D, González-Halphen D, Hallmann A, Hanikenne M, Hippler M, Inwood W, Jabbari K, Kalanon M, Kuras R, Lefebvre PA, Lemaire SD, Lobanov AV, Lohr M, Manuell A, Meier I, Mets L, Mittag M, Mittelmeier T, Moroney JV, Moseley J, Napoli C, Nedelcu AM, Niyogi K, Novoselov SV, Paulsen IT, Pazour G, Purton S, Ral J-P, Riao-Pachón DM, Riekhof W, Rymarquis L, Schroda M, Stern D, Umen JG, Willows R, Wilson N, Zimmer SL, Allmer J, Balk J, Bisova K, Chen C-J, Elias M, Gendler K, Hauser C, Lamb MR, Ledford H, Long JC, Minagawa J, Page MD, Pan J, Pootakham W, Roje S, Rose A, Stahlberg E, Terauchi AM, Yang P, Ball S, Bowler C, Dieckmann CL, Gladyshev VN, Green P, Jorgensen R, Mayfield S, Mueller-Roeber B, Rajamani S, Sayre RT, Brokstein P, Dubchak I, Goodstein D, Hornick L, Huang YW, Jhaveri J, Luo Y, Martínez D, Ngau WCA, Otillar B, Poliakov A, Porter A, Szajkowski L, Werner G, Zhou K, Grigoriev IV, Rokhsar DS, Grossman AR (2007) The *Chlamydomonas* genome reveals the evolution of key animal and plant functions. Science 318:245–250
Michod RE (2006) The group covariance effect and fitness trade-offs during evolutionary transitions in individuality. Proc Natl Acad Sci U S A 103:9113–9117
Michod RE, Viossat Y, Solari CA, Hurand M, Nedelcu AM (2006) Life-history evolution and the origin of multicellularity. J Theor Biol 239:257–272
Müller OF (1773) Vermium Terrestrium et Fluviatilium, seu Animalium Infusoriorum, Helminthicorum et Testaceorum, non Marinorum, Succincta Historia. Vol. 1
Nedelcu AM (2009a) Comparative genomics of phylogenetically diverse unicellular eukaryotes provide new insights into the genetic basis for the evolution of the programmed cell death machinery. J Mol Evol 68:256–268
Nedelcu AM (2009b) Environmentally induced responses co-opted for reproductive altruism. Biol Lett 5:805–808
Nedelcu AM, Michod RE (2004) Evolvability, modularity, and individuality during the transition to multicellularity in Volvocalean green algae. In: Schlosser G, Wagner GP (eds) Modularity in development and evolution. Oxford University Press, Oxford, pp 466–489
Nedelcu AM, Michod RE (2006) The evolutionary origin of an altruistic gene. Mol Biol Evol 23:1460–1464
Nedelcu AM, Borza T, Lee RW (2006) A land plant-specific multigene family in the unicellular *Mesostigma* argues for its close relationship to Streptophyta. Mol Bol Evol 23:1011–1105
Nedelcu AM, Driscoll WW, Durand PM, Herron MD, Rashidi A (2011) On the paradigm of altruistic suicide in the unicellular world. Evolution 65:3–20
Nishii I, Ogihara S, Kirk DL (2003) A kinesin, invA, plays an essential role in *Volvox* morphogenesis. Cell 113:743–753
Nozaki H (1983) Morphology and taxonomy of two species of *Astrephomene* in Japan. Jpn J Phycol 58:345–352
Nozaki H (1996) Morphology and evolution of sexual reproduction in the Volvocaceae (Chlorophyta). J Plant Res 109:353–361

Nozaki H (2003) Origin and evolution of the genera *Pleodorina* and *Volvox*. Biologia 58:425–431
Nozaki H (2014). Sexual Reproduction in animals and plants. In: Sawada H, Inoue N, Iwano M (eds) Sexual reproduction in animals and plants. Springer Japan, Tokyo, pp 215–227
Nozaki H, Coleman AW (2011) A new species of *Volvox* sect. Merrillosphaera (Volvocaceae, Chlorophyceae) from Texas. J Phycol 47:673–679
Nozaki H, Itoh M (1994) Phylogenetic relationships within the colonial Volvocales (Chlorophyta) inferred from cladistic analysis based on morphological data. J Phycol 30:353–365
Nozaki H, Krienitz L (2001) Morphology and phylogeny of *Eudorina minodii* (Chodat) Nozaki et Krienitz, comb. nov. (Volvocales, Chlorophyta) from Germany. Eur J Phycol 36:23–28
Nozaki H, Kuroiwa T (1992) Ultrastructure of the extracellular matrix and taxonomy of *Eudorina*, *Pleodorina* and *Yamagishiella* gen. nov. (Volvocaceae, Chlorophyta). Phycologia 31:529–541
Nozaki H, Ohtani S (1992) *Gonium sociale* (Volvocales, Chlorophyta) from Antarctica. Jpn J Phycol 40:267–271
Nozaki H, Itoh M, Watanabe MM, Kuroiwa T (1996) Ultrastructure of the vegetative colonies and systematic position of *Basichlamys* (Volvocales, Chlorophyta). Eur J Phycol 31:67–72
Nozaki H, Ott FD, Coleman AW (2006) Morphology, molecular phylogeny and taxonomy of two new species of *Pleodorina* (Volvocaceae, Chlorophyceae). J Phycol 42:1072–1080
Nozaki H, Yamada TK, Takahashi F, Matsuzaki R, Nakada T (2014) New "missing link" genus of the colonial volvocine green algae gives insights into the evolution of oogamy. BMC Evol Biol 14:37
Otsuka J (2008) A theoretical approach to the large-scale evolution of multicellularity and cell differentiation. J Theor Biol 255:129–136
Pickett-Heaps JD (1975) Green algae: structure, reproduction and evolution in selected genera. Sinauer Associates, Inc, Sunderland
Playfair GI (1915) Freshwater algae of the Lismore District: with an appendix on the algal fungi and Schizomycetes. Proc Linn Soc N S W 40:310–362
Pocock MA (1954) Two multicellular motile green algae, *Volvulina* Playfair and *Astrephomene*, a new genus. Trans R Soc S Afr 34:103–127
Pringsheim EG, Wiessner W (1960) Photoassimilation of acetate by green organisms. Nature 919–921
Prochnik SE, Umen J, Nedelcu AM, Hallmann A., Miller SM, Nishii I, Ferris PJ, Kuo A, Mitros T, Fritz-Laylin LK, Hellsten U, Chapman J, Simakov O, Rensing SA, Terry A, Pangilinan J, Kapitonov V, Jurka J, Salamov A, Shapiro H, Schmutz J, Grimwood J, Lindquist E, Lucas S, Grigoriev IV, Schmitt R, Kirk DL, Rokhsar DS (2010) Genomic analysis of organismal complexity in the multicellular green alga *Volvox carteri*. Science 329:223–226
Ratcliff WC, Denison RF, Borrello M, Travisano M (2012) Experimental evolution of multicellularity. Proc Natl Acad Sci U S A 109:1595–1600
Ratcliff WC, Herron MD, Howell K, Pentz JT, Rosenzweig F, Travisano M (2013) Experimental evolution of an alternating uni- and multicellular life cycle in *Chlamydomonas reinhardtii*. Nat Commun 4:2742
Richards RJ (1988) The moral foundations of the idea of evolutionary progress: Darwin, Spencer, and the neo-darwinians. In: Nitecki MH (ed) Evolutionary progress. University of Chicago Press, Chicago, pp 129–148
Rokas A (2008) The origins of multicellularity and the early history of the genetic toolkit for animal development. Annu Rev Genet 42:235–251
Schaap P (2011) Evolutionary crossroads in developmental biology: *Dictyostelium discoideum*. Development 138:387–396
Schaap P, Winckler T, Nelson M, Alvarez-Curto E, Elgie B, Hagiwara H, Cavender J, Milano-Curto A, Rozen DE, Dingermann T, Mutzel R, Baldauf SL (2006) Molecular phylogeny and evolution of morphology in the social amoebas. Science 314:661–663
Schlichting CD (2003) Origins of differentiation via phenotypic plasticity. Evol Dev 5:98–105
Sebé-pedrós A, Roger AJ, Lang FB, King N, Ruiz-Trillo I (2010) Ancient origin of the integrin-mediated adhesion and signaling machinery. Proc Natl Acad Sci U S A 107:10142–10147

Sebé-Pedrós A, Mendoza A de, Lang BF, Degnan BM, Ruiz-Trillo I (2011) Unexpected repertoire of metazoan transcription factors in the unicellular holozoan *Capsaspora owczarzaki*. Mol Biol Evol 28:1241–1254

Shaw WR (1894) *Pleodorina*, a new genus of the Volvocineæ. Bot Gaz 19:279–283

Sleigh MA (1989) Protozoa and other protists. Edward Arnold Limited, New York

Starr RC (1970) Control of differentiation in *Volvox*. Dev Biol 4:S59–S100

Stein JR (1959) The four-celled species of *Gonium*. Am J Bot 46:366–371

Stein JR (1965) On cytoplasmic strands in *Gonium pectorale* (Volvocales). J Phycol 1:1–5

Suga H, Chen Z, Mendoza A de, Sebé-Pedrós A, Brown MW, Kramer E, Carr M, Kerner P, Vervoort M, Sánchez-Pons N, Torruella G, Derelle R, Manning G, Lang BF, Russ C, Haas BJ, Roger AJ, Nusbaum C, Ruiz-Trillo I (2013) The *Capsaspora* genome reveals a complex unicellular prehistory of animals. Nat Commun 4:2325

Szathmáry E, Jordan F, Pál C (2001) Can genes explain biological complexity? Science 292:1315–1316

Ueki N, Nishii I (2009) Controlled enlargement of the glycoprotein vesicle surrounding a *Volvox* embryo requires the InvB nucleotide-sugar transporter and is required for normal morphogenesis. Plant Cell 21:1166–1181

Umen JG, Olson BJSC (2012) Genomics of volvocine algae. Adv Bot Res 64:185–243

van Leeuwenhoek A (1700) Part of a letter from Mr Antony van Leeuwenhoek, concerning the worms in sheeps livers, gnats, and animalcula in the excrements of frogs. Philos Trans R Soc London 22:509–518

Van Valen L (1973) A new evolutionary law. Evol Theory 1:1–30

Van de Berg WJ, Starr RC (1971) Structure, reproduction, and differentiation in *Volvox gigas* and *Volvox powersii*. Arch Protistenkd 113:195–219.

Velicer GJ, Vos M (2009) Sociobiology of the myxobacteria. Annu Rev Microbiol 63:599–623

Waddington CH (1969) Paradigm for an evolutionary process. In: Waddington CH (ed) Sketching theoretical biology: toward a theoretical biology, vol 2. Aldine Transaction, Chicago, pp 106–128

Wolpert L, Szathmáry E (2002) Multicellularity: evolution and the egg. Nature 420:745

Znachor P, Jezberová J (2005) The occurrence of a bloom-forming green alga *Pleodorina indica* (Volvocales) in the downstream reach of the River Malše (Czech Republic). Hydrobiologia 541:221–228

Emergence of *Ectocarpus* as a Model System to Study the Evolution of Complex Multicellularity in the Brown Algae

J. Mark Cock, Olivier Godfroy, Martina Strittmatter, Delphine Scornet, Toshiki Uji, Garry Farnham, Akira F Peters and Susana M Coelho

Abstract In contrast to the situation in land plants and animals, very little information is available concerning the molecular mechanisms underlying multicellular development in the brown algae. Historically, one of the reasons for this has been the lack of an effective model organism for the latter group that would permit the application of powerful genomic and genetic approaches to explore these processes. This situation has changed in recent years with the emergence of the filamentous brown alga *Ectocarpus* as a model organism. This chapter describes the genetic and genomic resources that are currently available for this organism and describes some

J. M. Cock (✉) · O. Godfroy · M. Strittmatter · D. Scornet · T. Uji ·
G. Farnham · A. F. Peters · S. M. Coelho
CNRS, UMR 8227, Integrative Biology of Marine Models, Station Biologique de Roscoff, Sorbonne Universités, UPMC Univ Paris 06, CS 90074, 29688 Roscoff cedex, France
e-mail: cock@sb-roscoff.fr

O. Godfroy
e-mail: godfroy@sb-roscoff.fr

M. Strittmatter
e-mail: strittmatter@sb-roscoff.fr

D. Scornet
e-mail: scornet@sb-roscoff.fr

T. Uji
e-mail: toshiki.uji@sb-roscoff.fr

G. Farnham
The Marine Biological Association of the United Kingdom, The Laboratory, Citadel Hill, Plymouth, Devon, PL1 2PB, UK

Plymouth Marine Laboratory, Prospect Place, The Hoe, Plymouth PL1 3DH, UK

A. F. Peters
Bezhin Rosko, 40 Rue des pêcheurs, 29250 Santec, France
e-mail: akirapeters@gmail.com

S. M Coelho
e-mail: coelho@sb-roscoff.fr

I. Ruiz-Trillo, A. M. Nedelcu (eds.), *Evolutionary Transitions to Multicellular Life*,
Advances in Marine Genomics 2, DOI 10.1007/978-94-017-9642-2_8

of the additional tools that are under development. Potential additional models that would provide access to the biological diversity within the brown algae are also discussed, with a particular focus on the evolution of multicellular complexity within this group.

Keywords Brown algae · Development · Ectocarpus · Genetic · Genomic · Model system · Multicellularity

Brown Algal Model Systems

Whilst brown algae as a group clearly exhibit complex multicellularity (see Chap. 16), the level of developmental complexity is variable across the group, ranging from species with relatively simple filamentous thalli to species with complex organs consisting of multiple tissue types. Recently evolved groups such as the Fucales and Laminariales (fucoid algae and kelps) exhibit the highest degree of developmental complexity, but the members of these groups are also large organisms with long life cycles that are difficult to cultivate under laboratory conditions. Experimental work on brown algae from the Fucales and Laminariales has therefore mostly involved either manipulation of *ex situ* material for short periods in the laboratory or experimentation under field conditions. Fucoid species have been used as models to study zygote polarisation and early embryogenesis. The large size of fucoid egg cells and the fact that fertilisation is external, and therefore easily observable under the microscope, have permitted very detailed characterisation of early developmental processes at the cellular level. These studies have led to several important advances, including elucidation of the physical determinants of zygote polarity, the relationship between initial cell polarisation and the first cell cycle, and the discovery of a role for the cell wall in determining cell identity (Berger et al. 1994; Bouget et al. 1998; Coelho et al. 2002; Corellou et al. 2001a, 2005; Kropf et al. 1988; Shaw and Quatrano 1996). However, our understanding of the molecular circuitry that regulates these cellular events is much less complete (Fowler et al. 2004; Corellou et al. 2001b), making it difficult to make meaningful comparisons with systems from other multicellular lineages such as green plants or animals. Until recently, progress on understanding brown algal developmental processes at the molecular level has been held back by several factors, including a lack of gene sequence information (genome data) and the absence of tools for the manipulation of gene function. The long life cycles of brown algae from the Fucales and Laminariales have also limited the scope for genetic analysis.

These limitations do not apply to all the brown algae. The filamentous brown alga *Ectocarpus* has been studied for many years and is emerging as the model of choice for the application of genomic and genetic approaches to diverse questions concerning the biology of this group of organisms (Charrier et al. 2008; Cock et al. 2011; Coelho et al. 2007, 2012a; Cock et al. 2010a). From a developmental point of view, *Ectocarpus* exhibits significantly less complexity than members of the Fucales and Laminariales, and this is true of the group Ectocarpales in general. However, recent phylogenies indicate that the Ectocarpales is a sister group to the Laminariales, the two lineages only having diverged about 100 million years ago (Silberfeld et al. 2010). It is therefore likely that the regulatory circuits that control development in members of the Ectocarpales and the Laminariales share many common features, despite the simpler bodyplans of the former.

Multicellular Development During the *Ectocarpus* Life Cycle

Like many brown algae, *Ectocarpus* has a haploid-diploid life cycle that involves alternation between two multicellular organisms, the sporophyte and the gametophyte (Fig. 1; Kornmann 1956; Müller 1964, 1967; Peters et al. 2008). Both sporophyte and gametophyte thalli consist of uniseriate, branched filaments. Gametophyte germlings are made up of a rhizoid and an upright filament, the latter consisting of cylindrical cells. The upright filament grows and branches to produce the mature thallus, which carries plurilocular gametangia in which the gametes are produced. The developmental program of the sporophyte is slightly more complex in that it produces a basal structure consisting of round and elongated cells before producing upright filaments. The upright filaments of the sporophyte resemble those of the gametophyte but are less profusely branched. The sporophyte upright filaments bear two types of reproductive structures, plurilocular sporangia containing mito-spores (which will germinate to produce clones of the parent sporophyte) and unilocular sporangia where a single meiotic event produces the meio-spores that are the initial cells of the gametophyte generation. Overall, during its life cycle *Ectocarpus* produces about eight different cell types (not including zoids; Fig. 1), significantly less than the 14 reported for kelps (reviewed in Bell and Mooers 1997).

Ectocarpus sporophytes can be derived from zygotes (i.e. formed by the fusion of two gametes) or can develop parthenogenetically from a gamete that has failed to find a partner of the opposite sex (in which case they are called partheno-sporophytes). Being derived from a single gamete, most partheno-sporophytes are haploid (Fig. 2, but see Bothwell et al. 2010). With *Ectocarpus*, therefore, it is possible to obtain both the gametophyte and sporophyte generations as haploid individuals, a feature that greatly facilitates genetic analysis of life-cycle-related developmental processes.

Genetic Analysis Using *Ectocarpus*

Ectocarpus has been used as a genetic model for several decades, earlier studies demonstrating Mendelian inheritance of a number of natural biological characters including sexuality, viral sequences that insert into the genome and membrane lipid composition (Müller 1967, 1991; Bräutigam et al. 1995; Müller and Eichenberger 1997). All stages of the life cycle can be grown in the laboratory and it takes about

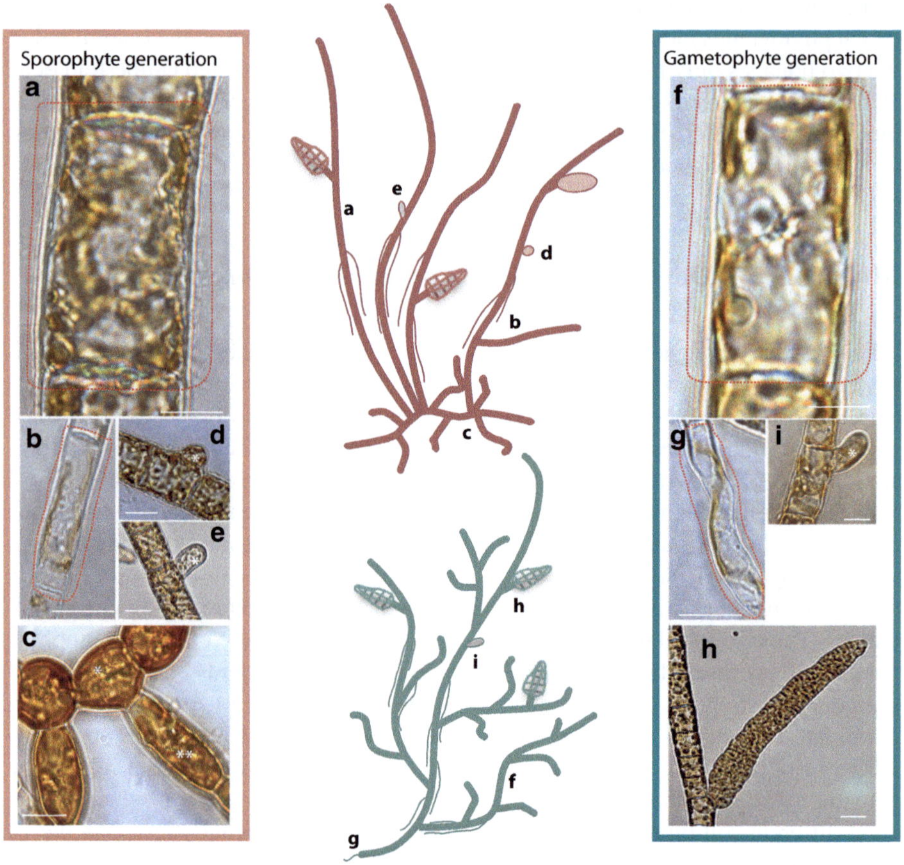

Fig. 1 Morphologically distinct cell types found in the sporophyte and gametophyte generations of the model brown alga *Ectocarpus*. The *left* and *right* panels show sporophyte and gametophyte cell types, respectively. **a** Upright filament cell (sporophyte), delineated with a *dotted red line*. **b** Rhizoid cell (sporophyte), delineated with a *dotted red line*. **c** Prostrate base cells, round *(asterisk)* and elongated *(double asterisk)*. **d** Unilocular sporangium initial cell (*asterisk*). **e** Plurilocular sporangium initial cell (*asterisk*). **f** Upright filament cell (gametophyte), delineated with a *dotted red line*. **g** Rhizoid cell (gametophyte), delineated with a *dotted red line*. **h** Mature plurilocular gametangium. **i** Plurilocular gametangium initial cell (*asterisk*). The central drawings indicate where each cell type occurs in the multicellular sporophyte (*brown*) and gametophyte (*green*) bodyplans

3 months to complete the sexual life cycle under these conditions. Individuals are raised in Petri dishes containing either natural or artificial seawater. Thalli usually become fertile when they are less than a centimetre in size. The small size of the thallus not only allows multiple individuals to be maintained in a small space, but also facilitates screens for genetic mutants, particularly when the screens are carried out during early development. Protocols have been developed for both ultraviolet

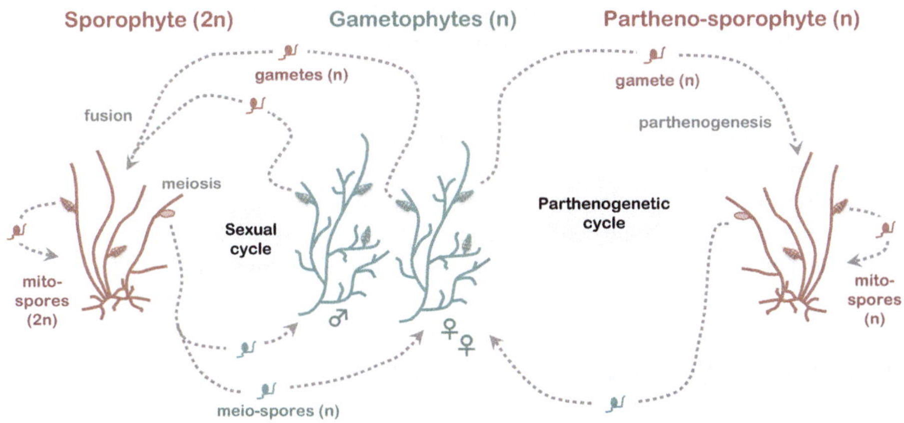

Fig. 2 Representation of the *Ectocarpus* life cycle showing the alternation between the diploid sporophyte and the haploid, dioicous gametophytes during the sexual cycle and the production of haploid partheno-sporophytes by gamete parthenogenesis (parthenogenetic cycle). Note that both the sporophyte and gametophyte generations exhibit multicellular development

(Coelho et al. 2011) and chemical (EMS and ENU) mutagenesis along with screening methodologies for developmental mutants. Classical genetic analysis of mutants is possible, with protocols having been developed both for carrying out genetic crosses and for the isolation of meiotic progeny (Coelho et al. 2012b, c). Using these approaches, large segregating populations can be generated for mapping experiments and the first mutant allele has recently been identified by positional cloning using these tools (unpublished results, see Chapter "Independent Emergence of Complex Multicellularity in the Brown and Red Algae"). Inbred lines can also be created by repeated crosses between siblings, although this is not required for many analyses because it is possible to work with haploid individuals.

There are also considerable genetic resources associated with *Ectocarpus*, including a collection of more than 2000 strains held at the Station Biologique de Roscoff (France). These strains represent worldwide diversity within the genus and include several collections of populations from single sites, providing access to information about local population structures and diversity. Strains can be stored under low light and low temperature conditions for at least 1 year before the culture needs to be refreshed. In addition, an alternative stock maintenance method based on cryopreservation has recently been developed (Heesch et al. 2012).

Ectocarpus as a Genomic Model

The small size of the *Ectocarpus* genome (214 Mbp) compared to those of most other brown algal species also represents an important advantage. The complete sequence of the genome is available and the transcribed regions have been characterised using data generated by several different approaches including Sanger expressed sequence tags (91,000), deep Illumina-base RNA-seq experiments, whole genome tiling arrays and deep Illumina sequencing of small RNAs (Cock et al. 2010b and unpublished data). An EST-based microarray has also been developed to allow near-genome-wide assessment of changes in transcript abundances (Dittami et al. 2009). A high quality genome reference sequence has been established based on these diverse transcriptomic data and on manual annotation of a large proportion of the genes in the genome. Genome data and genetic data have been combined to generate a sequence-anchored genetic map, which both provides a chromosome-scale assembly of the genome and represents an important resource for ongoing genetic analyses (Heesch et al. 2010).

Tools for the Analysis of *Ectocarpus* Gene Function

The major current bottleneck for *Ectocarpus* as a model system is the lack of tools to investigate gene function. Despite considerable investment, it has proved to be very difficult to develop a reliable genetic transformation protocol for this model organism. The recent demonstration that injection of double stranded tubulin RNA into *Fucus* zygotes led to disruption of the microtubule cytoskeleton (Farnham et al. 2013) suggests a possible alternative approach. The results of the *Fucus* experiments indicate that brown algae possess a functional RNA interference (RNAi) system and this hypothesis is supported by the presence of putative dicer and argonaute genes in the *Ectocarpus* genome (Cock et al. 2010b). Current work is aimed at adapting the *Fucus* protocol for use in *Ectocarpus*.

Another promising approach is the reverse genetic technique referred to as Targeting Induced Local Lesions in Genomes (TILLING). In this approach, a large population of individuals, each carrying hundreds of different mutations is screened for individuals carrying a mutation in a specific gene of interest by analysis of amplified polymerase chain reaction fragments (Kurowska et al. 2011). A recently completed pilot study in which a mutant population produced by UV and chemical mutagenesis was screened using a three-dimensional pooling and next-generation-sequencing-based approach indicates that this approach is feasible in *Ectocarpus* (unpublished results).

The availability of a tool that will allow gene function to be investigated experimentally, whether it be based on transformation, RNAi or TILLING, will represent an important step forward for *Ectocarpus* as a model organism. The combination of such a tool with existing resources and technologies, such as genome information, mutant screens and positional cloning, is expected to lead to significant progress in our understanding of diverse aspects of brown algal biology, including the developmental processes that underlie multicellular development in this lineage. It is also clear, however, that it will be important to complement experiment work using *Ectocarpus* with data from more developmentally complex brown algae. This will be required both to test the generality of information obtained using *Ectocarpus* and to investigate how developmental processes operate in more morphologically complex brown algae.

Additional Brown Algal Models

The development of an RNAi protocol for *Fucus* (Farnham et al. 2013) is likely to stimulate renewed interest in fucoids for brown algal research. The lack of a complete genome sequence for this genus remains a limiting factor, but an increasing amount of transcriptomic data is being generated (Coyer et al. 2009; Pearson et al. 2010) and this will be an important resource for future work. It is likely that *Fucus*-based research will continue to focus on early developmental events because of the problems associated with extended culture in the laboratory, but this should nonetheless provide access to many interesting questions relevant to the evolution of multicellularity.

As the most developmentally complex group among the brown algae, kelps are of particular interest for questions related to the emergence of complex multicellularity. Potential model species from within this group include *Undaria, Laminaria, Saccharina* and *Macrocystis* (Cock et al. 2012). *Saccharina* and *Macrocystis* are of particular interest because they are accessible to genetic analysis, whilst at the same time having a high potential for biotechnological and aquaculture applications (Li et al. 2007; Westermeier et al. 2010, 2011; Gutierrez et al. 2006). The *Saccharina* genome is estimated to be around 600 Mbp, whereas that of *Macrocystis* is larger (about 1 Gbp), comparable in size to those of fucoid species (Kapraun 2005; Phillips et al. 2011; Peters et al. 2004).

If an effective kelp model can be developed, it will be particularly interesting to compare processes related to multicellularity in this group with information obtained for *Ectocarpus*, both at the level of genome evolution and, where possible, from a mechanistic point of view. Analysis of the genome sequence of *Ectocarpus* has identified a number of features that may be related to the transition to multicellularity (see Chapter "Independent Emergence of Complex Multicellularity in the Brown and Red Algae"). It will be interesting to determine whether these features are also observed in a kelp genome, with perhaps more marked trends such as expansions of key gene families. In the longer term, it will also be of great interest to investigate how the developmental genes that are currently being identified in *Ectocarpus* function in the more morphologically complex kelps.

Summary

1. *Ectocarpus* has emerged as a model system for the brown algae, allowing the application of genetic and genomic approaches to this group of organisms.
2. The life cycle of *Ectocarpus* involves two multicellular stages, the sporophyte and gametophyte generations, with a total of about six different cell types.
3. Genomic and genetic resources that have been developed for *Ectocarpus* include a complete genome sequence, extensive transcriptomic data, protocols to produce and screen mutant populations, a genetic map, and a large collection of strains.
4. Methods being developed to analyse gene function in *Ectocarpus* include RNA interference and Targeting Induced Local Lesions in Genomes (TILLING).

5. In the future, extension of work carried out in *Ectocarpus* to more developmentally complex brown algal models, such as kelp species, should provide further insights into the emergence of complex multicellularity in this group.

References

Bell G, Mooers AO (1997) Size and complexity among multicellular organisms. Biol J Linn Soc 60:345–363

Berger F, Taylor A, Brownlee C (1994) Cell fate determination by the cell wall in early *Fucus* development. Science 263(5152):1421–1423

Bothwell JH, Marie D, Peters AF, Cock JM, Coelho SM (2010) Role of endoreduplication and apomeiosis during parthenogenetic reproduction in the model brown alga *Ectocarpus*. New Phytol 188(1):111–121

Bouget FY, Berger F, Brownlee C (1998) Position dependent control of cell fate in the *Fucus* embryo: role of intercellular communication. Development 125(11):1999–2008

Bräutigam M, Klein M, Knippers R, Müller DG (1995) Inheritance and meiotic elimination of a virus genome in the host *Ectocarpus siliculosus* (phaeophyceae). J Phycol 31(5):823–827

Charrier B, Coelho S, Le Bail A, Tonon T, Michel G, Potin P, Kloareg B, Boyen C, Peters A, Cock J (2008) Development and physiology of the brown alga *Ectocarpus siliculosus*: two centuries of research. New Phytol 177(2):319–332

Cock JM, Coelho SM, Brownlee C, Taylor AR (2010a) The *Ectocarpus* genome sequence: insights into brown algal biology and the evolutionary diversity of the eukaryotes. New Phytol 188(1):1–4

Cock JM, Sterck L, Rouzé P, Scornet D, Allen AE, Amoutzias G, Anthouard V, Artiguenave F, Aury J, Badger J, Beszteri B, Billiau K, Bonnet E, Bothwell J, Bowler C, Boyen C, Brownlee C, Carrano C, Charrier B, Cho G, Coelho S, Collén J, Corre E, Da Silva C, Delage L, Delaroque N, Dittami S, Doulbeau S, Elias M, Farnham G, Gachon C, Gschloessl B, Heesch S, Jabbari K, Jubin C, Kawai H, Kimura K, Kloareg B, Küpper F, Lang D, Le Bail A, Leblanc C, Lerouge P, Lohr M, Lopez P, Martens C, Maumus F, Michel G, Miranda-Saavedra D, Morales J, Moreau H, Motomura T, Nagasato C, Napoli C, Nelson D, Nyvall-Collén P, Peters A, Pommier C, Potin P, Poulain J, Quesneville H, Read B, Rensing S, Ritter A, Rousvoal S, Samanta M, Samson G, Schroeder D, Ségurens B, Strittmatter M, Tonon T, Tregear J, Valentin K, von Dassow P, Yamagishi T, Van de Peer Y, Wincker P (2010b) The *Ectocarpus* genome and the independent evolution of multicellularity in brown algae. Nature 465(7298):617–621

Cock JM, Peters AF, Coelho SM (2011) Brown algae. Curr Biol 21(15):R573–575

Cock JM, Arun A, Godfroy O, Macaisne N, Strittmatter M, Peters AF, Coelho SM (2012) Genomics of brown algae: current advances and future prospects. Genes Genomics 34:1–5

Coelho S, Taylor A, Ryan K, Sousa-Pinto I, Brown M, Brownlee C (2002) Spatiotemporal patterning of reactive oxygen production and Ca(2+) wave propagation in *Fucus* rhizoid cells. Plant Cell 14(10):2369–2381

Coelho S, Peters A, Charrier B, Roze D, Destombe C, Valero M, Cock J (2007) Complex life cycles of multicellular eukaryotes: new approaches based on the use of model organisms. Gene 406(1–2):152–170

Coelho SM, Godfroy O, Arun A, Le Corguillé G, Peters AF, Cock JM (2011) OUROBOROS is a master regulator of the gametophyte to sporophyte life cycle transition in the brown alga *Ectocarpus*. Proc Natl Acad Sci U S A 108:11518–11523

Coelho SM, Scornet D, Rousvoal S, Peters N, Dartevelle L, Peters AF, Cock JM (2012a) *Ectocarpus*: a model organism for the brown algae. Cold Spring Harb Protoc 2012:193–198

Coelho SM, Scornet D, Rousvoal S, Peters N, Dartevelle L, Peters AF, Cock JM (2012b) Genetic crosses between *Ectocarpus* strains. Cold Spring Harb Protoc 2012(2):262–265

Coelho SM, Scornet D, Rousvoal S, Peters NT, Dartevelle L, Peters AF, Cock JM (2012c) How to cultivate *Ectocarpus*. Cold Spring Harb Protoc 2012(2):258–261

Corellou F, Brownlee C, Detivaud L, Kloareg B, Bouget FY (2001a) Cell cycle in the *Fucus* zygote parallels a somatic cell cycle but displays a unique translational regulation of cyclin-dependent kinases. Plant Cell 13(3):585–598

Corellou F, Brownlee C, Kloareg B, Bouget FY (2001b) Cell cycle-dependent control of polarised development by a cyclin-dependent kinase-like protein in the *Fucus* zygote. Development 128(21):4383–4392

Corellou F, Coelho SM, Bouget FY, Brownlee C (2005) Spatial re-organisation of cortical microtubules in vivo during polarisation and asymmetric division of *Fucus* zygotes. J Cell Sci 118(Pt 12):2723–2734

Coyer JA, Hoarau G, Beszteri B, Pearson G, Olsen JL (2009) Expressed sequence tag-derived polymorphic SSR markers for *Fucus serratus* and amplification in other species of *Fucus*. Mol Ecol Resour 9(1):168–170

Dittami S, Scornet D, Petit J, Ségurens B, Da Silva C, Corre E, Dondrup M, Glatting K, König R, Sterck L, Rouzé P, Van de Peer Y, Cock J, Boyen C, Tonon T (2009) Global expression analysis of the brown alga *Ectocarpus siliculosus* (Phaeophyceae) reveals large-scale reprogramming of the transcriptome in response to abiotic stress. Genome Biol 10(6):R66

Farnham G, Strittmatter M, Coelho SM, Cock JM, Brownlee C (2013) Gene silencing in *Fucus* embryos: developmental consequences of RNAi-mediated cytoskeletal disruption. J Phycol 49:819–829

Fowler JE, Vejlupkova Z, Goodner BW, Lu G, Quatrano RS (2004) Localization to the rhizoid tip implicates a *Fucus distichus* Rho family GTPase in a conserved cell polarity pathway. Planta 219(5):856–866

Gutierrez A, Correa T, Muñoz, Santibañez A, Marcos R, Caceres C, Buschmann AH (2006) Farming of the giant kelp *Macrocystis pyrifera* in southern Chile for development of novel food products. J Appl Phycol 18:259–267

Heesch S, Cho GY, Peters AF, Le Corguillé G, Falentin C, Boutet G, Coëdel S, Jubin C, Samson G, Corre E, Coelho SM, Cock JM (2010) A sequence-tagged genetic map for the brown alga *Ectocarpus siliculosus* provides large-scale assembly of the genome sequence. New Phytol 188(1):42–51

Heesch H, Day JG, Yamagishi T, Kawai H, Müller DG, Küpper FC (2012) Cryopreservation of the model alga *Ectocarpus* (Phaeophyceae). Cryo Lett 33(5):327–336

Kapraun DF (2005) Nuclear DNA content estimates in multicellular green, red and brown algae: phylogenetic considerations. Ann Bot 95:7–44

Kornmann P (1956) Über die Entwicklung einer *Ectocarpus confervoides*-Form. Pubbl Della Stn Zoologica di Napoli 28:32–43

Kropf DL, Kloareg B, Quatrano RS (1988) Cell wall is required for fixation of the embryonic axis in *Fucus* zygotes. Science 239(4836):187–190

Kurowska M, Daszkowska-Golec A, Gruszka D, Marzec M, Szurman M, Szarejko I, Maluszynski M (2011) TILLING: a shortcut in functional genomics. J Appl Genet 52(4):371–390

Li X, Cong Y, Yang G, Shi Y, Qu S, Li Z, Wang G, Zhang Z, Luo S, Dai H, Xie J, Jiang G, Liu J, Wang T (2007) Trait evaluation and trial cultivation of Dongfang No. 2, the hybrid of a male gametophyte clone of *Laminaria longissima* (Laminariales, Phaeophyta) and a female one of *L. japonica*. J Appl Phycol 19:139–151

Müller DG (1964) Life-cycle of *Ectocarpus siliculosus* from Naples, Italy. Nature 26:1402

Müller DG (1967) Generationswechsel, Kernphasenwechsel und Sexualität der Braunalge *Ectocarpus siliculosus* im Kulturversuch. Planta 75:39–54

Müller DG (1991) Mendelian segregation of a virus genome during host meiosis in the marine brown alga *Ectocarpus siliculosus*. J Plant Physiol 137:739–743

Müller DG, Eichenberger W (1997) Mendelian genetics in brown algae: inheritance of a lipid defect mutation and sex alleles in *Ectocarpus siliculosus* (Ectocarpales, Phaeophyceae). Phycologia 36(1):79–81

Pearson GA, Hoarau G, Lago-Leston A, Coyer JA, Kube M, Reinhardt R, Henckel K, Serrão ET, Corre E, Olsen JL (2010) An expressed sequence tag analysis of the intertidal brown seaweeds *Fucus serratus* (L.) and *F. vesiculosus* (L.) (Heterokontophyta, Phaeophyceae) in response to abiotic stressors. Mar Biotechnol (NY) 12 (2):195–213

Peters AF, Marie D, Scornet D, Kloareg B, Cock JM (2004) Proposal of *Ectocarpus siliculosus* (Ectocarpales, Phaeophyceae) as a model organism for brown algal genetics and genomics. J Phycol 40(6):1079–1088

Peters AF, Scornet D, Ratin M, Charrier B, Monnier A, Merrien Y, Corre E, Coelho SM, Cock JM (2008) Life-cycle-generation-specific developmental processes are modified in the *immediate upright* mutant of the brown alga *Ectocarpus siliculosus*. Development 135(8):1503–1512

Phillips N, Kapraun DF, Gómez Garreta A, Ribera Siguan MA, Rull JL, Salvador Soler N, Lewis R, Kawai H (2011) Estimates of nuclear DNA content in 98 species of brown algae (Phaeophyta). AoB Plants 2011:plr001

Shaw SL, Quatrano RS (1996) The role of targeted secretion in the establishment of cell polarity and the orientation of the division plane in *Fucus* zygotes. Development 122(9):2623–2630

Silberfeld T, Leigh JW, Verbruggen H, Cruaud C, de Reviers B, Rousseau F (2010) A multi-locus time-calibrated phylogeny of the brown algae (Heterokonta, Ochrophyta, Phaeophyceae): investigating the evolutionary nature of the "brown algal crown radiation". Mol Phylogenet Evol 56(2):659–674

Westermeier R, Patiño DJ, Müller H, Müller DG (2010) Towards domestication of giant kelp (*Macrocystis pyrifera)* in Chile: selection of haploid parent genotypes, outbreeding, and heterosis. J Appl Phycol 22:357–361

Westermeier R, Patiño DJ, Murúa P, Müller DG (2011) *Macrocystis* mariculture in Chile: growth performance of heterosis genotype constructs under field conditions. J Appl Phycol 23:819–825

Part III
Theoretical Approaches

Evolutionary Transitions in Individuality and Recent Models of Multicellularity

Erik R. Hanschen, Deborah E. Shelton and Richard E. Michod

Abstract An evolutionary transition in individuality (ETI) is a fundamental shift in the unit of adaptation. ETIs occur through the evolution of groups of individuals into a new higher-level individual. The evolution of groups with cells specialized in somatic (viability) or reproductive functions has been proposed as a landmark of the unicellular to multicellular ETI. Several recent models of the evolution of multicellularity and cellular specialization have contributed insights on different aspects of this topic; however, these works are disconnected from each other and from the general framework of ETIs. While each of these works is valuable on its own, our interest in ETIs motivates an attempt to connect these models. We review the theory of ETIs along with these recent models with an eye towards better integrating insights from these models into the ETI framework. We consider how each model addresses key recurring topics, such as the importance of cooperation and conflict, life history trade-offs, multi-level selection, division of labor and the decoupling of fitness at the level of the group from the level of the cell. Finally, we identify a few areas in which conflicting views or unanswered questions remain, and we discuss modeling strategies that would be most suited for making further progress in understanding ETIs.

Keywords Cellular specialization · Division of labor · Evolution · Evolutionary transition in individuality (ETI) · Major transition · Modeling · Multicellularity

---

E. R. Hanschen (✉) · D. E. Shelton · R. E. Michod
Department of Ecology and Evolutionary Biology, University of Arizona,
1041 E. Lowell St., BioSciences West, Rm. 310, Tucson, AZ 85721, USA
e-mail: hanschen@email.arizona.edu

D. E. Shelton
e-mail: deborah.shelton@gmail.com

R. E. Michod
e-mail: michod@email.arizona.edu

I. Ruiz-Trillo, A. M. Nedelcu (eds.), *Evolutionary Transitions to Multicellular Life*,
Advances in Marine Genomics 2, DOI 10.1007/978-94-017-9642-2_9

Introduction

Why do groups of individuals evolve into new kinds of individuals? Answering this question is basic to understanding the origin of the hierarchy of life: genes, chromosomes, cells, cells within cells (eukaryotic cell), multicellular organisms, and societies. We refer to transitions among levels in the hierarchy of life as evolutionary transitions in individuality or ETIs. Building on previous work (Buss 1987; Maynard Smith 1988, 1991; Maynard Smith and Szathmáry 1995; Otto and Orive 1995), Michod and colleagues developed a framework for understanding ETIs, with special focus on the transition from unicellular to multicellular life. This framework involves five interrelated components: the evolution of cooperation (and conflict) in groups, multi-level selection, life history trade-offs, division of labor among the basic components of fitness (reproduction and viability), and, finally, decoupling of fitness of the two levels in the selection hierarchy (after which the fitness of the group is no longer a simple additive function of the fitnesses of the lower level units; Michod 1996, 1997, 1999, 2003, 2006, 2007; Michod and Roze 1999, 2001; Michod and Nedelcu 2003; Michod et al. 2006).

Recently, a number of interesting models have addressed the evolution of multicellularity and division of labor from several different perspectives. While each of these works is valuable on its own, our interest in ETIs motivates an attempt to integrate them, so we consider these models from the perspective of the ETI framework. Of course other useful and valid frameworks for the conceptual issues discussed exist, but for the purposes of this paper, we have restricted ourselves to the ETI framework as discussed below. As each of these models was developed for their own quite valid and compelling reasons, discussing how these models relate to the ETI framework is not meant as criticism. Rather, we wish to analyze these models to help us better understand the main issues relevant to ETIs. In this section we briefly review the ETI framework and the models upon which it is based. Then, in the next sections we discuss some of the more recent models and how they relate to the ETI framework as well as provide general suggestions for future work.

Overview of the ETI Framework

Because of cooperation, groups may function in ways that their members cannot. For example, a cell may not be able to swim and divide at the same time, but a group containing cells specialized at either swimming or cell division may undertake both functions simultaneously. In this way, groups of cooperating cells may break through the life history constraints governing life as a single cell. When cells start forming groups, selection operates at both the level of the cell and the level of the group and may lead to the evolution of cooperation among cells in the group. Cooperation in turn provides the opportunity for cheating and conflict among lower-level units, which must be mediated if cooperation is to be stable. Under certain conditions, conflict mediators evolve that lead to specialized reproductive and non-reproductive somatic cells. Division of labor among lower level units specialized in the basic fitness components of the group enhances the individuality of the group. Once cells are specialized in one of the necessary components of fitness, say reproduction or viability, they can no longer exist outside of the group and the fitness of the group is no longer the average fitness of the cells belonging to the group.

In summary, a working hypothesis for the basic steps in an ETI is as follows: (i) formation of groups, (ii) increase of cooperation within groups, (iii) cheating and conflict, (iv) conflict mediation leading to enhanced cooperation, (v) division of labor in the basic components of fitness leading to (vi) fitness decoupling and individuality of the group. The extent to which these steps apply in diverse empirical cases and the implications of empirical results are active areas of research.

The evolution of cooperation, the central problem of social biology, gains special significance during ETIs because altruism and other forms of cooperation lead to the transfer of fitness from the lower level (the costs of altruism) to the group level (the benefits of altruism). By "transfer of fitness" we do not mean to suggest that fitness is a conserved quantity; rather, we mean that a single change (e.g., the increase in cooperation) both decreases the fitness of the lower level and increases the fitness of the higher level. For example, in the additive model of altruism with costs to self $-c$ and benefit to group b, c and b are not equal nor are they typically related in magnitude (although in specific cases there may be some connection between the two).

Related to cooperation and altruism is specialization of group members in the two fitness components, reproduction and viability, of the group (Michod 2005, 2006). As already mentioned, when cells completely specialize at one of the two basic fitness components, they lose their overall fitness and capacity to function as evolutionary individuals if they existed outside the context of the group. By virtue of their specialization, cells have low cell-level fitness if they existed as single-celled individuals, while they contribute to increasing the fitness of the group. As a consequence, the fitness of the group is no longer the average of the individual fitnesses of the component cells. This is an example of a general principle promoting group cohesion: cell traits optimal in a group context may no longer be optimal outside of the group context; indeed group-selected traits may be deleterious at the cell level (Shelton and Michod 2014).

Energy, resources, and time expended on one fitness component often detract from another component, resulting in trade-offs among fitness components. Fitness trade-offs drive the diversification of life history traits in extant species (Stearns 1992; Roff 2002) but gain special significance during ETIs for three reasons (Michod 2006, 2007; Michod et al. 2006). First, specialization by cells in the fitness components of the group can be driven by cell-level fitness trade-offs. The curvature of the trade-off between survival and reproduction is known to be a central issue in life history evolution (Stearns 1992; Roff 2002). In the case of the origin of multicellularity, if the lower-level fitness trade-off is of convex curvature (positive second derivative) and assumptions are made about how lower-level fitness components relate to higher-level fitness components (see below), specialization by cells in the reproductive and viability fitness components (termed "germ-soma specialization" below) of the group will be an optimal group-level strategy (Michod et al. 2006).

Second, tradeoffs between fitness components at the cell level provide the basis for cooperation and division of labor in groups. Life history "trade-off genes" are genes that down regulate reproduction so as to enhance survival of the organism in stressful environments. Such genes can be co-opted to produce reproductive altruism in cell groups (somatic cells) through the shifting of their expression from a temporal (environmentally-induced) context into a spatial (developmental) context (Nedelcu and Michod 2006). For example, an important component of viability in the volvocine green algae is flagellar motility, but cell division and reproduction interfere with flagellar motility. In the unicellular members of this lineage, selection will optimize the allocation of time and energy to these two processes, but, in a group, cells that spend more time flagellated divide less frequently. As flagellar action of cells benefits group motility, cells with a greater propensity to remain actively flagellated are altruistic relative to cells that spend less time flagellated (because less flagellated cells reproduce more).

Third, as discussed below in Eq. 1, fitness tradeoffs can enhance the fitness of the group through a covariance effect by which group fitness is augmented beyond the average fitness of cells according to the covariance of cellular contributions to viability and fecundity (Michod 2006).

Models Related to the ETI Framework

The ETI framework described above is based on results from two different kinds of models: two-locus population genetic modifier models (Michod 1996, 1997, 1999, 2003, 2006; Michod and Roze 1997, 1999, 2001; Michod and Nedelcu 2003; Michod et al. 2003, 2006) and optimality models of division of labor (Michod 2006; Michod et al. 2006).

Two Locus Modifier Models

The two locus modifier models seek to explain how modifiers of development evolve in response to mutation and selection at a cooperation locus. Development involves the conversion of a propagule into an adult cell group via cell division. Propagules contain a cell or cells sampled from an adult group in the previous generation (or from several adult groups in the case of aggregation). Sex may occur in the case of single celled propagules that fuse with propagules from other groups to start a new group.

The cooperation locus has two alleles, C and D, which express cooperation and defection, respectively, among cells. Cooperation benefits other cells in the group at a cost to the cooperating cell. Defecting cells do not pay any cost, but receive the benefits of cooperating cells in their group. Benefit and cost are in terms of the cell death rate or cell division rate. During development of the adult group, there is recurrent mutation from C to D at each cell division (back mutation is ignored on the view that there are many more ways to lose a functional trait like cooperation than to gain it again). These defector mutations disrupt the functioning of the adult cell group by reducing the level of cooperation. Mutation increases the variance and opportunity for selection at the within-group or cell level. After the adult group is formed, a propagule is made. Depending on the parameters of development (mutation rate, cell replication and death rates, total number of cell divisions, the costs and benefits of cooperation, and the mode of propagule formation), a polymorphism may be maintained at the C/D locus by mutation selection balance. This polymorphism sets the stage for the evolution of modifiers of development assumed to be encoded by a second locus.

The second modifier locus affects the parameters or mode of development of the adult group and so affects the degree to which the propagule produced by an adult resembles the propagule that founded the group (a measure of group heritability). By molding development, modifier alleles affect group heritability and the capacity of the groups to reproduce themselves (Griesemer 2000). Why do these modifiers evolve and how do they lead to the capacity of a group to reproduce itself?

The mutation-selection balance equilibrium at the C/D locus implies that C alleles are fitter than D alleles, to compensate for mutation from C to D. Under certain conditions, alleles at the modifier locus evolve due to interaction with the fitter C allele. This has the effect of increasing the between-group variance and decreasing the within-group variance, thereby increasing the level of cooperation and the fitness of the group. Examples of conflict modifiers studied by this approach include germ-soma specialization, reduced mutation rate, policing, programmed cell death, passing the life cycle through a single-cell zygote stage, and fixed group size (reviewed in Michod 2003). By increasing the variance at the group level and decreasing the variance at the cell level, the modifiers lead to fitness decoupling between levels and enhance the capacity of group to reproduce itself. The capacity of a group to reproduce itself may be measured by the degree to which a group created by a propagule resembles the group the propagule came from. Alternatively, since the group is made from a propagule, and the two locus recurrence equations are in terms of the gene and genotype frequencies at the propagule stage, we may measure the capacity for reproduction and heritability as the degree to which the propagules produced by a group are similar to the propagule(s) that created the group.

Optimality Division of Labor Model

The evolution of specialization at the two basic components of fitness, reproduction and viability, was also studied using an optimality approach (Michod 2006; Michod et al. 2006). In these models, the phenotype of a cell is described by its effort at reproduction (fecundity) with the remainder of effort put into the viability. In the simplest case, the viability, V, and fecundity, B, of the group are assumed to be arithmetic averages of the cell efforts at the two fitness components, viability and fecundity, v and

b, respectively, or $V = \sum_i v_i/N$ and $B = \sum_i b_i/N$. This formulation assumes an initial isomorphism between fitness components at the two levels, because it assumes that the activities of the cell at cellular viability and cellular fecundity contribute, respectively, to group viability and group fecundity. Clearly, as cells become more specialized and integrated into the group, this isomorphism breaks down; but this assumption likely applies as cells first start joining (and leaving) groups. This formulation of group fitness also assumes that the two fitness components of the group, viability and fecundity, are first composed separately from cell properties, and then combined (multiplicatively as is appropriate for discrete generations) to generate the fitness of the group, W. Without this (or a related assumption), evolution of specialization at activities which trade-off with one another at the lower level would not be possible. Group fitness would be zero, if group fitness was composed directly out of cell fitness, and cells completely specialized in one fitness component or the other. The cell fitness (vb) of specialized cells is zero since one component (v or b) is zero. In this way, the group may break through the trade-off constraints imposed at the cell level. Finally, it is assumed that the total fitness at either level is the product of viability and fecundity, as is appropriate for cells or multicellular organisms with discrete generations.

One result of the model is the group covariance effect given in Eq. 1, which shows that the fitness of the cell group, W (taken as the product of V and B), is greater than the average fitness of member cells, $\bar{w} = \sum_i v_i b_i/N$, by an amount equal to the negative covariance of the fitness components at the cell level (viability, v, and fecundity, b).

$$W = VB = \bar{w} - Cov[v, b] \tag{1}$$

If the covariance between fitness components is itself negative, as it is when fitness components trade-off with one another, there is an enhanced fitness at the group level from what would be expected from the average fitness of cells. The covariance effect given in Eq. 1 translates the negative covariance of fitness components of group members into a benefit at the group level. Alternatively, if fitness components were to positively co-vary, then there would be a decrease in fitness of the group from that expected by the average cell fitness. High fitness for any unit at any level of organization requires a balance of fitness components at that level; in the explicit formulation here of discrete generations the components are multiplied together to give total fitness. The covariance effect translates a lack of balance at the lower cell level into an advantage at the group level, especially under conditions of convexity of the trade-offs. Convexity of the trade-off curvature allows for enhanced effectiveness at each fitness component for cells that specialize (Michod 2006; Michod et al. 2006).

The particular mathematical representation of the covariance effect given in Eq. 1 depends upon additivity of effects on the viability and reproduction components of fitness as described above. Additivity of fitness effects is the simplest assumption possible, nevertheless, the assumption of additivity of the contributions of cells to the viability of the group may be relaxed and the general points still hold (Michod et al. 2006).

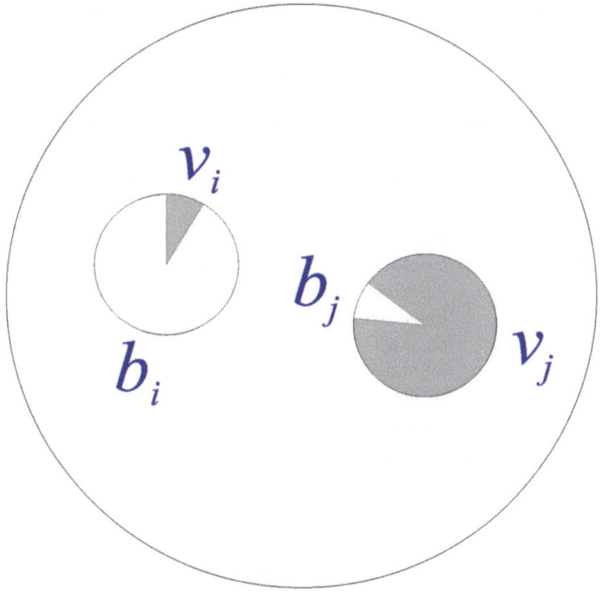

Fig. 1 Two cells specializing in different fitness components, reproduction (*white*) and viability (*grey*). Cell i specializes in reproduction, with reproductive effort b_i, with less effort put into viability functions, v_i. Cell j does the reverse. Alone they would each have low fitness, because they are unbalanced and high fitness requires a balance at the two components. However, together, they may constitute a good team and bring high fitness to the group. (From Michod (2007))

As illustrated in Fig. 1, what is required for the covariance effect is that, if one cell has a high reproductive effort (and hence a low viability, and a low cell fitness), this may be compensated for by another cell with high viability (and hence a low fecundity, and also a low cell fitness) (Michod et al. 2006). Consequently, even though each of these cells by themselves would have a low fitness, together they can bring a high fitness to the group, especially under conditions of convexity of the trade-off. In effect, such oppositely specialized cells complement one another and constitute a good and integrated "team" under conditions of convexity of the trade-off curve. This kind of joint effect contributes to fitness decoupling and integration of the group, and would not be possible if group fitness were the average of cell fitness.

During an ETI, fitness must be reorganized so that the individuals (e.g., cells) that were the focus of evolution and adaptation become components of a new higher-level individual (e.g., multicellular group) that is the new focus of evolution and adaptation. As already discussed, this involves two major processes. First, fitness and its variance is increased at the new group level and reduced at the lower level. Second, the previously-existing individuals, which are now members of the group, specialize in the fitness components (reproduction and viability) of the higher level group. This leads to the decoupling of fitness between the levels so that the fitness of the group may be quite high, while the fitness of the group members may be low or even non-existent were they to leave the group. The reorganization of fitness during ETIs may involve a number of cycles of cooperation, conflict and conflict mediation. Conflict through its mediation allows for more cooperation. This cycle of cooperation, conflict and conflict mediation drives ETIs.

Table 1 Summary of how each model incorporates five interrelated components of ETIs

	Cooperation in groups	Multilevel selection	Life history trade-offs	Division of labor	Fitness decoupling
Willensdorfer 2009	Implicit (see discussion in text)	None	Colony level (benefit and cost of soma)	Cells specialize on viability or fertility	None
Gavrilets 2010	Implicit Conflict between levels arises developmentally from mutation	Viability depends on colony trait, fertility depends on cell trait	Cell level (fertility and viability)	Cells specialize on viability or fertility	Present once division of labor evolves
Ispolatov et al 2012	Synergistic cooperation	Cell death rate depends on cell trait, cell birth rate depends on colony trait	Cell level (metabolic process A and B)	Cells specialize on a single metabolic process	None
Rueffler et al 2012	None	None	Colony level (between two tasks)	Cells specialize on a single task	None
Van dyken and Wade 2012a, b	Altruism	Implicit	Organism level (between tasks)	Organisms specialize on viability or fertility	Implicit (see discussion in text)

Recent Models

In addition to the models described above, which were built around the concepts associated with the ETI framework, several recent models have addressed various aspects associated with the evolution of multicellularity and cellular specialization. In Table 1 below we briefly relate the recent models considered here to the ETI framework presented in the previous section.

Willensdorfer 2009

Willensdorfer (2009) takes a colony-level optimality approach, defining colony fitness as the product of three functions. The first function represents the cost of soma, the second the cost of colony size, and the third the benefit of group living. Although not explicitly stated in the paper, by assuming costly somatic cells and germ cells that contribute less to group benefit function than do somatic cells, the model is based on cell-level trade-offs among fitness components. Willensdorfer is interested in when the benefit of soma to biomass production outweighs the cost of soma. Based on this condition, he predicts when somatic cells will evolve and what the proportion of somatic cells will be. Similar to Solari et al. (2006) and the optimality framework above, this model investigates the optimal proportion of soma given certain benefits and costs. Because this is a single-level model, Willensdorfer (2009) models the evolution of division of labor after an evolutionary transition to group living has occurred.

We see similarities and differences between Willensdorfer's model and the ETI framework (Table 1). We have already mentioned that fitness trade-offs are implicitly assumed and drive the division of labor in his model. In addition, cooperation between soma and germ cells is present. Multi-level selection is absent from this model because cell-level variation and cell-level selection have been explicitly removed. It is unclear how Willensdorfer's biomass fitness definition relates to more traditional fitness definitions and life history issues (r, R_0, invasion fitness, etc.). To try to understand fitness in this model from a life history perspective (say with fitness equal to the product of reproduction and viability), one could view the numerator of the first term (biomass of reproductive cells) of Willensdorfer's fitness definition as reproduction and the rest of the equation (allometric cost of size $\times$ benefit of multicellularity/total biomass) as viability (Eq. 2, Willensdorfer 2009). However this results in a viability term that is very difficult to interpret biologically. Novel definitions of fitness are interesting, but make it difficult to connect to other models or to the biological cases we wish to understand. In terms of the steps in an ETI, Willensdorfer is investigating how multicellular organisms evolve in the later stages after obligate coloniality has evolved.

Gavrilets 2010

Gavrilets (2010) decomposes fitness into reproductive and survival fitness components and defines a trade-off between them. He assumes four genes, two functional genes for each of the two fitness components and two regulatory loci, one for each fitness component. Investment in a functional locus increases cell fertility or cell viability, but negatively influences the other fitness component. The two regulatory loci determine to what extent cells will contribute to viability and reproduction. Thus, in order to evolve cellular differentiation in this model, a colony must increase investment in viability and fertility (via the functional loci) as well as regulation (via the regulatory loci) so every cell will specialize in either viability or fertility. Such cell specialization increases colony-level fitness by avoiding the cell level trade-off. Gavrilets studies how the shape of the trade-off affects the evolution of division of labor at reproductive and viability functions.

The life cycle consists of development, survival selection, and reproduction selection. During development, a single cell divides to found a colony and subsequent cell divisions result in a colony of a specified size. Mutation at all four loci occurs during cell division, which produces intracolony and intercolony variation. A set proportion of cells always develop as potential soma, regardless of whether regulation actually functionally distinguishes them from germ cells. During survival selection, a colony survives based on the average of its cell investment in survival. Lastly, during reproduction selection, cells reproduce with probability proportional to their investment in fertility. In this way, conflict between cell level fitness (which favors cell fecundity) and the colony level fitness (which favors a balance between cell fecundity and cell investment in colony viability) arises. Individual-based simulations are used to investigate how the shape of the tradeoff between viability and fecundity alters the evolution of division of labor and Gavrilets concludes that convex trade-offs are necessary for division of labor to evolve.

The model shares many of the properties of the ETI framework above: life history trade-offs, multi-level selection, cooperation, and reproductive division of labor (see Table 1). Life history trade-offs are manifested as functional trade-offs at the cell level (survival and reproduction). When this trade-off is convex, cooperation is observed as germ and somatic cells evolve. Multi-level selection is present as selection operates at the colony level through survival selection and at the cell level through fecundity selection on variation introduced through developmental mutation. Fitness decoupling is also present as the fitness of cells decreases (if they were outside the colony) as colony-level fitness increases. Lastly, division of labor in reproduction and survival components is clearly observed and basic to the transition to a multicellular individual.

Interestingly, some portions of Gavrilets' discussion suggest that we would be mistaken to see multilevel selection and cooperation/conflict as major elements of his model. For example, Gavrilets says (2010, p. 6), "In the model, cell differentiation and the division of labor are driven by individual selection maximizing the number of colony-producing offspring of a colony-producing cell. That is, the transition to individuality can be explained in terms of immediate selective advantage to individual replicators (Maynard Smith and Szathmáry 1998). Note that mutant cells that 'cheat' by having increased fertility within colonies will tend to lose in competition at the colony level after they develop their own colonies. Therefore, the conflict between individual and colony level selection is largely removed."

We understand the word "individual" in the above quote to be synonymous with "cell" or "cell-level" (except when he says "individuality," here he is talking about a multicellular individual). Given that interpretation, Gavrilets is making the claim that cell-level selection "drives" division of labor in this model. Apparently in support of this claim, Gavrilets points out that the success of "cheater" (high fertility) cells depends on how they are assorted into groups. Recall that this model assumes an extreme bottleneck: every colony originates from a single cell and eventually disintegrates into single cells. Recall also in this model that (i) a colony survives based on the average of its cells' investments in survival, (ii) cells reproduce proportionally to their cell-level investment in fertility, and (iii) there is a cell level trade-off between the two investments. Following Gavrilets' quote, consider a non-cheating cell that gives rise to a colony and a cheater that arises by mutation during development of that colony. Call the cells that descend from the original cell the F1 cells. We would expect the cheater genotype to be over-represented among F1 cells because a cheater does well at reproduction at the expense of its non-cheating colony-mates. Call the cells that descend from an F1 cheater cell and actually survive to begin the process of creating colonies of their own the $F2_c$ cells and cells which descend from a F1 non-cheater cell the $F2_{nc}$ cells. Our expectation of which genotype is doing better in the overall population ($F2_c$ and $F2_{nc}$) is not obvious—it depends on the fertility advantage at the cell level and the survival disadvantage at the group level. Because the F1 cheater genotypes produce poor-surviving colonies, their ability to make it into the F2 generation is compromised as are the colonies that result from them if they make it into the F2 generation. We agree with Gavrilets that the sorting effect of the single-cell bottleneck is an important aspect of his model. However, this point does not support the idea that cell-level selection "drives" the evolution of division of labor in his model. In order to see what traits cell-level selection favors, the relevant question is not the relative representation of cheater cells in the overall cell population in F2. That would be informative about the net effect of cell-level and group-level selection combined. Rather, the effect of individual selection should be assessed by asking what the effect of a trait change is within each group (or what its effect would be if the groups were dissolved). We think that Gavrilets' logic here—that cheater cells are not doing better overall in the F2 generation, therefore cell-level selection does not favor cheating—is an example of what others have called the "averaging fallacy" (Sober and Wilson 1998; Okasha 2006). Gavrilets' claim that individual-level selection is driving the results of his model notwithstanding, we see this model as clearly overlapping with the ETI framework with respect to having multilevel selection.

From these considerations, it is clear to us that levels of selection, cooperation and conflict are major elements of Gavrilets' model. In addition, there exist several group-level properties built into colonies from the start of the simulations, including the unspecified traits that allow the assumption of undifferentiated colonies as a starting point for the analysis, the developmental plasticity of cells to terminally differentiate, and the set proportion of germ and somatic precursor cells in a colony. In addition, groups of cells reproduce groups of cells through a group-level life cycle in the sense that a colony produces single cell propagules that develop into colonies.

Gavrilets sets the expression of the regulatory loci to be zero in the founding population thereby starting simulations with undifferentiated colonies. However, this initial condition is an unstable starting point. In the first few generations, the regulatory loci mutate away from zero, and germ and soma provide different contributions to colony-level fitness. Because undifferentiated colonies are unstable, Gavrilets has implicitly assumed the initial existence of division of labor and explained the subsequent evolution of more pronounced division of labor. A similar comment on Gavrilets' model was made by Ispolatov et al. (2012). Because initial colonies have division of labor and division of labor is the hallmark of evolutionary individuals, we view the initial colonies as already possessing properties of higher level individuals. For this reason, we see the model as addressing the latter stages of an ETI.

Ispolatov, Ackermann, and Doebeli 2012

Ispolatov et al. (2012) model the evolution of multicellularity in cyanobacteria where individual cells need to perform two incompatible metabolic processes such as oxygenic photosynthesis and anoxic nitrogen fixation. They assume a fitness cost to producing both metabolic products simultaneously. They utilize chemical mass action equations and assume an ephemeral group state in which cells aggregate into pairs that dissociate into cells. Cells are assumed to quickly alter their metabolic investment based on the paired group state and fitness landscape to increase cell fitness. Lastly, they model the evolution of a heritable trait, stickiness, which is representative of the evolution of multicellularity, as the amount of time a cell spends in a paired group state.

One innovation of this model is the changing dimensionality of the fitness landscape based on the solitary or paired state. When a cell is solitary, its fitness is a function of its investment in each metabolic process (two dimensions). When a cell is in a pair, its fitness is a function of its and its partner's investment in each metabolic process (four dimensions). This change in dimensionality changes a cell's fitness landscape from a single optimal peak when solitary to a saddle point when paired. When a saddle point is present in the fitness landscape, cells in a pair will differentiate to specialize on one metabolic product. The introduction of changes in dimensionality in fitness is important because it mathematically represents the increased complexity of multilevel selection in the group state. They find that multicellularity in the sense of the paired state, as a function of cellular stickiness, evolves when the cost of producing both metabolic products is high enough to make the mixed partial derivative of the fitness landscape negative (representing a saddle point) and the cost of being sticky is relatively low.

During their discussion, they emphasize that this model has several advantages absent in other models. Specifically, this model does not explicitly assume preexisting groups or differentiation (e.g., Willensdorfer 2009; Gavrilets 2010) and this allows the simultaneous evolution of groups (of size two) and division of labor. In this way, the pre-existence of developmental pathways or undifferentiated groups of cells is avoided. As much research investigates how differentiated colonies evolved from undifferentiated colonies (Willensdorfer 2008, 2009; Gavrilets 2010; Rueffler et al. 2012), a model which simultaneously evolves small colonies and division of labor is an explicit demonstration these processes are not necessarily separate.

This model has similarities and differences with the ETI framework (Table 1). When in a pair, cells independently alter their metabolic investment to maximize fitness. This leads to each cell specializing on a different metabolic process without apparent communication and cells share the metabolic products equally. In this way, metabolic adjustment appears to be synergistically cooperative, though nonaltruistic. This synergism causes cell fitness to increase when in a multicellular pair, but there is no discussion of group-level fitness so we found it difficult to determine whether fitness decoupling is occurring in their model. However Ispolatov et al. do not speak of cooperation, even though it is clear that synergistic cooperation is present in the paired state. Modeling cell interaction in this way represents a rudimentary form of the cellular integration that is critical to the evolution of multicellularity (Folse and Roughgarden 2010; Clarke 2011).

Multi-level selection is present in the Ispolatov et al. model, although they do not present it as such. Every cell experiences the same death rate and cells in groups have nearly identical birth rates (due to a small error term in metabolic investment). In comparison to solitary cells, cells in a pair have much higher fitness due to sharing of metabolic products and avoidance of the fitness cost associated with producing both products. Groups serve to produce population structure that affects the fitness of the cells. This context-dependent fitness means that group-level selection operates on pairs of cells to increase cell birth rate. However, the chemical mass action approach results in the lack of a canonical group level life cycle, in the sense that groups do not directly beget other groups.

Because of the lack of a canonical group life cycle, one interpretation is that cells, not multicellular groups, are the only evolutionary individuals in this model. However, this ignores the facts that group selection is occurring and group properties are evolving in the model. Although it is difficult to see this model as explaining the evolution of multicellularity as an ETI, it is novel in explaining the origin of group structure and properties (Table 1). The model has the novel feature of applying to the early stages of an ETI as well as showing the division of labor can evolve early in an ETI. While the evolution of division of labor is observed, reproductive division of labor is not addressed by the model.

With respect to division of labor, Ispolatov et al. (2012) model incompatible metabolic processes such as photosynthesis and nitrogen fixation in cyanobacteria, showing how non-reproductive division of labor may evolve. However, nitrogen fixing heterocysts in cyanobacteria do not reproduce (Kumar et al. 2010; Muro-Paster and Hess 2012). So while reproductive division of labor is present in cyanobacteria, it is not modeled by Ispolatov et al. (2012), which may call into question the applicability of their results both to cyanobacteria and to ETIs with division of labor in reproductive functions. The observation that reproductive division of labor tends to precede other kinds of functional specialization (Simpson 2012) indicates that there is something special about reproductive specialization; whatever this is, it is not being addressed by this model.

Rueffler, Hermisson, and Wagner 2012

Rueffler et al. (2012) develop a mathematical model with general assumptions in order to identify general principles about the conditions that favor the evolution of division of labor. Their starting point is a group of two modules. Each module is characterized by a trait value, which is initially constrained to be the same (i.e., the organism is initially made up of undifferentiated parts). The trait value determines how well the module performs on two tasks. Task performance is constrained by a trade-off, such that performance at both tasks cannot be optimized simultaneously by one module (or two undifferentiated modules). Fitness is taken to be an increasing function of performance of both tasks. Note that this is a single-level fitness concept like in Willensdorfer's model, so Rueffler et al. (2012) is modeling the evolution of division of labor after an evolutionary transition to group living. The group of modules is the only unit that has fitness. Fitness increases as the ability of groups of modules to perform tasks increases. Rueffler et al. assume no differentiation in the ancestral state and that trait values had evolved to a fitness maximum based on performance constraints. They then ask, given an ancestor with undifferentiated trait values, what conditions favor differentiation of the modules?

The main connection with the ETI theory, as Rueffler et al. (2012, p. E333) point out, is that the covariance effect described by Michod (2006) and Michod et al. (2006) can be seen as a special case of their observation that accelerating performance functions favor the evolution of differentiation. Michod (2006) and Michod et al. (2006) consider the trade-off between viability and fecundity as two contrasting basic categories of performance that combine multiplicatively to give fitness. Rueffler et al. consider a trade-off between performance of two tasks, meaning that increases in performance at one task come at the cost of decreases in performance of the other task. Performance at each task makes some positive contribution to fitness. They show that when the performance landscape is convex along the trait axes, specialization is favored. Specialization can lead to higher fitness in this case because the loss in performance at one task is due to one module's deviation from the (constrained) optimal trait value can be more than compensated for if the other module deviates in the other direction. Aside from the shape of the performance function, Rueffler et al. also find that positional effects (in which one module is less good at a particular task due to its position) and synergistic interactions (when the joint contribution of two differentiated modules exceeds the sum of their separate contributions) can favor the evolution of specialization. Because larger organisms a priori have more distinct areas, there are more positional effects to select for many cell types. Also, note that the greater the number of cell types, the higher the heterogeneity of the organism, hence also the highest positional effects will be present.

Rueffler et al.'s model highlights some commonalities underpinning the evolution of functional specialization in a variety of different kinds of units (e.g., genes, appendages, cells, etc.). Their framework is so general that it does not touch on issues that may be more specific to division of labor during an ETI. For example, the Rueffler et al. model does not address what (if anything) is special about reproductive division of labor. By leaving aside the issue of how the traits affect fitness, Rueffler et al. also leave aside the insight that basic life history trade-offs present at lower levels can disappear at higher levels while new trade-offs emerge at the higher level.

Van Dyken and Wade 2012a, b

In a series of papers, Van Dyken and Wade (2012a, b) propose a resource-based model for the evolution of different kinds of altruism. The evolution of altruism is fundamental to ETIs, so we include these papers here, even though they focus on the evolution of sociality and do not specifically address the evolution of multicellularity. Van Dyken and Wade assume discrete and non-overlapping generations for a population that is sub-divided into groups. They present fitness of a focal individual as the product of two factors: one factor represents the fitness that the individual would have given infinite resources and the second factor accounts for the effects of local resource availability and efficiency of use (2012a, their Eq. A6). Group-mates affect the local resource availability and resource use of the focal individual.

With respect to the evolution of division of labor, a major conclusion of Van Dyken and Wade's work is that local resource pressure, which depends both on the external/physical environment and the behavior of group-mates, determines which specialized tasks evolve. For example, abundant resources favor specializations on tasks such as nest defense whereas scarce resources favor the evolution of specialized foragers. We think Van Dyken and Wade's focus on the diverse ecological effects of altruistic behaviors begins to address an important shortcoming of some previous work on altruism. However, Van Dyken and Wade do not model the fundamental life history trade-off between viability and fecundity of a focal individual. They write: "$v_{ij} f_{ij}$ is individual *ij*'s asymptotic fitness; that is, its resource-independent probability of survival to maturity (viability, v_{ij}) times its maximum fecundity (f_{ij})" (Van Dyken and Wade (2012a, p. 2487). However, viability and fecundity are unconnected parameters in this model, not functions of more basic variables like reproductive effort or time spent at the two activities as is more commonly assumed in life history modeling. This is an interesting choice, and is probably appropriate for some cases. For example, an altruistic behavior such as alarm calling simply lowers the personal viability of the caller (and increases group fitness) without an increase in personal fecundity of the caller. However, in many other cases, a single behavior has contrasting effects on both personal viability and fecundity through a constrained resource such as time. For example, take a volvocine somatic cell that spends more time or energy beating its flagella than another cell. The fitness effect of this behavior would likely be a decrease in fecundity and an increase viability were it to live on its own. The behavior is costly (at the cell level) if the benefit of the behavior (increased cell-level viability) is insufficient to outweigh its costs (decreased cell-level fecundity). The cost of the behavior translates into a lack of balance in the two fitness components at the cell level. Survival-fecundity tradeoffs at the level of single individuals are common (see Roff and Fairbairn 2007), so considering this kind of situation is important. The viability and fecundity terms in Van Dyken and Wade are independent parameters so they do not capture the idea that behaviors can be costly at the lower level by creating an imbalance of investment in lower level fitness components.

Van Dyken and Wade's classification of four types of altruism depends on their fitness function and the types of exchanges between individuals that are allowed, in particular, the lack of any interdependence of viability and fecundity. They write: "Altruism is typically modeled as increasing a recipients' survival or fecundity. However, our model (Eq. 1) provides four different parameters that control fitness, each of which can be modified by altruism..." (Van Dyken and Wade 2012a, p. 2488).

Although they see this approach as advantageous, we question whether the hierarchical structure of the concepts involved is being ignored. That is, when one considers how increased resources increase a recipient's fitness, it becomes clear that this increase must be channeled through benefits to the recipient's fecundity, viability or both (assuming constant generation times). So it seems artificial to have four distinct ways in which altruistic behavior can benefit a partner (increased survival, increased fecundity, increased resources, increase resource use efficiency). Rather, why are there not two ways of affecting the partner's resource (total amount or efficiency of use), which the partner then channels into personal viability and/or fecundity? The typical meaning of viability and fecundity (probability of survival to reproduction and number of offspring conditioned on survival) is closely tied to the idea that both of these features are connected through their resource-dependency. By contrast, Van Dyken and Wade use viability and fecundity to refer to resource-independent characteristics of organisms. It is not clear how these resource-independent notions of viability and fecundity should be applied to classifying social interactions. The very idea of one individual helping another individual to reproduce seems to require an exchange of resources; offspring are made of resources. Yet Van Dyken and Wade's framework suggests that "fecundity altruism" occurs when one individual increases another's maximum, resource-independent fecundity. The usefulness of this way of parsing interactions is not yet clear.

In the Van Dyken and Wade approach, the "individual" fitness reflects both individual and group properties. ("Individual fitness was modeled as the physiological consequence of resource consumption in an environment composed of other consumers." (Van Dyken and Wade 2012b, p. 2498)). Specifically, the amount of crowding is a property of the group that affects the "individual" fitness (Van Dyken and Wade 2012a, p. 2487, their Eq. 2). The focus on this kind of fitness function (i.e., "individual" fitness that includes group effects) makes the multi-level nature of the processes that are affecting trait change implicit rather than explicit. So while it seems likely that the Van Dyken and Wade models are examples of the evolution of traits that decouple lower- and higher-level fitness, this result is not explicit and is not explicitly discussed in the papers. For example, a somatic cell in a colony decouples fitness at the two levels because this trait would be detrimental in a global population of cells but can be favored by selection among groups of cells. Thus, when somatic cells evolve, it is clear that the groups of cells with high fitness are not simply groups that contain the cells that would do best in a global population. Note that the realized, direct fitness of cells within a group (i.e., the kind of fitness in the Van Dyken and Wade papers) is not helpful here. Fitness is decoupled when the strictly single-level fitness that a cell would have declines and the fitness of the group increases. The same logic applies to all of the types of altruism that Van Dyken and Wade propose, suggesting that these models investigate the evolution of altruism after an evolutionary transition in individuality. One consequence of their work is the possibility that contrasting kinds of altruism can increase together via co-evolutionary niche construction (Van Dyken and Wade 2012b). With respect to division of labor, the ETI framework focuses on the lower-level trade-off between viability and fertility. However, Van Dyken and Wade explicitly assume that these two components of fitness are independent. Van Dyken and Wade (2012b, p. 2509) suggest that natural selection would act so as to "partition tasks into survival/fecundity and resource task specialists in proportions that most efficiently cope with negative environmental feedback at the colony level." This is an interesting and novel idea that certainly could be explored more with respect to the multicellularity ETI.

Discussion

We have discussed how each model relates to five components of ETIs: the evolution of cooperation in groups, multi-level selection, life history trade-offs, division of labor, and decoupling of fitness (see Table 1). While not all the models reviewed here cleanly relate to this ETI framework, none conflict with it, and they can all be understood as contributing new information to the different components. Of course, each of these works has a scope defined by their own objectives, so the observation that not all components are included in each model is not meant as a criticism. Two major themes have emerged in our analysis of these models, (i) the role of life history traits and life cycles in evolutionary transitions and (ii) what the models assume and purport to explain, in particular, whether groups are assumed to already have the properties we wish to explain.

Life History Traits and Life Cycles

The evolution of multicellularity involves the evolution of a group life cycle from the life cycles of single cells. Thus it is natural to consider life history traits and how they change and are reorganized at the two levels during this ETI. Ignoring the critical role life history traits play in defining fitness creates ambiguities in comparing and understanding definitions of fitness.

Rueffler et al. (2012), in the search for generality, did not assume explicit life history traits like survival and reproduction in their model; the detachment of this model from life history traits and reproductive specialization limits its application to ETIs. Van Dyken and Wade (2012a, b) do not explicitly model how resources affect viability and/or fecundity. They also ignore the most fundamental aspect of life history traits by assuming that viability and fecundity are independent and do not trade off in a focal individual. Because of these assumptions, it was difficult to relate their definition of fitness and types of altruists to the ETI framework, although their most general point, that resource use needs to be explicitly modeled, should be pursued in the context of ETIs.

We recognize the difficulty of defining fitness in a situation where the very unit of evolution and hence fitness is changing, but this is the challenge that must be met in understanding an ETI. When fitness is based on life history traits, a tractable conception of fitness results as the level of fitness changes. Novel definitions of fitness (such as Willensdorfer (2009) who defined fitness as biomass production) may be interesting, but make it difficult to connect these definitions to other models or to the biological cases we wish to understand. Care must be taken to ensure fitness definitions are biologically tractable and applicable.

Most of the recent models (Willensdorfer 2009; Gavrilets 2010; Rueffler et al. 2012; Van Dyken and Wade 2012a, b) along with the earlier modifier and optimality models, assume the initial existence of group-level life cycles. This leaves a critical gap in our understanding of an ETI. How might a group life cycle emerge from unicellular cycles? Life cycles can be thought of in terms of a few fundamental elements (Fig. 2). We consider the asexual life cycles of three different kinds of species of volvocine green algae—a diverse group of photosynthetic eukaryotes ranging from unicellular to complex multicellular forms (see Chapter "Volvocine Algae: From Simple to Complex Multicellularity"): *Chlamydomonas reinhardtii* (unicellular), *Basichlamys sacculifera* or *Tetrabaena socialis* (colonial with group size 2–4) and *Gonium pectorale* (colonial with group size 4–16). In *C. reinhardtii*, a parent cell (1c) grows and then (2c) divides through several rounds of synchronous cell divisions. Finally, offspring cells (3c) separate from the parent and from each other.

In the undifferentiated multicellular volvocine algae such as *G. pectorale*, we see the same three basic life cycle elements applying to groups of cells on the right-hand-side of Fig. 2. A group of cells, i.e. colony, (1g) grows; next, divisions occur (2g). Although divisions are at the cell level, each cell undergoes divisions while still being a group member. Whereas cell divisions in the cell cycle (2c) are immediately responsible for the number of offspring cells, cell divisions in the group cycle (2g) correspond to the number of cells in the adult group, that is, the adult group's body size.

The major elements of the asexual life cycle are similar in the unicellular *Chlamydomonas* and the multicellular *Gonium* (i.e. 1c-2c-3c is similar to 1g-2g-3g). However, note that these two cycles (Fig. 2, left side and right side, respectively) are distinct and share no states in common. How might evolution get from one to the other? The earliest-branching colonial volvocine species (*B. sacculifera* and *T. socialis*) have a hybrid pattern that may represent a transitional stage between a cell-level and group-level life cycle (Fig. 2, grey lines). In these species, among the simplest of known multicellular forms, a colony undergoes growth (1g), then the cells of the colony separate from each other (3c), and then the cells undergo cell division (2c). From a unicellular ancestor similar to *Chlamydomonas*, a *Basichlamys*-like life cycle could arise from a change in timing of the cell separation state. Step 3c (separation) fails to occur at the normal time (see connection between C and D'). Instead, cells separate later (see connection between E and B), after a period of growth within the colony. A mutation that causes temporary adherence of offspring cells to the mother cell wall material could create a *Basichlamys*-like cycle from a unicellular cycle. If so, then growth as a colony was the first aspect of a fully colonial life cycle to evolve in volvocine algae. This would suggest that the effects of colony-level traits (e.g. colony cell number, overall colony size) on growth could form the basis for specifically group-level selection very early in the

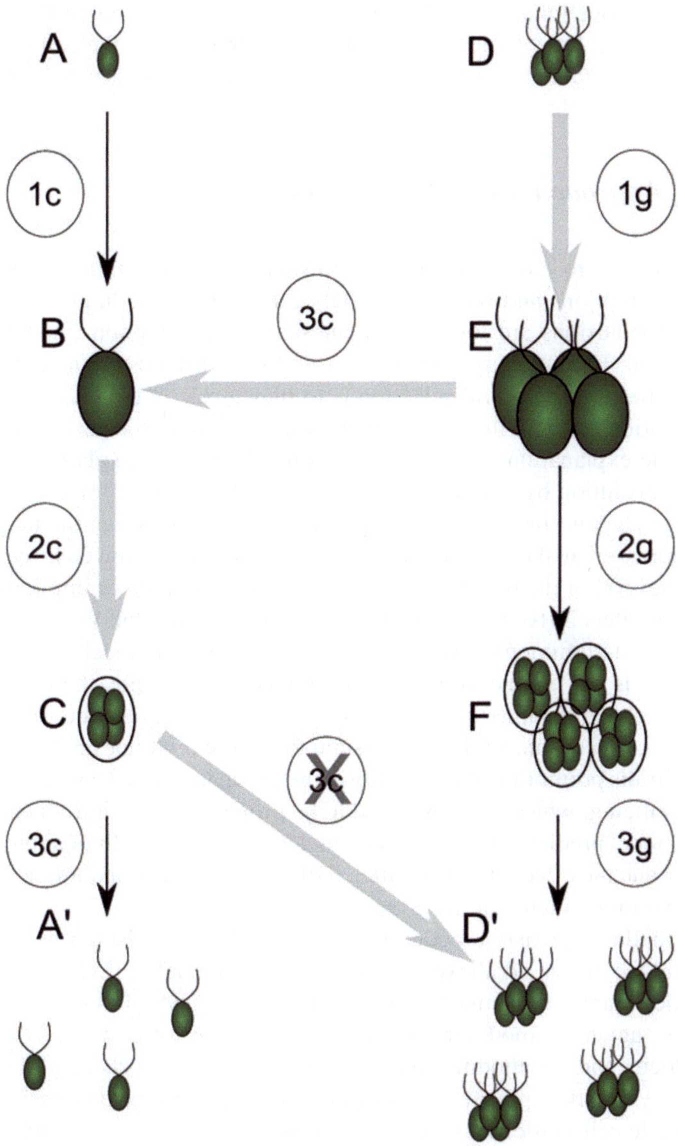

Fig. 2 *Left side*: The three major elements of a single-celled asexual cycle. (*1c*) growth, (*2c*) division, (*3c*) separation. *Right side*: The same three elements are applicable in the same order in a simple colonial (group) life cycle (e.g. as seen in *G. pectorale*). Note the cell and group life cycles are distinct and do not have states in common. *Gray lines*: The simplest colonial volvocine algae (*B. sacculifera* and *T. socialis*) show a hybrid asexual cycle in which growth (*1g*), but not division (*2c*) or separation (*3c*), occurs while cells are in groups. Throughout this figure, the number of rounds of division is shown as two. In reality, this value can vary based on external conditions

transition to multicellularity. Modeling these issues in such simple life cycles can help us understand how the level of adaptation changes from the cell to the group during a transition in individuality (Shelton 2013; Shelton and Michod 2014).

Modeling Approaches and Challenges

The second theme to emerge from our analysis regards the relationship between the initial assumptions and what the model seeks to explain. In any mathematical model, the conclusions are logical outcomes of the assumptions, some of which may be implicit. Evolutionary transitions in individuality pose unique problems to the modeler, because if natural selection is to play a role in the transition, then the units of evolution are both the explanandum (phenomenon to be explained) and the explanans (the explanation of the phenomenon). We seek to explain a new higher-level unit of evolution by making assumptions about the lower level units and how they interact. The multilevel selection approach determines when the magnitude of selection increases at the higher level and decreases at the lower level. We have argued that several of the models studied here either explicitly or implicitly use this approach (Gavrilets 2010; Ispolatov et al. 2012; Van Dyken and Wade 2012a, b).

The issue of assuming pre-existing group properties in a model meant to explain group features is not unique to the recent models, for example, the modifier models discussed above assume the prior existence of group-level traits such as group structure and cooperation. With these assumptions, the modifier models studied the evolution of traits promoting individuality at the group level, such as germ line segregation and policing, which were assumed to be properties of an introduced modifier allele. The model predicted the conditions under which the modifier allele increased in frequency and the effect of the modifier allele on levels of cooperation, heritability of fitness, and individuality at the group level.

Similar challenges exist in modeling division of labor. As Ispolatov et al. (2012) point out and we discussed above, Gavrilets' (2010) model of division of labor initially assumes many of the antecedents of division of labor. To help us understand this, consider again the modifier models in which there were two loci, the cooperate/defect locus and the modifier locus. The modifier locus changed an aspect of the life history or development, such as mutation rate, germ soma division of labor, policing, single cell bottlenecks, or genetic control of group size, which in turn increased the fitness of the already more fit C allele. Consequently, the evolution of division or labor at the modifier locus and its effect on individuality emerged out of an interaction between the two loci. The modifier approach allows us to track the genetic and selective factors involved in the evolution of division of labor.

Gavrilets (2010) sets the proportion of soma precursor cells arbitrarily to be 25 % and the conditions for which division of labor evolves via the regulatory loci are determined. However we do not know whether or in what way these conditions depend upon the arbitrary figure of 25 % somatic cell precursors (although simulations were also done with 75 % somatic precursors). A modifier approach would assume that the proportion of soma precursors was encoded by a third locus that was initially set at 0 %. In this case, the evolution of division of labor would emerge out of an interaction between the regulatory loci and the precursor locus and not as a result of a specific preordained frequency of soma precursors.

These challenges highlight the need for careful interpretation of the model. Here, we have discussed how the interpretation of a model can lead to substantial differences in understanding the causal factors underlying the evolution of multicellularity. The clearest examples of this were Gavrilets' suggestion that selection solely at the level of cells can drive the evolution of division of labor in multicellular groups and Ispolatov et al.'s suggestion that multicellularity can evolve without cooperation or increased cellular integration.

Future Directions

The papers we have reviewed here are largely focused on the later stages of ETIs, specifically the evolution of division of labor (Step v, Table 1). One exception to this was Ispolatov et al. (2012), which focused on initial group formation and increasing synergistic cooperation (Steps i and ii). Previous ETI models largely focused on conflict mediation, division of labor among fitness components and fitness decoupling (Steps iii–vi). In comparison, there has been little recent research on group formation and the evolution of group life cycles as discussed in the previous section.

In modifier models, the life cycle is summarized in terms of heritability of group traits but a more explicit treatment of the components of the life cycle and how they are reassembled at the level of the group has not yet been achieved. This represents a critical gap in the theory given the emphasis on life history traits in the ETI framework. Recent empirical (e.g. Rainey and Rainey 2003; Ratcliff et al. 2012, 2013), conceptual (Libby and Rainey 2013) and theoretical work (Shelton 2013; Shelton and Michod 2014) studying the evolution of group life cycles will provide ideas for this future work to address. Furthermore, we suspect a better theoretical understanding of group-level life cycles will integrate the ETI framework with the multi-level selection (MLS) framework (Damuth and Heisler 1988; Okasha 2006), in which Okasha has defined the evolution of individuality as the transition among different kinds of MLS (MLS1 and MLS2) (Okasha 2006). This is an important step to integrating the large body of research on MLS1 and MLS2 with the ETI framework.

We also suggest that modeling non-canonical systems would be valuable for our understanding of ETIs. Many recent models are implicitly or explicitly modeling a canonical model system (multicellular development from a single celled propagule). While it is important to understand this form of multicellularity, there are other, albeit rare, pathways to multicellularity. These pathways deserve mathematical investigation to more fully appreciate what factors predispose a system to evolve towards one kind of multicellularity or another and why certain kinds of multicellularity are rare. We envision models investigating multicellularity via aggregation (such as the slim mold, *Dictyostelium*; Chapter "The Evolution of Developmental Signalling in

Dictyostelia from an Amoebozoan Stress Response") and species with unique life histories (such as the red alga, *Porphyra*). Such models may investigate the evolutionary consequences of alternative pathways. Major differences, such as the difference between groups formed by aggregation ("coming together", Tarnita et al. 2013) and groups formed by reproductive products staying together ("staying together", Tarnita et al. 2013), may relate to the generalized step (i) group formation differently. This distinction may also affect the timing of step (iii) cheating and conflict, as relatives that stay together may be less prone to conflict than genetically distinct individuals that aggregate. Ispolatov et al. (2012) is the only model reviewed here which models multicellularity via aggregation rather than via development. As such, this model serves as an important first step to understanding alternative pathways to multicellularity.

Finally, we find it interesting that each paper has taken a different mathematical approach, made different biological assumptions, and has made new contributions to our understanding of the evolution of multicellularity. This suggests that our theoretical understanding of the evolution of multicellularity is far from complete. It seems clear that continued theoretical exploration of this subject would be highly valuable to our understanding of multicellularity and ETIs more generally.

Summary

1. An evolutionary transition in individuality (ETI) is a fundamental shift in the unit of selection, when a group of individuals become a new, higher-level individual. ETIs are thought to be responsible for the origin of the hierarchy of life: genes, chromosomes, cells, cells within cells (eukaryotic cell), multicellular organisms, and societies.
2. We discuss five interrelated components of ETIs: the evolution of cooperation in groups, multi-level selection, life history trade-offs, division of labor among the basic components of fitness, and decoupling of fitness of the new, higher-level unit from the fitnesses of the lower-level units.
3. We review the two-locus modifier models and optimality models that provide the basis for the ETI framework emphasizing the role of lower-level fitness trade-offs.
4. We discuss how six recently published papers (Willensdorfer 2009; Gavrilets 2010; Ispolatov et al. 2012; Rueffler et al. 2012; Van Dyken and Wade 2012a, b) relate to the five components of the ETI framework.
5. We identify life history traits and life cycles as critical considerations in understanding ETIs that have been missing in some recent mathematical models. We show how group life cycles may emerge from cell life cycles in the volvocine green algae model system.
6. Incorporating recent papers into the ETI framework reveals several less understood aspects of ETIs, which we suggest as directions for future research, including the evolution of group life cycles, how alternative frameworks relate to ETIs, and the need to model non-canonical systems.

Acknowledgements This research was supported by the National Science Foundation under Grant No. DEB-0742383 to R. E. Michod, the National Science Foundation under Grant No. DGE-0654435 to the University of Arizona, and the National Aeronautics and Space Administration under Grant No. NNX13AH41G to R. E. Michod. We would like to thank J. D. Van Dyken, M. Doebeli, A. M. Nedelcu, and two anonymous reviewers for their helpful comments on the manuscript.

References

Buss LW (1987) The evolution of individuality. Princeton University, Princeton

Clarke E (2011) The problem of biological individuality. Biol Theory 5:312–325

Damuth J, Heisler IL (1988) Alternative formulations of multilevel selection. Biol and Philos 3:407–430

Folse HJ III, Roughgarden J (2010) What is an individual? A multilevel selection perspective. Q Rev Biol 85:447–472

Gavrilets S (2010) rapid transition towards the division of labor via evolution of developmental plasticity. PLoS Comput Biol 6:e1000805

Griesemer J (2000) The units of evolutionary transition. Selection 1:67–80

Ispolatov I, Ackermann M, Doebeli M (2012) Division of labour and the evolution of multicellularity. Proc Biol Sci 279:1768–1776

Kumar K, Mella-Herrera RA, Golden JW (2010) Cyanobacterial heterocysts. Cold Spring Harb Perspect Biol 2:a000315

Libby E, Rainey PB (2013) A conceptual framework for the evolutionary origins of multicellularity. Phys Biol 10:035001

Maynard Smith J (1988) Evolutionary progress and levels of selection. In: Nitecki MH (ed) Evolutionary progress. University of Chicago Press, Chicago, pp 219–230

Maynard Smith J (1991) A Darwinian view of symbiosis. In: Margulis L, Fester R (eds) Symbiosis as a source of evolutionary innovation. MIT Press, Cambridge, pp 26–39

Maynard Smith J, Szathmáry E (1995) The major transitions in evolution. W.H. Freeman, San Francisco

Michod RE (1996) Cooperation and conflict in the evolution of individuality. II. Conflict mediation. Proc: Biol Sci 263:813–822

Michod RE (1997) Cooperation and conflict in the evolution of individuality. I. Multi-level selection of the organism. Am Nat 149:607–645

Michod RE (1999) Darwinian dynamics, evolutionary transitions in fitness and individuality. Princeton University Press, Princeton

Michod RE (2003) Cooperation and conflict mediation during the origin of multicellularity. In: Hammerstein P (ed) Genetic and cultural evolution of cooperation. MIT Press, Cambridge, pp 261–307

Michod RE (2005) On the transfer of fitness from the cell to the multicellular organism. Biol Philos 20:967–987

Michod RE (2006) The group covariance effect and fitness trade-offs during evolutionary transitions in individuality. Proc Natl Acad Sci U S A 103:9113–9117

Michod RE (2007) Evolution of individuality during the transition from unicellular to multicellular life. Proc Natl Acad Sci U S A 104(1):8613–8618

Michod RE, Nedelcu AM (2003) Cooperation and conflict during the unicellular-multicellular and prokaryotic-eukaryotic transitions. In: Moya A, Font E (eds) Evolution: from molecules to ecosystems. Oxford University Press, Oxford, pp 195–208

Michod RE, Roze D (1997) Transitions in individuality. Proc: Biol Sci 264:853–857

Michod RE, Roze D (1999) Cooperation and conflict in the evolution of individuality. III. Transitions in the unit of fitness. In: Nehaniv CL (ed) Mathematical and computational biology: computational morphogenesis, hierarchical complexity, and digital evolution. American Mathematical Society, Providence, pp 47–92

Michod RE, Roze D (2001) Cooperation and conflict in the evolution of multicellularity. Heredity 81:1–7
Michod RE, Nedelcu AM, Roze D (2003) Cooperation and conflict in the evolution of individuality IV. Conflict mediation and evolvability in *Volvox carteri*. BioSystems 69:95–114
Michod RE, Viossat Y, Solari CA, Hurrand M, Nedelcu AM (2006) Life history evolution and the origin of multicellularity. J Theor Biol 239:257–272
Muro-Pastor AM, Hess WR (2012) Heterocyst differentiation: from single mutants to global approaches. Trends Microbiol 20:548–557
Nedelcu AM, Michod RE (2006) The evolutionary origin of an altruistic gene. Mol Biol Evol 23:1460–1464
Okasha S (2006) Evolution and the levels of selection. Oxford University Press, Oxford
Otto SP, Orive ME (1995) Evolutionary consequences of mutation and selection within an individual. Genetics 3:1173–1187
Rainey PB, Rainey K (2003) Evolution of cooperation and conflict in experimental bacterial populations. Nature 425:72–74
Ratcliff WC, Denison RF, Borrello M, Travisano M (2012) Experimental evolution of multicellularity. Proc Natl Acad Sci U S A 109:1595–1600
Ratcliff WC, Herron MD, Howell K, Pentz JT, Rosenzweig F, Travisano M (2013) Experimental evolution of an alternating uni- and multicellular life cycle in *Chlamydomonas reinhardtii*. Nat commun 4:2742
Roff DA (2002) Life history evolution. Sinauer Associates, Sunderland
Roff DA, Fairbairn DJ (2007) The evolution of trade-offs: where are we? J Evol Biol 20:433–447
Rueffler C, Hermisson J, Wagner GP (2012) Evolution of functional specialization and division of labor. Proc Natl Acad Sci U S A 109:E326–E335
Shelton DE (2013) The shifting role of cell division during an evolutionary transition to multicellular-level individuality. Doctoral Dissertation, University of Arizona
Shelton DE, Michod RE (2014) Group selection and group adaptation during a major evolutionary transition: insights from the evolution of multicellularity in the volvocine algae. Biological Theory 9:452–469
Simpson C (2012) The evolutionary history of division of labor. Proc: Bio Sci 279:116–121
Sober E, Wilson DS (1998) Unto others: the evolution and psychology of unselfish behavior. Harvard University Press, Cambridge
Solari CA, Kessler JO, Michod RE (2006) A hydrodynamics approach to the evolution of multicellularity: flagellar motility and germ-soma differentiation in volvocalean green algae. Am Nat 167:537–554
Stearns SC (1992) The evolution of life histories. Oxford University Press, Oxford
Tarnita CE, Taubes CH, Nowak MA (2013) Evolutionary construction by staying together and coming together. J Theor Biol 320:10–22
Van Dyken JD, Wade MJ (2012a) Origins of altruism diversity I: the diverse ecological roles of altruistic strategies and their evolutionary responses to local competition. Evolution 66:2484–2513
Van Dyken JD, Wade MJ (2012b) Origins of altruism diversity II: runaway coevolution of altruistic strategies via "reciprocal niche construction". Evolution 66:2498–2513
Willensdorfer M (2008) Organism size promotes the evolution of specialized cells in multicellular digital organisms. J Evol Biol 21:104–110
Willensdorfer M (2009) On the evolution of differentiated multicellularity. Evolution 63:306–323

Multicellular Life Cycles as an Emergent Property in Filamentous Bacteria

Valentina Rossetti and Homayoun C. Bagheri

Abstract Multicellularity is an integral part of all organisms that grow to be larger than microscopic scales and is a necessity for complex morphologies. Hence, a central question is: what are the conditions that can lead to the evolution of multicellular development? Here, we outline a theoretical framework that serves as basis to understand the interactions that can lead to the evolution of multicellular life cycles in simple filamentous organisms. By assuming the prior evolution and existence of filamentous multicellularity, and not considering *a priori* selective advantages, we explore the extent to which intrinsic processes such as cellular birth and death rates can drive the development of such multicellular organisms. The chapter offers an overview of our mathematical setup and of the validation experiments in natural populations of filamentous bacteria. These studies show the presence of a common pattern in terms of filament growth, which provides a hypothesis for the emergence of primitive multicellular life cycles in simple organisms.

Keywords Emergent property · Life-history traits · Streptococci · Cyanobacteria · Heterotrophs

Introduction

Multicellularity characterizes all macroscopic organisms such as fungi, plants and animals, reaching an order of magnitude of 10^{14} cells and several hundred cell types in humans (Arendt 2008). The transition from single cells to multicellular individuals occurred several times during the history of life, in both prokaryotic and eukaryotic

V. Rossetti (✉) · H. C. Bagheri
Institute of Evolutionary Biology and Environmental Studies,
University of Zurich, Zurich, Switzerland
e-mail: valentina.rossetti@uzh.ch

V. Rossetti
Anthropological Institute and Museum, University of Zurich, Zurich, Switzerland

H. C. Bagheri
e-mail: bagheri@ieu.uzh.ch

I. Ruiz-Trillo, A. M. Nedelcu (eds.), *Evolutionary Transitions to Multicellular Life*,
Advances in Marine Genomics 2, DOI 10.1007/978-94-017-9642-2_10

lineages (Buss 1987; Rokas 2008; Schirrmeister et al. 2011). However, multicellularity first evolved among bacteria. According to fossil records, multicellular types of organization can be potentially traced back to more than 3.2 billion years ago (Schopf 2006) and correspond to filamentous bacteria. Later in the history of our planet, cyanobacteria have been among the most abundant species exhibiting multicellularity as reflected in fossils belonging to the proterozoic era (Schopf 2006; Schirrmeister et al. 2012). Bacteria can hence serve as model organisms for investigating the conditions that can lead to the evolution of multicellularity (Claessen et al. 2014).

Several studies focused on the shift of the unit of selection linked to the transition from unicellularity to multicellularity. What may be beneficial in terms of reproduction and survival of a single cell could at the same time be detrimental to the fitness of a multicellular aggregate (Michod and Roze 2001; Nedelcu and Michod 2006; Ispolatov et al. 2012, Chapters ("Evolutionary Transitions in Individuality and Recent Models of Multicellularity", "The Evolutionary Ecology of Multicellularity: The Volvocine Green Algae as a Case Study"). However, situations can arise in which cells share their secreted products in order to provide nutrients to the whole aggregate, and may forego "selfish" reproduction. Linked to these studies are those on the main risk associated to cooperation in a multicellular environment, namely the presence of cheater cells that express a selfish behavior. Defecting from cooperation can provide selective advantage over competitors (Boomsma and Franks 2006; Sachs et al. 2004; Axelrod and Hamilton 1981; Strassmann et al. 2000), but can ultimately lead a population to its collapse ("tragedy of the commons", (Hardin 1968)).

Another category of studies considers the potential selective advantages provided by the increased size of multicellular individuals. It has been hypothesized that big organismal size can enhance feeding efficiency, improve dispersal and predator avoidance (Berleman and Kirby 2009; Bonner 1974; Dworkin 1972; Hahn and Hoefle 1998; Solari et al. 2006; Boraas et al. 1998; Ratcliff et al. 2013).

In addition to the "fitness oriented" approaches mentioned above, one can look at the ecological processes and first principle "ground rules" that can affect multicellularity and the evolution of the developmental programs associated with it. More importantly, one can embark on such an approach without an a priori assumption of selective advantage. This implies a study of the basic underlying mechanisms that regulate the growth of a multicellular organism at the cellular level. Given this perspective, we make two main assumptions. The first is that we consider multicellularity as possible, i.e. that the organismal machinery to produce multicellular filaments is available to the individual cell. Second, the organism is viewed as a collection of attached cells governed by the processes of cell birth and death. The interplay of these two processes is expected to drive the growth of the organism at its different life stages. Differences in birth and death rates at the cellular level can hence potentially be responsible for differences in the degree of multicellularity reached by different organisms. We offer a new perspective on the evolution of multicellularity as an emergent property instead of driven by a selective advantage. The size of a multicellular organism, viewed as the number of cells, can be interpreted as population size, where the population coincides with the organism itself. Under this perspective, we study multicellularity in filamentous bacteria within the frame of more general ecological theories for population growth.

Filamentous bacteria represent an ideal case study to test such hypotheses. Thanks to the simple morphology of filamentous bacteria, birth and death coincide with the addition of one cell and with the breakage of a chain of cells, respectively. In natural populations, one can easily observe that the length of filaments across species and within species during an entire lifecycle is not constant. For example, filamentous cyanobacteria can contain a variable number of cells, from few units up to hundreds (Komarek and Anagnostidis 2005). Environmental conditions such as temperature, solar irradiation and nutrient concentrations have been identified as factors determining the mean size (filament length) of different cyanobacterial species (Kamp et al. 2008; Kruskopf 2006; Wu et al. 2005). Moreover, filaments can break because of external mechanical stress, lytic processes initiated by pathogens (van Hannen et al. 1999; Weinbauer 2004) or programmed cell death (Adamec et al. 2005; Berman-Frank et al. 2004; Daft and Stewart 1973; Lewis 2000; Ning et al. 2002).

Filament length in terms of number of cells can be interpreted also as the degree of multicellularity, and the changes in length as the ability of a species to maintain a multicellular state. There are not many theoretical studies addressing the distribution of filament lengths and the population dynamics leading to shorter or longer filaments. In a recent study (Rossetti et al. 2011) we studied a mathematical model to simulate such dynamics in filamentous multicellular bacteria and we validated it in cyanobacteria as well as in heterotrophic species. The model did not assume selective advantages for multicellularity. Rather, it considered multicellular filaments as a population of cells governed by birth and death rates, as commonly assumed in classical ecological theory for population dynamics.

Based on advances documented in our previous publications (Rossetti et al. 2013; Rossetti et al. 2011), in the following sections we first illustrate a mathematical model that provides the theoretical framework for the study of the emergence of multicellular life cycles. Subsequently we present the validation of the model in bacterial filamentous strains living in two different habitats, namely aquatic and oral bacteria.

A Theoretical Framework for the Emergence of Multicellular Life Cycles

As commonly assumed in ecological models for population dynamics, the growth of a population is governed by birth and death rates that depend on the population size. In the case of filamentous bacteria, the population is represented by the set of cells forming the filaments. Hence, the population size is the total number of cells that form all the filaments in a population. Birth and death rates are those of the cells, and given any time point, all the cells in the population are assumed to have the same birth and death rates. Furthermore, these rates are a function of the total number of

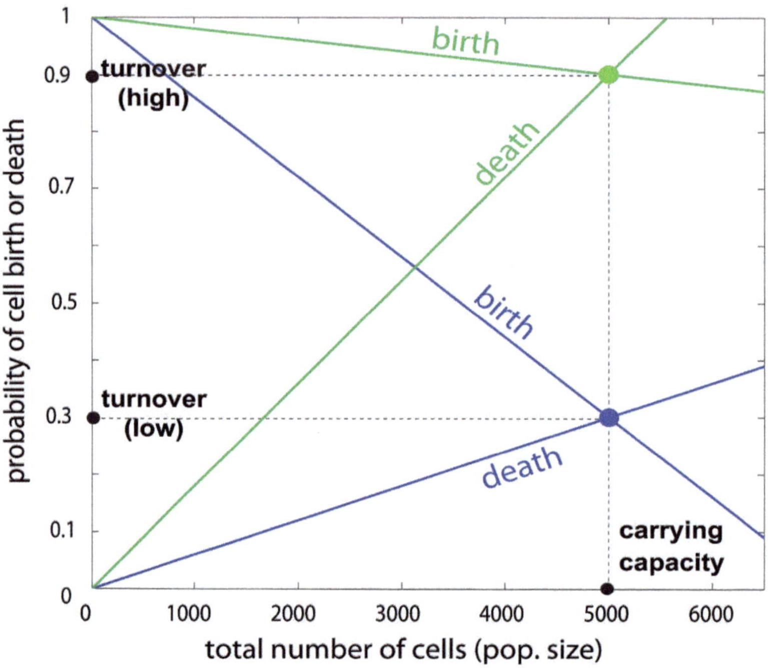

Fig. 1 Schematic representation of linear birth and death rates that depend on population size. Birth and death rates equal when the population is at its carrying capacity. The corresponding value of the rates is here referred to as turnover. The same carrying capacity can be achieved with different turnovers, depending on the slope of the linear birth and death rates

cells in the population (i.e. the population size). The shape of such a function can be linear or nonlinear. As a first approximation in our studies, we assumed that birth and death rates depend linearly on the population size. Figure 1 schematically depicts this framework. These linear functions cross at a specific value of the population size, namely at its carrying capacity. At this point, the birth rate equals the death rate, both having a value here referred to as the turnover rate. In the model, the turnover is a central variable that can help us study the differences in filament length according to the birth and death rates. This is because as shown by the pair of functions in Fig. 1, a population can achieve the same carrying capacity with different turnover rates.

We investigated the question whether differences in turnover rate can influence the length of filaments during their life cycle. To achieve this, we set the birth and death rate functions as follows:

$$\text{birth rate} = \frac{\theta - 1}{N^*} N + 1 \tag{1}$$

$$\text{death rate} = \frac{\theta}{N^*} N \tag{2}$$

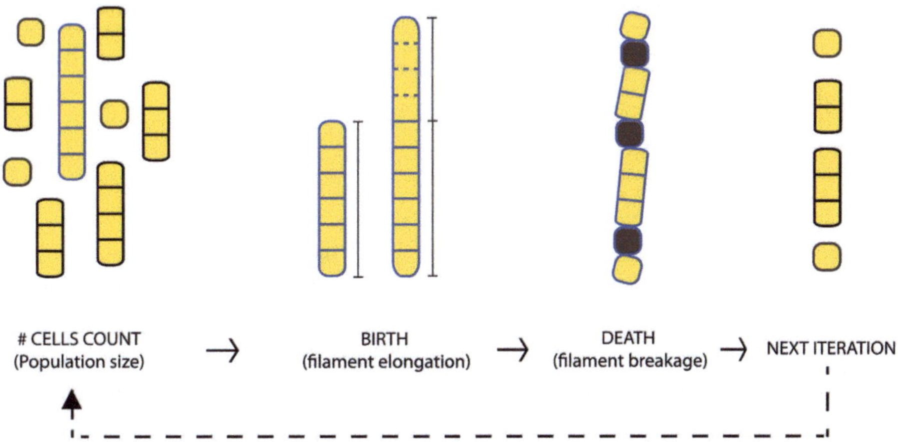

Fig. 2 Illustration of the main steps of the algorithm to simulate the dynamics of a population of filaments governed by density dependent birth and death rates

where N is the population size, θ is the turnover rate and N^* is the population size at carrying capacity. We then implemented a computer algorithm that simulates the dynamics of a population of undifferentiated multicellular individuals. The algorithm is designed as a loop of iterations, each consisting in distinct steps, as illustrated in Fig. 2. Each of the iterations can be interpreted as a round of cell births and deaths in the population. Since the birth and death functions depend on the population size, the first step is the cell count in the population. Accordingly, birth and death rates are computed and the filaments are elongated and broken in smaller units. The same steps are then repeated on the resulting set of filaments, until the population reaches its carrying capacity.

We monitored the mean filament length at every iteration and we tried different turnover rates. As measure of fitness, we used the growth rate of the population. From the results, we first observed that the growth curve of the population was the same for all tested turnover rates. This indicates that strains with different turnovers can have the same fitness.

We then found that regardless of the turnover, the average filament length follows a cyclic pattern, whereby filaments are short in the beginning, reach a maximum length during the exponential phase, and become shorter again when carrying capacity is reached (Fig. 3, panels A and B). The presence of this short-long-short pattern indicates that the simple interplay of birth and death rates at the cellular level can lead to the emergence of multicellular cycles. Moreover, depending on the turnover, these cycles differ in terms of average length achieved during growth. During the exponential phase, filaments of populations with low turnover were able to elongate much more than filaments with high turnover rate. Similarly, at carrying capacity, filaments in low turnover populations were on average longer than the high turnover variants.

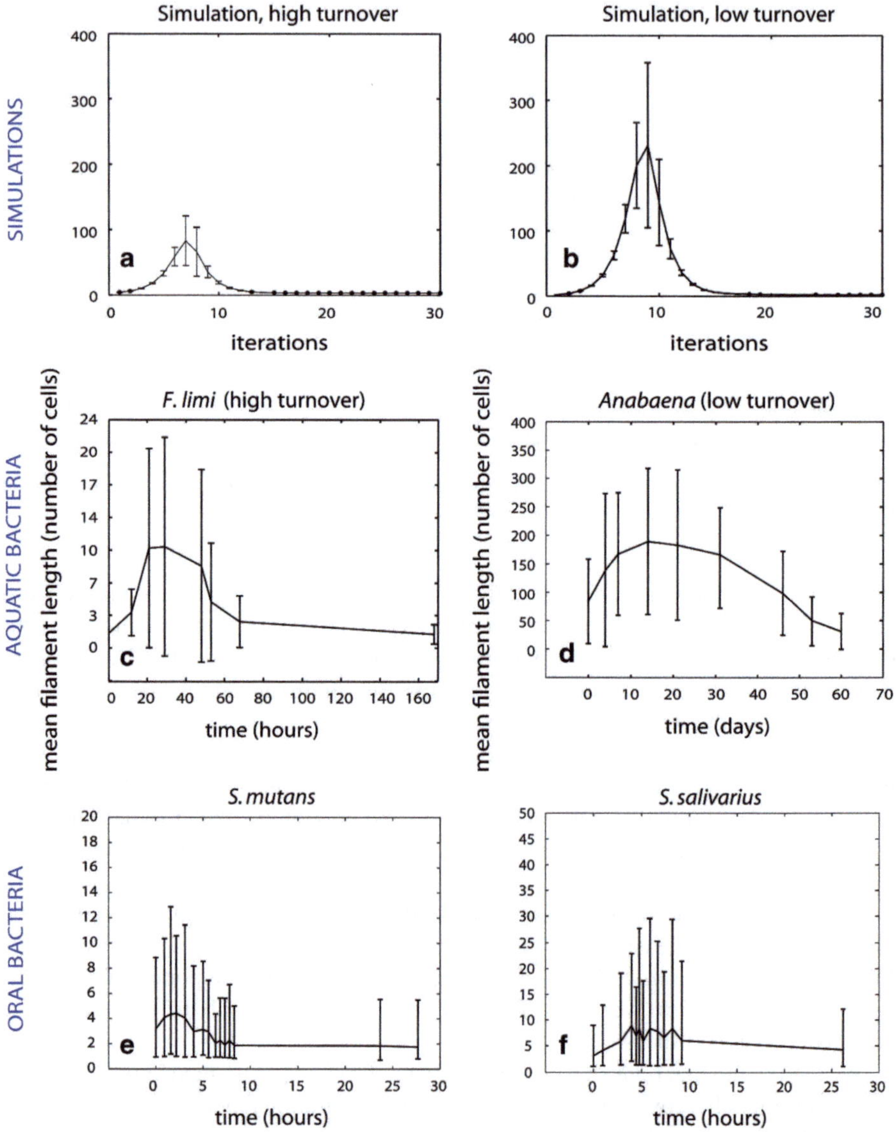

Fig. 3 Mean filament length of selected simulation and experimental cases. In every plot, the mean filament length in terms of number of cells is plotted against the time unit (iterations for the simulations, hours or days for the bacteria). *Lower* and *upper* edges of the error bars indicate the 2.5th and 97.5th percentiles respectively. Qualitatively, it is possible to observe a cyclic pattern of filament length across species, as well as in the simulated populations. In the cases of high turnover, the filaments are generally shorter than those of the low turnover populations

The results obtained with the model indicate that differences in fitness are not the sole explanation for the modulation and evolution of multicellular life cycles in filamentous bacteria. In fact, density dependent birth and death rates of the cells can play a fundamental role, as turnover can affect the transient and carrying capacity phases. Given that birth and death rates are intrinsic properties of every organism, the multicellular life cycle results as an emergent property. The next section shows the results obtained by testing these hypotheses in natural bacterial populations.

Validation in Aquatic and Oral Bacteria

In addition to cyanobacteria, multicellular filamentous bacterial species can be found in many environments, such as freshwater, oceans, soil and extreme habitats. In the human body, the oral cavity is inhabited by several bacterial species. Among the most abundant oral bacteria are members of the genus *Streptococcus*, and many streptococcal groups are filamentous and multicellular (Paerl et al. 2000; Rosen et al. 2007; Tannock 1999; Whitton and Potts 2000; Kolenbrander 2000). For our studies, we chose two marine cyanobacterial species (*Nostoc muscorum* and *Anabaena variabilis*), two marine heterotrophic strains (*Rudanella lutea* and *Fibrella aestuarina*) and four oral streptococcal strains (*S. mutans, S. salivarius, S. anginosus* and *S. oralis*). Bacteria were grown in batch cultures and monitored along their life cycle. At regular time intervals, pictures of the bacterial population were taken and analyzed to collect the mean length of the filaments.

In the theoretical model, the turnover rate was the key parameter determining the length of the filaments. However, turnover rate is not an easily measurable quantity in natural populations, hence we established a mathematical correlation between turnover and generation time of the used species. Our formula predicts that the generation time G of a natural bacterial population is a function of the turnover rate θ according to

$$G(\theta) = \frac{(1-\theta)}{\theta} \tag{3}$$

Equation 3 allows for a comparison of theoretical and experimental results, and states that low and high turnover rate correspond to long and short generation time respectively. In the case of our species, cyanobacteria have a very long generation time, and hence were considered as low turnover species. Heterotrophs instead have generally faster generation times and hence high turnover.

The experimental results (Fig. 3, C–F) supported the predictions from simulations (Fig. 3, A and B). A short-long-short pattern of average length was recognizable in cyanobacteria, marine heterotrophs and in oral streptococci. In all these cases, the filaments are short in the beginning, while the average length increases significantly during the growth phase, only to shorten again as it reaches the carrying capacity. Cyanobacteria, having a long generation time (hence with lower turnover), reached much greater length than the heterotrophs, which had a faster generation time (hence

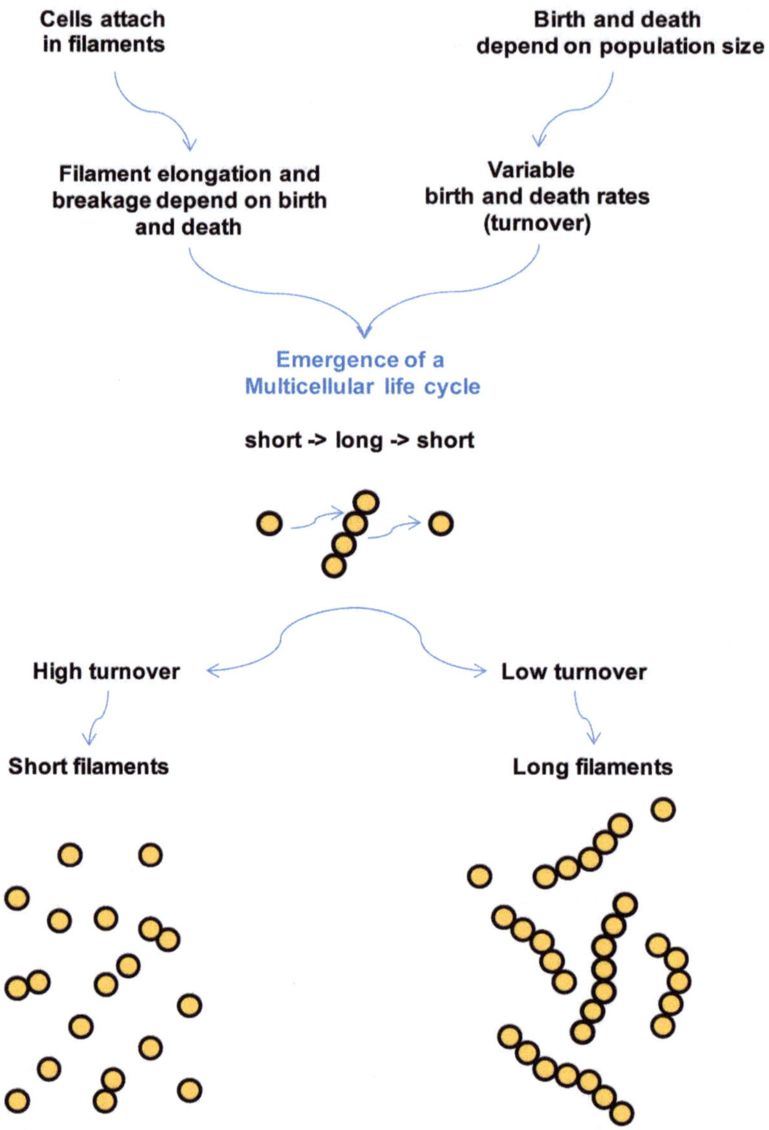

Fig. 4 Schematic overview of the main conclusion. Differences in cellular turnover (corresponding to generation time) are intrinsic properties of the organisms. These can lead to an emergent multicellular life cycles in filamentous species. The size of the multicellular organism can be regulated by the generation time, which depends on the turnover

higher turnover), both during the growth phase and at carrying capacity. Within the Streptococci, the phenotypic diversity was also relevant. *S. salivarius* showed greater elongation than the other three strains. *S. oralis* and *S. anginosus* were in contrast characterized by very short filaments at every growth phase. Oral streptococci are generally linked to a variety of oral diseases such as caries and abscesses. Interestingly, the strain with the longest filaments (*S. salivarius*) is usually associated with oral health instead of oral diseases.

Conclusions

Our studies argue that density-dependent processes at the cellular level can induce a cycle between single-celled and multicellular stages at the filament level (Fig. 4). Theory and experiments both show that due to differences in parameters affecting life history traits, namely birth and death rates (and hence generation time), bacteria can cover a wide spectrum of filament lengths. This prediction was tested in photoautotrophic and heterotrophic species, suggesting that the observed pattern can be common to different genera. This leads to the conclusion that the predominance of multicellularity in the life cycle can be achieved to different degrees, and that most importantly, this is not necessarily determined by morphology-dependent differences in fitness. Here, the factors affecting multicellularity are turnover rate and generation time. Growth rate, considered as an indicator of fitness, can be the same for different turnovers, as shown by the simulations.

The life history traits governing filament length are intrinsic properties of every living organism. The mulitcellular life cycle described here automatically arises from the interplay between ecology and the filamentous nature of the bacteria, and can hence be considered as an emergent property. Within this perspective, the evolution of a simple developmental program can have its roots in the interaction between ecology and life history.

Summary

1. The emergence of multicellular life cycles is studied in a theoretical framework that does not imply any *a priori* selective advantage of multicellularity—rather, it considers changes in birth and death rates at the cellular level as drivers of multicellular cycles in simple organisms.
2. The theoretical model predicts that according to different birth and death rates of cells, simple filamentous organisms can cover a wide spectrum of lengths in terms of number of cells, and this is not necessarily a consequence of morphology-dependent differences in fitness.
3. Birth and death rates of cells are related to the generation time of natural bacterial populations using a new mathematical formula. This allows for a validation of the model in aquatic and oral bacteria, which shows that the predicted patterns are common to different genera.

References

Adamec F, Kaftan D, Nedbal L (2005) Stress-induced filament fragmentation of *Calothrix elenkinii* (Cyanobacteria) is facilitated by death of high-fluorescence cells. J Phycol 41(4):835–839
Arendt D (2008) The evolution of cell types in animals: emerging principles from molecular studies. Nat Rev Genet 9(11):868–882
Axelrod R, Hamilton WD (1981) The evolution of cooperation. Science 211(4489):1390–1396
Berleman JE, Kirby JR (2009) Deciphering the hunting strategy of a bacterial wolfpack. FEMS Microbiol Rev 33(5):942–957
Berman-Frank I, Bidle KD, Haramaty L, Falkowski PG (2004) The demise of the marine cyanobacterium, Trichodesmium spp., via an autocatalyzed cell death pathway. Limnol Oceanogr 49(4):997–1005
Boomsma JJ, Franks NR (2006) Social insects: from selfish genes to self organisation and beyond. Trends Ecol Evol 21(6):303–308
Bonner JT (1974) On development: the biology of form. Harvard University Press, Cambridge
Boraas ME, Seale DB, Boxhorn JE (1998) Phagotrophy by a flagellate selects for colonial prey: A possible origin of multicellularity. Evol Ecol 12(2):153–164. doi:10.1023/a:1006527528063
Buss LW (1987) The evolution of individuality. Princeton University Press, Princeton
Claessen D, Rozen DE, Kuipers OP, Sogaard-Andersen L, van Wezel GP (2014) Bacterial solutions to multicellularity: a tale of biofilms, filaments and fruiting bodies. Nat Rev Microbiol 12(2):115–124. doi:10.1038/nrmicro3178
Daft MJ, Stewart WDP (1973) Light and electron-microscope observations on algal lysis by bacterium CP-1. New Phytol 72(4):799–808
Dworkin M (1972) The myxobacteria: new directions in studies of procaryotic development. Crit Rev Microbiol 1(4):435–452
Hahn MW, Hoefle MG (1998) Grazing pressure by a bacterivorous flagellate reverses the relative abundance of comamonas acidovorans PX54 and Vibrio Strain CB5 in Chemostat Cocultures. Appl Environ Microbiol 64(5):1910–1918
Hardin G (1968) The tragedy of the commons. Science 162:143
Ispolatov I, Ackermann M, Doebeli M (2012) Division of labour and the evolution of multicellularity. Proc R Soc Lond B Biol Sci 279(1734):1768–1776. doi:10.1098/rspb.2011.1999
Kamp A, Roy H, Schulz-Vogt HN (2008) Video supported analysis of Beggiatoa filament growth, breakage, and movement. Microb Ecol 56(3):484–491
Kolenbrander PE (2000) Oral microbial communities: biofilms, interactions, and genetic systems. Annu Rev Microbiol 54:413–437. doi:10.1146/annurev.micro.54.1.413
Komarek J, Anagnostidis K (2005) Susswasserflora von Mitteleuropa 19/2. Cyanoprokaryota, 2. Teil: Oscillatriales. Spektrum, Heidelberg
Kruskopf M (2006) Growth and filament length of the bloom forming *Oscillatoria simplicissima* (Oscillatoriales, Cyanophyta) in varying N and P concentrations. Hydrobiologia 556(1):357–362
Lewis K (2000) Programmed death in bacteria. Microbiol Mol Biol Rev 64(3):503–514
Michod RE, Roze D (2001) Cooperation and conflict in the evolution of multicellularity. Heredity 86(Part 1):1–7
Nedelcu AM, Michod RE (2006) The evolutionary origin of an altruistic gene. Mol Biol Evol 23(8):1460–1464. doi:10.1093/molbev/msl016
Ning SB, Guo HL, Wang L, Song YC (2002) Salt stress induces programmed cell death in prokaryotic organism Anabaena. J Appl Microbiol 93(1):15–28
Paerl HW, Pinckney JL, Steppe TF (2000) Cyanobacterial-bacterial mat consortia: examining the functional unit of microbial survival and growth in extreme environments. Environ Microbiol 2(1):11–26
Ratcliff WC, Pentz JT, Travisano M (2013) Tempo and mode of multicellular adaptation in experimentally evolved *Saccharomyces cerevisiae*. Evolution 67(6):1573–1581. doi:10.1111/evo.12101

Rokas A (2008) The origins of multicellularity and the early history of the genetic toolkit for animal development. Ann Rev Genet 42(1):235–251

Rosen DA, Hooton TM, Stamm WE, Humphrey PA, Hultgren SJ (2007) Detection of intracellular bacterial communities in human urinary tract infection. PLoS Med 4(12):1949–1958

Rossetti V, Ammann TW, Thurnheer T, Bagheri HC, Belibasakis GN (2013) Phenotypic diversity of multicellular filamentation in oral streptococci. Plos One 8(9). doi:10.1371/journal.pone.0076221

Rossetti V, Filippini M, Svercel M, Barbour AD, Bagheri HC (2011) Emergent multicellular life cycles in filamentous bacteria owing to density-dependent population dynamics. J R Soc Interface 8(65):1772–1784. doi:10.1098/rsif.2011.0102

Sachs JL, Mueller UG, Wilcox TP, Bull JJ (2004) The evolution of cooperation. The Quart Rev Biol 79(2):135–160

Schirrmeister BE, Antonelli A, Bagheri HC (2011) The origin of multicellularity in Cyanobacteria. BMC Evol Biol 11(1):45

Schirrmeister BE, Dalquen DA, Anisimova M, Bagheri HC (2012) Gene copy number variation and its significance in cyanobacterial phylogeny. BMC Microbiol 12. doi:10.1186/1471-2180-12-177

Schopf JW (2006) Fossil evidence of Archaean life. Philos Trans R Soc Lond B Biol Sci 361(1470):869–885

Solari CA, Kessler JO, Michod RE (2006) A hydrodynamics approach to the evolution of multicellularity: flagellar motility and germ-soma differentiation in Volvocalean Green Algae. Am Nat 167(4):537–554

Strassmann JE, Zhu Y, Queller DC (2000) Altruism and social cheating in the social amoeba *Dictyostelium discoideum*. Nature 408(6815):965–967

Tannock GW (1999) Analysis of the intestinal microflora: a renaissance. Anton Leeuw Int J G Mol Microbiol 76(1–4):265–278

van Hannen EJ, Zwart G, van Agterveld MP, Gons HJ, Ebert J, Laanbroek HJ (1999) Changes in bacterial and eukaryotic community structure after mass lysis of filamentous cyanobacteria associated with viruses. Appl Environ Microbiol 65(2):795–801

Weinbauer MG (2004) Ecology of prokaryotic viruses. FEMS Microbiol Rev 28(2):127–181

Whitton BA, Potts M (eds) (2000) The ecology of cyanobacteria. Their diversity in time and space. Springer

Wu H, Gao K, Villafane VE, Watanabe T, Helbling EW (2005) Effects of solar UV radiation on morphology and photosynthesis of filamentous cyanobacterium *Arthrospira platensis*. Appl Environ Microbiol 71(9):5004–5013

The Evolutionary Ecology of Multicellularity: The Volvocine Green Algae as a Case Study

Cristian A. Solari, Vanina J. Galzenati and John O. Kessler

Abstract The volvocine green algae in the order Volvocales are an ideal model-system for studying the unicellular-multicellular transition since they comprise an assemblage of lineages featuring varying degrees of complexity in terms of colony size, colony structure, and cellular specialization. Here, we have investigated the size-related advantages that might have caused single-celled volvocine algae to start living in groups, and the possible reasons for the evolution of cellular differentiation as group size increased, which created multicellular volvocine algae with germ-soma separation. Primordial cell clusters might have benefited from decreased predation, increased nutrient uptake, nutrient storage, and enhanced motility capabilities. We have tested these hypotheses by analyzing previous data on motility and growth rates in Volvocales. We have also compared the growth rates of the unicellular *Chlamydomonas reinhardtii* and *Gonium pectorale*, a 1–16 celled volvocine alga, at different nutrient concentrations, and measured predation rates on these species using the phagotrophic euglenoid *Peranema trichophorum*. Our analyses support the hypothesis that predation was an important selective pressure for the origin of multicellularity, but found no evidence that increased motility and nutrient uptake were advantages for the first cell groups. The extra-cellular matrix necessary for cell clustering might have been later co-opted for nutrient storage. With regards to cellular differentiation, we review a model inspired by the Volvocales that explains the dynamics of the transition to germ-soma differentiation as size increases. We found that flagellar motility constraints and opportunities were important driving forces for germ-soma separation in this group. We argue that germ-soma separation in Volvocales evolved to counteract the increasing costs of larger multicellular colonies.

C. A. Solari (✉) · V. J. Galzenati
Laboratorio de Biología Comparada de Protistas,
IBBEA, CONICET-UBA, Universidad de Buenos Aires,
C1428EHA Buenos Aires, Argentina
e-mail: casolari@bg.fcen.uba.ar

V. J. Galzenati
e-mail: vaninagalzenati@yahoo.com.ar

J. O. Kessler
Department of Physics, University of Arizona, Tucson, AZ 85721, USA
e-mail: jottok@gmail.com

I. Ruiz-Trillo, A. M. Nedelcu (eds.), *Evolutionary Transitions to Multicellular Life*,
Advances in Marine Genomics 2, DOI 10.1007/978-94-017-9642-2_11

Keywords Cost of reproduction · Germ-soma differentiation · Life-history evolution · Motility · Multicellularity · Nutrient storage · Nutrient uptake · Predation · Size-related advantages · Volvocales

Introduction

Multicellularity evolved multiple times independently, from quite different unicellular ancestors (Chapter "Timing the Origins of Multicellular Eukaryotes Through Phylogenomics and Relaxed Molecular Clock Analyses"). As Grosberg and Strathmann (2007) state in their extensive review on the subject, the frequent origination and general spread of multicellularity suggest that selection pressures that favor multicellularity are common and that genetic and developmental obstacles related to this transition are relatively easy to overcome. Recent experiments applying strong artificial selection on unicellular yeast *Saccharomyces cerevisiae* and the unicellular green algae *Chlamydomonas reinhardtii* have been successful at evolving multicellular strains from unicellular ones in just a few generations (Ratcliff et al. 2012, 2013).

It is generally assumed that size-related advantages are responsible for the evolution of multicellularity. For unicellular organisms, general constraints such as the decrease in the surface to volume ratio set an upper limit to their size. Given these constraints, the aggregation of mitotic products held together by a cohesive extra-cellular material enabled organisms in certain lineages to increase in size by increasing cell number (instead of cell size). These primordial groups of undifferentiated cells might have benefited from decreased predation, as well as having novel opportunities to increase nutrient uptake, nutrient storage, and motility capabilities, create a buffered internal environment, share metabolites between cells, to name a few (see Grosberg and Strathmann (2007) for a more thorough review of the size-related advantages).

Size-related advantages can benefit both viability (e.g., in terms of predation avoidance and higher motility) and fecundity (e.g., higher number or quality of offspring), the two basic fitness components. Nevertheless, a large size can also become costly, both in terms of viability (e.g., increased need for local resources) and fecundity (e.g., increased generation time). As size increases, such costs can reach a point where the fitness of the emerging multicellular individual might be negatively affected. Consequently, to maintain levels of fitness that allow for further increase in size, the benefits have to be increased and/or the costs have to be reduced. In the early multicellular organisms, cellular differentiation was one of the solutions to increase the benefits and reduce the costs.

In the first group of cells that stayed together and formed the simplest colonies, all cells retained both vegetative and reproductive functions and remained undifferentiated. Natural selection acted on these colonial/undifferentiated organisms and generated more complex forms with cellular differentiation—that is, with cells specialized in vegetative (i.e., soma) and reproductive (i.e., germ) functions. Germ-soma separation helps to create the emergence of a higher level of individuality since both

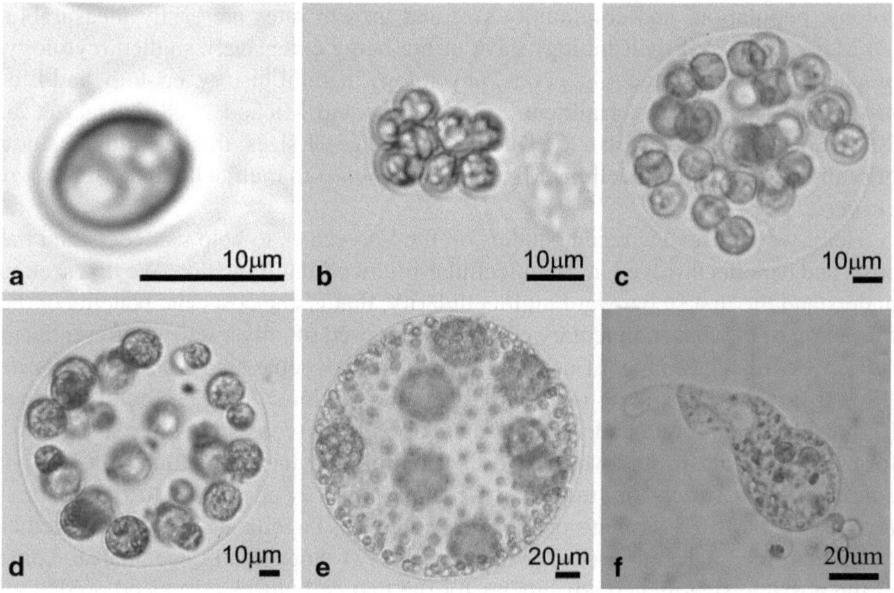

Fig. 1 A selection of the volvocine green algae arranged according to organism size, and the phagotroph *Peranema trichophorum*. Unicellular *Chlamydomonas reinhardtii* (**a**), undifferentiated *Gonium pectorale* (**b**) and *Eudorina elegans* (**c**), the soma-differentiated *Pleodorina californica* (**d**) and germ–soma differentiated *Volvox carteri* (**e**). When two cell types can be recognized (**d** and **e**), the smaller are the somatic cells, and the larger are the germ cells. *Peranema trichophorum* (**f**) trapped in 0.5 % concentration agar. The spheroids inside *P. trichophorum* are *C. reinhardtii* cells being digested; the clear spheroids outside are the expelled vacuoles after digestion

cell lines depend on each other for the success of the whole organism. As cells specialize in the different fitness components (i.e., fecundity and viability), they relinquish their autonomy in favor of the group, and, as a result, fitness and individuality are transferred from the cell to the group level.

The volvocine green algae are an ideal model system for studying the transition from unicellular to multicellular life, since they comprise an assemblage of lineages featuring varying degrees of complexity in terms of colony size, colony structure, and cell specialization (Fig. 1; e.g., Koufopanou 1994; Kirk 1998; Solari et al. 2006a; Herron and Michod 2008; Coleman 2012; Chapter "Volvocine Algae: From Simple to Complex Multicellularity"). These freshwater bi-flagellated organisms range from unicellular species such as *Chlamydomonas*, to colonies composed of 4–64 cells with no cellular differentiation, e.g., *Gonium*, *Pandorina*, *Eudorina*, to multicellular individuals comprising 1000–50,000 cells with complete germ-soma separation, e.g., *Volvox*. Within this lineage, cellular differentiation has evolved independently multiple times (e.g., Nozaki et. al. 2006; Herron and Michod 2008, Coleman 2012). The multicellular *Volvox* genus, which contains species with germ-soma differentiation, is polyphyletic.

Volvocales exhibit a number of features that make them especially suitable for empirical studies: (1) Asexual, monoclonal populations are easily obtained. (2) Populations are easily grown in a well-defined freshwater mineral medium. (3) Cell, colony, population, and community size and growth rates are easily measurable. (4) Many aspects of their biology have or are being extensively studied (cytology, biochemistry, development, genetics, physiology, natural history, ecology and life-history; e.g., Kirk 1998, Hallman 2006, Herron and Michod 2008, Nozaki et al. 2006, Solari et al. 2006a, b). (5) Due to their range of sizes, they enable the study of scaling laws (from unicellular 10^0 *Chlamydomonas* to multicellular $\sim 10^4$ *Volvox barberi*).

Here, we will review recent studies on the Volvocales to help us understand the costs and benefits involved in the unicellular-multicellular transition. We first seek to understand the first step toward multicellularity; that is, the costs and benefits of the formation of the first rudimentary colonies composed of only a few undifferentiated cells. We will center most of our analysis on the volvocine alga *Gonium pectorale* (Fig. 1b), which depending on environmental conditions it can be found in either its unicellular (as single-celled individuals) or colonial (consisting of up to 16 cells) form. This organism has worldwide distribution and has been found on all five continents in a variety of streams and ponds (Fabry et al. 1999). In the colonial form, *G. pectorale* is a plate of undifferentiated bi-flagellated cells held together by an extra-cellular matrix (ECM; Coleman 2012). Specifically, we analyze some of the possible size-related advantages for the first cell clusters: reduced predation, increased nutrient uptake, enhanced nutrient storage, and increased motility.

Our second goal is to examine the transition from cell clusters of undifferentiated cells to multicellular organisms with cellular differentiation. Over the last years, many features such as motility and metabolism have been studied in detail in the Volvocales (e.g., Short et al. 2006, Solari et al. 2006a, b, 2011). In these organisms, the constraints and opportunities of flagellar motility as colony size increased might have been the major driving force in the transition to multicellularity and germ-soma separation. We will review a model inspired by the volvocine green algae that explains the dynamics involved in the unicellular-multicellular transition using life-history theory and allometry (Solari et al. 2013). The model analyzes the two fitness components (fecundity and viability) and compares the fitness of hypothetical colonies of different sizes with varying degrees of cellular differentiation to understand the general principles that underlie the evolution of multicellularity. Some of the allometric relationships that have been derived on the motility and metabolism studies in this algae group will help us illustrate the model, shedding some light on the transition to cellular differentiation.

The First Steps Toward Multicellularity

Motility

Volvocales are negatively buoyant (i.e., denser than water) and need flagellar beating to avoid sinking. They are found in quiet, standing waters of transient vernal puddles, water bodies such as rice paddies, or in permanent lakes when thermal stirring stops and the lake becomes stratified (Reynolds 1984; Kirk 1998). Thus, Volvocales need motility to stay afloat, to control their position in the water column, and to reach light and nutrients. In these still environments higher motility capabilities might give a competitive advantage. For example, Sommer and Gliwicz (1986) found that *Volvox* colonies migrated vertically several meters at night, presumably in search of higher phosphorous concentrations.

The upward swimming (V_{up}) and sedimentation (V_{sed}) velocities were measured for Volvocales of different size and complexity (Solari et al. 2006a). It was shown that in Volvocales swimming speeds increase with colony cell number (N; $V_{up} \propto N^{0.27}$). However, when we look at where *G. pectorale* (Fig. 1b), a species that forms a cluster of few undifferentiated cells, fits in the derived allometric relationship we find that its swimming speed is below the linear regression line, even outside the 95 % confidence interval (Fig. 2). Furthermore, when swimming speeds in synchronized populations at 1 and 3 h into the light cycle (Solari et al. 2006a) and in full and diluted medium (Solari et al. 2011) was measured, the average upward swimming speed of *G. pectorale* colonies was even lower than that of the unicellular *C. reinhardtii*. In Solari et al. (2006a), *G. pectorale* upward swimming speeds were 32.5 μm/sec vs. 38 μm/sec for *C. reinhardtii* (most *G. pectorale* colonies in this experiments were 8-celled), and in Solari et al. (2011), *G. pectorale* upward swimming speeds were 33 μm/sec in full and 37 μm/sec in diluted medium vs. 43 μm/sec and 54 μm/sec for *C. reinhardtii* (most *G. pectorale* colonies in this experiments were 4-celled). Moreover, the plate colony design of *G. pectorale* did not slow sinking (Fig. 2). Sedimentation speeds were higher for *G. pectorale* compared to *C. reinhardtii* in Solari et al. (2006a; 9.5 μm/sec vs. 1.4 μm/sec) and in Solari et al. (2011; 10 μm/sec in full and 8.9 μm/sec in diluted medium vs. 6.5 μm/sec in full and 3.7 μm/sec in diluted medium).

G. pectorale swims awkwardly. The plate formed by the cells irregularly rotates and is not capable of keeping a straight trajectory line as *C. reinhardtii* or other more complex Volvocales are capable of (personal observations). In more complex 16–64-celled volvocine species like *Eudorina elegans* (Fig. 1c), upward swimming speeds increase at least twofold (82 μm/sec, Solari et al. 2006a). But, this species has increased structural complexity, producing a spherical structure made up of extra-cellular matrix (ECM). Each of the cells in a colony of *E. elegans* is found at the surface of the ECM, with its two flagella oriented outwards. The external ECM spherical structure allows a symmetric organization and the separation of cells within the colony. This innovation makes a better design for coordinated collective flagellar swimming.

In the case of *G. pectorale*, the simple modifications for colony swimming seem not to be enough to increase motility over that of the unicellular counterpart. The flagellar apparatus of *G. pectorale* tends to be less modified than in other more complex volvocine species (Gerisch 1959; Greuel and Floyd 1985; Hoops 1997). The cells at the center of a *G. pectorale* 12-celled plate retain the flagellar symmetry present in unicells, whereas the ones at the periphery do not. When isolated, the central cells can swim forward as *Chlamydomonas* cells do, but the outer cells spin in the same place (Gerisch 1959).

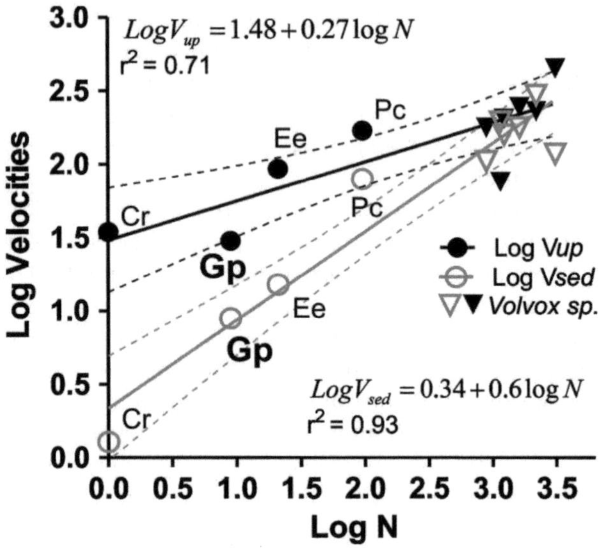

Fig. 2 V_{up} and V_{sed} as a function of N. Only the first measurements of the time series from Solari et al. (2006a) were used for the interspecies allometric analysis of newly hatched synchronized colonies (1 h after algae hatched). The average V_{up} and V_{sed} for each species were measured on the same population as detailed in Solari et al. (2006a). *Cr Chlamydomonas reinhardtii*, *Gp Gonium pectorale*, *Ee Eudorina elegans*, *Pc Pleodorina californica*. *Volvox* species are *Volvox carteri* grown at 600 foot-candles (*fc*); *V. carteri* grown at 1000 fc; *Volvox obversus*; *Volvox tertius*; *Volvox aureus*; *Volvox rousseletii*

Due to the several steps that are needed to enhance motility, we find it difficult to conclude that motility was one of the size-related advantages that were responsible for the evolution of the first cell clusters in this lineage. To have directional motility for a new cell cluster, changes in the flagellar beating pattern would be needed for at least some of the cells in the group, but then colony design must be improved for directional swimming and increased speed (as occurs in *Eudorina* colonies).

Nutrient Uptake

Microorganisms live in a world of low Reynolds number regime ($Re < 1$, Guyon et al. 2001). $Re = RV\rho_w/\eta < 1$, where R can be the volvocine unicell or colony radius, V the swimming or sedimentation speed, η the viscosity of water, and ρ_w the density of water. Even for a hypothetical large *Volvox* colony swimming at a considerable speed, $Re < 1$ (e.g., $Re = 0.25$ if $R = 0.05$ cm, $V = 0.05$ cm/sec, $\eta = 10^{-2}$g/sec cm, $\rho_w = 1$ g/cm^3). In this regime, motion is dominated by friction, flows are linear, time is reversible, and transport is dominated by diffusion. The transport of dissolved nutrient molecules to the surface of an organism is proportional to the molecular concentration gradient. The magnitude of this gradient depends both on the remote concentration of the molecules and on the rate at which those molecules are absorbed by the organism. These insights are the (correct) basis of conventional wisdom regarding the dynamics of microorganisms.

The situation changes radically when many closely spaced, moving structures, such as flagella arrayed on the surface of an algal colony, collectively generate flows. Vigorous boundary layer stirring and the flow associated with swimming can greatly increase transport rates by advection and mixing of molecules. The relative importance of these processes can be evaluated by a ratio of time constants for diffusion ($t_{diff} = L^2/D$) and advection ($t_{adv} = L/V$) called the Peclet number ($Pe = LV/D$), where V is the fluid or swimming speed, L is a characteristic length such as the organism's radius, and D the diffusion coefficient of an important molecule (e.g., $D = 2 \times 10^{-5} cm^2/sec$ for O_2). If $Pe < 1$, a strong interaction between flagellar beating and increased nutrient uptake can be ruled out, whereas if $Pe > 1$, advection is important (Solari et al. 2006b). Collective flagellar beating increases the speed (V) and distances (L) of the flows, so Pe may be > 1 for large volvocine colonies.

For example, $Pe < 1$ for *C. reinhardtii* (Fig. 1a), but in the large *Volvox* spheroids (Fig. 1e) with $\sim 10^3$–10^4 flagellated somatic cells, $Pe >> 1$ (For O_2 transport, $Pe = 0.25$ and 40 for *C. reinhardtii* and *V. carteri* respectively; Solari et al. 2006b). Therefore, advection might not be that important for nutrient uptake in a unicellular flagellated organism, but might be important for nutrient uptake in a flagellated colonial one.

If we analyze the case of *G. pectorale* or any similar colonial species with few cells, even if we assume L to be 40 μm for a plate of 16 cells, $Pe = 1$ for O_2 transport ($V \sim 5 \times 10^{-3}$cm/sec, $D = 2 \times 10^{-5}$ cm^2/sec). At the scale of a group of only a few cells, a preliminary simple analysis shows no evidence that collective flagellar beating might increase nutrient uptake.

If flagella in colonies are generating fluid flows to enhance nutrient uptake, then conditions of nutrient deprivation might trigger changes, a phenotypic plasticity response, in the motility apparatus to mitigate such stress. Complementing the low Pe calculation on *G. pectorale*, it was reported in Solari et al. (2011) that two *Volvox* species with thousand of cells made an investment into collective flagellar beating during nutrient deprivation, but under those same conditions *C. reinhardtii* and *G. pectorale* did not.

In short, it is difficult to see how the first cluster of cells in the Volvocales might have benefited from increased nutrient uptake. Circumstantial evidence is also consistent with this conclusion. *G. pectorale* did not have higher growth rates compared to unicellular *C. reinhardtii* when populations were synchronized in bubbling medium and transferred every day (Solari et al. 2006a), or when population growth was measured in the exponential phase (Solari et al. 2011), or had higher abundance when populations were maintained at equilibrium in semi-continuous batch cultures at six temperatures (personal observations, unpublished data).

Storage

It was hypothesized by Bell (1985) and later discussed by Kirk (1998) that the ECM used to hold together the cells in multicellular volvocine algae might also function as storage for nutrients such as phosphorous. Once algae populations deplete the resources in a lake or pond, their growth and reproduction might depend on their capability to absorb and store these essential nutrients. It has been shown that population growth rates might be related more to the phosphorus concentration within the organisms than to the external concentrations (e.g., Rhee 1973). A *Chlamydomonas* unicell quickly takes up phosphorous from the environment and accumulates it as polyphosphates, but a multicellular volvocine alga has the potential to accumulate polyphosphates also in the ECM, decreasing the inhibition of the transport process and adding storage. The ECM in colonial Volvocales is mainly composed of glycoproteins and has developed from the simpler cell wall of the *Chlamydomonas*-like unicellular ancestor (Hallman 2006). It was shown that the ECM in Volvocales contains polyphosphates rich in phosphorous, and that the amount of polyphosphates accumulated in the ECM is proportional to the concentration of phosphate available in the medium (Werner et al. 2007). Nevertheless, there is still no evidence that these nutrients are used for growth when they are not available in the medium (Kirk 1998, Coleman 2012).

To evaluate if nutrient deprivation affects colonial species less than unicellular ones due to increased storage in the ECM, we have performed a preliminary analysis with populations of *C. reinhardtii* and *G. pectorale* cultured in full and 10^{-1} diluted medium (methods in Section "Conclusions"). In the 7 days that the experiment lasted, the population growth rate decreased in the diluted medium for both species compared to the full medium, but the decrease was steeper in *C. reinhardtii* populations (Fig. 3). When performing a multiple additive regression analysis, after taking into account the differences in growth rate between species and between the diluted and full medium, we found a positive interaction between the diluted treatment and *G. pectorale* (model $r^2 = 0.87$, all parameters $p < 0.0001$), supporting the hypothesis that additional nutrient storage might be helping the colonial species.

Additional support for this hypothesis is the break up of colonies observed in the experiment we performed (Fig. 3) as well as in older *G. pectorale* stock cultures (personal observation). On both cases where nutrients are scarce, most of the *G. pectorale* population is formed by unicells. We speculate that the lack of nutrients might be forcing colonies to use the ECM as a nutrient source, which would break up the colonies and would not allow new colonies to form.

These results, although indirect and preliminary, are consistent with the idea that additional nutrient storage in the ECM might benefit colonial species when nutrients are scarce. Detailed experiments with more *Chlamydomonas* and *Gonium* strains in different nutrient conditions are needed. In short, we find it plausible that the ECM necessary to form a cell cluster might have been co-opted to function as storage of nutrients.

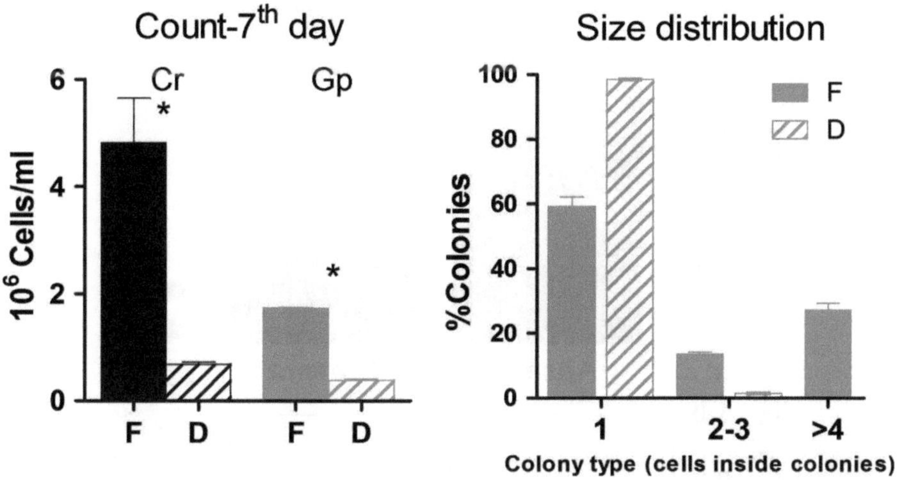

Fig. 3 Population size (cells/ml) of *C. reinhardtii* (*Cr*) and *G. pectorale* (*Gp*) and colony size distribution of *G. pectorale* populations after seven days in full (*F*) and 10^{-1} diluted (*D*) medium. Although the population size is significantly lower for both species after seven days in the diluted compared to the full medium, it decreased proportionally less in *G. pectorale*. B- Colony size distribution of *G. pectorale* populations after seven days in full and diluted medium. The population in the full medium has a higher proportion of colonies with more cells

Predation

Grosberg and Strathmann (2007) argued that phagotrophic microorganisms consuming unicellular prey were probably the most important selective pressure exerted for the origin of multicellularity. In different experiments with unicellular prey (e.g., *Scenedesmus acutus*, *Chlorella vulgaris*) colony formation was favored in the presence of predators (Boraas et al. 1998; Lurling and Van Donk 2000). Here we performed similar experiments where *C. reinhardtii* and *G. pectorale* populations were grown with the phagotrophic colorless euglenoid flagellate *Peranema tricophorum* (Fig. 1f; Chen 1950; Chang 1966; Breglia et al. 2006). *P. tricophorum* has worldwide distribution and is found in freshwater puddles, ponds, and lakes. This organism can feed on organic material, but is also a voracious predator of microbes such as bacteria, protists, and yeasts.

To measure how predation pressure could affect the evolution of multicellularity, we performed a short 10 h experiment to measure predation rates and a long three-week experiment to track the changes in the colony size distribution of *G. pectorale* populations (methods in Section "Summary"). Since there is no reproduction in the light period, the short 10 h experiment allowed us to measure predation rates. We found that the predation rate was much higher for *C. reinhardtii* than for *G. pectorale* (4.23 ± 0.40 vs. 1.08 ± 0.26 cells/h per *P. tricophorum*). Since on average *G. pectorale* cells were larger than *C. reinhardtii* ones (6.47 ± 0.05 vs. 5.45 ± 0.14 μm of cell radius at the end of the experiment), it was not possible to find out in this

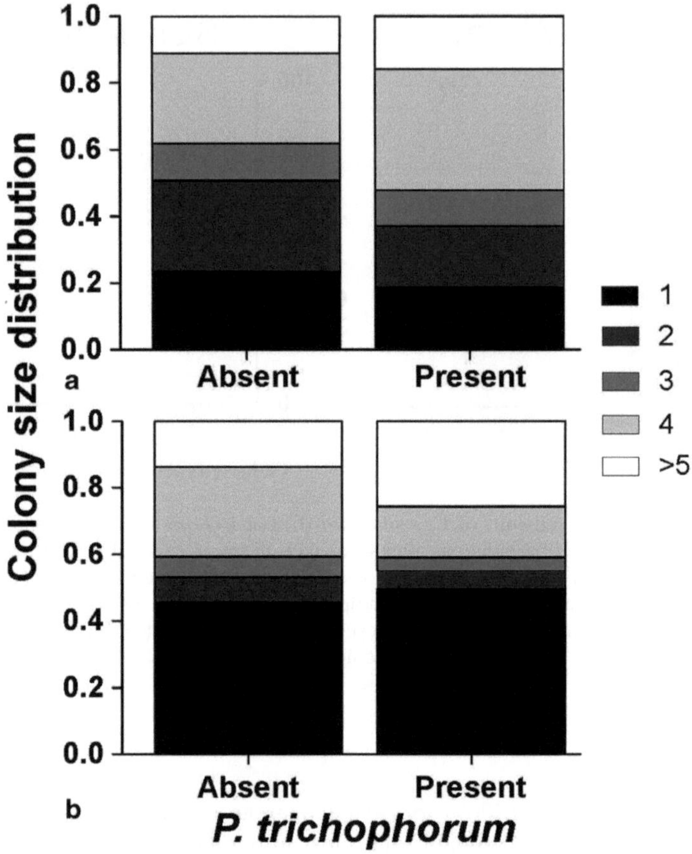

Fig. 4 Contigency analysis on the colony size distribution of the *G. pectorale* population with and without the presence of *P. tricophorum* for 10 h (**a**) and 3-weeks (**b**). In both cases there was an increase in the proportion of colonies with more than five cells

experiment if the lower predation rate on *G. pectorale* was due to cell size, colony formation, or both.

To decouple the effect of cell size and colony formation on predation, we compared the colony size distribution between the *G. pectorale* populations that were grown with and without the presence of *P. tricophorum*. Our contingency analysis on the 10 h experiment shows an increase in the proportion of colonies with 4 cells or more when *P. tricophorum* was present, and a decrease in the proportion of unicells and colonies with less than 4 cells (Chi-square $p < 0.0001$, Fig. 4a). For the duration of the three week long experiment, the contingency analysis also shows an increase in the proportion of colonies with more than 5 cells, but also an increase in the proportion of unicells compared to the control (Chi-square $p < 0.0001$, Fig. 4b).

When analyzing in detail the evolution of the proportion of unicells, 2–3-celled, 4-celled, and 8-celled colonies for the three weeks, we found that the largest difference in favor of 8-celled colonies compared to the control, and the decrease of the 2–3 and 4 celled colonies, agreed with the peak of the predator population (Fig. 5). Note that the proportion of 8-celled colonies was always higher in the population when *P. trichophorum* was present.

In short, our results agree with previous evidence, showing that predation can be a strong selective pressure for the origin of the first cell clusters. When the predator was present, colonial *G. pectorale* had a lower predation rate than unicellular *C. reinhardtii*, and there was an increase in the proportion of larger colonies in *G. pectorale* populations.

Conclusion

We have considered some of the size-related advantages that have been previously proposed for the transition from unicells to simple cell clusters using as a model system the volvocine green algae, and in particular the 1–16 celled colonial *G. pectorale*. The Volvocales give a rare insight into the costs and benefits of the transition from unicellular to the first, most basic colonial organisms. We found additional evidence supporting predation as one of the main selective pressures for the origin of multicellularity. We think that the ECM necessary for the evolution of multicellularity might have been co-opted for storage and used as a nutrient source, but good direct evidence is lacking. Based on our results and theoretical analysis, we found no evidence that increased motility and nutrient uptake were advantages of the first cell clusters in Volvocales.

The Evolution of Cellular Differentiation

Life-history Evolution

The fitness of any evolutionary unit can be understood in terms of its two basic components: fecundity and viability. In unicellular individuals, the same cell must be involved in both fitness components, typically these components being separated in time. However, in multicellular organisms, under certain circumstances, cells may specialize in one component or the other, the result being a division of labor, leading to the differentiation of germ and soma. The evolution of a specialized and sterile soma can increase viability and indirectly benefit fecundity (e.g., increasing nutrient uptake) but, all things being equal, must directly cost fecundity by reducing the number of cells producing offspring. On the other hand, the evolution of a specialized germ will benefit fecundity (by reducing the generation time and/or increasing the quality of offspring), but must directly cost viability by reducing the number of cells participating in viability-related functions.

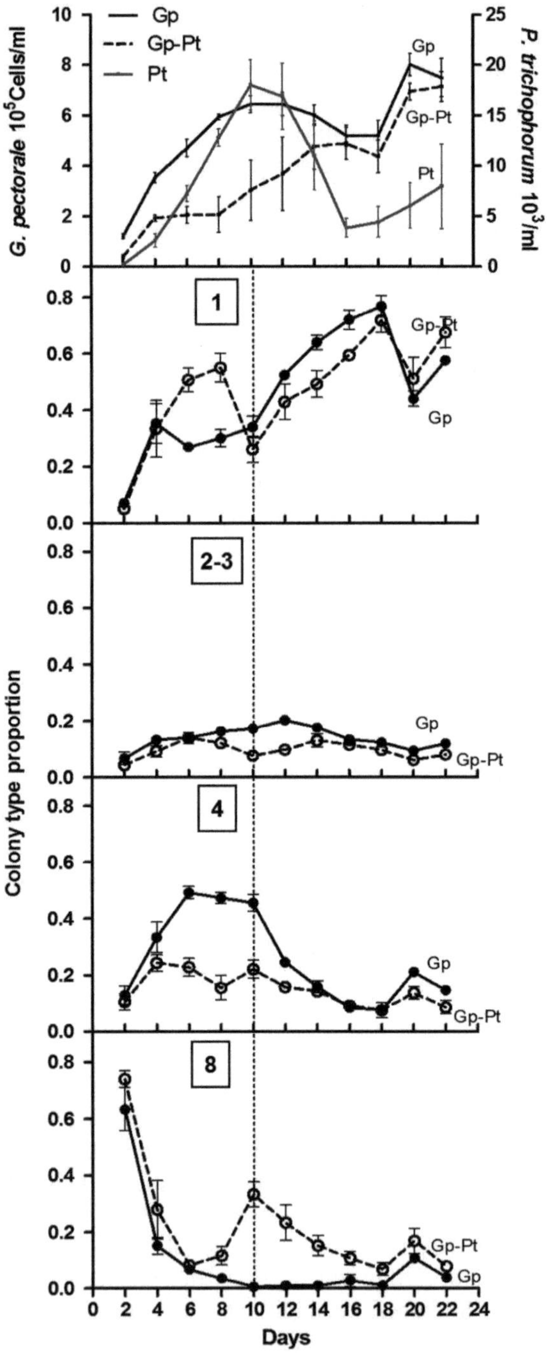

Fig. 5 Time-series analysis of the evolution of the total cells/ml of *G. pectorale* with (*Gp-Pt*) and without (*Gp*) the predator *Peranema trichophorum* (*Pt*) present, and the changes in proportion of its different colony sizes (*1*, *2–3*, *4*, *8*). In general, the larger colonies are favored (8-celled) in detriment of the 2–3 and 4-celled colonies when the predator is present, having the larger proportion change at the peak of the predator population (a *pointed line* was drawn at 10 days to aid the eye)

The various trade-offs between viability and fecundity are reflected in the variety of life-history traits among extant multicellular lineages. We argue that the evolution of germ-soma separation and the emergence of individuality and increased complexity at a higher level during the unicellular-multicellular transition can be also a consequence of these trade-offs. By using the volvocine green algae as a model system we show that the evolution of soma might be the expected outcome of reducing the cost of reproduction in order to realize the benefits associated with increasing size. As size increases further, the viability and fecundity benefits can be better achieved via the increase in specialization of germ and soma, and as a result, increased levels of complexity are achieved. In short, we suggest that the emergence of higher levels of complexity during the unicellular-multicellular transition can be a consequence of life history evolution.

The Model

Fecundity

The trade-offs between fecundity, viability, and size have been recently studied in Volvocales (Short et al. 2006; Solari et al. 2006a, b), and a model was recently developed based on life-history evolution (Solari et al. 2013). Some of the allometric relationships that have been derived in these studies on the Volvocales are used to illustrate this model.

Several aspects of the Volvocales' life history fit the assumptions of the model developed by Solari et al. (2013); asexual reproduction, discrete generation time, two cell types, fixed cell number throughout development, and the autocolony life cycle, in which each reproductive cell in the colony grows to produce offspring with the same cell number (Kirk 1998). In Volvocales, population growth is achieved via asexual reproduction; they only go through the sexual phase to produce resistant spores when conditions for survival are not met. Volvocales have discrete generation time; when new colonies hatch, the mother colony disintegrates. *Volvox* colonies with a differentiated germ and soma have only two cell types. Since the number of cells in Volvocales is determined by the number of cleavage divisions that take place during embryonic development, cell number is not augmented by accretionary cell divisions after juveniles hatch (Kirk 1997).

If an exponential growth model is used for the growth of undifferentiated cells, then $m = m_o e^{rt}$, where m_o is the initial size of the cells in a newly produced colonies, r the intrinsic growth rate for such cell and t time. As the number of cells n in the colony increases, the size m to which each cell within the colony grows, and the number of divisions d it performs to produce a daughter colony of the same type increase. Since each cell/embryo grows by a factor of n to produce daughter colonies of the same type (i.e., the autocolony process), also the size $m = nm_o$, where n is the number of cells in the colony. If we solve for cell number, $n = e^{rt}$, and if we solve for time, $t = Ln(n)/r$.

Assuming that cell division d is instantaneous, then t equals generation time T ($T = Ln(n)/r$). Since colonies have discrete generation time, the per-generation fecundity Ro of the group of cells or colony is equal to the number of cells in the colony since all cells produce daughter colonies of the same type: $Ro = n = 2^d$. For example, in a 16-cell colony ($n = 16$), each cell grows by a factor of n and undergoes 4 divisions ($d = 4, n = 2^d$) to develop a daughter colony with 16 cells with initial size m_o. By assuming discrete generation time, Ro can also be written in a simple way as a function of the fecundity rate λ, $Ro = \lambda^T$ (Stearns 1992). Since $n = e^{rT}$ and $Ro = n$, then

$$Ro = n = r^{rT} = \lambda^T, or \lambda = e^r \tag{1}$$

Since the size term n gets cancelled out, organisms of different sizes with a simple autocolony life cycle have the same fitness (i.e., fecundity rate λ). Equation 1 shows that size does not matter if the intrinsic cell growth rate r is not size-dependent.

But if these colonies invest in soma and a proportion s of cells in the colony becomes sterile, then the fecundity $Ro = n(1 - s)$ and the fecundity rate $\lambda = e^r(1 - s)^{1/T}$. Since $1/T = r/Ln\ (n)$, and $Ln(n) = dLn2$, then $1/T = r/dLn2$ and

$$\lambda = e^r(1 - s)^{1/T} \rightarrow \lambda = e^r\left[(1 - s)^{r/Ln2}\right]^{1/d} \tag{2}$$

Equation 2 shows that investing in soma (s) decreases the fecundity rate (λ), but it also shows that the negative effect of soma decreases and dilutes as colony size increases (Fig. 6). An increase in colony size (i.e., n, thus, the number of divisions d; $d = Log_2(n)$) decreases the exponent of the proportion $(1 - s)$, thus increasing the fecundity rate. In short, regardless of the costs and benefits that size might have on fitness, larger size gives a direct scaling benefit to cellular differentiation by decreasing its cost on the fecundity rate.

However, the intrinsic growth rate r, on which the fecundity rate greatly depends (Eq. 2), is not a constant, but is dependent on the supply and demand of resources, which in turns depends on size and cellular differentiation. It was envisioned in Solari et al. (2013) that the resources needed to produce the next generation, the cost of reproduction C, is proportional to the total number of cells a colony has to produce for the next generation. C greatly increases with size for undifferentiated colonies since all cells produce colonies ($C \sim n^2$), but this cost can be eased and shifted to a larger size by increasing the proportion of sterile somatic cells ($C \sim n^2(1 - s)$).

On the other hand, the acquisition of resources B needed to grow and reproduce is performed by the sterile somatic cells that lose reproductive functions (ns) and the undifferentiated reproductive cells that retain those functions ($n(1 - s)(1 - g)$), $B = ns + n(1 - s)(1 - g)$. Parameter g goes from no specialization ($g = 0$), meaning that reproductive cells retain full vegetative functions, to full specialization in reproductive functions ($g = 1$), meaning that reproductive cells lose all vegetative functions. In short, soma specialization can have an indirect benefit to the colony by decreasing the reproduction costs (the demand for resources C) and a direct benefit by helping the reproductive cells acquire resources for their metabolic needs (the supply of resources B).

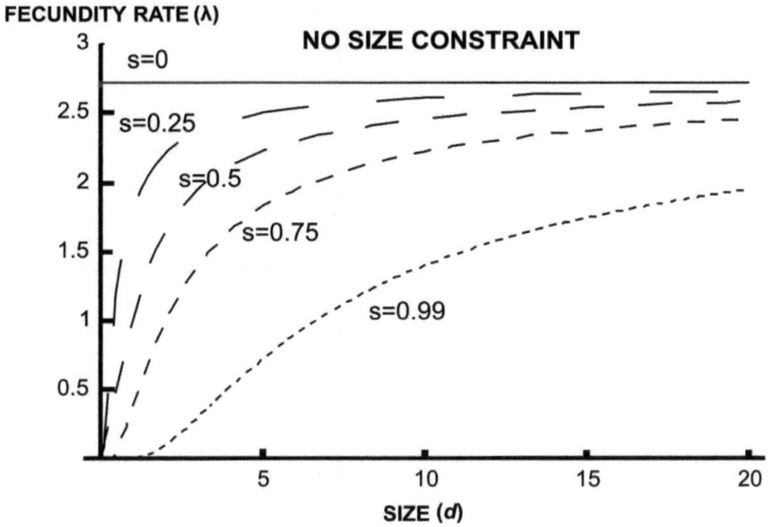

Fig. 6 The fecundity rate of colonies with no size dependency (the cell growth rate rate $r = 1$ remains constant as size increases) as a function of size d ($\text{Log}_2\ n$) for different proportions of somatic cells s

The ratio between the supply B and demand C of resources (B/C_r) is used as the factor that limits the intrinsic growth rate r of reproductive cells/embryos as colonies increase in size. Thus, r becomes,

$$\begin{aligned} &\textit{if}\, B/C_r \geq 1,\ r \approx r_o \\ &\textit{if}\, B/C_r < 1,\ r \approx r_o B/C_r \end{aligned} \qquad (3)$$

where r_o is the growth rate of a cell with no size constraints on its metabolic rate such as a unicellular *Chlamydomonas*. Equation 2 is a stepwise function since it is assumed that r_o is the maximum possible rate for an undifferentiated cell. If $B/C_r \geq 1$, supply meets the demand of resources, so cells grow at their maximum possible rate. If $B/C_r < 1$, the supply of resources does not meet the demand and limits r (Fig. 7a).

By using Eq. 3 in Eq. 2, the effects of size and cellular specialization on the fecundity rate can be evaluated. Fecundity rate curves of hypothetical colonies form peaks that shift to a larger size as the proportion of somatic cells s increases (Fig. 7b). For a fixed proportion of somatic cell s there is a colony size that optimizes the fecundity rate, this optimal size increasing as s increases. Depending on colony size and the proportion of somatic cells, colonies with specialized germ cells might have higher fecundity rates than colonies with non-specialized reproductive cells.

What happens in Volvocales with the supply and demand dynamics as size increases, and how does this affect the growth rates of colonies? We know that flagella are used for self-propulsion, but collective flagellar beating also serves to enhance the molecular transport of nutrients as explained in the previous Section "The First

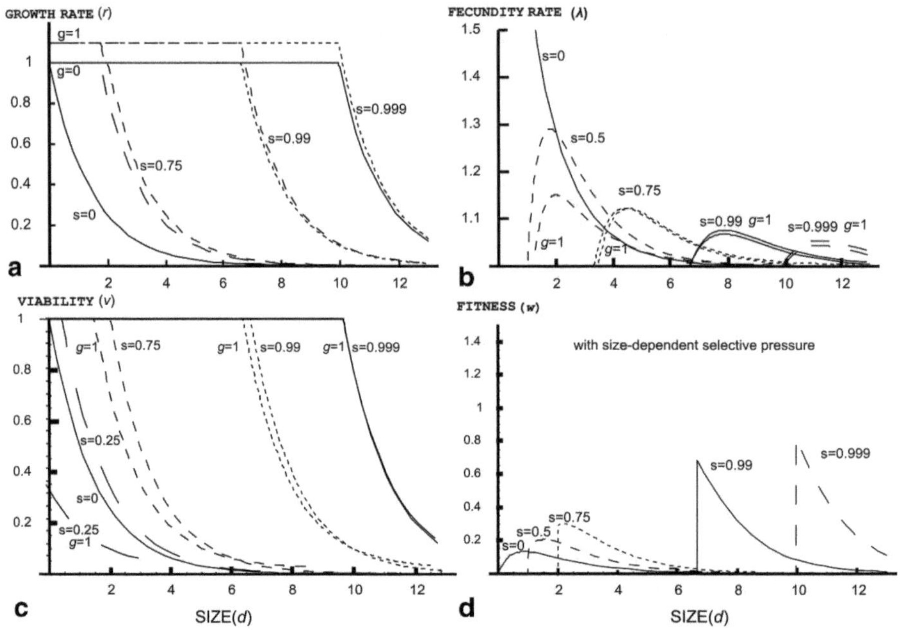

Fig. 7 **a** The reproductive cell growth rate r as a function of size d ($\text{Log}_2\ n$) for different proportions of somatic cells s (the unicellular growth rate $r_o = 1$). Plot of r vs d with ($g = 1$, $u_g = 0.1$) and without ($g = 0$) germ specialization. Germ specialization allows an overall increase of the growth rate regardless of d and s. **b** The fecundity rate of colonies λ as a function of size d ($\text{Log}_2\ n$) for different proportions of somatic cells s (the unicellular growth rate $r_o = 1$). Plot of λ vs. d with ($g = 1$, $u_g = 0.1$) and without ($g = 0$) germ specialization. For each proportion of somatic cells s there is a size that optimizes the fecundity rate λ. As s increases the fecundity rate peaks shift to larger size. Colonies with specialized germ cells and a low proportion of somatic cells s have lower fecundity rate peaks than colonies with no specialized reproductive cells since only a low proportion of cells contribute to the supply of resources. As s increases the peaks of the colonies with specialized germ cells may be higher since the vegetative functions are met by the somatic cells and specialized reproductive cells have higher growth rates. **c** The viabiltiy of colonies v as a function of size d ($\text{Log}_2\ n$) for different proportions of somatic cells s. Plot of v vs. d with ($g = 1$) and without germ specialization ($g = 0$). In general, the increase in the cost of reproduction C decreases viability as size increases; an increase in the proportion of somatic cells shifts the constraint to a larger size. Germ specialization decreases viability since specialized reproductive cells do not contribute to the viability of the colony, but this is compensated as the proportion of somatic cells s increases. **d** Fitness w with size-dependent mortality added $p = 1z(n)^{-\mu}$ ($z = 1, \mu = 0.25$). The fitness of undifferentiated colonies is depressed by the size-dependent selective pressure, thus, larger colonies with germ-soma differentiation have higher fitness than smaller undifferentiated ones

Steps Towards Multicellularity" (Short et al. 2006; Solari et al. 2006b). Previously calculated thresholds and scaling relationships in Volvocales can be used for nutrient uptake (B) to calculate cell growth rates (r) for the fecundity (λ) in the model.

Volvocales show a diffusive bottleneck as colonies increase in size (Short et al. 2006). When the demand for essential molecules exceeds the diffusive current, metabolism is constrained. This bottleneck can be circumvented by the increased advection generated by the flagellated cells arrayed at the surface of the colony. It was shown in Short et al. (2006) that the absorption rate of nutrients in organisms with a spherical design such as the Volvocales is $I_a \sim RPe^{1/2}$, where R is colony radius and Pe is the Peclet number. Here, the Peclet number can be expressed in terms of the velocity U of the flow generated by the flagellated cells, the sphere (colony) diameter ($2R$), and a diffusion constant D for a molecule such as O_2 ($Pe = 2RU/D$).

R and U can be expressed as a function of cell number and proportion of somatic and germ cells; it was shown that $U \propto R$ (Solari et al. 2006a; Short et al. 2006). Taking into account these relationships, it was shown in Solari et al. (2013) that the supply of nutrients to the colonies $I_a \sim B$ (the supply variable on the model). On the other hand, the demand of resources depends on the number of cells in the colony and the total number of cells it has to produce for the next generation (the cost of reproduction $C = n^2(1 - s)$). Therefore, the ratio that constraints the growth rate r in the Volvocales becomes

$$B/C_r \sim bB/cC \quad (4)$$

where b and c would be the normalization constants for the absorption rate and the metabolic demand of the unicell, respectively (e.g., *C. reinhardtii* in Volvocales). As size increases in Volvocales, colonies have to invest in somatic cells to increase advection at the surface of the colony (bB) and decrease the metabolic demand (cC; the dynamics is shown in Fig. 7a). If $cC > bB$ then the growth rate is limited by an insufficient inflow of nutrients via diffusion, therefore decreasing the fecundity rate (Fig. 7b).

Viability

Viability gives the proportion of colonies that will survive to reproduce the next generation. Viability goes from 0 (no survival) to 1 (100 % survival). Since somatic cells specialize in vegetative functions, they contribute to viability functions such as motility, while totally specialized germ cells spend all their energy in reproductive (i.e., fecundity) related functions, therefore decreasing the viability of the colony. Solari et al. (2013) also modeled viability as the ratio between the contribution B and the cost C of the cells in the colony to survival. Figure 7c shows the viability rates of colonies as a function of germ-soma specialization and size. Colonies that invest in germ specialization decrease their viability since they do not contribute to viability, but that decrease can be compensated by increasing the proportion of somatic cells.

Due to the importance of motility for survival in these algae, motility of colonies can be used as a proxy for viability, and assume that colonies that sink strongly compromise their survival (viability). In Volvocales, the flagellated cells are used for self-propulsion, to avoid sinking, and to reach light and nutrients (Hoops 1997; Kirk 1998; Koufopanou 1994; Solari et al. 2006a). The contribution to motility for self-propulsion and to avoid sinking depends on the flagellar force F generated by the biflagellated cells n_F arrayed at the colony surface (both somatic and undifferentiated reproductive cells). It was previously shown that in Volvocales $F \propto {n_F}^{3/4}$ (Solari et al. 2006a), and that $n_F = B$ (again the supply equation in the model; Solari et al. 2013).

The cells in Volvocales are denser than water, therefore, the downward gravitation rate of colonies depends on the number of cells and the somatic to reproductive cell ratio (again, the cost of reproduction $C = n^2(1-s)$; Solari et al. 2013). The viability (motility) constraint becomes,

$$v \approx b_v B^{3/4}/c_v C \tag{5}$$

where B is composed by all the cells performing motility functions, C is composed by all the cells in the colony, and b_v and c_v are the normalization constants for the flagellar force and the downward gravitation rate of the unicell respectively (e.g., *C. reinhardtii*). If $v < 1$, the negative gravitational force of the colony ($c_v C$) is higher than the flagellar force generated by the flagellated cells for propulsion ($b_v B^{3/4}$), which makes colonies sink. To avoid sinking and be able to swim to acquire resources, as volvocine colonies increase in cell number, they have to invest in a higher proportion of somatic cells to increase the flagellar force and decrease the negative gravitational force (Fig. 7c).

Fitness

The overall fitness w of colonies is the product of their fecundity λ and viability v rates, $w = \lambda v$. If we compare the fitness of colonies for specific sizes, increased soma differentiation (i.e., increased ratio of somatic to reproductive cells) is favored as colony size increases. By adding a size-dependent selective pressure $p = 1z(n)^{-\mu}$ to the model, where z is the mortality coefficient and μ the size-dependent scaling exponent, the fitness of colonies becomes $w = \lambda v p$. Figure 7d shows how larger colonies with increased germ-soma differentiation can now have a higher fitness than unicells and undifferentiated colonies. The size-dependent selective pressure p could represent predation pressure (e.g., size thresholds for zooplankton filter feeders; Porter 1977; Morgan 1980) or resource availability (e.g., migration capabilities through the water column to get nutrients; Sommer and Gliwicz 1986), where small colonies have a higher predation rate or lower resource availability than larger ones.

Conclusion

Using life-history theory and allometry, Solari et al. (2013) have produced a model inspired by the volvocine green algae that describes the dynamics of the emergence of germ-soma differentiation as size increases in multicellular organisms. The results show that the cost of reproducing an increasingly larger group has likely played an important role in the evolution of complexity and individuality in the transition to multicellularity. As selective pressures first pushed multicellular organisms to increase in size (i.e., predation), the costs of reproducing an increasingly larger group also increased, having negative effects on their fitness. At some threshold size, fitness decreased dramatically and overcoming this threshold might have required the separation of reproductive and vegetative functions between two cell types, which resulted in increased complexity.

Germ-soma separation was one of the solutions that Volvocales used to deal with this problem. Other innovations such as a different life cycles, changes in geometry, enhanced storage capabilities, etc, have surely also helped emerging multicellular individuals in other lineages decrease their costs associated with size (in algae there are many examples; Graham and Wilcox 2000). Nevertheless, we argue that to a large extent in Volvocales, and probably to some extent in other lineages, germ-soma differentiation was a solution (in parallel to other ones) to counteract these increasing costs, helping to create a new level of individuality. Once germ-soma differentiation evolved, this trait by itself can add selective advantages such as increased motility in Volvocales.

This model shows first that the cost of investing in soma decreases with size, regardless of any size constraint or benefit (Eq. 2). Second, for lineages such as the Volvocales, soma specialization can benefit the colony indirectly by decreasing reproduction costs and directly by helping the reproductive cells acquire resources for their metabolic needs and increase motility. Third, as the ratio of somatic to reproductive cells increases, a specialized germ cell can easily evolve since the vegetative functions are taken care of by the somatic cells, benefiting the colony's fecundity. Moreover, a higher somatic to reproductive cell ratio increases the motility (viability) of colonies (from ~ 200 μm/sec in *V. carteri* to up to ~ 800 μm/sec in *V. barberi*; Solari et al. 2008). In Volvocales, the ratio of somatic to reproductive cells increases with colony cell number, in agreement with the experiments results and modeling (Kirk 1998; Koufopanou 1994; Short et al. 2006; Solari et al. 2006a, b, 2013).

The trade-offs between fecundity, viability (motility), and size recently studied in Volvocales show in detail how metabolic (B/C_r) and motility constraints (v) as colonies increase in size might be strong enough to push the organism design to cellular specialization and higher complexity. The higher cost of reproducing a larger organism was probably an important driving force for the evolution of cellular differentiation and individuality during the transition to multicellularity in Volvocales, and probably to some degree in all extant multicellular lineages. Once germ-soma separation evolves, further specialization and differentiation might counteract increasing costs associated with larger size, but can also by itself give a selective advantage by increasing the fecundity and/or viability of the larger organism.

Summary

1. The volvocine green algae are an ideal model system for studying the unicellular-multicellular transition since they comprise an assemblage of lineages featuring varying degrees of complexity in terms of colony size, colony structure, and cellular specialization.
2. Using the Volvocales we have investigated what size-related advantages caused single cells to start living in groups, and why cellular differentiation evolved as size increased, creating multicellular organisms with germ-soma separation.
3. Our analysis supports the hypothesis that predation was an important selective pressure for the origin of multicellularity, but found no evidence that increased motility and nutrient uptake were advantages for the first cell clusters in Volvocales.
4. We think that the extra-cellular matrix necessary to form cell groups might have been co-opted for nutrient storage in this group.
5. Using life-history theory and allometry, Solari et al. (2013) have produced a model inspired by the volvocine green algae that describes the dynamics of the emergence of germ-soma differentiation as size increases in multicellular organisms.
6. The results of the model show that, if there is a selective advantage to increase in size, the cost of reproducing an increasingly larger group has likely played an important role in the evolution of complexity and individuality in the transition to multicellularity.
7. The trade-offs between fecundity, motility (viability), and size recently studied in Volvocales show in detail how metabolic and motility constraints as colonies increase in size might be strong enough to push the organism design to cellular specialization and higher complexity.
8. Once germ-soma separation evolves, increased specialization by itself might counteract increasing costs, but also give further selective advantages (e.g., increased motility) to the multicellular organism.

Methods

General

For all experiments, cell and colony counting was manually done with samples fixed with formalin in a microscope counting chamber (Neubauer hemocytometer). Axenic conditions of the experiment were checked for bacterial contamination at the beginning and at the end by plating on 1 % agar plates with bacteria growth medium (LB). For cell size calculations, digital photos were taken and analyzed with Image J software. For statistical analysis JMP software was used (SAS Institute, Cary, NC, USA). Variance reported and bars in figures are standard errors.

Nutrient Storage Experiment

To check for differences in growth rates between media with different nutrient concentrations, we grew the algae cultures in full and 10^{-1} diluted medium. We performed the experiment with unicellular *C. reinhardtii* (UTEX89, Fig. 1a) and 1–16 celled colonial *G. pectorale* (UTEX LB826, Fig. 1b). The populations for inoculation were kept in exponential phase in 10 ml test tubes with standard *Volvox* medium (SVM; Kirk and Kirk 1983), and illuminated by homogeneous cool white light (~10,000 lx) in a daily cycle of 16 h light (at 25 °C) and 8 h in the dark (at 23 °C). From these populations, 10^4 cells/ml were inoculated with individually wrapped sterile serological plastic pipettes under laminar flow hood in 125 ml Erlenmeyer flasks with 20 ml of sterile SVM (in *Gonium* cells inside colonies were counted to reach 10^4 cells/ml). The flasks were set at the same temperature, medium, and light conditions outlined above. The experiment lasted seven days and each species had three replicates; samples were taken everyday for cell counting. A multiple additive linear regression analysis was performed using indicator variables to account for factors such as species and medium treatments (full or diluted).

Predation Experiments

Peranema trichophorum (Fig. 1f) was purchased from Carolina Biological Supply Co. (Burlington, NC; Fig. 1f). Monoaxenic cultures were established using the method described by Saito et al. (2003). *P. trichophorum* cells were washed in 0.01 % Knop solution with a mixture of three antibiotics (160 μg/ml streptomycin, 160 μg/ml kanamycin, and 280 μg/ml penicillin) and fed with axenic *C. reinhardtii*. Once *P. trichophorum* was bacteria-free, we were able to sub-culture them on SVM with axenic *C. reinhardtii* and *G. pectorale* as a food source. We keep *P. trichophorum* stock cultures in SVM, illuminated by homogeneous cool white light (~ 3,000 lx) in a daily cycle of 12 h light and 12 h dark at 25 °C. To keep stock cultures we inoculate an initial concentration of 10^3 *P. trichophorum* cells/ml and 10^5 cells/ml of the axenic culture of the volvocine algae. In 7–14 days *P. trichophorum* cells deplete all the algae prey, its population growing to 2–4 × 10^4 cells/ml from the initial concentration; at this point the axenic population of *P. trichophorum* are retransferred or used for experiments.

We performed a short term experiment to measure predation rates and a long term experiment to track the changes in the colony size distribution of *G. pectorale* populations. At 3 h into the light period, in the same flasks and conditions outlined above for the nutrient storage experiment, we inoculated 4 × 10^3 *P. trichophorum* cells/ml and, as a food source, 10^5 cells/ml of axenic *C. reinhardtii* and *G. pectorale*. We counted the initial and final cell concentration of the six replicates. The difference between the initial and final cell concentration was divided by the predator concentration to get the predation rate (cells/h) per predator. Digital photos were taken to record cell and colony size.

In the 3-week long experiment, axenic *G. pectorale* were grown with and without *P. tricophorum* added in the culture. *G. pectorale* were inoculated as outlined above; when the predator was present, 4×10^2 *P. trichophorum* cells/ml were added. For the experiment, the flasks were set at 25 °C with the same medium and light outlined above. Four replicates were set up for each treatment. The experiment cultures were semicontinuous, diluted every two days by 10 % with fresh sterilized medium. The aliquot removed (2 ml) was then used for measuring cell and colony number and size.

Acknowledgements We would like to thank Rick Michod and Ray Goldstein for all the discussions throughout this last decade, for their support, and for being an integral part of some of the work that is reviewed here. We would like to thank Benjamin Basanta for helping in the culturing and isolation of *Peranema trichophorum*. Without his help it would have been impossible to perform the predation experiments.

References

Bell G (1985) The origin and early evolution of germ cells as illustrated by the Volvocales. In: Halvorson H, Monroy A (ed) The origin and evolution of sex. A. R. Liss, Inc., New York, pp 221–256

Boraas ME, Seale DB, Boxhorn JE (1998) Phagotrophy by a flagellate selects for colonial prey: a possible origin of multicellularity. Evol Ecol 12:153–164

Breglia SA, Slamovits CH, Leander BS (2006) Phylogeny of phagotrophic euglenids (Euglenozoa) as inferred from Hsp90 gene sequences. J Eukaryot Microbiol 54(1):86–92

Chang SL (1966) Observations on the organelles, movement, and feeding of *Peranema trichophorum*. Trans Am Microsc Soc 85:29–45

Chen YT (1950) Investigations of biology of *Peranema trichophorum*. Q J Microsc Sci 91:279–308

Coleman AW (2012) A comparative analysis of the Volvacaceae (*Chlorophyta*). J Phycol 48:491–513

Fabry S, Kohler A, Coleman AW (1999) Intraspecies analysis: comparison of ITS sequence data and gene intron sequence data with breeding data for a worldwide collection of *Gonium pectorale*. J Mol Evol 48:94–101

Gerisch G (1959) Die Zelldifferenzierung bei Pleodorina californica Shaw and die Organisation der Phytomonadinenkolonien. Arch Protistenkd 104:292–358

Graham LE, Wilcox LW (2000) Algae. Prentice Hall, Engelwood Cliffs

Greuel BT, Floyd GL (1985) Development of the flagellar apparatus and flagellar orientation in the colonial green alga *Gonium pectorale* (Volvocales). J Phycol 21:358–371

Grosberg RK, Strathmann RR (2007) The evolution of multicellularity: a minor major transition? Ann Rev Ecol Evol Syst 38:621–654

Guyon E, Hulin JP, Petit L et al (2001) Physical hydrodynamics. Oxford University Press, New York

Hallmann A (2006) The pherophorins: common, versatile building blocks in the evolution of extracellular matrix architecture in Volvocales. Plant J 45:292–307

Herron MD, Michod RE (2008) Evolution of complexity in the volvocine algae: transitions in individuality through Darwin's eye. Evolution 62:436–451

Hoops HJ (1997) Motility in the colonial and multicellular Volvocales: structure, function, and evolution. Protoplasma 199:99–112

Kirk DL (1997) The genetic program for germ-soma differentiation in *Volvox*. Ann Rev Gen 31:359–380

Kirk DL (1998) *Volvox*: molecular-genetic origins of multicellularity and cellular differentiation. Cambridge University Press, Cambridge
Kirk DL, Kirk MM (1983) Protein synthetic patterns during the asexual life cycle of Volvox carteri. Dev Biol 96:493–506
Koufopanou V (1994) The evolution of soma in the Volvocales. Am Nat 143:907–931
Lurling M, Van Donk E (2000) Grazer-induced colony formation in *Scenedesmus*: are there costs to being colonial? Oikos 88:111–118
Morgan NC (1980) Secondary production. In: Le Cren ED, Lowe-McConell RH (eds) The functioning of freshwater ecosystems, IBP 22. Cambridge University Press, Cambridge, pp 247–340
Nozaki H, Ott FD, Coleman AW (2006) Morphology, molecular phylogeny and taxonomy of two new species of *Pleodorina* (*Volvoceae*, *Chlorophyceae*). J Phycol 42:1072–1080
Porter KG (1977) The plant-animal interface in freshwater ecosystems. Am Sci 65:159–170
Ratcliff WC, Denison RF, Borrello M et al (2012) Experimental evolution of multicellularity. Proc Natl Acad Sci U S A 109:1595–1600
Ratcliff WC, Herron MD, Howell K et al (2013) Experimental evolution of an alternating uni- and multicellular life cycle in *Chlamydomonas reinhardtii*. Nat Commun 4:2742
Reynolds CS (1984) The ecology of freshwater phytoplankton. Cambridge University Press, Cambridge
Rhee GY (1973) A continuous culture study on phosphate uptake, growth rate and polyphosphate in *Scenedesmus* sp. J Phycol 9:495–506
Saito A, Suetomo Y, Arikawa M et al (2003) Gliding movement in *Peranema trichophorum* is powered by flagellar surface motility. Cell Motil Cytoskelet 55:244–253
Short MB, Solari CA, Ganguly S et al (2006) Flows driven by flagella of multicellular organisms enhance log-range molecular transport. Proc Natl Acad Sci U S A 103:8315–8319
Solari CA, Kessler JO, Michod RE (2006a) A hydrodynamics approach to the evolution of multicellularity: flagellar motility and the evolution of germ-soma differentiation in volvocalean green algae. Am Nat 167:537–554
Solari CA, Ganguly S, Kessler JO et al (2006b) Multicellularity and the functional interdependence of motility and molecular transport. Proc Natl Acad Sci U S A 103:1353–1358
Solari CA, Michod RE, Goldstein RE (2008) Volvox barberi, the fastest swimmer of the Volvocales (Chlorophyceae). J Phycol 44:1395–1398.
Solari CA, Drescher K, Ganguly S et al (2011) Flagellar phenotypic plasticity in volvocalean algae correlates with Péclet number. J R Soc Interface 8:1409–1417
Solari CA, Kessler JO, Goldstein RE (2013) A general allometric and life-history model for cellular differentiation in the transition to multicellularity. Am Nat 181:369–380
Sommer U, Gliwicz ZM (1986) Long-range vertical migration of Volvox in tropical lake Cahora Bassa (Mozambique). Limnol Oceanogr 31:650–653
Stearns SC 1992. The evolution of life histories. Oxford University Press, Oxford
Werner TP, Armhein N, Freimoser FM (2007) Inorganic polyphosphate occurs in the cell wall of Chlamydomonas reinhardtii and accumulates during cytokinesis. BMC Plant Biol 7:51 doi:10.1186/1471-2229-7-51

Cells Acting as Lenses: A Possible Role for Light in the Evolution of Morphological Asymmetry in Multicellular Volvocine Algae

John O. Kessler, Aurora M. Nedelcu, Cristian A. Solari and Deborah E. Shelton

Abstract The volvocine green algae have been extensively used to address various questions related to the evolution of multicellularity and cell differentiation, in terms of the genetics, developmental constraints, and underlying selective forces specific to this group. More recently, physical characteristics of the environment and of the emerging multi-celled entities have also been considered as potential contributors to the evolution of multicellularity in this lineage. However, the role of light in the evolution of multicellularity—beyond its direct photosynthetic role—has not been explored. The objectives of this work are (1) to show that algal cells, in both unicellular and multicellular algae, concentrate incident light, and (2) to suggest that this concentrated light might have contributed to the evolution of multicellularity in volvocine algae. We show that single algal cells can act as lenses and concentrate light from a remote source (e.g., the Sun) into beams, by a combination of standard refractive imaging of transmitted light and diffractive Arago-Poisson imaging of the light surrounding the cells. In the spheroidal multicellular volvocine algae, the peripheral cells facing the Sun can concentrate incident sunlight towards the interior of the colony. We suggest that the evolution of morphological asymmetries associated with the anterior-posterior polarity exhibited by multicellular spheroidal volvocine algae may have been influenced by this phenomenon. Whether the effect of these light beams is still important to extant spheroidal volvocine algae remains to be

J. O. Kessler (✉)
Physics Department, University of Arizona, Tucson, AZ 85721, USA
e-mail: jottok@gmail.com

A. M. Nedelcu
Department of Biology, University of New Brunswick,
Fredericton, NB, Canada
e-mail: anedelcu@unb.ca

C. A. Solari
Laboratorio de Biología Comparada de Protistas, IBBEA, CONICET-UBA,
Universidad de Buenos Aires, C1428EHA Buenos Aires, Argentina
e-mail: casolari@bg.fcen.uba.ar

D. E. Shelton
Department of Ecology and Evolutionary Biology, University of Arizona,
Tucson, AZ 85721, USA
e-mail: deborah.shelton@gmail.com

I. Ruiz-Trillo, A. M. Nedelcu (eds.), *Evolutionary Transitions to Multicellular Life*,
Advances in Marine Genomics 2, DOI 10.1007/978-94-017-9642-2_12

investigated. Future experiments are also needed to assess the relative contributions of the two light concentrating mechanisms by algal cells.

Keywords Light beams · Focus · Refraction · Diffraction · *Pleodorina* · *Volvox* · Volvocine · Volvocales · Morphological asymmetry · Development · Photoreceptors · Evolution of multicellularity

Introduction

During the transition from unicellular to multicellular life a series of general life properties (immortality, totipotency), cellular processes (cell growth and division) and fitness components (survival and reproduction) had to be re-organized to reflect the emergence of the new multicellular individual (Michod and Nedelcu 2003; Nedelcu and Michod 2004). In addition, the ways in which cells interacted with the physical environment must have changed when previously free-living unicellular organisms became part of a group; for instance, their surface area in direct contact with the environment became smaller, which would have limited nutrient and gas exchanges. Furthermore, the fraction of incident light scattered or absorbed by those cells of a group that are in the path of illumination is lost to the cells located in the shadow region. Overall, the evolution of multicellularity in various lineages is thought to have been influenced by a combination of developmental constraints, life history trade-offs and adaptations to the physical environment that were inherited from their unicellular ancestors (e.g., King 2004; Grosberg and Strathmann 2007; Knoll 2011, Chapter 11 "The Evolutionary Ecology of Multicellularity: The Volvocine Green Algae as a Case Study"). Here, we argue that in addition to such factors, passive outcomes of physical interactions between single-celled organisms and their environment might have become biologically significant to cells in a group or multicellular context, and could have influenced the evolution of multicellular-specific traits. Specifically, we show that cells acting as lenses concentrate incident light by both refraction and diffraction. We suggest that the resulting focused light could have contributed to the evolution of morphological asymmetry in a group of multicellular green algae in the order Volvocales, known as the volvocine algae.

Hypothesis and Objectives

Many algal cells are spheroidal—that is, they have a somewhat irregular, close-to-spherical shape. Algae are generally known (and studied) for their ability to utilize incident sunlight for photosynthesis and for directional locomotion (phototaxis). However, algal cells can also reflect and refract some of the incident illumination. Apparently, there is no serious mention in the literature of the fact, shown below, that algal cells act as miniature lenses that can concentrate, and focus into a beam, the fraction of light transmitted through them. Furthermore, the light passing nearby a cell is concentrated by diffraction into an additional beam. The anisotropy characteristic of these beams of light, i.e., their generally downward direction, due to their source in the sky above, may have been biologically significant during the early evolution of volvocine algae.

The chief objectives of this contribution are to (i) demonstrate the existence of this light focusing effect by algal cells; and (ii) speculate on its role in the evolution of volvocine algae and their morphology, in the hope that future research will demonstrate that some of these suggestions are valid. Below, we first introduce the volvocine algae, with an emphasis on the traits that are relevant to our discussion on the role of beams of sunlight in the evolution of these algae. Then, we demonstrate the principle using a single-celled dinoflagellate and several multicellular volvocine algae. Lastly, we suggest ways in which the effect of this focused light, always incident from "above", could have influenced the evolution of traits associated with morphological asymmetry in volvocine algae.

The Volvocine Algae

Volvocine algae comprise both unicellular and multicellular species with various numbers of cells and distinct morphological traits (Fig. 1, see Chapter 7 "Volvocine Algae: From Simple to Complex Multicellularity"). *Chlamydomonas* are single-celled algae that use their two flagella for motility, and a light-sensitive organelle (the eyespot) to orient themselves towards light. In their asexual phase, most multicellular (i.e., colonial) members have a morphology involving *Chlamydomonas*-like cells embedded on or within a spheroidal transparent gel-like structure known as the extracellular matrix (ECM). In *Eudorina* species, all 16–32 cells are located at the periphery of the spheroid (Fig. 1b); the cells are equal in size and potential—that is, they are flagellated during the first part of the life cycle, then they each start growing, divide and develop into daughter colonies that detach and repeat the cycle. *Pleodorina* species are initially morphologically similar to *Eudorina* (except they generally possess a larger number of cells; i.e., 64–128 cells). However, in these species cells grow at different rates, creating an anterior-posterior gradient of cells of different sizes and potential (the anterior pole is defined with respect to the direction of swimming); the largest cells (which become the reproductive cells that produce new daughter colonies) develop at the posterior end of the spheroid (Fig. 1c). In the most-studied volvocine species, *Volvox carteri*, up to 2000 small flagellated somatic cells are distributed on the surface of the spheroid, while up to 16 large reproductive cells (known as gonidia) are found inside the colony (Fig. 1d). Each gonidium will grow many-fold in volume and will then undergo a series of cell divisions to form an embryo that develops into a fully-formed daughter colony while still within the mother colony (Fig. 1e) whose peripheral somatic cells are visible as dots in the background.

Overall, all the spheroidal volvocine species exhibit an anterior-posterior polarity that is reflected in the differential expression of a number of traits. For instance, flagella have different orientations (due to the rotation of their basal bodies during

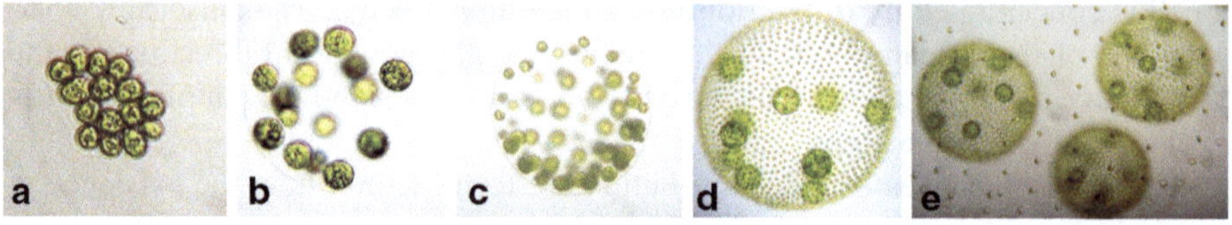

Fig. 1 A subset of multicellular volvocine green algae that show a progressive increase in cell number, volume of extracellular matrix per cell and division of labor between somatic and reproductive cells. **a** *Gonium pectorale*. **b** *Eudorina elegans*. **c** *Pleodorina californica*. **d** Young *Volvox carteri* spheroid. **e** *V. carteri* daughter spheroids (with their own somatic cells and gonidia) are seen below the somatic cell sheet of the mother colony. Where two cell types are present, the smaller cells are the somatic cells, whereas the larger cells are the reproductive cells

embryonic development) depending on the location of the cell in the colony. This organization ensures a coordinated beat that results in the rotation of the spheroid around its anterior-posterior axis and the movement of the spheroid with the anterior pole in the direction of swimming. The size of the eyespot also differs as a function of each cell's location relative to the anterior pole of the spheroid, with cells located at the anterior pole of the spheroid having larger (and more sensitive to light) eyespots relative to the cells at the posterior pole (Kirk 1998, p. 230). In *V. carteri*, the placement and size of somatic cells is also not uniform over the entire surface of the colony. Specifically, especially in the adult colonies, the somatic cells at the anterior pole of the spheroid are smaller and more dispersed relative to the cells at the posterior pole. Furthermore, the reproductive cells are mostly located in the posterior half of the spheroid.

Consequently, all these traits define a morphological asymmetry correlated with the anterior-posterior polarity of these algae. In mature *Pleodorina* and *Volvox* spheroids, this asymmetry results in the center of gravity being located below the geometric center of the spheroid, due to the posterior, off-center location of reproductive cells. Thus, in still water, and without directional photic stimulus, these colonies have an up-down orientation consistently correlated with the colonies' anterior-posterior polarity. This equilibrium orientation of the colonies is the consequence of gravitational torque acting on the asymmetry of the distribution of mass (Kessler 1986). Such an asymmetry might confer some selective advantages by causing gravitaxis, the propensity of the colonies to swim generally upwards, towards sunlight—for energy, and towards the air-water interface—for gas exchanges, even when the environmental information for "which way is up" is absent or ambiguous. The sparse distribution of somatic cells at the anterior pole might also allow for more light to reach the gonidia located at the posterior pole. Nevertheless, the evolutionary and mechanistic basis for establishing these morphological asymmetries, especially the positioning of the gonidia toward the posterior pole in both *Pleodorina* and *Volvox* species, is not known.

Focusing of Incident Parallel Light by Algal Cells

Demonstration of Principle

Theoretical Framework

Light incident on a transparent spheroidal object whose index of refraction is even slightly greater than that of its surroundings is deflected toward its axis (e.g., Hecht 1987), forming a region of increased intensity. That focus can be sharp or extended, depending on the specific geometry of the object, which is acting as a lens. The distance of focus, or strongest convergence, depends on the ratio of the indexes of refraction n(in)/n(out) = Q. When Q = 1 there is no difference between "in" and "out", hence there is no convergence. When Q > 1, there is convergence at a distance below the lens. That distance is a function of Q. When $Q = 1 + a$, where $a << 1$, the focus is far from the surface. At larger Q the focal distance is shorter.

Parallel light from the Sun shines on the cells located at the anterior pole of a spheroidal volvocine colony. Light enters the cells—whose index of refraction is n(c), from water—which has an index of refraction n(w). The upper surface of the cells refracts the light, causing it to converge at a distance

$$F = Rn(c)/[n(c) - n(w)]$$

from that surface (Hecht 1987), where R is the radius of curvature of the cell's upper surface. The bottom surface of the cell also has a focusing effect. That effect is smaller since the ECM's index is probably nearly equal to n(c). The magnitude of the refractive indices varies with temperature, composition and wavelength of light. For an estimate, one may set n(w) = 1.33 and 1.39 < n(c) < 1.43 (Fiorani et al. 2008; Spizzichino et al. 2011 for *C. reinhardtii*). Assuming the radius of a cell is 5 μm, one obtains the value of F to be 70–100 μm. Although this estimate is based on idealized simple geometry that ignores scattering, it indicates that even with considerable aberrations, this light can be focused within a colony, and possibly on the cells near the posterior pole, if the diameter of the colony is around 70–100 μm. This is indeed the case for mature colonies such as *Pleodorina* and developing *V. carteri* daughter colonies (discussed below).

In addition to this refraction of incident light, cells can also diffract the light that passes near their edges. A spheroidal opaque object that absorbs, diffusely scatters or reflects the incident light casts a shadow in the region of the lost light. The light that passes near the edge of that object produces a diffraction pattern in the shadow region. That diffraction pattern is due to the wave nature of light. The bright beam of light at its center, seen as a spot of light on a screen that intercepts it, is named for its discoverers, Arago and Poisson (Harvey and Forgham 1984); for additional discussion and examples see Appendix. An opaque sphere that acts as a diffracting lens can image the Sun and produce an Arago "spot" (Sommargren and Weaver 1992). Using the actual distance from Earth and the diameter of the Sun, Sommargren and Weaver (1992) calculated that the image that forms 500 mm behind a sphere of 20 mm has a diameter of 4.7 mm. If we extrapolate to an algal cell, the image of the Sun forming at a distance of 100 μm behind the cell is 1 μm.

The two light-concentrating phenomena discussed above are independent. However, in some circumstances, the two beams—i.e., the one that can be understood using standard "geometrical optics" and the diffracted "Arago Spot" beam, can act simultaneously (see Fig. 8 in the Appendix).

Experimental Evidence

Below, we are using the single-celled dinoflagellate *Alexandrium fundyense* and several multicellular volvocine algae to demonstrate these light-focusing effects of cells. Although we cannot separate the two effects, our observations prove that cells can concentrate light below their surfaces at various focusing distances, depending on the physical characteristics of each cell and of their surroundings.

As with any microscopical observation of a cell, there is a set of nearby positions of the objective lens that best images the cell wall, flagella and internal organelles. Slight movement of the objective produces a slight change of the location where focus is sharp. Moving well away from that imaging location, and away from the incident illumination, a bright, usually featureless circular spot appears. Depending on the distance between the objective and the algal cell, that spot is sharply or fuzzily edged. The spot is actually a section through the beam of light emerging below the illuminated cell acting both as an approximately spherical lens and as a diffracting object—where "below" is appropriate for illumination from above. In a sunny outdoor setting, that beam of light forms the image of the sun. This assertion can be proven with an inverted microscope, which uses light coming from above and an objective positioned below the stage that supports the cell. When one removes the usual condenser lenses located in the path of the illumination toward the specimen, the light beam that emerges below the cell forms the image of the illuminating featureless finely ground glass plate below the light bulb. To prove the fact of imaging by the algal cells, one can insert below that illuminator arrangement a narrow slit cut into an opaque mask. The cell ought to image the slit. The observed length of the concentrated beam, with or without slit, is often tens of micrometers. Why state "often"? Because the focusing distance of the refracted light depends on the value of Q for every case. Observed focusing distances can range from about 100 μm to ca. 20–25 μm or less (see below).

First, we used the marine dinoflagellate *A. fundyense* as a test organism, because of its large cells, its dense contents (Fig. 2a) and the small value of *a* (likely due to similar refraction indexes between the cell and its saline surroundings)—as inferred from the relatively remote imaging of the slit. The elongated beam of concentrated light observed below the algal cell, with boundary ranging from sharp to fuzzy depending on the position between the observing objective and the cell, is an extended image of the light source (Fig. 2b). The imaging by the cell of the slit (Fig. 2c), effectively located at infinity compared with the distance between the microscope objective and the cell, is proof of the assertion that the cell focuses remote sources of light such as

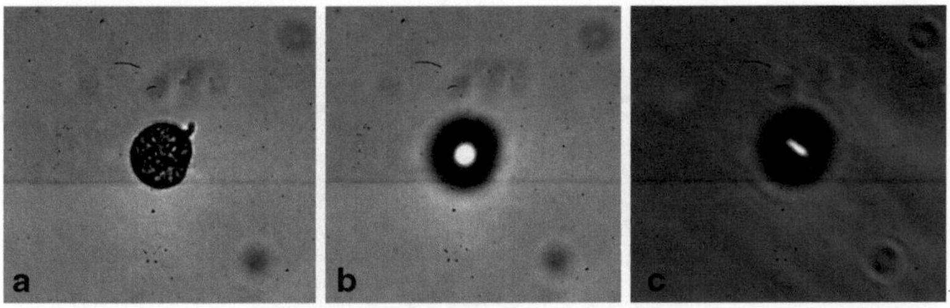

Fig. 2 Light concentrating action by the dinoflagellate *A. fundyense*, observed with an inverted microscope. **a** The microscope objective is positioned to visualize the cell's interior and boundary. **b** The objective is now lowered by 100 μm below its position in **a**; it produces an image of the ground glass diffuser plate located just below the microscope lamp. The quoted distance of 100 μm is derived from a subjective decision concerning optimum sharpness. Moving the objective further down maintains the image, with increasing size, progressively lower brightness and lower sharpness. The image is actually a cross section of the beam of light emerging from the cell. **c** The light now emerges from a slit cut into an opaque mask, placed below the diffuser plate; it is focused by the cell into an image of the slit. The diameter of the cell is 33 μm; the diameter of the circular image spot is 11 μm. The background in **c** is darker than in **a** and **b** because most of the illuminating light was blocked by the mask

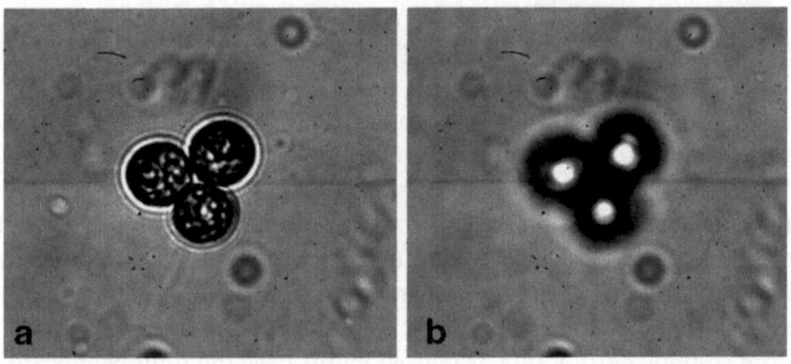

Fig. 3 Three adherent cells of *Gonium pectorale*. **a** light is focused on the cells and their components. **b** shows the image of the light source (the ground glass disk), focused by the cells; focusing distance is 40 μm (i.e., the objective was moved 40 μm downwards, below the image in **a**, away from the light source)

the Sun, simulated by the image of the circular ground glass plate. The image of the slit can be remarkably sharp when the distance from the microscope objective to the algal cell is appropriately adjusted (at ca. 100 μm; Fig. 2c).

To prove that similar effects can be observed in the freshwater volvocine algae, we first used *Gonium pectorale*, a simple colonial volvocine alga composed of 8–16 cells arranged in a convex plate (Fig. 1a). Figure 3 shows the light focusing effect of three *G. pectorale* cells detached from a colony. Next, we used a spheroidal species—*Pleodorina starrii*. In the adult *Pleodorina* colonies, cells at the anterior pole remain small and do not reproduce (Fig. 1b). Figure 4 shows the light focusing effect of the small somatic cells.

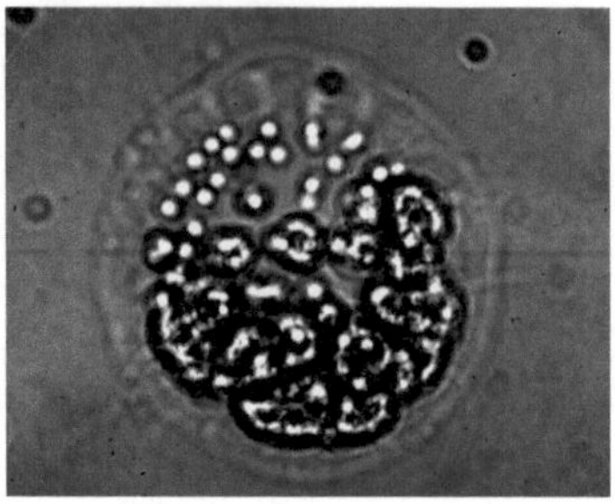

Fig. 4 *Pleodorina starrii* showing the light-focusing effect of the small anterior cells. Each cell produces a concentrated beam of illumination, seen here in cross section as a *circular bright spot*, the image of the ground glass source of light. The colony, seen *sideways*, is attached to the transparent surface of a Petri dish located on the stage of the microscope. The diameter of the colony is ca. 95 μm

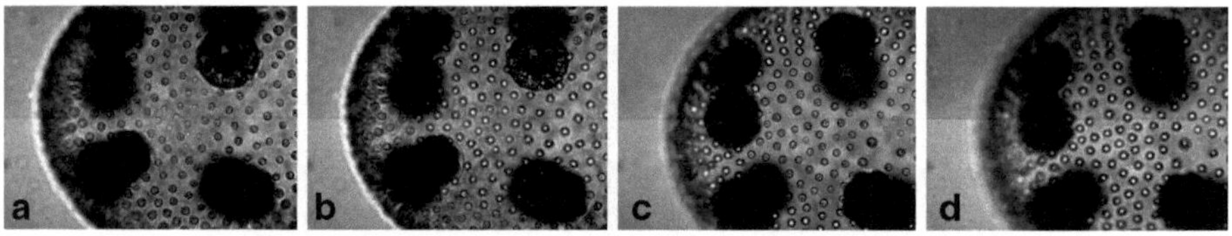

Fig. 5 The light-focusing effect of peripheral somatic cells in a young *V. carteri* colony (the *larger* and *darker* cells are gonidia). The colony was immobile between a transparent substrate and cover glass. Judging by the spatial distribution of gonidia, the anterior pole of the colony is located on the *right* of the images. **a** The focus is on the *top* somatic cells. **b** The ground glass disk that provides the illumination is imaged by the *top* somatic cells (the objective of the microscope was 20–30 μm below its position in *panel a*); the light beams produced by the *top* somatic cells are projected inside the colony. **c** The focus is on the *bottom* somatic cells. **d** The image as in *panel b* but now showing the bottom somatic cells acting as lenses; in this case the light beams are projected outside the colony

We applied the same procedure to young *V. carteri* spheroids, in which the somatic cells act as individual lenses. Figure 5 shows the focusing action of the peripheral somatic cells situated at the top (panel B) and bottom (panel D) of the colony. The sharpest bright spots are located 20–30 μm below the focal distance between the microscope objective and the cell. Note that the curvature of the colony's surface causes the distance between the cells and the objective to vary, most strongly near the edges, and thus the focusing is unequal. For instance, in Fig. 5b, the upper most somatic cells (in the center of the image) generate the light beams, while the cells near the edge (left side of image) are closer to the microscope objective than the central cells and do not produce the light spots. In Fig. 5c, the effect is opposite since the image shows the cells at the bottom surface of the sphere; in this case, the central cells are nearer to the microscope objective (recall that the objective is below the specimen) than the cells near the left edge which are positioned further up from the objective.

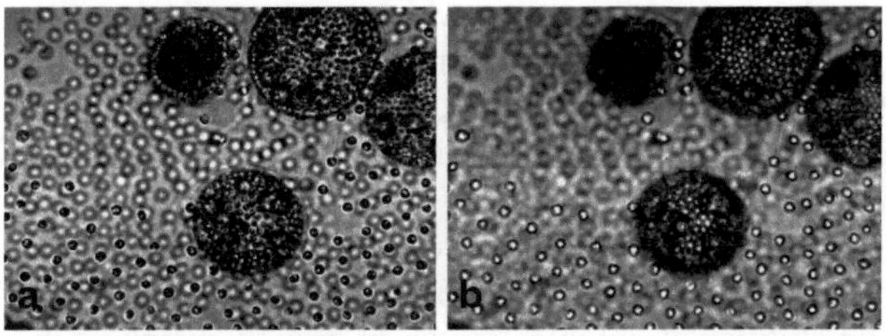

Fig. 6 Focusing of incident light by the *bottom* somatic cells of a *V. carteri* mother colony and by the *bottom* somatic cells of the daughter colonies maturing while embedded in vesicles within the mother's ECM. The young gonidia, visible within the daughter colonies and illuminated by the light beams emerging from the *upper* somatic cells, become the third generation. The *bottom* somatic cells are pictured because the images of the *top* somatic cells are occluded by the closely spaced *bottom* somatic cells in the daughter colonies. The *top* somatic cells of the parent show up as out of focus cells surrounding the *bright spots* of the light beam focused by them. **a** The microscope is focused on the somatic cells, mother and daughter colonies. **b** The microscope objective has been moved 10 μm downward to show focusing by the somatic cells of the daughters. The "best" focus of the much larger mother's somatic cells' beams of light is at 20–30 μm down relative to position in *panel a*. The diameters of the two most central daughter colonies are 82 μm and 98 μm

As mentioned previously, the length of the beam of illumination, concentrated by the focusing action of algal cells, varies depending on the difference between the mean index of refraction of the cells and their surroundings. For *V. carteri*, typical length of a concentrated beam was 10–30 μm below the somatic cell that focused it. Distances between light-focusing somatic cells and targets that the beams could influence ought to be in that range. But distances between the peripheral somatic cells of a mature *V. carteri* and its gonidia are typically 100 μm. However, the distances between the somatic cells of developing daughter colonies and their gonidia (Fig. 1e) are within this range. Figure 6 shows that somatic cells of the daughter spheroids inside the mother colony can also focus light from above into beams that can reach the newly formed gonidia within the daughter colonies.

Implications and Applications

As shown in Figs. 2, 3, 4, 5, we established that single algal cells can act as tiny lenses. Cells of different species of algae have various values of n(in). Various growth media, e.g., fresh or brackish water, and the ECM of different species can also have different values of n(out). Consequently, the light-focusing effects of various species will be different.

In the volvocine algae, such as *Pleodorina* and *V. carteri*, in the context of peripheral cells embedded in the ECM, and perhaps also in contact with the aqueous medium on the outside, the values of Q and R will vary as the colonies mature. Young spheroidal volvocine algae grow in size by the deposition of ECM. The indices of refraction of the ECM and of the cells depend on their molecular content. Thus they can also vary as the organism matures, which results in the location of convergence of incident light also varying as a function of the stage of growth. The observed variations in focal distance at various stages and for different species of the Volvocales could be used to measure temporal variations in the implied molecular content of the cells and the ECM.

To address whether the light focusing effect of individual peripheral cells has any biological consequences in the extant species one might compare the morphology and phototactic abilities of colonies grown in growth chambers in unstirred media illuminated only from below, or only from above, with colonies grown in stirred media in conventional, uniformly illuminated growth chambers. We might find a phenotypic plastic response and developmental changes if such effects are relevant in colonies such as *Pleodorina* or in young daughter colonies of *V. carteri*, where the distances between cells are in the range of the light-focusing capabilities of the peripheral cells.

Nevertheless, although such effects might not have any biological relevance in the extant species, it is possible that they influenced the early evolution of multicellularity in volvocine algae. Indeed, in contrast to single-celled organisms, cells in the emerging multicellular organisms were likely affected by the light beams produced by their neighboring cells. It is conceivable that cells responded to these new "signals", and if the responses were adaptive they could have become fixed. Below we suggest ways in which the light-focusing effect of cells might have influenced the evolution of morphological traits in volvocine algae.

Light-Focusing and the Evolution of Morphological Asymmetry

What role might the concentration of solar illumination by algal cells have played in the evolution of morphological asymmetry in the spheroidal volvocine algae, including the posterior positioning of gonidia and the differences in cell and eyespot size among peripheral cells? Could this light-focusing effect represent a unique contribution, beyond so many other conventionally considered factors, to the evolution of multicellularity and cell specialization in these algae? And if so, how did it influence their evolution?

Rather than providing a small additional quantity of photosynthetic energy, it seems likely that the light focused inside the spherical colony could act via photoreceptors. The actions triggered or steered by photic signals require very low light intensity. As the colonies swim or are rotated by their fluid environment, the signaling light strikes any one target for only a very short interval (Drescher et al. 2010). That is adequate, even desirable, for signaling but likely quite ineffective for photosynthesis.

Photoreceptors are generally involved in directional movement in response to light. However, their roles extend to a greater range of cellular activities, including modulating gene expression and protein synthesis, regulating photosynthetic activities, controlling cell division, cell fate and differentiation (Ozawa et al. 2009; Beel et al. 2012; Kianianmomeni and Hallmann 2014). In *V. carteri*, the initiation of cell divisions during embryogenesis and the final steps of cell differentiation are regulated by light, and these processes appear to be mediated by photoreceptors (Kirk and Kirk 1985; Kirk 1998). A large number of photoreceptors have been described in *C. reinhardtii* and *V. carteri* (Ebnet and Fischer 1999; Kianianmomeni et al. 2009; Kianianmomeni and Hallmann 2014). Interestingly, in *V. carteri*, the distribution and relative abundance of photoreceptors are different between the two cell types, which suggests that they interpret light signals using different signaling pathways. For instance, the animal-type rhodopsin VR1 is expressed only in gonidia, where it might be involved in regulating photosynthetic activities (Ebnet and Fischer 1999); other photoreceptors are expressed only in somatic cells (Kianianmomeni and Hallmann 2014). Of particular interest is phototropin—a blue light receptor that in *C. reinhardtii* not only regulates the expression of channelrhodopsins (light-gated ion channels in the eyespot area; Kianianmomeni et al. 2009) and the eyespot size, but is also involved in cell size regulation (see Kianianmomeni and Hallmann 2014 for discussion and references). As in *V. carteri* the phototropin gene is highly expressed in somatic cells. It has been suggested that a connection between this photoreceptor and the mechanism of cell size regulation during development could exist (Kianianmomeni and Hallmann 2014). Furthermore, *V. carteri* possesses four histidine kinase rhodopsins, which belong to a two-component signal transduction system that in prokaryotes is involved in response to stress stimuli and control of cell division and cell growth (Schaller et al. 2011). In *V. carteri*, they could produce cAMP or cGMP (which are secondary messengers in a variety of cellular processes) in a light-dependent manner, which can activate transcription factors controlling cell cycle regulation and cell differentiation during development (Kianianmomeni and Hallmann 2014).

Although little is known about the composition and role of photoreceptors in volvocine algae other than *C. reinhardtii* and *V. carteri*, it is likely that light and photoreceptor-mediated signaling pathways were also important in the early multicellular volvocine algae. Thus, we suggest that the evolution of specific aspects related to the morphological asymmetries of larger volvocine algae might have been influenced by the light focusing properties of spheroidal cells, via photoreceptor-mediated signaling pathways. The two aspects discussed below include (i) the different growth rates and reproduction potential among the cells in a *Pleodorina* colony (with the cells at the anterior pole losing their reproductive potential, Fig. 1c) and (ii) the location of the reproductive cells in the posterior pole of *Volvox* spheroids (Fig. 1d). Interestingly, anterior-posterior polarity mutants have been described in *V. carteri*; since they affect some but not all polarity-related traits (e.g., the distribution of gonidia in the posterior half of the spheroid but not the graded differentiation of somatic cells in terms of cell and eyespot size), it has been suggested that these traits are under separate genetic controls (Kirk 1998).

The mechanism responsible for the differential growth rates of cells within *Pleodorina* colonies is not known. Likewise, little is known about the factors determining the posterior location of gonidia in *Volvox*. In *V. carteri* (and related species that share the same developmental program, as defined in Desnitski 1995) the distinction between gonidia and somatic cells is apparent early during embryogenesis as a result of a series of asymmetric cell divisions that set apart larger cells to become gonidia. However, these asymmetric divisions occur only in the anterior half of the embryo, which, following a process of inversion (turning itself inside-out), becomes the posterior half of the adult. This restriction of asymmetric divisions to the anterior half of the embryo explains why the gonidia are located in the posterior half of the spheroid. However, what determines the difference in cell division patterns between the two hemispheres of the cleaving embryo is not known. We suggest that these aspects now stably associated with the anterior-posterior polarity of *Pleodorina* and *Volvox* species could have evolved from initially plastic traits induced by light signals focused by the cells at the anterior pole of the mother spheroids, and that acted as morphogenetic signals affecting the development of the daughter spheroids. Below we present a possible evolutionary scenario relating the light focusing effect discussed above to the evolution of morphological asymmetry in volvocine algae; specifically, the differential growth rate in *Pleodorina* and the distribution of gonidia at the posterior pole of *Volvox* colonies.

First, one can envision a *Gonium*-like convex plate of cells (Fig. 1a) that "accidentally" turned into a spherical structure at the end of embryogenesis (such as in the extant species, *Pandorina*), and whose cell walls eventually expanded into an ECM (such as in *Eudorina*; Fig. 1b). In addition to providing size benefits (presumably to escape predators) and potential nutrient storage (see Chapter 11 "The Evolutionary Ecology of Multicellularity: The Volvocine Green Algae as a Case Study"), the transparent ECM could have also allowed the optimal action of the cells' concentrated sun-beams on neighboring cells. To form these light-beams the cells should not be in direct contact. As the number of cells/lenses and size of the ECM and the spheroid increased, the dynamics of these light beams is likely to have changed, and it is conceivable that they eventually reached cells on the opposite side of the spheroid. At this point, we envision that some aspects of anterior-posterior polarity (e.g., the gradient in flagellar orientation) had already been established, as is seen in modern-day *Eudorina* colonies. Depending on each cell's location with respect to the pre-existing aspects of anterior-posterior polarity, these beams could have differentially affected other cells in the spheroid. If these beams were perceived as photoreceptor-mediated signals acting on photosynthetic activities, it is possible that they increased the growth rates of cells at the posterior half of the colony, resulting in a morphology similar to that observed in the current *Pleodorina* species (Fig. 1c). Furthermore, the light beams would have affected more the top half (facing the sun) of the growing and dividing reproductive cells. As, after the inversion of the embryo, the top side of the developing embryo becomes the bottom side of the adult, the asymmetric effects of these beams could have been indirectly transmitted to the offspring. In other words, the phenotype of the offspring would have been already "set" to follow the developmental pattern of the mother. Initially, this could have simply involved the asymmetric distribution of photoreceptors, transcripts, photosynthetic products or any other components that might have accumulated unequally in the growing reproductive cell in response to the additional light signal produced by the cells above acting like lenses and focusing the light on the anterior pole of the growing reproductive cells. The asymmetric distribution of cellular components is known to be responsible for the differential fate of cells during early embryogenesis in other systems (for instance, the asymmetric distribution of germ granules determines germ fate in most animals; Voronina et al. 2011).

In some species, these light beams might have also affected the cell division pattern in the developing embryos, with cells at the anterior pole dividing asymmetrically and producing large cells that, after inversion, became located at the posterior pole of the embryo. As colonies without asymmetric divisions became larger (for instance, to avoid predators; see Chapter 11 "The Evolutionary Ecology of Multicellularity: The Volvocine Green Algae as a Case Study"), it is conceivable that the distance between the somatic cells and gonidia increased as the colony matured (the mother colony grows by the accumulation of ECM), to the point that the convergence point of the refracted light became far above the gonidia. Consequently, these beams might not have been able to influence the development of gonidia in the adult colony. However, the daughter colonies, while growing in the mother colony, have cells that are at the appropriate distances to affect each other (Figs. 1e and 6). Thus, the light beams focused by the anterior cells of the daughter colonies themselves had the potential to influence the further development of their anterior-posterior polarity (in terms of the gradient in somatic cell and eyespot size) and possibly the development of the next generation by establishing the polarity of the gonidia, which will determine the pole that will undergo asymmetric divisions in the next generation. This could have involved the unequal accumulation of photoreceptors (or their transcripts)—e.g., channelrhodopsins and phototropins in the somatic cells, and volvoxrhodopsin in gonidia, as discussed earlier.

If the effects of the cell-induced light beams we envisioned above provided a fitness advantage, it is conceivable that these environmentally-induced plastic responses could have become fixed via genetic assimilation. Genetic assimilation occurs when a phenotype initially induced by an environmental stimulus becomes constitutively expressed, so that the original stimulus is no longer required for its expression (Waddington 1953; Crispo 2007). While phenotypic plasticity is thought to have been important for the evolution of cellular division of labor in early multicellular lineages (Pigliucci et al. 2006; Gavrilets 2010), it has not yet been implicated in the evolution of morphological asymmetry. Furthermore, although it has been suggested that adaptive responses to the environment that evolved in in unicellular lineages have been co-opted for multicellular-specific traits (such as signaling and cell differentiation; Nedelcu and Michod 2006; Ritchie et al. 2008; also see Chapter 21 "The Evolution of Developmental Signalling in Dictyostelia from an Amoebozoan Stress Response"), the co-option of passive outcomes resulting from the interactions between cells and their environment has not received as much attention (but see Schlichting 2003).

Conclusion

Volvocine algae have been used to address various questions related to the evolution of multicellularity and cell differentiation, in terms of the genetics, developmental constraints, and underlying selective forces specific to this group (see Chapters 7 and 11 "Volvocine Algae: From Simple to Complex Multicellularity" and "The Evolutionary Ecology of Multicellularity: The Volvocine Green Algae as a Case Study"). More recently, it has been proposed that physical characteristics of the environment as well physical aspects associated with the emerging multicellular groups (e.g., increased mass, drag) are also likely to have contributed to the evolution of specific multicellularity traits in volvocine algae (see Chapter 11 "The Evolutionary Ecology of Multicellularity: The Volvocine Green Algae as a Case Study"). However, the role of light—outside its photosynthetic context, in the evolution of multicellularity has not been explored. Here, we argue that light and the optical properties of approximately spherical cells might have contributed to the evolution of specific traits associated with the morphological asymmetries observed in these spheroidal multicellular algae. As single algal cells can act as miniature lenses that concentrate incident illumination, it seems likely that the concentration of incident light by one portion of an emergent multicellular organism onto another part of itself could have been biologically significant.

As far as we know, volvocine algae grown in laboratories settings, with agitation of the growth medium and approximately uniform illumination do exhibit the morphological asymmetries mentioned. However, the role of light (especially in terms of its directionality) for some of these traits has not been specifically addressed. It is possible that these lights beams might provide some morphogenetic information when they reach germ cells and daughter colonies, at least in natural environments.

Summary

1. Single algal cells act as lenses that concentrate illumination incident on them, producing beams of light that may vary in length, depending on the geometry of the cell and the difference between the index of refraction of the cell and its surroundings.
2. When illuminated from above, a downward beam of light emerges from each of the peripheral cells embedded in the extracellular matrix of multicellular Volvocales. The beams that originate from the cells facing the sun penetrate the ECM; we suggest that such beams may have influenced the evolution of some morphological asymmetries associated with the anterior-posterior polarity in these algae.
3. The evolved spheroidal geometry and remarkably transparent ECM are well suited to intermittent photic signaling associated with slight changes of the colonies' axes as they swim, providing potential information in a changing environment.

4. The function of somatic cells acting as lenses concentrating sunlight into downward beams that might operate as morphogenetic information when they reach gonidia and daughter colonies is an additional property, besides motility and phototaxis, of non-reproductive somatic cells.

Materials and Methods

Algal Strains and Culturing

Cultures of the dinoflagellate *A. fundyense* (kindly provided by Jennifer Alix; NOAA, Northeast Fisheries Science Center) were cultured in unshaken flasks of f/2 seawater medium (Guillard and Ryther 1962), modified for dinoflagellates and adjusted to a salinity of 28 ppt. Flasks were in a growth chamber kept at 18°C with a diurnal light/dark regime of 12/12 h (~ 600 foot candles).

The volvocine strains used in this study were: *G. pectorale* (UTEX LB82), *P. starrii* (NIES 1361), and *V. carteri* (strain EVE; kindly provided by Dr. D. Kirk, Washington University). Cultures were grown in gently shaken flasks containing standard *Volvox* medium (SVM; Starr and Zeikus 1993). Flasks were placed in a growth chamber kept at 26°C with a diurnal light/dark regime of 16/8 h (~ 600 foot candles).

Microscopy Observations

The images and focusing data are taken from 33 and 50 frames per second sequences obtained with a Phantom camera, version 630, manufactured by Vision Research Inc., attached to an Olympus IM inverted microscope via an Olympus camera-attaching tube that contained a 2.5 × Olympus photo-relay lens. Objectives used were 10 ×, 20 × and long working distance 40 ×. Depending on objective lenses used, the distances between the front of the objectives and the algae being imaged, as well as the imaged cross sections of the beams of focused light emerging from them, were a few millimeters or less. All the standard light-concentrating lenses that normally illuminate the object being examined were removed, leaving empty space between the object and a circular ground glass light diffusing plate situated below the microscope lamp. The light passing through the ground glass plate illuminates the object, e.g. an alga, on the microscope stage. The distance between that plate and the algal object being imaged was approximately 34 cm, effectively at infinity, i.e., 340 times the distance between that object and the imaging objective, assuming 1 mm for that distance. The "infinite" distance between the Sun and a swimming alga outdoors is thus simulated by the closer yet effectively infinite distance between the source of diffuse light and the algal object on the microscope stage. The circular spot of light observed at a section of the beam that emerges from an algal cell acting as a lens (Figs. 2, 3, 4, 5, 6) is the image of the circular ground glass, the source of light. This assertion was proven by placing below the ground glass of an opaque sheet of paper, into the central region of which was cut a narrow linear opening, a slit. The illumination emerging from that slit was then focused by the cell into an emergent beam with slit-shaped cross section, shown in Fig. 2c. Summarizing, we have developed a method for demonstrating formation by an algal cell of an image of a remote source of light, e.g. the Sun, an illuminated circular ground glass, or a thin rectangular source of light.

Maintaining the circular ground glass plate as the source of light, the distance between microscope objective and algal object can be chosen so as to image a cross section of the alga's interior (e.g., Figs. 2a and 3a). When that distance is increased, the image of the cell's interior structures become fuzzy; increasing that distance further, a bright circular spot, the image of the light source, appears within the remaining outline of the cell (e.g., Figs. 2b and 3b). The spot remains visible for a few to many tens of micrometers, depending on the optical properties of the focusing algal cell and its surroundings, as discussed in the text. If the cell were a "perfect" lens, there would be a focus at a sharply defined distance from the lens. Aberrations due to imperfections of shape, variation of the index of refraction within the cell, and other irregularities are the inferred cause of the extended "focus" which implies that there is an extended distance of action for the concentrated beam of light. The method for estimating that extended distance was to move the objective in the down-beam direction, away from the focus on the cells' interior, by turning the fine-focus dial, by 10 μm steps. The interval over which the spot remained reasonably sharp is therefore reported as a range. The beam of diffracted light that produces the "bright spot of Arago and Poisson" is superimposed on the refracted beam. It modifies the total length of concentrated illumination.

Appendix

Figure 7 demonstrates the spot of diffracted light that occurs behind approximately circular opaque objects, namely, powdered graphite. The black background permits a clear view of the bright spot behind the small particles. The larger, odd shaped graphite fragments are too irregular to generate a bright spot by addition of wavelets. In Figure 8 we used 21 μm diameter polystyrene spheres (Bangs Labs) to demonstrate complex focusing and diffraction. The interior of these spheres is surely less complex than the interior of algal cells. The images show focusing of the incident light in apparently two distinct steps, followed by a diffraction pattern that is similar to the diffraction patterns observed with the Arago-Poisson effect (Sommargren and Weaver 1990; Kolodziejczyk et al. 2002). It is too difficult to show unambiguously the contribution of the Arago-Poisson diffractive imaging to the light being concentrated by the algal cells, because of their internal structure, surface irregularities and deviations from sphericity. Both experimental and theory investigations are currently underway. In particular, we have shown that by restricting the area of the illumination source, the beam propagating the image of the light source becomes longer

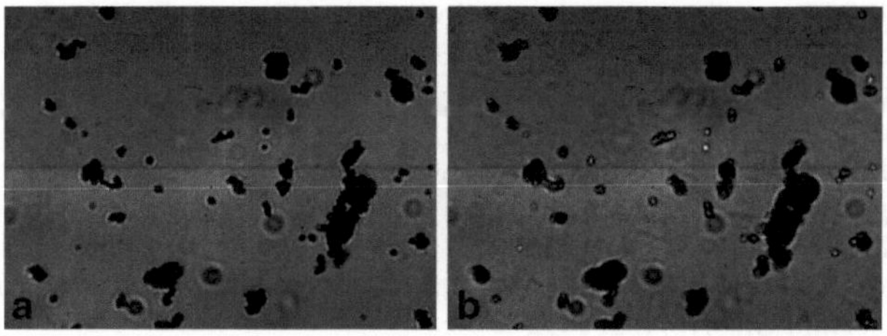

Fig. 7 Graphite powder viewed from below with a 40 × objective. Light is incident from above. **a** In *panel a* all graphite particles appear black because they do not transmit light. A diffraction *ring* is seen outside the particles. **b** In *panel b* the objective was moved downwards by a few micrometers. The smaller particles show the Arago *bright spot* that is produced by diffracted light originating near the particles' edge. The larger particles do not show the spot because their irregular surface eliminates coherent addition of wavelets

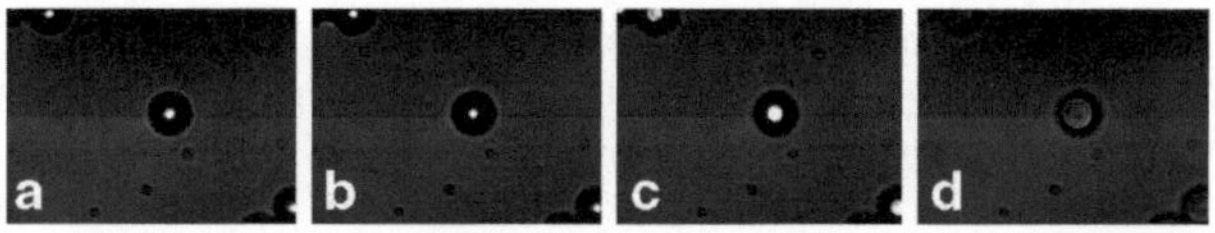

Fig. 8 The light emerging below a 21 μm diameter polystyrene sphere surrounded by water, and illuminated from above. The images are obtained with a 40 × objective being moved away from the sphere. *a, b* and *c* are in sequence. The *light spot* is a section through the concentrated beam emerging from the sphere. For image in *panel d*, showing a diffraction pattern, the objective was lowered a few micrometers away from its position for image in *panel c*

and its cross section increases with distance, as would be expected from the Arago effect. These experiments, indicating the dependence on geometry of the source, are preliminary to investigate imaging of the Sun by algal cells.

Acknowledgements We thank Agneta Persson (Department of Biological and Environmental Sciences, Göteborg University, Göteborg, Sweden) for much helpful advice and information on dinoflagellates. The dinoflagellate culture and media ingredients were kindly provided to us by Jennifer Alix (NOAA, Northeast Fisheries Science Center). Jeremiah Hackett (Department of Ecology and Evolutionary Biology, University of Arizona) provided growth chamber space and lab equipment for our dinoflagellate cultures. Rick Michod (Department of Ecology and Evolutionary Biology, University of Arizona) provided materials, space, and the volvocine algal cultures. JOK also wishes to acknowledge the very helpful conversations with Peter Evennett and Luis Cisneros. AMN acknowledges support from NSERC.

References

Beel B, Prager K, Spexard M et al (2012) A flavin binding cryptochrome photoreceptor responds to both blue and red light in *Chlamydomonas reinhardtii*. Plant Cell 24:2992–3008

Crispo E (2007) The Baldwin effect and genetic assimilation: revisiting two mechanisms of evolutionary change mediated by phenotypic plasticity. Evolution 61:2469–2479

Desnitski A (1995) A review on the evolution of development in *Volvox*-morphological and physiological aspects. Eur J Protistol 31:241–247

Drescher K, Goldstein R, Tuval I (2010) Fidelity of adaptive phototaxis. Proc Natl Acad Sci U S A 107:11171–11176

Ebnet E, Fischer M (1999) Volvoxrhodopsin, a light-regulated sensory photoreceptor of the spheroidal green alga *Volvox carteri*. Plant Cell 11:1473–1484

Fiorani L, Maltsev V, Nekrasov V et al (2008) Scanning flow cytometer modified to distinguish phytoplankton cells from their effective size, effective refractive index, depolarization, and fluorescence. Appl Opt 47:4405–4412

Gavrilets S (2010) Rapid transition towards the division of labor via evolution of developmental plasticity. PLoS Comput Biol 6:e1000805

Grosberg R, Strathmann R (2007) The evolution of multicellularity: a minor major transition? Annu Rev Ecol Evol Syst 38:621–654

Guillard R, Ryther J (1962) Studies of marine planktonic diatoms I. *Cyclotella nana* (Hustedt) and *Detonula confervacea* (Cleve) Gran. Can J Microbiol 8:229–239

Harvey J, Forgham J (1984) The spot of Arago: new relevance for an old phenomenon. Am J Phys 52:243–247

Hecht E (1987) Optics, 2nd edn. Addison Wesley, Reading

Kessler JO (1986) The external dynamics of swimming micro-organisms. In: Round FE, Chapman DJ (eds) Progress in Phycological Research. Biopress, Bristol 4:257–307

Kianianmomeni A, Hallmann A (2014) Algal photoreceptors: in vivo functions and potential applications. Planta 239:1–26

Kianianmomeni A, Stehfest K, Nematollahi G et al (2009) Channelrhodopsins of *Volvox carteri* are photochromic proteins that are specifically expressed in somatic cells under control of light, temperature, and the sex inducer. Plant Physiol 151:347–366

King N (2004) The unicellular ancestry of animal development. Dev Cell 7:313–325

Kirk D (1998) *Volvox*: molecular-genetic origins of multicellularity. Cambridge University Press, Cambridge

Kirk M, Kirk D (1985) Translational regulation of protein synthesis, in response to light, at a critical stage of *Volvox* development. Cell 41:419–428

Knoll A (2011) The multiple origins of complex multicellularity. Annu Rev Earth Planet Sci 39: 217–239

Kolodziejczyk A, Jaroszewicz Z, Henao R et al (2002) An experimental apparatus for white light imaging by means of a spherical obstacle. Am J Phys 70:169–172

Michod R, Nedelcu A (2003) On the reorganization of fitness during evolutionary transitions in individuality. Integr Comp Biol 43:64–73

Nedelcu A, Michod R (2004) Evolvability, modularity, and individuality during the transition to multicellularity in volvocalean green algae. In: Schlosser G, Wagner G (eds) Modularity in development and evolution. University of Chicago Press, Chicago, pp 466–489

Nedelcu A, Michod R (2006) The evolutionary origin of an altruistic gene. Mol Biol Evol 23:1460–1464

Ozawa S-I, Nield J, Terao A et al (2009) Biochemical and structural studies of the large Ycf4-photosystem I assembly complex of the green alga *Chlamydomonas reinhardtii*. Plant Cell 21:2424–2442

Pigliucci M, Murren C, Schlichting C (2006) Phenotypic plasticity and evolution by genetic assimilation. J Exp Biol 209:2362–2367

Ritchie A, Van Es S, Fouquet C et al (2008) From drought sensing to developmental control: evolution of cyclic AMP signaling in social amoebas. Mol Biol Evol 25:2109–2118

Schaller G, Shiu S-H, Armitage J (2011) Two-component systems and their co-option for eukaryotic signal transduction. Curr Biol 21:R320–330

Schlichting C (2003) Origins of differentiation via phenotypic plasticity. Evol Dev 5:98–105

Sommargren G, Weaver H (1990) Diffraction of light by an opaque sphere 1. Description and properties of the diffraction pattern. Appl Optics 29:4646–4657

Sommargren G, Weaver H (1992) Diffraction of light by an opaque sphere 2. Image formation and resolution considerations. Appl Optics 31:1385–1398
Spizzichino V, Fiorani L, Lai A et al (2011) First studies of pico- and nanoplankton populations by a laser scanning flow cytometer. J Quant Spectrosc Radiat Transf 112:876–882
Starr R, Zeikus J (1993) UTEX—the culture collection of algae at the University of Texas at Austin 1993 list of cultures. J Phycol 29:1–106
Voronina E, Seydoux G, Sassone-Corsi P et al (2011) RNA granules in germ cells. Cold Spring Harb Perspect Biol 3:a002774
Waddington C (1953) Genetic assimilation of an acquired character. Evolution 7:118–126

In Silico Transitions to Multicellularity

Ricard V. Solé and Salva Duran-Nebreda

Abstract The emergence of multicellularity and developmental programs are among the major problems of evolutionary biology. Traditionally, research in this area has been based on the combination of data analysis and experimental work on one hand and theoretical approximations on the other. A third possibility is provided by computer simulation models, which allow to both simulate reality and explore alternative possibilities. These *in silico* models offer a powerful window to the possible and the actual by means of modeling how virtual cells and groups of cells can evolve complex interactions beyond a set of isolated entities. Here we present several examples of such models introducing different components required to generate and evolve multicellular systems. Each one illustrates the potential for artificial modeling of the transition to multicellularity.

Keywords Complexity · Embodiment · Artificial life · Gene networks

The Physics of Multicellularity

The transition to multicellularity is tied to the emergence of interactions among previously isolated cells. As a problem in complexity (Anderson 1972) we could say that a multicellular system defines a level of organization whose global properties cannot be reduced to the properties of the individual units. This statement is relevant for many reasons. First, because the presence of an evolutionary innovation necessarily requires the cooperation between previously unrelated units (Schuster 1996). Once such interactions are in place, a network of connected elements needs to be considered in order to understand what is now at work. In early phases predating the

R. V. Solé (✉)
Santa Fe Institute, Santa Fe, NM, USA
e-mail: ricard.sole@upf.edu

R. V. Solé · S. Duran-Nebreda
ICREA-Complex Systems Lab, Institut de Biologia Evolutiva-UPF-CSIC, Barcelona, Spain

S. Duran-Nebreda
e-mail: salvador.duran@upf.edu

I. Ruiz-Trillo, A. M. Nedelcu (eds.), *Evolutionary Transitions to Multicellular Life,*
Advances in Marine Genomics 2, DOI 10.1007/978-94-017-9642-2_13

transition to complex multicellular life forms, the network involved cell-cell as well as cell-substrate interactions.

In dynamical terms, the transition required the emergence of cooperation among elements, which share a common space where they relate to each other and respond to environmental changes in a concerted manner. Most theoretical and computational approaches to this problem do not take into account the fact that these systems are formed by physical objects and it might seem not so relevant when dealing with the generic mechanisms associated to cooperative dynamics. As usual, the level of detail that is used in a model scales with the type of question we wish to answer. If we search for general principles pervading the appearance of cooperative aggregates, general models considering population dynamics and gene interactions are enough.

But multicellularity also provides the first steps towards developmental programs and a preceding step to other major innovations. Such innovations are always associated to novelties in the ways cells and tissues interact. The spatial arrangement of cells and the diversity potential provided by space, and thus an explicit introduction of spatial degrees of freedom, is essential.

Meaningful models of evolutionary dynamics of multicellular systems need to consider the role of generic physical mechanisms of morphogenesis that are not the result of complex regulatory processes. In this context, physical forces including gravity, adhesion or diffusion, and their generative potential, are considered (Newman and Comper 1990; Goodwin 1994). The interplay between these mechanisms allows for spontaneous pattern formation through segregation of cell types, differential cell growth and mortality. Some of these generic, pattern-forming mechanisms likely predate the early history of both pre-cellular and multicellular life forms (Forgacs and Newman 2005; Solé et al. 2007; Solé 2009), along with others controlled by genetic circuits.

Using some of these mechanisms, an evolutionary model of form can be constructed. Moreover, since some of these mechanisms seem to strongly constrain the repertoire of potential structures that can be generated, they also offer a powerful framework to understand the origins of convergent designs (Alberch 1980). In this context, as pointed out by John Tyler Bonner, simple explanations based on mathematical and computational models can help to grasp the principles of multicellular organization (Bonner 2001; Forgacs and Newman 2005). As noticed in Newman and Bhat (2008) the interplay between physical constraints and genetic regulatory mechanisms has been traditionally overlooked in most studies, with few exceptions (see for example Eggenberger 1997; Coen et al. 2004; Cummings 2006; Doursat 2008; Kaandorp et al. 2008).

Although physics and embodiment are usually discussed in the context (or at the level of) organisms or tissues, there is another level of embodiment that requires attention: the external world, whose fluctuations and properties influence the repertoire of adaptations that can be available. Thus, other factors playing a role in the early stages of multicellularity, including the ecological context and the physics of the environment should also be taken into account.

This chapter considers several *in silico* models illustrating different aspects of the emergence of multicellularity, involving increasing complexity but also different

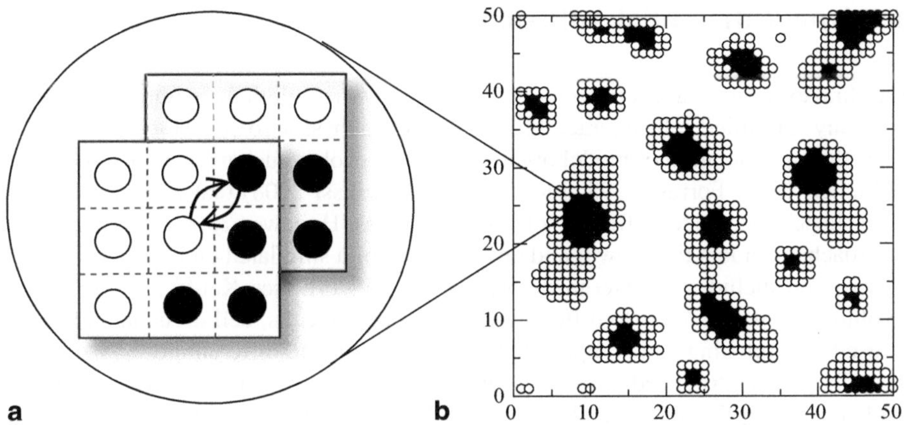

Fig. 1 Pattern formation through differential adhesion. Here a system composed by two types of cells (*open* and *filled circles*) are initially scattered at random on a two-dimensional lattice. Cells move and can flip their location with a neighboring cell **a** provided that the final energy is reduced (*Box 1*). Eventually, a stable arrangement of ordered cell assemblies **b** is obtained. In this example, the adhesion parameters satisfy J(1,2) > (J(1,1) + J(2,2)) = 2 and J(1,1) > J(2,2)

factors that favor cell aggregation, cell differentiation, spatial patterning or even coupled evolution between multicellular aggregates and ecological dynamics. All illustrate the generative potential of the role played by the explicit embodiment that pervades the interactions among virtual cells.

Box 1. Modelling Cell Sorting Under Differential Adhesion

The dynamics of cell sorting can be easily modelled by considering a discrete lattice Ω. Each site(i, j) (Fig. 1) is characterized by a "state" S(i, j) which can be 0 if the site is empty and either 1 or 2 if occupied by two possible "cell" states. Cells can be more or less prone to remain together and also might tend to either avoid or attach to the external medium. Each cell can interact with only eight nearest neighbors. Let J(S(i, j), S(k, l)) indicate the energy associated to the interactions between the sites (i, j) and (k, l). Cells move to a neighboring position by switching the two local states provided that the final state has a smaller energy. This can be obtained using a function

$$\mathcal{H} = \sum_{i,j} \sum_{k,l} (1 - \delta_{s(i,j)s(k,l)}) \, J(s(i,j), s(k,l))$$

where the sum is performed only over nearest pairs. Here $\delta_{min} = 1$ when m = n and zero otherwise, and thus the term $1 - \delta_{min}$ just discards pairs of sites with identical states. More sophisticated approaches were developed by Glazier and co-workers, involving "cells" composed of connected sites defining a specific cell, whose shapes can change in realistic ways (Graner and Glazier 1992; Glazier and Graner 1993).

Evolving Differential Adhesion

Our first example is a discrete model of evolution where digital creatures composed by many simplified cells interact through discrete adhesion forces. Hogewegs model considers a population of model organisms that is evolved using a genetic algorithm (Mitchell 1998; Forrest 1993; Floreano and Mattiussi 2008) that allows to search over shape space under different selection pressures (Hogeweg 2000a, b). Hogewegs approach considers the growth and development of a simulated embryo. The model description includes an internal boolean gene network (see Kauffman 1993), the evolution of which leads to different adhesion among cells, cell division and death caused by stretching and compression, cell migration and differentiation.

Adhesion is introduced using very simplified but effective physical models (Graner and Glazier, 1992; Glazier and Graner 1993; Savill and Hogeweg 1997; Podgorski et al. 2007) and is one of the main players influencing the evolutionary dynamics of these virtual metazoans and their potential for diversification, consistent with the role played by development in the context of morphological radiations (Eble 2003). Cell adhesion can easily promote the spatial organization of an initially disorganized, mixed group of cells (Steinberg 1963, 1978; Steinberg and Takeichi 1994). This is a very robust, repeatable and predictable mechanism of organization that amplifies initial disorder experienced by a mixed set of cells that move in space and aggregate with other cells under differential adhesion. The preferential adhesion mechanism is responsible for the sorting of cells in space, in a way that corresponds to the behavior of immiscible fluids (Forgacs and Newman 2005) (Fig. 1).

As pointed out by Hogeweg (b), differential cell adhesion is regulated by the gene network affecting cell behavior and the communication between cells through cell-cell interactions. The model considers different types of fitness functions but the only strong pressure is directed to maximizing the diversity of cell types and thus there is no explicit search for special spatial arrangements or predefined developmental programs. Hogewegs work revealed the existence of a neutral landscape of possible phenotypes that pervades the punctuated nature of transitions (Hogeweg b; see also Fontana and Schuster 1998). Long periods of stasis are characterized by slow increases in fitness as small variations in phenotype are achieved. Selection for diverse gene expression patterns is used (see also Solé et al. 2003). The number of cell types is a good measure of complexity, which is known to increase through metazoan evolution (Carroll 2001, 2005; Valentine et al. 1994). Increases in cell type number provide a high potential for further evolution of anatomical and functional complexity, essentially through division of labor and the formation of specialized tissues (Maynard-Smith and Szathmáry 1995) (Fig. 2).

Since the imposed selection pressure is rather generic, no special constraints are posed on the way genes interact and influence cell arrangements; no particular, predefined architectures and developmental plans are favored. The model is able to evolve complex forms, and in the process of evolving them, different remarkable changes take place. Complex shapes and some familiar ways of obtaining them (such as tissue engulfing, budding, etc) appeared and complex interactions between apoptosis or migration emerged. As pointed out in Hogeweg (2000a), morphogenesis itself emerges as a byproduct of optimization for cell diversity. It is worth noting that

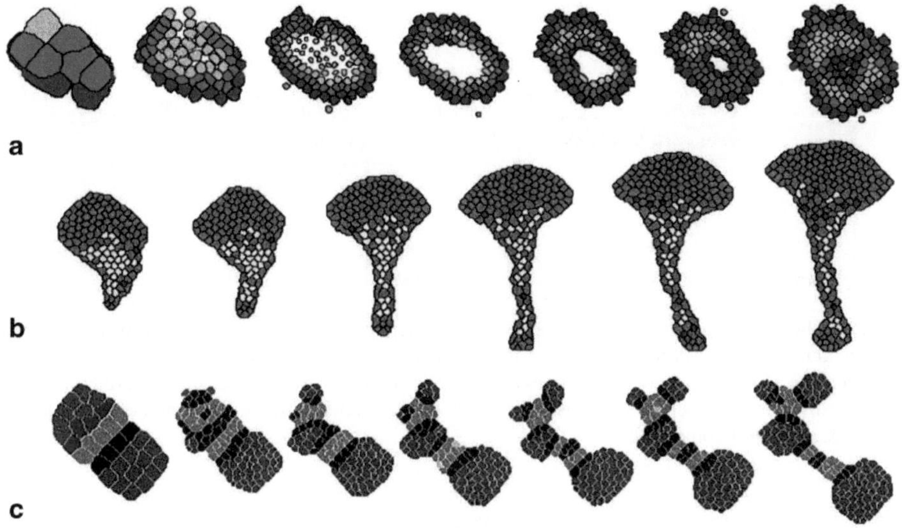

Fig. 2 Some examples of the developmental programs obtained in Hogeweg's in silico experiments involving the evolution of cell adhesion and differentiation (see Hogeweg 2000a). Differential cell adhesion leads to cell migration and tissue remodeling, including intercalation of cells and cell layers. Here we show three different outcomes of the evolution process, where selection for maximal cell differentiation produces, as a side effect, morphogenetic processes and pattern formation. Among the developmental processes observable, we find: **a** cell migration and engulfing. **b** budding and elongation and **c** cell death and re-differentiation

other works involving cell type richness as a fitness function favor the explosion of pattern forming motifs as soon as a threshold of genetic complexity is reached (Solé et al. 2003).

Multicellularity for Free

One of the most relevant and striking features of the transition to multicellularity is the fact that it took place multiple times in different lineages in the history of life (King 2004; Bonner 2001, Niklas and Newman 2013). For several authors, this suggests that there is a certain component of inevitability in this process (Buss 1987; Grossberg and Strahmann 2007), because cooperation and specialization are such powerful innovations that will inevitably result in the convergent evolution of multicellularity. In this vein, if Hogewegs model proves that selecting for cell diversification can lead to the co-selection of unexpected, emergent properties and behaviors, the model by Kaneko and colleagues (Kaneko and Yomo 1999), discussed below, shows that even when there is no selection at work, cells can easily drift into multicellular phenotypes including differentiation and spatial patterning.

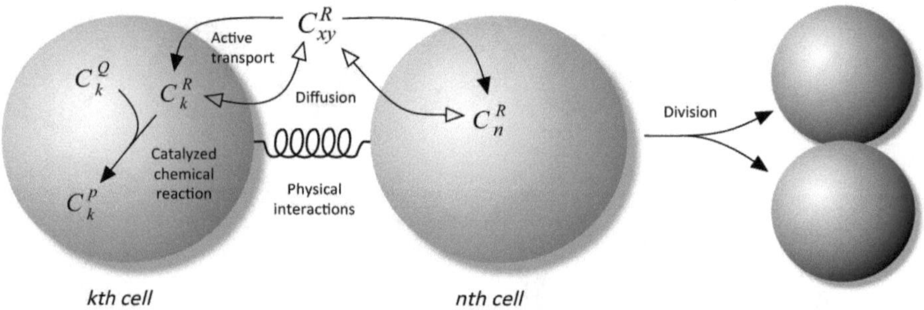

Fig. 3 Schematic representation of Kaneko's model, depicting the internal cell dynamics (catalyzed chemical reactions governed by sets of differential equations) as well as cell-cell interactions (diffusion and active transport of metabolites). The accumulation of key species up to a fixed threshold leads to division and stochastic allocation of the molecules between the new daughter cells. Redrawn from (Kaneko and Yomo 1999)

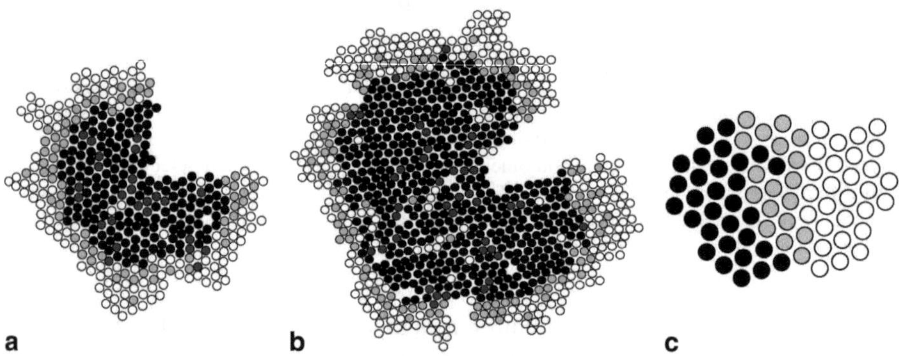

Fig. 4 Some of the multicellular organisms obtained by Kaneko's model (redrawn from Furusawa and Kaneko 1998; Furusawa and Kaneko 2003), the different shades of *grey* represent different cell types determined by mean metabolite concentration. Here we show the evolution of an organism after the removal of a quarter of the cluster (**a**), which leads to the regeneration of the previous ring pattern by means of differential cell growth and differentiation (**b**). Other natural occurring developmental forms are also observed, such as *stripes* (**c**)

As in the last example, the model starts with a single cell embryo that can develop into simple yet hallmark structures (see Fig. 4), just by allowing cell-cell and cell-environment interactions. The former are simulated by non-preferential adhesion between close cells and the latter by transport and processing of chemical species present in the virtual environment, which cells use to grow and divide. The internal workings of the cells are modeled by sets of coupled ordinary differential equations that describe autoregulatory transcription factor networks (see box 2). Remarkably, Kanekos work shows that even if the internal networks, the initial state of the cell and the environment composition are randomized, a significant fraction of all possible cases develops consistently into virtual organisms with life-like features (Fig. 3).

Box 2. Dynamical Networks of Differentiation
Internal dynamics Cellular states are defined by the set of concentration of each species, namely $C_k^1(t), \ldots, C_k^l(t)$ for the concentration of the l species in the *kth* cell at time t. A reaction matrix with notation $W(R, P, Q)$ is constructed, each position taking the value 1 if there exists a reaction from chemical R to chemical P catalyzed by Q and 0 otherwise. The Concentration change in this simple set of three species reads:

$$\partial C_k^P / \partial t = W(R, P, Q) C_k^R \left(C_k^Q\right)^2$$

$$\partial C_k^R / \partial t \quad = -W(R, P, Q) C_k^R \left(C_k^Q\right)^2$$

Cell-Environment dynamics The environment is simulated by a lattice of equal sized patches, each one of them characterized by C^1 $(x, y, t), \ldots$ C^l (x, y, t). Two different modes of material transfer are considered, passive diffusion and active transport.

Cell division and death Division is considered to be the direct consequence of the accumulation of key, predefined chemicals inside the cell.

Kaneko and colleagues observed that from the subset of configurations displaying cell growth and intermediate connectivity (case study in Kaneko and Yomo 1999 contains nine paths for a system of 32 chemical species), most led to periodic or quasiperiodic changes in metabolite concentrations inside the cells, analogous to natural cell cycles observed in both unicellular and multicellular organisms. Beyond a certain population threshold, synchronization and phase stabilization appeared among the cells in the same cluster, the first steps towards cooperation and collective action in cell societies. Further increase in the numbers within the ensemble enables the divergence of the mean chemical concentrations and the cycle itself (i.e. the differentiation of the cells). These new trajectories in the phase diagram were stabilized by mutually reinforcing metabolite exchanges between neighboring cells.

This process of generation of variants from previously equal individuals, dubbed isologous diversification, provides a solid testbed to study both the community effect (Carnac and Gurdon 1997) and positional information (Wolpert 1969) theories. Moreover, after taking a first round of differentiation, cells could subsequently change into previously unavailable cell types, thus creating a tree-like hierarchical structure observed in natural developmental lineages. Some of this virtual organisms displayed other key features of living systems like robustness to noise during development (Kaneko and Yomo 1999), resistance to injury through regeneration of spatial patterning (Furusawa and Kaneko 1998) and even life cycles exhibiting senile/proliferative stages (Furusawa and Kaneko 1998).

In conclusion, Kanekos model demonstrates that even in the absence of evolution or selection, valid unicellular genotypes have the potential to create complex, emergent phenomena given that size thresholds can be surpassed. Whether this phenotypical changes suppose an increase in fitness is not of relevance here, but the feasibility to become multicellular and display potential task allocation as a side effect of cell-cell interactions. This results articulate a clear connection to Kauffmans work (Kauffman 1987) and the realization that some of the high order features observed in natural systems can arise not as a result of natural selection but the unavoidable properties of systems with high epistatic connectivity.

Evolving Multicellular Aggregates

Multicellularity has been a recurrent novelty in the story of life and some clues to its origins can be found (at least at the functional level) in living unicellular systems, such as bacteria or yeast. Many unicellular species display multicellular traits (Shapiro 1998; Bonner 2001) associated to the presence of signals that provide the source of coherent population responses. As a consequence of these responses (mainly to stress signals) multicellular aggregates can form and display some degree of specialization and/or differentiation. Following these observations, a very promising approach to study the feasibility of the transition to multicellularity is the use of artificial selection in natural systems.

This strategy was put forward in a recent set of experiments (Ratcliff et al. 2012), in which the authors sequentially subcultured *S. cerevisiae* cells with the fastest sedimentation in order to force the selection of cooperating aggregates. Yeast is a specially interesting candidate to explore the potential first steps of the evolution of multicellularity due to the fact that it already presents some pre-adaptations thought to be relevant in this major transition in evolution (Maynard-Smith and Szathmáry 1995; Szathmáry 1994) and its biology and multicellular states are well enough described so that new emergent phenomena can be easily distinguished from them. Remarkably, after just 60 selection events in a timescale much shorter than previously thought the so-called snowflake phenotype appeared consistently in all cultures. These are roughly spherical clusters of cells formed not by aggregation but by defective separation of cells after division (Fig. 5).

The clonal formation of the clusters ensures limited conflict of interest among the elements of the new multicellular individual, as discussed elsewhere (see Grossberg and Strahmann 2007; Bonner 2001; Michod 2000). Later on, the authors studied the role played by cellular fate in cluster reproduction. It was found that clusters did not reproduce through events associated to single cells but instead involved a group-level set of events that led to the generation of a propagule. This was achieved through a division of labor in the form of the active control of apoptosis, which caused the asymmetrical splitting of the cluster once it reached a threshold size.

In order to test alternative explanations to some of the phenomena observed particularly the presence of a group-level reproduction and its relation to division of labor

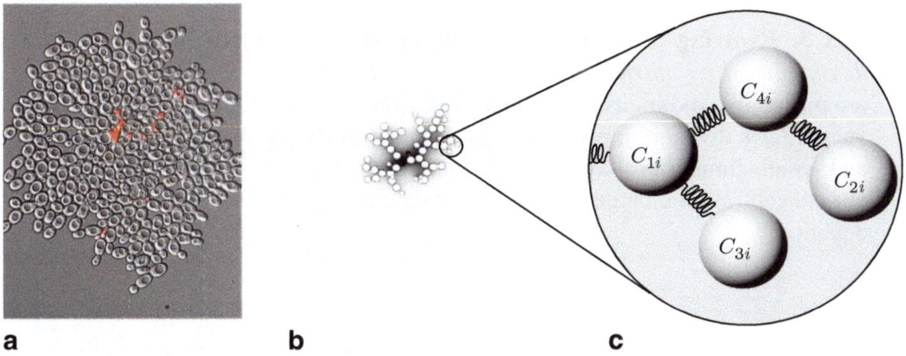

Fig. 5 The path from experimental results to in silico modeling. In **a** clusters of snowflake yeast with some apoptotic cells dyed in *red* in the *center* (from Ratcliff et al. 2012), breackage of a large cluster which generates juvenile propagules during the course of the simulation **b** and schematic representation of cell physics in the model **c** as described in *box 3* (from Duran-Nebreda and Solé 2014)

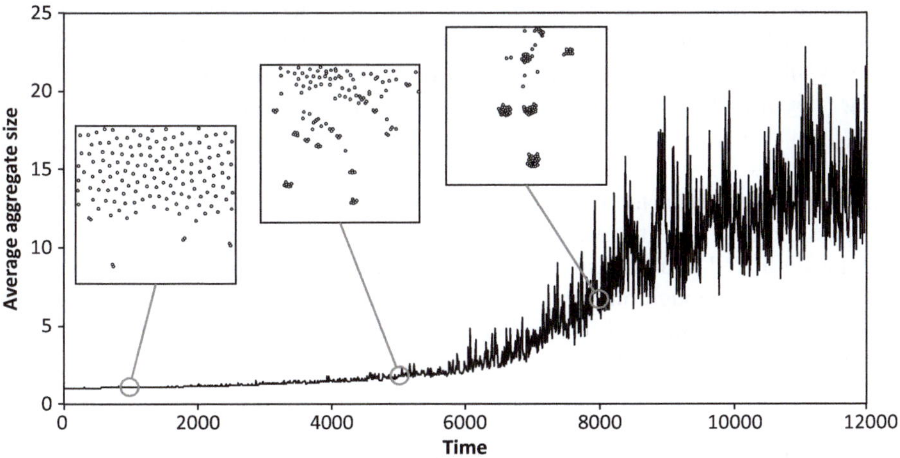

Fig. 6 In silico evolution of aggregates under the nutrient depletion model (Duran-Nebreda and Solé 2014). Average cluster size increases over time and stabilizes after 10^4 simulation steps (mean for 10 replicate experiments). The maximum aggregate size is dependent on the nutrient concentration fixed at the beginning of the simulation (data not shown)

and test other potential scenarios for the rise of multicellular ensembles, a simple embodied computational model was created (Duran-Nebreda and Solé 2014). In this model, yeast cells are represented as point elements in a bidimensional lattice, which can fail to separate correctly after division, thus remaining attached by a spring. The cells movement is modeled by a biased random-walk. No explicit genetic network is implemented, instead cells inherit the mothers parameters with small deviations. Two causes of cell death are tested, apoptosis as well as a simpler alternative based on the depletion of resources (Fig. 6).

Box 3. Evolving Multicellularity Under Size-Dependent Selection *Cell death caused by nutrient depletion* To take into account this process, cells are placed in a bidimensional lattice that holds nutrient. Cells have an energy value M_{ni}, a division threshold M^c_{ni}, a counter on the number of divisions Δ_{ni} and an attachment probability to daughter cells once they divide p_{ni}. Nutrient concentration change in the finite element R_{ij} is given by the following equation:

$$^{\partial R_{ij}}/_{\partial t} = D\nabla^2 R_{ij} - \rho\theta_{ij} R_{ij}$$

The heaviside function θ_{ij} is used to indicate the presence or absence of cells in that particular patch of the lattice (so $\theta_{ij} = 1$ if a cell is present and zero otherwise). The energy change for C_{ij} is:

$$^{\partial M_{ij}}/_{\partial t} = \rho R_{ij} - \beta_c M_{ij}(1 + \kappa\Delta_{ni})$$

Where β_c represents the maintenance costs. If the energy value of a particular cell reaches its division threshold, a new cell is created and the original energy value is split asymmetrically between the cells. Conversely, cells die if: $M_{ni} \leq \delta_c$, where δ_c is the energy limit cells can withstand.

The conclusions extracted from the simulations draw a slightly different picture on the possible first steps towards multicellularity. In this model, physical constraints previously linked to a decrease in fitness for the aggregates namely, the added difficulty to attain enough nutrients to survive caused by limited diffusion in the core are shown to work in favor of reproductive fitness, debilitating the cluster structure and promoting splitting after a certain size threshold is achieved.

Physical Forces and Ecological Scales

An important component of a consistent theory of multicellularity, particularly in relation to the emergence of not just cooperation, but also developmental programs, requires considering community ecology in embodied models. Such models introduce physics and spatial interactions, and they naturally incorporate some selection forces, since the explicit physics introduce strong constraints to the potential forms and multicellular aggregates that can be obtained. If cell aggregates explicitly move, adhere to substrates, develop cooperation through cell-cell shared nutrients and resist external fluctuations, the evolution of adhesion and other variable features will occur under well-defined selection pressures. If these changes take place in a physical environment where available resources spread and are consumed, ecosystem level processes might take place, and some particular structures, such as budding, coherent multicellular shapes, differentiated aggregates or even life cycles could emerge.

Moreover, potential changes in grazing efficiencies and the rise of predators can trigger arm races tied to changes in developmental programs, as it is likely the case for the transition between the Ediacaran and Cambrian biotas (Marshall 2006, Fedonkin 2007, Erwin and Valentine 2013).

In order to incorporate all these components, we need to build a complex simulation framework able to capture the essential physics, the presence of a population of interacting artificial agents and mechanisms of evolving the parameters that weight different metabolic and adhesion properties. Such type of model belongs to the tradition of so-called artificial life approaches (Ray 1991, Langton 1991, 1995; Sipper 1995; Adami 1998), which involve the study of artificial life-like systems in artificial environments (along with a wet version associated to the construction of living systems using genetic engineering techniques).

Box 4. Modeling Ecology, Physics and Evolution
The CHIMERA model was build as an advanced framework introducing artificial cells as particles in a physical world where Newtonian forces, along with selection pressures, genes and metabolism are taken into account.

Cells and particles Our starting point is a population of single-cell organisms, where each cell in the initial population is identical. Cells and particles are simulated with rigid bodies moving within a fluid-like environment. A cell (particle) has spherical geometry with radius R_i, mass M_i, spatial position r_i and velocity $v_i = \partial r_i/\partial t$. The motion of a cell is described by the standard second law:

$$M_i {}^{\partial v_i}/_{\partial t} = F_i$$

Cell movement is obtained by numerical integration of the Newton equations. Cell velocity at time $t + \Delta t$ is thus: $v_i(t + \Delta t) = v_i\ (t) + F_i/M_i \Delta t$ where Δt is the size of the integration step, and F_i the total force acting on M_i.

Cell-substrate adhesion Attachment of cells to surfaces may provide a favorable microenvironment for cell aggregates to develop. If $\mathrm{D}(r_i)$ indicates the cell-to-wall distance, when a cell with adhesion strength to the substrate $J_i^f > 0$ is closer than a given adhesion range, we attach a spring connecting the cell with the wall (Fig. 6).

Cell-cell adhesion Cells can form aggregates by attaching to other cells. Each cell has an intrinsic probability J_i^c to create a new adhesion link. Given two close cells located at r_i and r_j, we will set an adhesion string connecting them with probability $(J_i^c + J_j^c)/2$. The adhesion force to any cell is the sum of forces contributed by all the active cell-cell adhesion springs.

Computational models, which can reproduce realistic scenarios or completely ignore them, provide the ideal framework to explore the generative potential of an evolving set of rules allowing structures to emerge through time. Following this view, an embodied model of evolution, the so-called CHIMERA model (Solé and Valverde 2013)

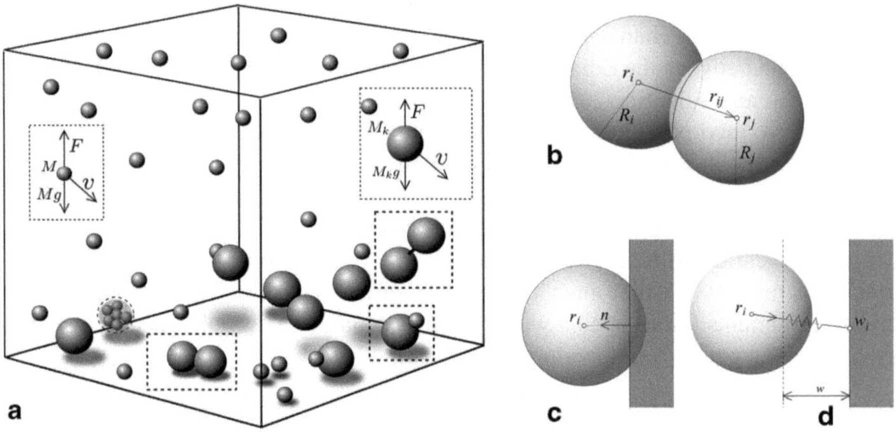

Fig. 7 Basic scheme of the components of the CHIMERA model (Solé and Valverde 2013). The system (**a**) is confined within a rigid *cube*. Nutrient particles fall from the top layer experiencing physical forces. Cells also experience the same forces, as described by Newton's laws (see *Box* 4). Additionally, both cells and particles get degraded. Cells can interact with the boundaries of the system as well as between them. Cells increase in mass every time they collide with a nutrient particle if they have the right internal metabolism. We also display the mechanical forces acting between cells (**b**) and the interactions between cells and the boundaries (**c-d**). In both cases, adhesion forces stabilize interactions within some range, but interpenetration is forbidden

was introduced as a way of including Newtonian physics, fluctuations, evolution and ecology in a unified fashion.

The model was intended as an approach to the pre-Mendelian universe, which can be approached by studying the interplay between physical forces such as gravity, diffusion and adhesion and generic pattern-forming mechanisms. In its simplest version (Solé and Valverde 2013) Chimera considers a cubic world involving a fluid like medium with gravity and turbulence (Fig. 7) where an initial set of identical cells exploit one of a number of potential energy particles falling from the upper side of the cube. A set of rules is then used to evolve the system:

1. Movement: both particles and cells experience both a gravitational and a fluctuating velocity field (the later associated to turbulence). Particles are removed from the system with some probability.
2. Each cell carries a given set of internal parameters and variables: they have a given size and mass and they have a list of possible particle types that they can take and the efficiency of the grazing for each particle type.
3. Cells can attach to the surface of the walls with some probability. When they are, a spring is used to properly define the physical interaction. Another adhesion probability is used for cell-cell adhesion. At the beginning all are set to zero.
4. If a cell interacts (collides) with a given particle, it ingests it if the efficiency for metabolizing that type of particle is non-zero. If taken, the mass of the particle gets transformed into mass of the cell.

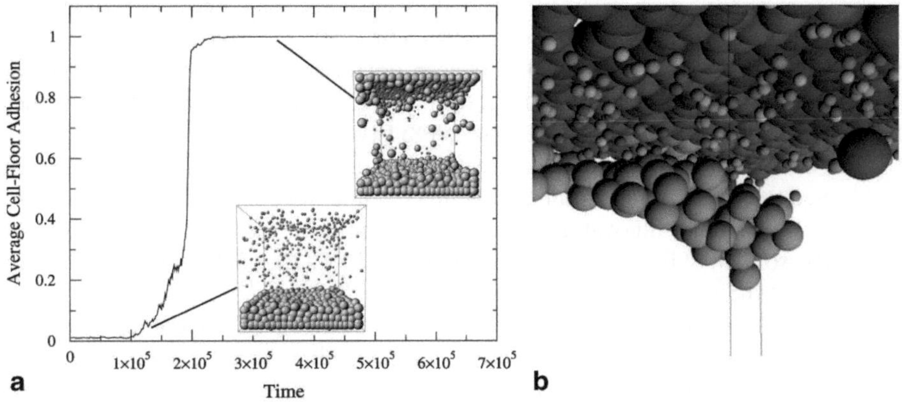

Fig. 8 **a** in the basic, newtonian CHIMERA model, the evolution of the system under enough nutrient levels drives the population from a few layers placed at the bottom to an inverted system where cells occupy and adhere to the *upper* wall. This occurs thanks to a broad distribution of efficiencies (i. e. evolved generalists) exploiting all resources, together with an increased adhesion between cells and specially between cells and surfaces. The main plot shows a fast growth in cellfloor adhesion towards its maximum value. Two snapshots of the system are also shown, before and after the transition. In **b** an example of an evolved multicellular aggregate is shown. This was obtained by using a more detailed implementation of the cell-cell interactions that allow aggregates to emerge Solé and Valverde 2013)

5. Once a maximum cell size is reached, the cell splits into two daughter cells. Moreover, if the cell goes below a minimal value, it dies and it disintegrates. Detritus particles are also allowed to be part of the nutrient intake of cells.
6. Each time a cell divides, mutations can occur in the daughter cell. Metabolic efficiencies and adhesion rates can change.

The model is able to display complex forms of pre-multicellularity in terms of loose aggregates of cooperating cells. These aggregates evolve adhesion rates, both between cells but specially in relation to the physical substrate. Two major trends take place here. From the point of view of grazing efficiency, individual cells tend to become generalists: their metabolism changes in order to exploit all types of nutrient particles, although at the price of being less efficient. Secondly, in order to gather more particles, there is an advantage in providing a larger cell surface, which is possible provided that cells increase their attachment probability (adhesion force) and occupy available space on the lateral walls. This tendency, once starts, is rapidly amplified and we can see (Fig. 8a) that eventually the cells discover the source of energy and occupy it. This is done by an increase of cell-floor adhesion but also by increasing cell-cell adhesion and (when fluctuations are large) cooperation. Also increasing cell-cell adhesion and (when fluctuations are large) cooperation.

Eventually, the flow of particles is almost shut down except for cell mortality and the fall of cells. Cell death creates detritus particles, which are also exploited with low efficiency. But once a critical amount of detritus gets accumulated at the bottom, a new community of specialized detritivores emerges. This new ecosystem with two layers is stable over time and represents a dramatic example of how ecosystem engineering emerges: the feedbacks between organisms and environment trigger a control of the later by the first, with a deep reorganization of energy flows (Jones et al. 1994, 1997; Hastings 2006; Erwin 2008; Erwin and Tweedt 2012).

The outcome of the simulation reveals how a microscopic process of evolution associated to cell adhesion, combined with a community adaptation to the environment leads to a major innovation. Using a very simple physics (Palsson and Othmer 2000; Palsson 2008; Ericson 2005; Sandersius and Newman 2008) where cells have volumes but their behavior is closer to point particles does not allow the formation of complex aggregates (Box 4). However, a more sophisticated and realistic definition of forces and spatial interactions allows the formation of multicellular aggregates (Fig. 8b) thus suggesting that much more can be obtained even under these simplified scenarios (Solé and Valverde 2013).

Combinatorial Explosions and the Cambrian Conundrum

As a final example, we consider the problem of how complex and diverse spatial pattern of gene expression (and thus cell types) can emerge as a consequence of gene networks in development. Specifically, we consider an abstract model of pattern formation where a one-dimensional digital organism is composed of C cells each carrying the same gene network.

Box 5. Gene Networks in Development

A simple, Boolean model of pattern formation can be implemented by defining a set of N = G + H genes interacting through a one-dimensional domain involving C cells (Solé et al. 2003). G genes interacting within the cell, whose state at a given time t will be indicated as $g_i^j(t)$, where i = 1, ..., G is gene number and j = 1, ..., C is cell number. The second term in the equation refers to generically labeled microhormones and their state will be indicated as $h_i^j(t)$. Hormones can receive inputs from any of the first G units, but they can only make output to genes in other cells. Two matrices will be required, indicated by W = (Wik) and C = (Cik), defining interactions among the G genes and between genes and hormones, respectively.

The basic set of equations of our gene network model read:

$$g_i^j(t+1) = \Phi\left[\sum_{l=1}^{G} W_{il}\, g_l^j\,(t) + \sum_{k=1}^{H} C_{ik}\delta(h_k^{j+1}(t), h_k^{j-1}(t))\right]$$

where d(x, y) = 1 if x + y > 0 and zero otherwise (i.e. an "OR" function). Similarly, hormones receive inputs only from inside the cell,

$$h_i^j(t+1) = \Phi\left[\sum_{l=1}^{G} W_{il}\, g_l^j(t)\right]$$

with additional, specific equations for the boundaries. The function $\Phi(x)$ is a threshold function, i. e. $\Phi(x) = 1$ if $x > 0$ and zero (inactive) otherwise.

The gene network includes both gene-gene interactions within the cell and between cells. In other words, we take into account regulatory interactions taking place within each cell together with cell-cell communications through given signals (to be called hormones). These models have been extensively used since the early days of theoretical biology (Kauffman 1987) and provide a simple way of approaching the problem of defining cell types and thus multicellular assemblies.

In its simplest form, we can define a gene network in terms of a set of n genes whose states g_i are confined to two possibilities, namely $g_i \in \{0,1\}$. This binary approximation implies that genes are ON-OFF elements, which of course is a simplified picture. The effect of gene g_i on gene g_j is given by a weight W_{ij}, which is positive in case of activation and negative in case of repression. No interaction is given by $W_{ij} = 0$. For simplicity, a discrete space of weights is used, namely $W_i \in \{-1, 0, +1\}$. In this way, a full exploration of the potential set of states can be performed. The state of a gene will change as a consequence of its interactions with other genes. This state is updated in discrete time units following:

$$g_i^j(t+1) = \Phi\left[\sum_{l=1}^{n} W_{il} g_l^j(t)\right]$$

which essentially tells us that the gene will become active or inactive if the global input acting on it is positive or negative, respectively. These networks can generate extremely simple (say, all genes inactive or active) or very complex dynamics (when chaotic changes occur). But for some ranges of connectivities, it leads to a rich diversity of stable states. If a cell type T is identified as a string S_T of active and inactive genes, namely $S_T = (g_1, \ldots, g_n)$ (Kauffman 1987) we have a potential of 2_n alternative strings. We can expand this formalism to take into account space, if multiple cells are also taken into account (Box 5). As shown in Fig. 9a, this is easily implemented and a spatial pattern can be described, for each gene, in terms of its expression level (0 or 1) in different cells. A detailed analysis of this type of model reveals that some types of patterns are easily found (Solé et al. 2003; Munteanu and Solé 2008; Tusscher and Hogeweg 2011.). This is the case of regular stripe patterns, which is a rather common one.

What is the generative potential of a given gene network complexity level? More precisely, if we start from a small network with G = 1 genes and H = 1 hormones and expand its number, what are the consequences? These questions are relevant when we consider the early events that shaped the gene networks affecting developmental processes in the history of multicellular life. Once again, thinking in the

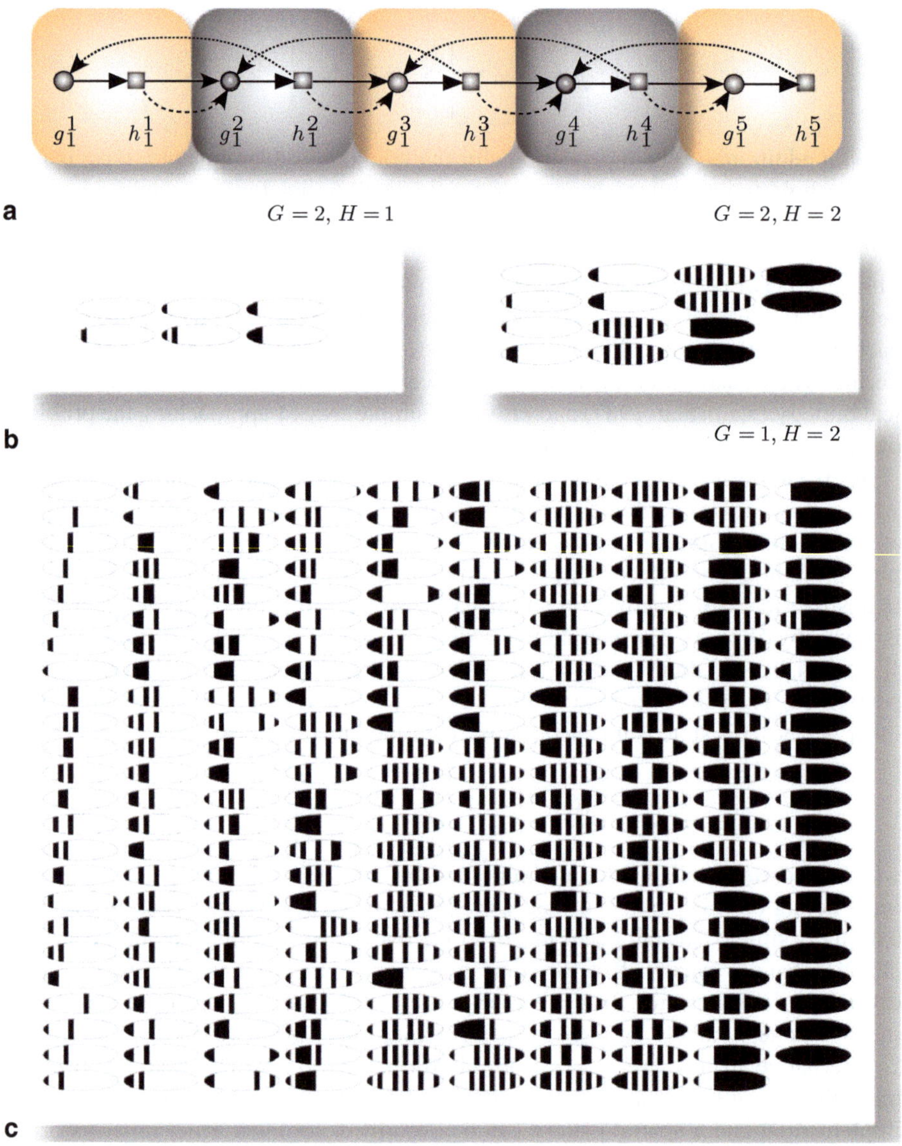

Fig. 9 Transitions from simple to complex (diverse) patterns in a minimal model of gene network model of biological pattern formation. The model considers a population of digital organisms composed by a linear chain of C cells, each carrying the same genetic network (see *Box 5*). The complexity of this network, measured in terms of the number of genes G associated to internal switches and the number of cell-cell signals H determines the amount of patterns that can be achieved through evolution. In **a** *dark* and *light* cells indicate the high and low expression of one of the genes, respectively. In Figs. (**b–c**) the complete set of patterns generated in this way is shown for different sets of genes. **b** For H = 2, G = 1 or H = 1, G = 2 only a few patterns are accessible. **c** Once we have H = G = 2, all possible patterns can be reached

Cambrian explosion of life, several factors can be considered (Marshall 2006; Erwin and Valentine 2013). These include external, abiotic factors as well as internal ones and they are likely to interact among them. But it would be interesting to know, even under some basic abstract model, if some particular elements can play a key role in promoting sudden changes in the amount of achievable morphologies. In order to analyze this problem, we use the presence of a genotype-phenotype mapping Ω, namely

$$\Omega : W \to \{\Phi\}$$

between the set of matrices W and the corresponding set of patterns $\{\Phi\}$. In other words, for a given network W_a, the arrangement of ON and OFF genes defining a stable pattern $P_a{}^*$ can be written as $P_a^* = \Phi(W_a) \in \{\Phi\}$.

Using our basic model, it is possible to explore the space of potential phenotypes through single mutations in the genotype, as defined by the matrix of gene-gene interactions. Each step in the simulation, we evolve the system by changing single elements in the matrix. Patterns that are not stable are discarded and a change in the matrix is accepted if the diversity of cells within the organism is increased (or at least remains the same). This movement in sequence space is known as an adaptive walk (Kauffman and Levin 1987).

Pattern-forming gene networks display sharp thresholds affecting their combinatorial potential. For small numbers of elements, i. e. when $H+G<4$ and $H, G \leq 2$ the range of possible spatial patterns is rather limited (Fig. 9b) but once the critical number $H=G=2$ is reached, all patterns become accessible (Solé et al. 2003). This is a very interesting finding, since it provides a possible logical explanation for the rapid diversification of developmental paths when genetic complexity thresholds are crossed. Such phase transition phenomenon. Along with other influences, a small increase in regulatory complexity can account for a sudden jump in the achievable diversity of developmental pathways.

Although these results are obtained from a toy model of regulatory interactions and ignore other pattern-forming factors, such as tissue organization, morphodynamic processes or cell division, sorting and apoptosis, the basic conclusions are likely to be robust: a relatively small increase in underlying genomic complexity can lead to rich morphogenetic potential (Marshall 2006). In earlier models of evolution on fitness landscapes (Niklas 1994) high diversity is linked to the presence of multiple optima on a morphological landscape. If such optima are easily reached, a diverse range of structures is expected to be obtained. An interesting feature of the space of spatial patterns defined by the gene network model is that it displays neutrality: large, neutral networks percolate sequence space allowing for efficient exploration of the phenotype space. The structure of this pattern forming network space is actually very similar to other found in RNA folding (see Solé et al. 2003; Munteanu and Solé 2008). This result tells us something important here. As soon as we reach the critical threshold of network complexity, not only multiple patterns become accessible. The intrinsic evolvability of the system is also very high.

Discussion and Prospects

In silico models of evolutionary change should be a natural component of our exploration of macroevolutionary patterns and the tempo and mode of evolutionary transitions. Despite their limitations, they offer, along with experimental dynamics using microbial populations (Lenski and Travisano 1994; Elena and Lenski 2003) what no other approach can: an opportunity to recreate the past and how complexity developed over time. Here we have summarized the outcomes of different models of artificially evolved "organisms". Although they are all far from "realistic", the previous results reveal a great generative potential implicit in the simple rules. In all cases, multicellular complexity experiences increases or even jumps and some remarkable results can be highlighted. These include, for example, the emergence of some ontogenetic processes resulting from an evolutionary algorithm searching for diverse cell types. Such processes typically incorporate cell-cell interactions that provide the capacity for tissue reorganization and shape changes together with cell differentiation. But even cellular and ecological scales can become related once evolving adhesion provides the exploratory potential for community level processes to unfold. This connection between such disparate scales provides a novel way of re-considering the problem of hierarchies in evolution (Eldredge 1985; McShea 2001).

Some more sophisticated models have been created able to evolve complex creatures with multiple connected components. These models involve a more or less detailed physical context, both in terms of the elements used to describe the virtual creature and the physics of the environment (Sims 1994). In these models, the final outcome often reminds us of some type of living creature. However, an essential difference is at work: in the evolved artificial creatures there is no developmental program at work and thus there is no genotype-phenotype mapping. This is no minor drawback, since developmental programs are the essential component required to properly understand and model evolutionary paths. Over the last years, novel approaches to this problem incorporating some type of morphogenetic rules are being considered (Doursat 2008; Jin 2011).

In the CHIMERA framework, our artificial creatures are autogenic engineers (Jones et al.1994): they change their environment mainly via their own physical structures. The success of our model might be due to the complete set of key components that we allow to interact freely. By using space, diverse ecosystems can be build through spatial segregation of subpopulations. By allowing simple components of pattern formation or aggregate generation it is possible to introduce simple forms of cooperative dynamics. By embedding the virtual creatures within an ecosystem where physics plays a role, selection pressures restrict the repertoire of cellular aggregates that can be formed.

Future work should address the potential for generating complex structures perhaps similar to the Ediacaran fauna and test the role played by both internal and external innovation triggers. The first includes for example the emergence of predators and the resulting arm races, which are known to be a major player in expanding morphological complexity. The second deals with extinctions caused by geological or astronomical shocks, which deeply altered communities and whole ecosystems. The aftermath of the extinction provides a unique lens to see different evolutionary processes in action. Such recovery patterns have been studied both from field data and modeling (Benton and Twitchett 2003; Erwin 1998, 2001; Solé et al. 2002; Chen and Benton 2012; Yedid et al. 2012) and offer an additional test for studying how organismal and ecological complexity react to stress.

As a final point, it is worth mentioning that another avenue to address the dawn of multicellular systems is provided by synthetic biology (Benner and Sismour 2005; Solé et al. 2007; Cheng and Lu 2012) which is considered by some researchers as the wet version of artificial life. By engineering unicellular systems, it is possible to build novel forms of cell-cell communication and thus create (and perhaps re-create) novel forms of multicellular assemblies, able to perform novel functions and even complex computations (Regot et al. 2011; Macia et al. 2012; Chuang 2012). Given the potential offered by genetic engineering techniques to alter the logic of cell-cell exchanges, we have a unique opportunity of exploring the landscape of transitions from uni- to multicellular forms of organization.

Summary

In this chapter we have begun to explore some key aspects of the origin of multicellularity through the unique perspective of theoretical models, namely:

1. The role of embodiment and accurate physical descriptions at different levels -subcellular, cellular and environmental- and their role in shaping cell to cell interactions.
2. The impact of emergent properties of ensembles of cells in engineering niches and building ecosystems.
3. The ease of acquiring multicellular phenotypes by genetic drift or "simple", purely physical, selective pressures.
4. The existence of sudden jumps or transitions in available complexity as the number of subcellular components -or basic toolkit genes- increases.

Acknowledgements We thank the members of the Complex Systems Lab for useful discussions. This work has been supported by grants of the James McDonnell Foundation, the Botin Foundation and by the Santa Fe Institute.

References

Adami C (1998) Introduction to artificial life. Springer, New York
Alberch P (1980) Ontogenesis and morphological diversification. Am Zool 20:653–667
Anderson P (1972) More is different. Science 17:393–396
Benner SA, Sismour AM (2005) Synthetic biology. Nat Rev Genet 6:533–543

Benton MJ, Twitchett RJ (2003) How to kill (almost) all life: the end- Permian extinction event. Trends Ecol Evol 18:358–365

Bonner JT (2001) First signals: the evolution of multicellular development. Princeton University Press, Princeton

Buss (1987) The evolution of individuality. Princeton University Press, Princeton

Carnac G, Gurdon JB (1997) The community effect in Xenopus myogenesis is promoted by dorsalizing factors. Int J Dev Biol 41:521–524

Carroll, SB (2005) Endless forms most beautiful: the new science of evo devo and the making of the Animal Kingdom. W. W. Norton and Co, New York

Chen Z-Q, Benton MJ (2012) The timing and pattern of biotic recovery following the end-Permian mass extinction. Nat Geosci 5:375–383

Cheng AA, Lu TK (2012) Synthetic biology: an emerging engineering discipline. Annu Rev Biomed Eng 14:155–178

Chuang JS (2012) Engineering multicellular traits in synthetic microbial populations. Curr Opin Chem Biol 16:370–378

Coen E, Rolland-Lagan AG, Matthews M, Bangham JA, Prusinkiewicz P. (2004) The genetics of geometry. Proc Natl Acad Sci U S A 101:4728–4735

Cummings F (2006) On the origin of pattern and form in early metazoans. Int J Dev Biol 50:193–208

Doursat R (2008) Organically grown architectures: creating decentralized, autonomous systems by embryomorphic engineering. In: Wrtz RP (ed) Organic computing. Springer, Berlin, pp 167–200

Duran-Nebreda S, Solé RV (2014) Emergence of multicellularity in a model of cell growth , death and aggregation under size-dependent selection. J R Soc Interface 6(12):102 20140982

Eble (2003) Developmental morphospaces and evolution. In: Crutchfield JP, Schuster P (eds) Evolutionary dynamics. Oxford University Press, New York

Eggenberger P (1997) Evolving morphologies of simulated 3d organisms based on differential gene expression. In: Husbands P, Harvey I (eds) Fourth European conference. Artificial life. MIT Press, Cambridge, pp 205–213

Eldredge N (1985) Unfinished synthesis: biological hierarchies and modern evolutionary thought. Oxford University Press, Oxford

Elena SF, Lenski RE (2003) Microbial genetics: evolution experiments with microorganisms: the dynamics and genetic bases of adaptation. Nat Rev Genet 4:457–469

Ericson C (2005) Real-time collision detection. Morgan Kaufmann, Burlington

Erwin DH (1998) The end and the beginning: recoveries from mass extinctions. Trends Ecol Evol 13:344–349

Erwin DH (2001) Lessons from the past: biotic recoveries from mass extinction. Proc Natl Acad Sci U S A 98:5399–5403

Erwin DH (2008) Macroevolution of ecosystem engineering, niche construction and diversity. Trends Ecol Evol 23:304–310

Erwin DH, Tweedt S (2012) Ecological drivers of the Ediacaran-Cambrian diversification of metazoa. Evol Ecol 26:417–433

Erwin DH, Valentine J (2013) The cambrian explosion: the construction of animal biodiversity. Roberts and Company, Atlanta

Fedonkin MA (2007) The rise of animals: evolution and diversification of the kingdom animalia. JHU Press, Baltimore

Floreano D, Mattiussi C (2008) Bio-inspired artificial intelligence. MIT Press, Cambridge

Fontana W, Schuster P (1998) Continuity in evolution: on the nature of transitions. Science 280(5368):1451–1455

Forgacs G, Newman SA (2005) Biological physics of the developing embryo. Cambridge University Press, Cambridge

Forrest S (1993) Genetic algorithms: principles of natural selection applied to computation. Science 261:872–878

Furusawa C, Kaneko K (1998) Emergence of multicellular organisms with dynamic differentiation and spatial pattern. Artif Life 4(1):79–93
Furusawa C, Kaneko K (2003) Robust development as a consequence of generated positional information. J Theor Biol 224:413–435
Goodwin BC (1994) How the leopard got its spots. Princeton University Press, Princeton
Glazier JA, Graner F (1993) Simulation of the differential adhesion driven rearrangement of biological cells. Phys Rev E Stat Phys Plasmas Fluids Relat Interdiscip Topics 47:2128–2154
Graner F, Glazier JA (1992) Simulation of biological cell sorting using a two-dimensional extended Potts model. Phys Rev Lett 69:2013–2016
Grossberg RK, Strahmann RR (2007) The evolution of multicellularity: a minor major transition? Annu Rev Ecol Evol Syst 38:621–654
Hastings A (2006) Ecosystem engineering in space and time. Ecol Lett 10:153–164
Hogeweg P (2000a) evolving mechanisms of morphogenesis: on the interplay between differential adhesion and cell differentiation. J Theor Biol 203:317–333
Hogeweg P (2000b) Shapes in the shadow: evolutionary dynamics of morphogenesis. Artif Life 6:85–101
Jin Y (2011) Morphogenetic robotics: an emerging new field in developmental robotics. IEEE Trans Syst Man Cybern 41:145–160
Jones CG, Lawton JM, Shachak M (1994) Organisms as ecosystem engineers. OIKOS 69:373–370
Jones CG, Lawton JM, Shachak M (1997) Positive and negative effects of organisms as physical ecosystem engineers. Ecology 78:1946–1957
Kaandorp JA, Blom JG, Verhoef J, Filatov M, Postma M, Müller WEG (2008) Modelling genetic regulation of growth and form in a branching sponge. Proc R Soc B 275:2569–2575
Kaneko K, Yomo T (1999) Isologous diversification for robust development of cell society. J Theor Biol 199:243–256
Kaufmann SA (1993) The origins of order: self-organization and selection in evolution. Oxford University Press, New York
Kauffman SA, Levin S (1987) Towards a general theory of adaptive walks on rugged landscapes. J Theor Biol 128:11–45
King N (2004) The unicellular ancestry of animal development. Dev Cell 7:313–325
Langton CG (1991) Life at the edge of chaos. In Artificial life II. Addison Wesley, Boston
Langton GG (ed) (1995) Artificial life: an overview. MIT Press, Cambridge
Lenski RE, Travisano M (1994) Dynamics of adaptation and diversification: a 10,000-generation experiment with bacterial populations. Proc Natl Acad Sci U S A 91:6808–6814
Macia J et al (2012) Distributed computation: the new wave of synthetic biology devices. Trends Biotechnol 30(6):342–349
Marshall CR (2006) Explaining the Cambrian explosion of animals. Annu Rev Earth Planet Sci 34:355–384
Maynard Smith J Szathmáry E (1995) The major transitions in evolution. Oxford University Press, Oxford
McShea DW (2001) The hierarchical structure of organisms: a scale and documentation of a trend in the maximum. Paleobiology 27:405–423
Michod RE (2000) Darwinian dynamics: evolutionary transitions in fitness and individuality. Princeton University Press, Princeton
Mitchell M (1998) An introduction to genetic algorithms. Bradford Books, MIT Press, Cambridge
Munteanu A, Solé RV (2008) Neutrality and robustness in evo-devo: emer gence of lateral inhibition. PLoS Comput Biol 4:e1000226
Newman SA, Baht R (2008) Dynamical patterning modules: physicogenetic determinants of morphological development and evolution. Phys Biol 5:015008
Newman SA, Comper WD (1990) 'Generic' physical mechanisms of morphogenesis and pattern formation. Development 110:1–18
Newman SA, Forgacs G, Mller GB (2006) Before programs: the physical origination of multicellular forms. Int J Dev Biol 50:289–299

Niklas KJ (1994) Morphological evolution through complex domains of fitness. Proc Natl Acad Sci U S A 91:6772–6779
Niklas KJ, Newman SA (2013) The origins of multicellular organisms. Evol Dev 15:41–52
Palsson E (2008) A 3–D model used to explore how cell adhesion and stiffness affect cell sorting and movement in multicellular systems. J Theor Biol 254(1):1–13
Palsson E, Othmer HG (2000) A model for individual and collective cell movement in *Distyostelium discoideum*. Proc Natl Acad Sci U S A 97:10448–10453
Podgorski T et al (2007) Fingering instabilities of a reactive micellar interface. Phys Rev E 76:016202
Ratcliff WC, Denison RF, Borrello M, Travisano M (2012) Experimental evolution of multicellularity. Proc Natl Acad Sci U S A 109:1595–1600
Ray TS (1991) An approach to the synthesis of life. In: Langton C, Taylor C, Farmer D (eds) Artificial life II. Westview Press, Boulder, pp 371–408
Regot S et al (2011) Distributed biological computation with multicellular engineered networks. Nature 469:207–211
Sandersius SA, Newman TJ (2008) Modelling cell rheology with the subcellular element model. Phys Biol 5:015002
Savill NJ, Hogeweg P (1997) Modeling morphogenesis: from single cells to crawling slugs. J Theor Biol 184:229–235
Schuster P (1996) How does complexity arise in evolution?. Complexity 2:22–30
Shapiro JA (1998) Thinking about bacterial populations as multicellular organisms. Annu Rev Microbiol 52:81–104
Sims K (1994) Evolving 3D morphology and behavior by competition. Artif Life 1:353–372
Sipper M (1995) Using artificial life using a simple, general cellular model. J Artif Life 2:1–35
Solé RV (2009) Evolution and self-assembly of protocells. Int J Biochem Cell Biol 41:274–284
Solé and Valverde (2013) Before the endless forms: embodied model of transition from single cells to aggregates to ecosystem engineering. PLoS ONE 8(4): e59664
Solé RV, Montoya JM, Erwin DH (2002) Recovery after mass extinction: evolutionary assembly in large-scale biosphere dynamics. Phil Trans Roy Soc Lond B 357:697–707
Solé RV, Fernandez P, Kauffman SA (2003) Adaptive walks in a gene network model of morphogenesis: insights into the Cambrian explosion. Int J Dev Biol 47:685–693
Solé RV, Munteanu A, Rodriguez-Caso C, Macia J (2007) Synthetic protocell biology: from reproduction to computation. Phil Trans R Soc Lond B 362:1727–1739
Steinberg MS (1963) Reconstruction of tissues by dissociated cells. Science 141(3579):401–408
Steinberg MS (1978) Cell–cell recognition in multicellular assembly: levels of specificity. Symp Soc Exp Biol 32:25–49
Steinberg MS, Takeichi M (1994) Experimental specification of cell sorting, tissue spreading, and specific spatial patterning by quantitative differences in cadherin expression. Proc Natl Acad Sci U S A 91(1):206–209
Szathmáry E (1994) Toy models for simple forms of multicellularity, soma and germ. J Theor Biol 169:125–132
Tusscher KH, Hogeweg P (2011) Evolution of networks for body plan patterning; interplay of modularity, robustness and evolvability. PLoS Comput Biol 7:e1002208
Valentine JW et al (1994) Morphological complexity increase in metazoans. Paleobiology 20(2):131–142
Wolpert L (1969) Positional information and the spatial pattern of cellular differentiation. J Theor Biol 25:1–47
Yedid G, Stredwick J, Ofria CA, Agapow PM (2012) A comparison of the effects of random and selective mass extinctions on erosion of evolutionary history in communities of digital organisms. PLoS One 7:e37233

Part IV
Genomics Insights

A Comparative Genomics Perspective on the Origin of Multicellularity and Early Animal Evolution

Mansi Srivastava

Abstract The genetic basis of the origin of animal multicellularity and the subsequent diversification of complex animal forms has been a long-standing question in biology. In the past decade, the genomes of species representing early-branching animal lineages and close unicellular relatives of animals were sequenced, providing an unprecedented wealth of data from these understudied phyla. This chapter focuses on comparative genomic analyses of four animal lineages, cnidarians, ctenophores, placozoans, and sponges, and of two unicellular lineages, choanoflagellates and filastereans. These studies revealed striking conservation of gene structure and genomic organization among animals, and uncovered deep evolutionary origins of the genetic circuits underlying biological processes essential for animal biology, including cell cycle control, cell growth, programmed cell death, and specialized cell types. Genomic analyses therefore allow us to infer that all extant animals have descended from an ancestor with a complex genome that encoded a vast majority of the gene content responsible for biological processes in vertebrates. Strikingly, genomes of animals that lack specialized cell types such as muscles and neurons encode the molecular machinery required for the function of these cell types. Thus, the genomic events by which the ancestral animal genome gave rise to gross differences in animal morphology remain unknown. The genome sequences described here will enable future functional genetic studies of anciently-evolved genes in early-branching animal lineages and their unicellular relatives to decipher the evolution of animal body plans.

Keywords Animal evolution · Comparative genomics · Cnidarians · Placozoans · Sponges · Multicellularity · Morphological complexity

M. Srivastava (✉)
Whitehead Institute for Biomedical Research, Cambridge, MA, USA
e-mail: mansi@wi.mit.edu

I. Ruiz-Trillo, A. M. Nedelcu (eds.), *Evolutionary Transitions to Multicellular Life*, Advances in Marine Genomics 2, DOI 10.1007/978-94-017-9642-2_14

Introduction

Biologists have long sought to explain the origins of animals and the astounding diversity in their morphologies. The first step towards understanding the basis of morphological variation is to identify and categorize different animal forms. The majority of animal (metazoan) phyla are bilaterally symmetric and fall in a group called the "Bilateria", which are characterized as having a primary body axis with head and tail ends, a dorsal-ventral axis with a spatially restricted central nervous system, three germ layers, true muscle, and an epithelial gut. The four other (non-bilaterian) animal lineages (i.e., cnidarians, ctenophores, placozoans, and sponges), diverged from bilaterians early in animal evolution (i.e., they are early-branching relative to lineages within the bilaterian clade) (Brusca and Brusca 2002; Hyman 1940; Fig. 1a). These lineages have been considered by some to be morphologically "simple" because they lack many bilaterian features. For example, cnidarians and ctenophores have a primary body axis (but lack an obvious secondary axis) and poorly organized nervous systems; sponges and placozoans have amorphous adult forms with six to ten different cell types but lack muscle or neural cells. Non-bilaterian animals are, in turn, more complex than closely related non-animal lineages, which are unicellular. A major question in understanding the origins of animal diversity focuses on identifying the genetic underpinnings of the apparent differences in morphological complexity between bilaterians and non-bilaterians, and between animals and their non-animal relatives.

Phenotypes result from interactions between the genotype and the environment, and thus, a major determinant of animal multicellularity and form is the underlying genome. With the advent of whole genome sequencing technology, it has become relatively easy to ascertain the entire genotype of an organism. A simple hypothesis can be laid out—perhaps the genomes of unicellular species are simpler than the genomes of animals, and non-bilaterian genomes are simpler than bilaterian ones by some measure. The twenty-first century brought us the sequencing and analysis of the complete genomes of cnidarians (*Nematostella vectensis*, *Hydra magnipapillata*, and *Acropora digitifera*), ctenophores (*Mnemiopsis leidyi*), placozoans (*Trichoplax adhaerens*), sponges (*Amphimedon queenslandica* and *Oscarella carmela*), choanoflagellates (*Monosiga brevicollis* and *Salpingoeca rosetta*), and filastereans (*Capsaspora owczarzaki*), with more non-bilaterian metazoan genomes continuing to join this list (Fig. 1a and Table 1; Putnam et al. 2008; Chapman et al. 2010; Shinzato et al. 2011; Srivastava et al. 2008, 2010; Nichols et al. 2012; King et al. 2008; Fairclough et al. 2013; Suga et al. 2013; Ryan et al. 2013). This chapter will explore how these genome sequences, through comparative analyses with the human genome and other bilaterian genomes, have revealed striking features of the last common ancestor of all animals. Broadly, these features pertain to (1) the structure of the genome itself, and (2) the molecular determinants of biological processes and cell types encoded in the genome. After exploring these aspects of the comparative genomics analyses, the question of how these genomes weigh-in on the evolution of complexity will be considered.

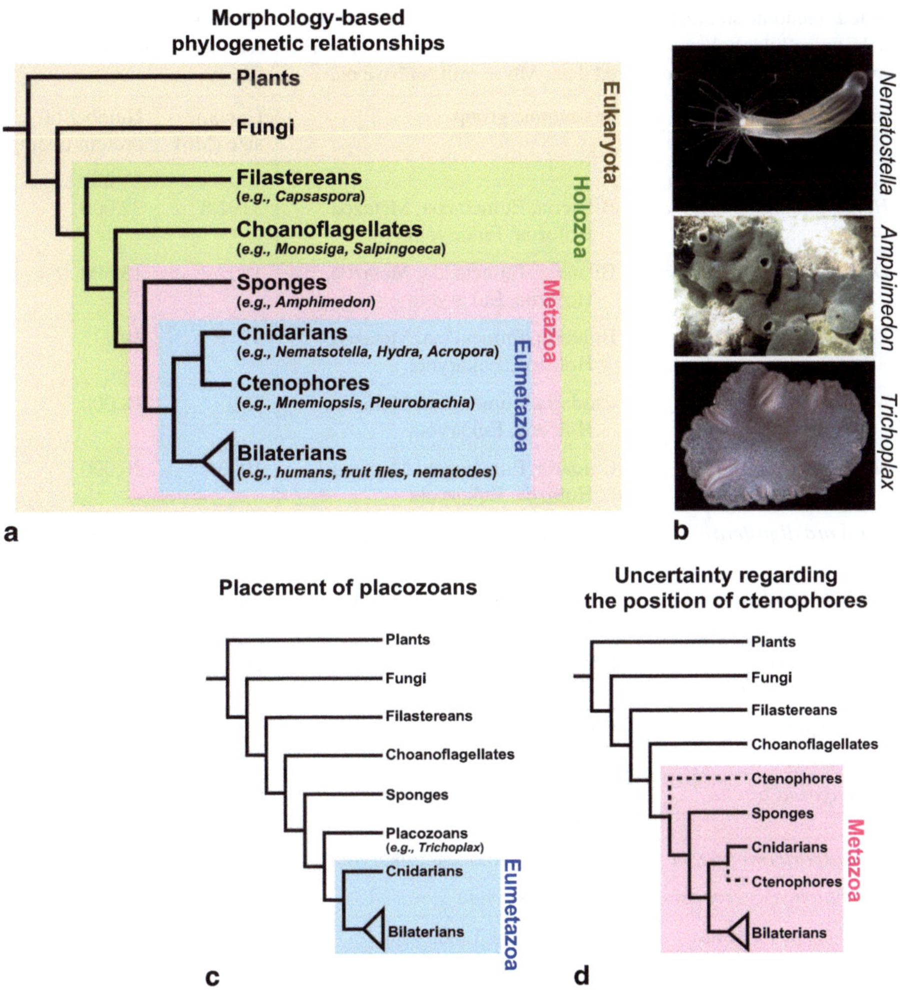

Fig. 1 Non-bilaterian animals with sequenced genomes and their phylogenetic relationships. **a** A schematic tree representing the morphology-based view of phylogenetic relationships of sponges, cnidarians, and ctenophores to bilaterally symmetric animals. Ctenophores and cnidarians form a monophyletic lineage (Coelenterata), which together with bilaterians comprises the "Eumetazoa". Sponges form the earliest-diverging animal lineage, with filastereans and choanoflagellates as the closest unicellular cousins of animals. Filastereans, choanoflagellates, and animals are "Holozoan" lineages, which exclude other eukaryotes such as plants and fungi. **b** Bright field images of *Nematostella vectensis*, *Trichoplax adhaerens*, and *Amphimedon queenslandica* courtesy of Mark Martindale, Ana Signorovitch, and Bernie Degnan, respectively. **c** Schematic representation of phylogenetic relationships used as the reference tree for this chapter. Though there is some outstanding debate regarding the placement of placozoans, many phylogenetic analyses using genome-wide sequence data have recovered *Trichoplax* as a sister lineage to cnidarians and bilaterians. Cnidarians and bilaterians will be referred to as eumetazoans in this chapter (Hyman 1940). **d** Analyses of genome-scale sequence data recover ctenophores as sister to cnidarians (congruent with the morphology-based view) or as the earliest-branching animal lineage. Alternative positions for ctenophores are indicated by *dashed lines* in this schematic tree. The implications of these alternative scenarios for nervous system evolution are shown in Fig. 5. The question of whether ctenophores are also a eumetazoan species remains to be answered and will not be addressed in this chapter

Table 1 Genome size and number of protein-coding genes for various bilaterians, non-bilaterian animals, unicellular holozoans, and other eukaryotes. These numbers are estimates that are continually revised based on newly generated data. Mb = million base pairs

Species	Taxonomic group	Genome size (Mb)	Number of protein-coding genes
Homo sapiens	Bilateria, Eumetazoa, Metazoa, Holozoa, Eukaryota	3101.8	22,000
Drosophila melanogaster	Bilateria, Eumetazoa, Metazoa, Holozoa, Eukaryota	180	13,600
Caenorhabditis elegans	Bilateria, Eumetazoa, Metazoa, Holozoa, Eukaryota	97	19,735
Nematostella vectensis	Cnidaria, Eumetazoa, Metazoa, Holozoa, Eukaryota	450	18,000
Hydra magnipapillata	Cnidaria, Eumetazoa, Metazoa, Holozoa, Eukaryota	1000	20,000
Acropora digitifera	Cnidaria, Eumetazoa, Metazoa, Holozoa, Eukaryota	420	23,700
Trichoplax adhaerens	Placozoa, Metazoa, Holozoa, Eukaryota	98	11,514
Amphimedon queenslandica	Porifera, Metazoaa, Holozoa, Eukaryota	167.1	30,327
Monosiga brevicollis	Choanoflagellata, Holozoa, Eukaryota	41.6	9171
Salpingoeca rosetta	Choanoflagellata, Holozoa, Eukaryota	55	11,629
Capsaspora owczarzaki	Filasterea, Holozoa, Eukaryota	28	8657
Saccharomyces cerevisiae	Fungi, Eukaryota	12.1	5863
Arabidopsis thaliana	Embryophyta, Eukaryota	125	25,498

Early-Branching Animal Lineages And Unicellular Relatives with Sequenced Genomes

The cnidarian phylum is defined by a sac-like body plan with a single "oral" opening, two epithelial tissue layers, the presence of numerous tentacles, a nerve net, and the characteristic stinging cells (cnidocytes, literally, "nettle cells") that give the phylum its name. These animals typically have a poorly organized nervous system and lack definitive muscle (though they have epithelial cells with myofibers). The starlet sea anemone *Nematostella vectensis* has emerged as a powerful cnidarian model system, with easy laboratory culture and methods for investigating gene expression and function (Fig. 1b; Genikhovich and Technau 2009).

Ctenophores, or comb jellies, are a phylum of gelatinous marine organisms that have rows ("combs") of cilia that beat synchronously for locomotion (Brusca and Brusca 2002). Like cnidarians, ctenophores also have a decentralized nervous system, but have muscle cells. The ctenophore *Mnemiopsis leidyi* has become accessible for biological experimentation.

Sponges (Phylum: Porifera) have amorphous sessile adult forms that are organized to filter-feed from water circulated by the action of flagellated cells called choanocytes. Sponges have multiple differentiated cell-types, yet tissue-level organization, obvious body axes, muscles, and neurons are lacking (Brusca and Brusca 2002). *Amphimedon queenslandica*, a haplosclerid demosponge (sponges with skeletons made of siliceous spicules and spongin fibers), has become a leading model system for experimental studies in sponges (Fig. 1b). The year-round availability of embryos and larvae makes *Amphimedon* an ideal system for studying the functions of genes important in animal development.

Placozoans (literally, "flat animals") are small (1–2 mm) disc-shaped animals with four to six cell types organized in two ciliated epithelial layers that sandwich a layer of multi-nucleate fiber cells, with the bottom surface of the animal acting as a temporary extraorganismal gastric cavity (Grell 1971b; Schierwater 2005). Placozoans have no evident axis of symmetry, they lack identifiable nerves and muscle, and a complete sexual life cycle has never been observed (Grell 1971a; Schierwater 2005). Placozoans are found in tropical and subtropical oceans, and the only named species, *Trichoplax adhaerens* F.E. Schulze, can be maintained in the laboratory (Fig. 1b).

In addition to non-bilaterian animal lineages, closely related non-animal, unicellular outgroups are essential for understanding the transition to animal multicellularity. The closest single-celled relatives of animals are the choanoflagellates (Lang et al. 2002; Shalchian-Tabrizi et al. 2008; Ruiz-Trillo et al. 2008), which are characterized by a single flagellum surrounded by a collar of actin filaments and morphologically resemble sponge choanocytes. The genomes of *Monosiga brevicollis*, which has a solitary lifestyle, and of the colony-forming *Salpingoeca rosetta* have been sequenced (King et al. 2008; Fairclough et al. 2013). Animals and choanoflagellates, together with two other lineages, filastereans and ichthyosporeans, form a monophyletic group referred to as the "Holozoa" (Fig. 1b; see Chapters. "Choanoflagellates: Perspective on the Origin of Animal Multicellularity" and "Filastereans and Ichthyosporeans: Models to Understand the Origin of Metazoan Multicellularity" for more details on these groups). The genome of the filasterean *Capsaspora owczarzaki* (Suga et al. 2013) has also generated significant inferences for the origins of animal genes and biological pathways.

The majority of non-bilaterian genomes featured in this chapter were sequenced using whole genome shotgun sequencing, which obtains nucleotide sequence data for an entire genome without any *a priori* knowledge of genome sequence and structure from that species. Briefly, genomic DNA is fragmented to different lengths and sequenced from both ends of each fragment. Based on sequence overlap, these "reads" are assembled into contiguous stretches ("contigs"). Paired reads from long-insert libraries allow these contigs to be assembled into "scaffolds" that have missing sequence of known length. Though the relative arrangement of these scaffolds on chromosomes remains unknown, *ab initio* gene modeling and transcriptome sequencing provide a comprehensive view of the protein-coding gene content in these genome assemblies (Yandell and Ence 2012).

Bilaterian genomes have a wide range of sizes (e.g., the human genome is estimated to have about 3100 million base pairs, whereas the *C. elegans* genome has only 97 million base pairs). The sequenced non-bilaterian animal genomes range from 98 megabases (Mb) in *Trichoplax* to ~1000 Mb in *Hydra* (Table 1). Among non-bilaterian genomes, *Trichoplax* and *Amphimedon* encode the smallest (~11,500) and largest (~30,000) numbers of proteins respectively (Table 1). There appears to be no correlation between genome size and the number of protein-coding genes predicted in animal genomes. For example, the *Amphimedon* genome, at an expected 210 Mb and about 30,000 genes, is comparable in size to the *Drosophila melanogaster* genome (180 Mb), which is predicted to have 16,000 genes. Though the sea anemone genome is much smaller than the human genome (450 Mb vs. 3100 Mb), the number of predicted genes is comparable (18,000 for *Nematostella* and 22,000 for the human genome). However, the three non-animal holozoan genomes appear to be smaller (28–55 Mb) and to encode fewer proteins (8657–11,627) than most animal genomes (Table 1).

Phylogenetic Relationships of Non-Bilaterians Based on Whole-Genome Sequence

A robust phylogenetic framework of the relationships of animals is necessary to understand the evolution of the biological pathways encoded in their genomes. Cnidarians and ctenophores have long been thought to comprise a monophyletic clade, the Coelenterata, the sister-group to the Bilateria (Brusca and Brusca 2002; Fig. 1a). Cnidarians and ctenophores, together with bilaterians, are often referred to as "eumetazoans", a term originally coined to encompass all animals except sponges (Hyman 1940). Based on comparisons of morphological characters, sponges are considered to be the earliest-diverging animal phylum (Fig. 1a). Though the phylogenetic position of placozoans was unresolved, it had been proposed that placozoans may resemble the primitive animal (Butschli 1884; Schierwater 2005). Analyses of whole-genome sequence have confirmed some of these relationships, but have also proposed a new view of early animal divergences.

Prior to the availability of whole-genome sequence, sequence-based phylogenetic analyses placed placozoans as secondarily simplified cnidarians, a sister group to bilaterians, a sister group to cnidarians and bilaterians, or the earliest-branching animal lineage (for a review see (Philippe et al. 2011)). The whole-genome sequences of cnidarians, a placozoan, and a sponge, provided a near-complete set of nuclear protein-co ding genes to address the question of how early-branching animal lineages relate to each other. These whole-genome data were applied in two types of analyses, one using only species with complete genome sequences resulting in very little missing data (Srivastava et al. 2010), and the second using a large number of species, many of which have relatively few EST sequences available, producing incomplete data matrices (Hejnol et al. 2009; Philippe et al. 2009). Both methods recovered placozoans as a sister-lineage to the traditional eumetazoan group that includes cnidarians and bilaterians (Fig. 1c). The debate on the position of placozoans continues, as researchers generate new kinds of data and analyze it in novel ways—for example, an analysis of a mixed dataset (combining morphology with molecular data) recovered all early-branching animal lineages (including placozoans) as a sister clade to bilaterians (Schierwater et al. 2009) and an analysis of ribosomal genes recovered placozoans as a sister-group to sponges (Nosenko et al. 2013). However, caveats to these alternative positions have been reported and the placement of placozoans as sister to cnidarians and bilaterians continues to find support in many analyses (Philippe et al. 2011; Nosenko et al. 2013). Based on the original use of the term "Eumetazoa" to include all animals except sponges, placozoans could be considered a eumetazoan lineage. However, the inclusion of placozoans within the Eumetazoa is currently under debate, and therefore, for the purpose of clarity in this chapter, the term "Eumetazoa" will refer to the grouping of cnidarians and bilaterians, with placozoans placed as the sister-lineage to eumetazoans (an outgroup to both cnidarians and bilaterians) (Fig. 1c).

Though ctenophores were considered to be eumetazoans, their position relative to other animals is now less well-resolved (Nosenko et al. 2013). Genome-scale sequence data continue to recover two well-supported positions—one as a sister group to cnidarians (Philippe et al. 2009) and the other as the sister group to all other animals (Dunn et al. 2008; Ryan et al. 2013; Fig. 1d). The genome sequences of ctenophores, including the recently published *Mnemiopsis leidyi* genome sequence (Ryan et al. 2013), will be essential resources for final resolution of animal relationships. The significance of these alternate positions for ctenophores will be considered at the end of this chapter. Genome-wide phylogenetic analyses support the placement of choanoflagellates as the sister lineage to animals, with filastereans (another holozoan) forming an outgroup to both choanoflagellates and animals (Fig. 1a).

Ancestral Genome Organization, Gene Structure, and Gene Content

Gene-modeling algorithms used to predict protein-coding genes in non-bilaterian genomes revealed intron-exon structures as well as the relative positions of genes on chromosomes. To understand the impact of comparative genomics using non-bilaterian genomes, it is instructive to first consider what was known about conservation of these features before non-bilaterian animal genomes were sequenced. By studying patterning of the anterior-posterior (AP) axis in fruit flies and vertebrate embryos, developmental biologists discovered that despite stark differences in morphology, embryonic development in these species utilizes certain shared molecular pathways. Hox genes, which cause homeotic transformations of segment identity in *Drosophila* and are required for patterning of the AP axis of fruit flies and vertebrates, were found to be expressed in tandem domains along the AP axis in both fruit fly and mouse embryos, and to be encoded in tandem (forming the Hox complex) in the genomes of the two species (Akam 1995). This striking similarity allowed biologists to infer that the molecular mechanism of AP patterning by the Hox complex must have been present in the approximately 550 million year old common ancestor of fruit flies and mice (the last common bilaterian ancestor). Other processes, *e.g.*, dorsal-ventral (DV) axial patterning, were also found to have similar molecular underpinnings in flies and vertebrates (De Robertis and Sasai 1996).

However, besides the Hox gene complex, few aspects of genome organization or intron-exon structure were known to have ancient bilaterian ancestry before non-bilaterian genomes were sequenced. The paradigm of comparing developmental processes across phyla to understand the direction of evolutionary change can be extended to features such as gene structure, genome organization, and gene content to understand early events in animal genome evolution. These approaches and results are discussed next.

Conservation of Introns

Though they are excluded from the final protein, introns regulate cellular biology as sources of noncoding RNA or by allowing alternative splicing. Orthologous genes from different animals, i.e., genes that have descended from a common ancestral gene present in the genome of the ancestor, can be identified to compare the locations and phases of their introns. Vertebrate genes were found to be intron-rich relative to their counterparts in invertebrate model organisms such as *Drosophila* and *Caenorhabditis elegans*, suggesting an important role for introns in the biological differences between vertebrates and invertebrates. However, similar comparisons of human genes with their *Nematostella* counterparts yielded very different results—within alignable regions, nearly 81 % of human introns were found in the same position and phase in *Nematostella* (Putnam et al. 2008; Sullivan et al. 2006). Introns shared by sea anemones and humans represent introns that were present in the cnidarian-bilaterian ancestor. Strikingly, *Drosophila* and *C. elegans* have lost 50–90 % of these ancestral introns.

With greater sampling of non-bilaterian genomes, a catalog of introns shared by all eukaryotic organisms, by all metazoans, and by placozoans and eumetazoans was built. Sponges, sea anemones, placozoans, and humans have all retained ancestral introns in high proportions (70–90 %) (Fig. 2a). Thus, maintaining ancestral gene structures appears to be a broad feature of animal genomes, a finding that would have been elusive without increased sampling of animal genomes at great evolutionary distances from humans. The large numbers of introns shared by humans and non-bilaterian species also suggests that the latter could serve as better models for studying the functions of introns in human genes better than traditional models such as flies and nematodes.

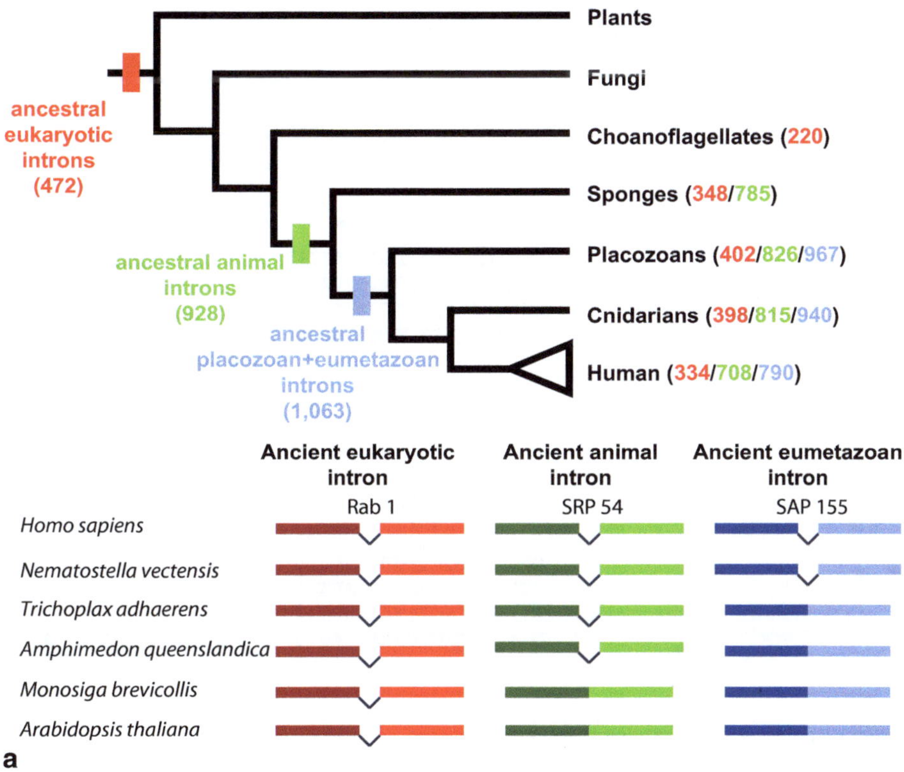

Fig. 2 Conservation of gene structure and genome organization. **a** Numbers of ancestral eukaryotic (*red*), ancestral metazoan (*green*), and ancestral placozoan + eumetazoan (*blue*) introns retained by extant animal lineages are indicated on a phylogenetic tree. Examples of various ancestral introns are shown in schematic form where *colored bars* represent exons and *chevrons* represent introns. Modified from Putnam et al. (Putnam et al. 2008). **b** A schematic example illustrating the evolution of genes on four chromosomes in an ancestral organism. Genes on different chromosomes are indicated with different colors, with genes on the same chromosome represented as different shades of that color. Immediately upon speciation, *i.e.*, the formation of two distinct lineages, orthologous genes (identifiable by the same color and shade) are in the same chromosomal locations in the genomes of the two descendant species. Therefore, in a dot plot, where each *dot* represents the position of a pair of orthologous genes on chromosomes from the two species plotted on the x and y axes, the dots should fall along the diagonal. Over time, the chromosomes re-arrange, altering the positions of genes relative to each other and scrambling the positions of genes on the dot plot. If there is conserved synteny, *i.e.*, genes are largely still present in the same chromosomal context despite shuffling of gene order, dots of the same color appear clustered in the dot plot. Off-diagonal dots represent genes that are removed from their original syntenic context as a result of inter-chromosomal translocations. Formally, this signature of ancestral linkage groups can be detected as a significant enrichment of orthologous genes compared to a null model where genes present on these chromosomal segment pairs are not orthologous. **c** A dot plot showing positions of orthologous genes on human chromosomal segments and *Trichoplax* scaffolds. As in the schematic schematic illustration in b, regions with a high density of dots are significantly enriched for orthologous gene pairs, evidencing conserved synteny between these two species (Srivastava et al. 2010)

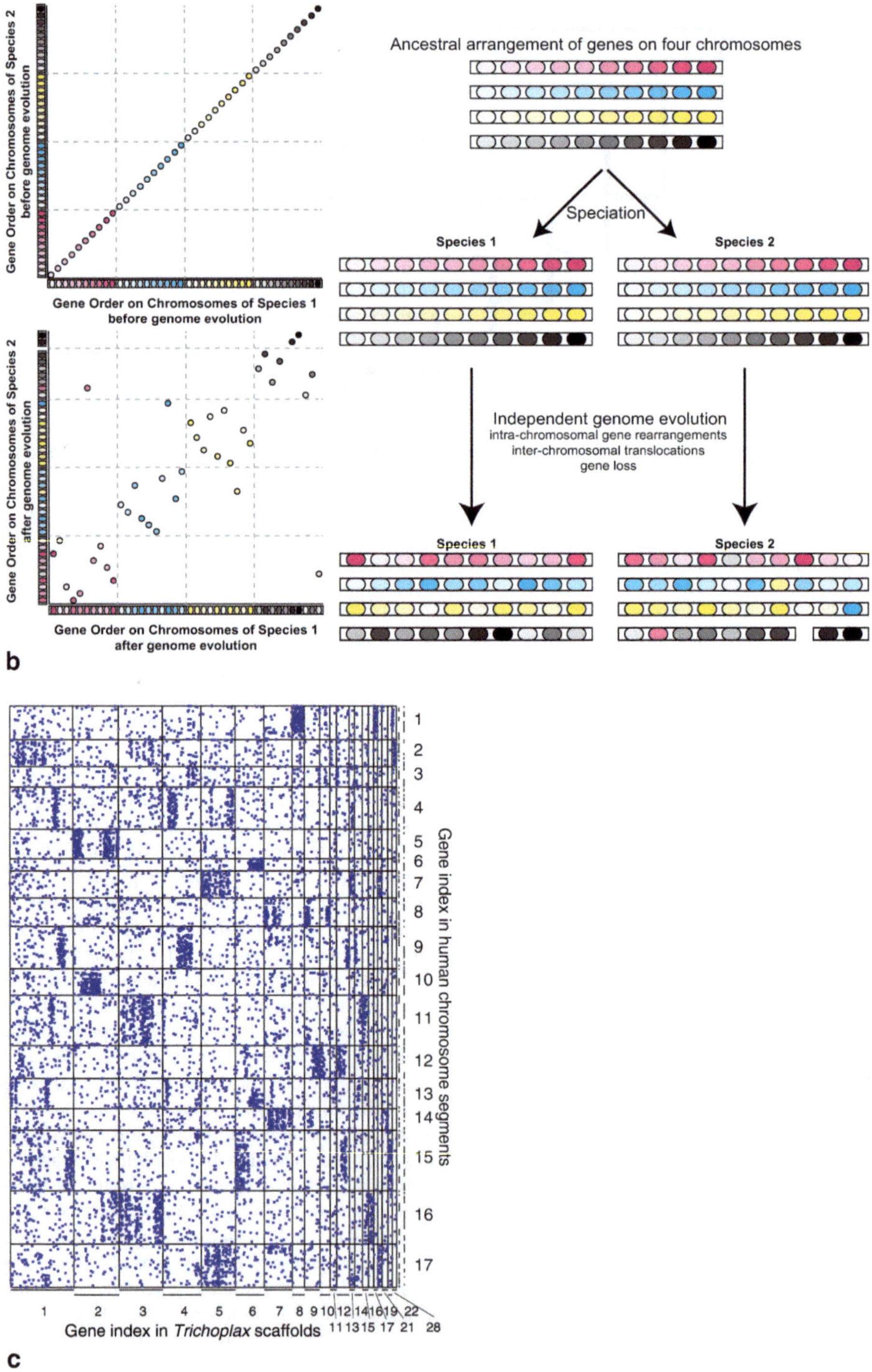

Fig. 2 (continued)

Conservation of Synteny

As populations evolve, their underlying genomes undergo many changes including site mutations and chromosomal rearrangements. Upon speciation, chromosomal segments can translocate to different locations in the two descendant species, resulting in a different order of genes on chromosomes and in the movement of genes across chromosomes (Fig. 2b). The observation of conserved synteny, literally "same ribbon/band", indicated the presence of orthologous genes in the same chromosomes in two species, though not necessarily in the same order or orientation. The human genome displays genome-wide conserved synteny with other vertebrates and cephalochordates (Jaillon et al. 2004; Putnam et al. 2007). Genes that are linked in the human genome and in the cephalochordate genome were likely present on the same chromosome in the common ancestor of humans and cephalochordates. However, syntenic orthologous genes were not detected at a genome-wide scale between humans and more distantly related animals such as fruit flies or *C. elegans*. Therefore, before the availability of non-bilaterian genomes, it was unclear whether gene linkages in the human genome pre-dated the chordate ancestor.

Conserved linkage groups (*i.e.*, sets of genes that are present on the same chromosome segments in multiple genomes) between the assembled scaffolds of non-bilaterian animal genomes (sea anemone, placozoan, and sponge) and human chromosomes were identified using statistical analysis. The first step in this analysis is to identify the locations of genes that have single orthologs in both the human genome as well as in a non-bilaterian animal genome. These locations map a pair of orthologous genes (each member of the pair belongs to one of the two species) to a position on a segment of human chromosome and to a position on an assembled scaffold in the non-bilaterian genome. A statistical test is then used to test whether the number of orthologous gene pairs identified between a given segment-scaffold pair is enriched over the number predicted by a null model where the genes on the segment-scaffold pair are drawn from a random sampling of genes shared between the two species. The *Nematostella*, *Trichoplax*, and *Amphimedon* genomes all revealed many scaffold-segment pairs that are significantly enriched for orthologous genes (Putnam et al. 2008; Srivastava et al. 2008, 2010). Orthologous genes in segment-scaffold pairs can be visualized in a "dot plot" (Fig. 2b), such as the one between human and *Trichoplax* segments shown in Fig. 2c (each dot represents the locations of orthologs of the same gene in the two species). If the linkage of orthologous genes was scrambled instead of being conserved between the two species, the dots would be randomly distributed across this plot. However, the regions of the graph with clustered dots indicate that many genes are linked on the same chromosomal segment in both species (these regions represent segment-scaffold pairs that are enriched for orthologous genes in statistical tests).

The methods described here have identified extensive conservation of synteny. The 40 human chromosome segments that show conserved synteny with *Nematostella* cover half of the human genome. Each of the 21 longest gene-rich *Trichoplax* scaffolds contains segments with a significant concentration of orthologs on one or more human chromosome segments (Fig. 2c). Eighty-three of the longest *Amphimedon* scaffolds contain segments with a significant concentration of orthologs in fifteen ancestral linkage groups inferred for the cnidarian-bilaterian ancestor. Scaffold-segment pairs that share a significant excess of orthologous genes likely represent regions of the two modern genomes (a non-bilaterian animal genome and the human genome) that descended from the same chromosome in the common ancestor. Thus, it appears that many genes that were linked together on the same chromosome in the ancestral animal continue to remain linked in modern-day animals.

It is remarkable that this conservation of synteny can be identified despite independent chromosomal rearrangement and scrambling of gene order over millions of years of independent evolution in extant animal lineages. The fruit fly and *C. elegans* genomes have undergone substantial rearrangement and have lost these ancestral gene linkages. These results suggest that the neutral rate of chromosomal rearrangements is low in non-bilaterian animal lineages and in humans, and with enough time, this signal of ancient gene linkages may be eliminated. Alternatively, selection may have acted to maintain syntenic groups of genes in non-bilaterian animals and humans; however, the functional relevance of this conservation remains unclear.

Hox genes lie in one of the conserved linkage groups thus identified. It appears that in the sea anemone-human ancestral linkage group that contains Hox genes, 225 other genes were also present, and their descendants remain linked in modern sea anemones and humans (Chourrout et al. 2006; Putnam et al. 2008; Ryan et al. 2007). Thus, compared to the few instances of conserved synteny in bilaterians (e.g., the Hox complex) that were known previously, the sequencing of non-bilaterian genomes has revealed much more extensive conservation of synteny in animals.

Ancestral Gene Content

The molecular functions of proteins encoded in genome sequence can be inferred by searching for functional sequence motifs such as those identified in the PANTHER and PFAM databases or by identifying homologous proteins with known functions in other species via BLAST. Sequence similarity discovered in this manner guides general classification of proteins from newly sequenced genomes into large protein families. However, for uncovering the gene content associated with most known biological pathways, it is important to establish true orthology (Koonin 2005). For example, a protein (e.g., Wnt1) in the *Nematostella* genome may show high similarity to Wnt ligands when BLAST is used to search a protein database, indicating that it is a Wnt family member. However, a detailed phylogenetic analysis is required to demonstrate that the human and *Nematostella* Wnt1 proteins descended from a single protein in the last common ancestor of these two animals (*i.e.*, they are true orthologs) (Kusserow et al. 2005). Combining BLAST-based annotation and phylogenetic methods, it has been shown that vast numbers of gene families previously thought to be unique to bilaterian animals are present in the genomes of *Nematostella*, *Trichoplax*, and *Amphimedon* (see for example, (Kusserow et al. 2005; Ryan et al. 2006; Adamska et al. 2007; Simionato et al. 2007; Putnam et al. 2008; Srivastava et al. 2008, 2010)). The metazoan ancestor was thus endowed with a "toolkit" of genes that is shared by many of its extant descendants, despite variations in their level of morphological complexity. It is possible that these toolkit genes allowed early animals to accomplish functions essential to multicellularity.

Hallmarks of Animal Multicellularity

The evolution of multicellularity poses special challenges to cells because they have to coordinate their actions with those of other cells. The solutions to these challenges can be considered hallmarks of multicellular life, and include intercellular controls on cell division, growth, and death; coordinated cell-cell and cell-matrix adhesion; specification of differentiated cell types, and processes for distinguishing self from non-self to maintain immunity. The molecular underpinnings of many of the hallmarks of animal multicellularity are known from studies in fruit flies, nematodes, and vertebrates, and their orthologs in the genomes of non-bilaterian animal lineages have been identified. These analyses provide insight into the sequential assembly of the molecular machinery underlying the hallmarks of animal multicellularity. Many of these processes are discussed in other chapters of this book (for example, see Chapter "Developmental Signalling and Emergence of Animal Multicellularity" for a discussion of developmental signaling pathways); this chapter will focus on two aspects—(1) control of cell cycling, growth, and death, and (2) specialized cell types.

Cell Cycling, Growth, and Death

In unicellular organisms, single cells complete their life cycles by making independent decisions on division, growth, and death. However, once in a multicellular context, such as at the dawn of animal life, cells began to coordinate their actions to work together for the success of the multicellular organism. Did the genomes of early animals acquire new proteins, or did they reuse ancient eukaryotic molecular machinery, to regulate these processes?

Cell Cycle

Cell cycle control is an anciently-evolved process that allows organisms (single- and multi-celled) to respond to stress (*e.g.*, lack of nutrients, DNA damage), and many of the proteins for cell cycle progression are conserved among eukaryotes. For example, cyclins and cyclin dependent kinases (CDKs) are at the core of cell cycle progression in all eukaryotes—different subfamilies of each of these proteins are upregulated for different phases during cell cycling to mediate downstream changes. Analysis of non-bilaterian animal genomes revealed that the modern-day vertebrate cell cycle is mediated by some recently-evolved regulators (relative to the eukaryotic ancestor), including modifications to the ancient cyclin-CDK system, that are novel to the holozoan, metazoan, or eumetazoan lineages (Fig. 3a). The CDK4/6 subfamily

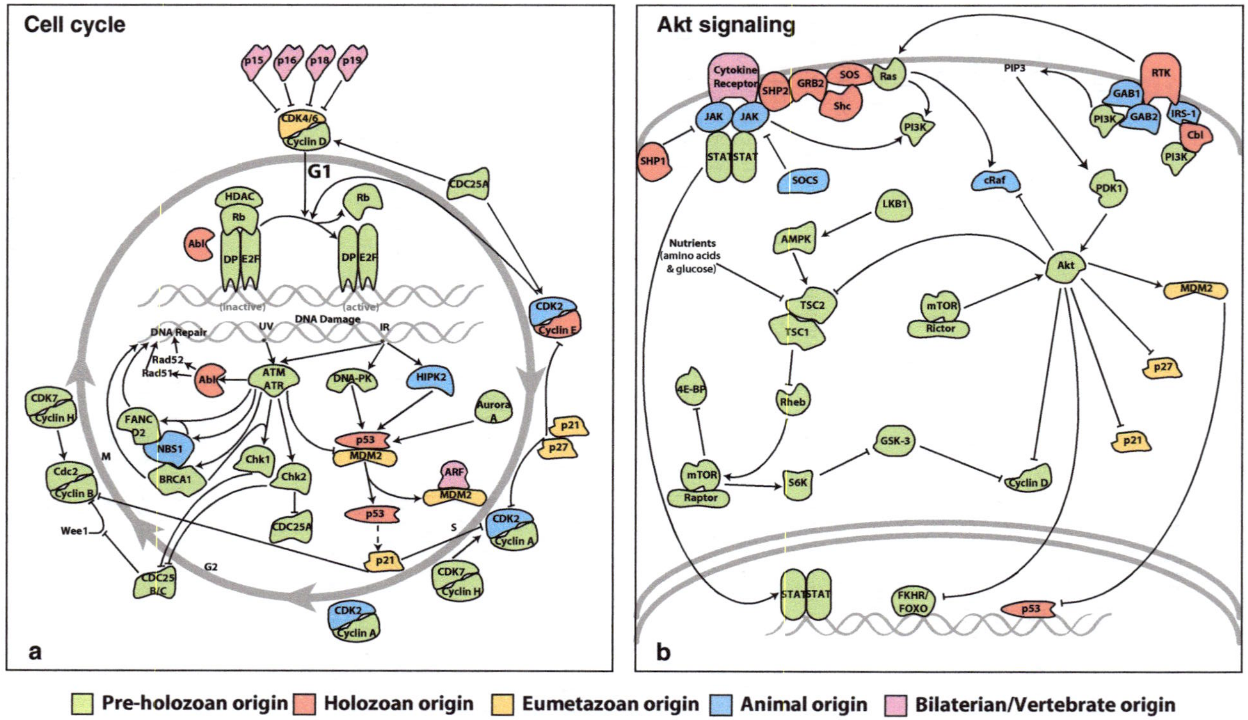

Fig. 3 Evolution of cellular pathways. Schematic representations of regulatory networks of proteins involved in the vertebrate (**a**) cell cycle, (**b** and **c**) cell growth, and (**d**) cell death. The proteins are color-coded to indicate the ancestral organism in which they are inferred to have first appeared. Upon the sequencing of the *Capsaspora* genome, orthologs for several components of Warts-Hippo signaling that were previously thought to be metazoan-specific proteins were found to be more anciently-evolved, in the holozoan ancestor. Whereas many components of cell cycle and cell growth pathways are anciently-evolved eukaryotic proteins, the apoptotic machinery has many components that were newly evolved in early animal evolution. (Adapted from Srivastava et al. 2010)

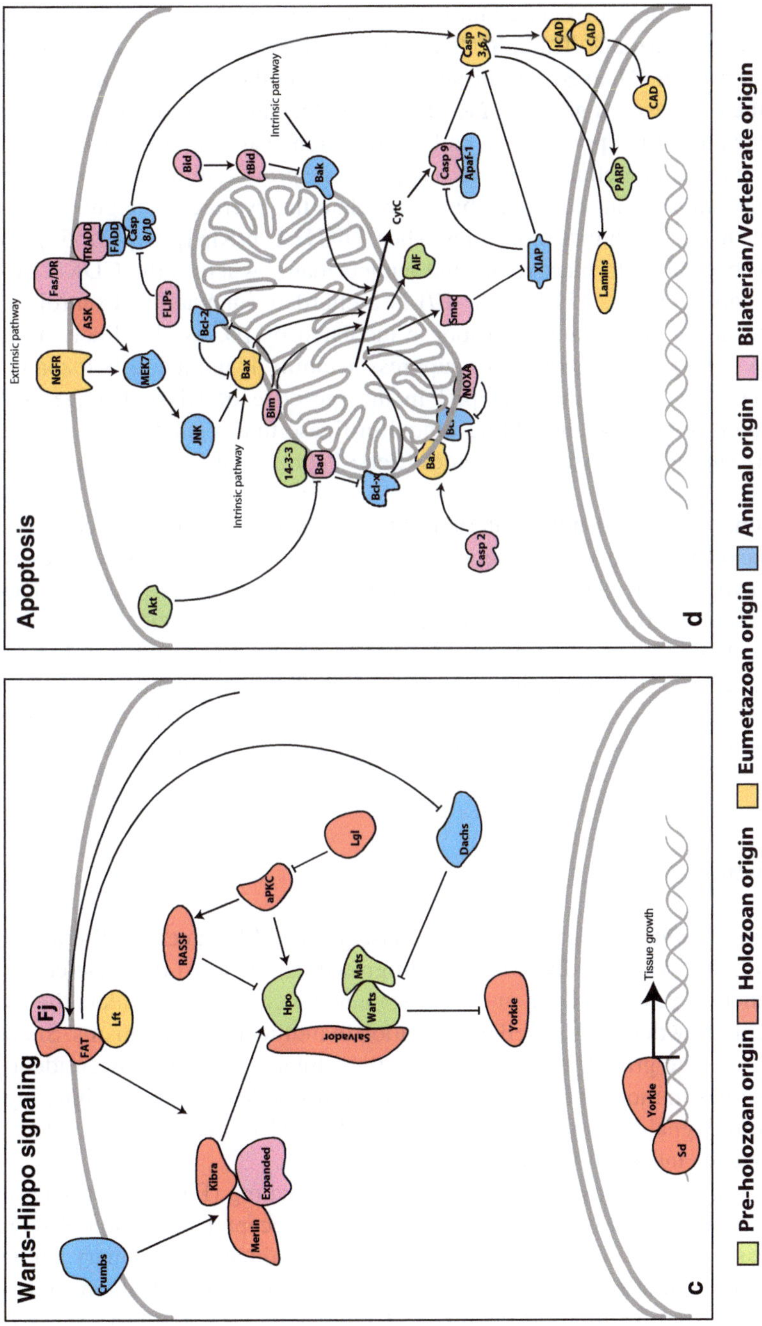

Fig. 3 (continued)

(a G1-specific CDK) is unique to eumetazoans and placozoans and the CDK2 subfamily of the ancient Cdc2 kinase family (required for the G1 to S transition) appears to be unique to metazoans. Though bonafide Cyclin E orthologs (also associated with the transition from G1 to S) are missing from choanoflagellate genomes, they are found in the *Capsaspora* genome, *i.e.*, they are holozoan innovations.

From the unicellular perspective, cell proliferation results directly in reproductive success, but in a multicellular context, inappropriate proliferation can be detrimental to the organism. Animal cells are able to regulate the cycling of neighboring cells through the use of inhibitory molecules. The CDKN1/Cip/Kip (p21/p27/p57) family appears to be unique to placozoans and eumetazoans, whereas the CDKN2/INK4 family of inhibitors (p. 15, 16, 18, 19) is a chordate innovation. Control of the cell cycle via the DNA damage response mediated by p53 is also likely a recent innovation—the p53/p63/p73 tumor suppressor family is unique to holozoans, the HIPK kinase that phosphorylates p53 in the presence of DNA damage is only found in metazoans, and the Mdm2 ubiquitin ligase that regulates p53 is specific to eumetazoan (and placozoan) genomes (Sebe-Pedros et al. 2011; Srivastava et al. 2010).

Functional studies in non-bilaterian animal lineages will be needed to verify whether novel metazoan components of the cell cycle acquired these roles when they first evolved, or whether they were co-opted for cell cycle functions later in animal evolution. For example, the absence of some key cell cycle regulators (*e.g.*, p21/p27 or Mdm2) from the *Amphimedon* genome raises the question of how these missing functions, such as the inhibition of cell proliferation by other cells, are accomplished in sponge cells. In this instance, detailed study of the sponge cell cycle can potentially help us understand functions of these recently-evolved cell cycle regulators.

Cell Growth

The growth of multicellular animals is a consequence of both cell proliferation and cell growth. Whereas cell division and cell growth are coupled in single-celled organisms such as yeast, external and developmental signals can modulate the extent to which cell growth results in cell proliferation in multicellular organisms. Six pathways that regulate growth in response to extracellular signals have been identified through studies in model organisms (Receptor tyrosine kinase (RTK) signaling via RAS, insulin signaling via the PI3K pathway, Rheb/Tor, cytokine-JAK/STAT, the Myc oncogene, and Warts/Hippo).

The first three pathways are mediated by Akt signaling (Fig. 3b). Though most components of this pathway are ancient, the insulin receptor substrate (IRS-1) and receptor associated proteins Gab1/Gab2 are novel to animals. Preliminary studies in *Hydra* suggest that insulin signaling may control growth and cell death in cnidarians (Lasi et al. 2010a). Cytokine receptors and Janus kinase (JAK) also appear to be unique to animals, suggesting that the control of growth by the JAK/STAT pathway was invented in the metazoan lineage (Fig. 3b). Myc is a holozoan innovation and core features of its molecular functions are preserved in choanoflagellates (Young et al. 2011). However, a role for non-bilaterian Myc homologs in growth control remains to be established.

The components of the sixth cell growth pathway, mediated by the Wart-Hippo-Mats complex, are encoded in the *Capsaspora* genome and their potential functions have been investigated in some detail. Warts, Hippo, and Mats are anciently-evolved eukaryotic proteins that operate in a molecular pathway that is preserved in animals and fungi. Though the module is preserved, the net outcomes are different—in *Drosophila* the pathway limits cell proliferation and tissue growth, but in yeast it enables cell division. Hippo (Mst in mammals) autophosphorylates and then phosphorylates Salvador (WW45 in mammals), Warts (Lats in mammals) and Mats (Mob in mammals) (Reddy and Irvine 2008). In bilaterians, Salvador facilitates the phosphorylation of Warts by Hippo, and Warts subsequently phosphorylates and inactivates the transcription factor Yorkie (Yki), which normally acts in conjunction with Scalloped (Sd) to mediate tissue growth. Several upstream regulators of the Warts-Hippo cassette in bilaterians such as Merlin, Kibra, Lgl, and aPKC were not found in the *Monosiga* genome, and therefore were thought to be unique to animals. However, the recent sequencing of the *Capsaspora* genome uncovered deeper ancestry for these proteins, making them holozoan innovations (Fig. 3c; Sebe-Pedros et al. 2012). Salvador, Scalloped and Yorkie are encoded in the *Capsaspora, Monosiga*, *Amphimedon*, *Trichoplax*, and *Nematostella* genomes.

Overexpression of Yki and Sd orthologs from *Capsaspora* in *Drosophila* eye imaginal discs resulted in overgrowth of eye tissue, recapitulating the outcome of overexpressed *Drosophila* Sd-Yki. Overexpression of *Capsaspora* Hippo in flies resulted in the opposite phenotype, smaller eyes, similar to the effect of fly Hippo overexpression (Sebe-Pedros et al. 2012). This suggests that these distant orthologs of the *Drosophila* Sd-Yki complex can substitute for the endogenous fly proteins and have enough sequence/structural similarity to affect the same downstream pathways. Congruently, *Capsaspora* Sd-Yki led to increased expression of a known Warts-Hippo signaling target gene in flies, Diap1. *Capsaspora* Hippo also phosphorylated *Capsaspora* Yorkie. Thus, it appears that core Warts-Hippo pathway components are functionally connected in a distant holozoan relative of animals.

The downstream effects of this pathway in *Capsaspora* remain unknown, however. Functional studies in this species will reveal whether the Warts-Hippo pathway has an ancestral role in regulating cell growth and proliferation. New regulators of the Warts-Hippo pathway appeared concurrently with the appearance of animal multicellularity. For example, the unconventional myosin Dachs, which is inhibited by Fat (a holozoan innovation) and inhibits the activity of Warts, belongs to a novel myosin subfamily in animals. The FERM-domain protein Expanded, which regulates Hippo in *Drosophila*, is only found in bilaterians. Thus, it is possible that the ancient Warts/Hippo/Mats cassette may have been co-opted in a tumor suppressor role in animals by coming under the control of proteins novel to the animal lineage.

Cell Death

In a multicellular context, many processes (such as embryonic development) require the sacrifice of some cells for the success of the organism. In vertebrates, programmed cell death proceeds via two pathways, one is intrinsic to the cell and the other is activated in response to extrinsic cues.

The intrinsic pathway is regulated by members of the Bcl-2 family (the pro-apoptotic subfamilies Bak, Bax, and, Bok, and the anti-apoptotic Bcl-2/Bcl-X) (Fig. 3d). Bcl-2/Bcl-X and Bak appear to be metazoan-specific, whereas Bax and Bok are missing in the *Amphimedon* genome. This pathway results in the permeabilization of the outer mitochondrial membrane, which signals through a series of caspases, a metazoan-specific family of cysteine aspartyl proteases, to mediate apoptosis. Several regulators of Bcl-2 and caspases, namely the BH3-only proteins (Bid, Bim, and NOXA) are bilaterian novelties, though BH3-only proteins with potential roles in apoptosis and unclear homology to bilaterian BH3-only proteins were identified recently in *Hydra* (Lasi et al. 2010b). It is unknown whether placozoans and sponges lack this additional layer of regulation entirely, or whether they utilize other proteins for that purpose.

In the extrinsic pathway, external signals that lead to apoptosis are typically detected by death domain-containing transmembrane receptors belonging to the tumor necrosis factor receptor (TNFR) family. Bonafide nerve growth factor receptor (NGFR) proteins with death domains are found in placozoan, cnidarian and bilaterian genomes. Classic death TNFRs (*i.e.*, Fas, DR4, DR5, and TNFR1) evolved later in bilaterian evolution.

In contrast to many components of the extrinsic pathway, the intrinsic pathway consists of relatively anciently-evolved eukaryotic and metazoan proteins, making the latter a likely candidate for the original process for programmed cell death in animals.

Specialized Cell Types

As cells unite in a multicellular entity, they can specialize their functions to optimize the survival of the organism. Animals have many cell types with differentiated functions such as neurons, muscles, excretory cells, epithelial cells, and digestive cells. The molecular determinants of some of these cell types are known from bilaterians and therefore can be investigated in non-bilaterian animal genomes. To understand the diversification of animal form, it is particularly instructive to focus on cell types that are missing in early-diverging species to understand how molecular pathways encoded in genomes correlate with the morphological appearance of a new cell type. Sponges and placozoans are believed to lack epithelial cells and neurons (Brusca and Brusca 2002; Grell 1991). The next two sections will explore the presence of orthologs of genes responsible for the function of these bilaterian cell types in the *Amphimedon* and *Trichoplax* genomes.

Epithelia

Epithelial tissues cover the outer surface of animals and line compartments within an animal, providing varying degrees of physiological barriers to solutes. Cnidarian and bilaterian tissues display an epithelial grade of organization (Brusca and Brusca 2002). Three characteristics define epithelia: (a) apical-basal polarity of cells, (b) cell-cell adhesion via belt-like junctions (adhesive or occluding), and (c) cell-ECM adhesion to an underlying basal lamina. Genes for these processes/structures are known from studies of epithelia in *Drosophila* and vertebrates. Epithelial cells with junctions and underlying basement membrane can be identified based on morphology by transmission electron microscopy. Though cell-cell junctions are found in placozoans in ultrastructural studies, cells with a basal lamina, or any ECM, have not been identified (Grell 1991). The presence of a true epithelium has been highly debated in sponges, but cell-cell junctions are present and a basal lamina has been reported from some sponge groups (Leys et al. 2009).

The *Trichoplax* and *Amphimedon* genomes encode most orthologs of the Par, Crumbs, and Discs Large complexes that are required for epithelial cell polarity (Fahey and Degnan 2010; Fig. 4a). Among cell-cell adhesion mechanisms, all components of adhaerens junctions (cadherins, alpha-catenin, delta-catenin) are present in placozoans and sponges (Srivastava et al. 2008, 2010; Nichols et al. 2012). However, whereas the *Trichoplax* genome encodes Neuroglian and Neurexin IV, two components of septate junctions, the sponge genome is missing these as well as the third component, Contactin. As with cnidarians, placozoan and sponge genomes both appear to be missing tight junction components (Claudin and Occludin). ECM components of the basal lamina such as Collagen IV Nidogen, and Perlecan are encoded in placozoans, but not found in the *Amphimedon* genome. Clear orthologs of another set of ECM proteins, Laminin alpha, beta, and gamma, are present in the *Trichoplax* genome, whereas the *Amphimedon* genome encodes laminin-like proteins with variant underlying domain composition. Among the cell adhesion proteins present in both sponges and placozoans, some such as cadherins, integrins, and certain laminins have ancient holozoan origins (Abedin and King 2008; Suga et al. 2013).

The absence of septate junction and basal lamina proteins in *Amphimedon* is consistent with the inability to find these structures in EM studies in demosponges. However, septate junctions were observed in a calcareous sponge (Ledger 1975) and a basal lamina has been reported in homoscleromorph sponges (Boute et al. 1996; Nichols et al. 2006). A type IV collagen protein is found in the homoscleromorph sponge *Pseudocorticium jarrei*, and was reported to be located in a basal lamina underlying the surface epithelium (Boute et al. 1996). Thus, if sponges are a monophyletic clade, it is possible that, lacking septate junctions and basement membranes, *Amphimedon* represents a derived state for sponges. However, it has been proposed that calcareous and homoscleromorph sponges may have diverged more recently than demosponges, raising the possibility that sponges are a paraphyletic group (Sperling et al. 2007). This would imply that the absence of molecular determinants of epithelia

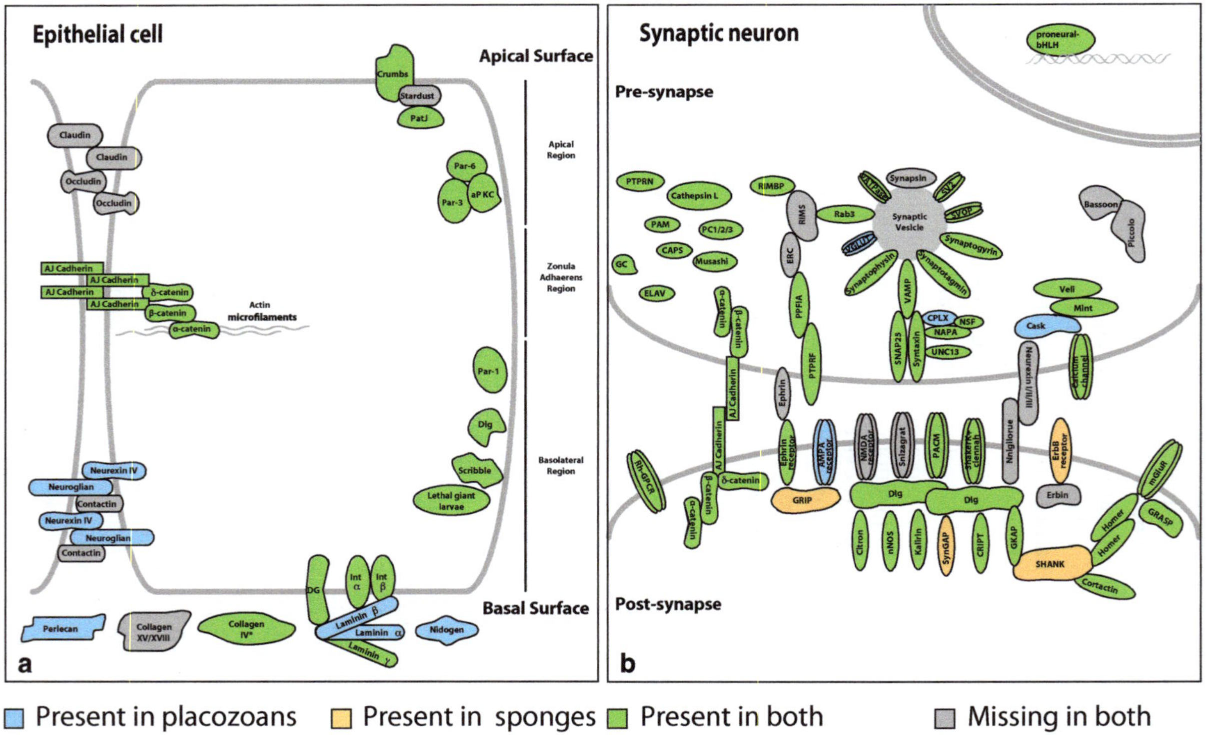

Fig. 4 Evolution of proteins for specialized cell types. Schematic representations of protein interactions required for the function of (**a**) epithelia (**b**) synaptic neurons, two cell types missing in sponges and placozoans. Proteins encoded in the *Trichoplax* genome are in *blue*, those encoded only in the *Amphimedon* genome are in *yellow*, and those encoded by both genomes are in *green*. *Asterisk*: Although type IV collagen is not encoded in the *Amphimedon* genome, it is present in homoscleromorph sponges (Boute et al. 1996). (Adapted from Srivastava et al. 2010)

in *Amphimedon* reflects the absence of these proteins in the ancestral animal instead of reflecting secondary loss in demosponges.

The question of whether the ancestral animal had true epithelial cells will be answered when the continuing debate about sponge monophyly is answered using new phylogenetic approaches. Analysis of the recently sequenced genome of the homoscleromorph sponge, *Oscarella carmela*, will be important for deciphering both the question of sponge phylogenetics and of differences in gene content between different groups of sponges (Nichols et al. 2012).

The absence of a clear basal lamina in *Trichoplax,* despite the presence of its molecular determinants, remains to be explained. It is possible that other life history stages of this species, that are yet to be observed, contain true epithelia. It is also possible that *Trichoplax* has recently lost the basal lamina, despite retaining the genes underlying this structure in other animals.

Neurons

Cnidarians, ctenophores, and bilaterians have morphologically identifiable neurons and use neural conduction to rapidly transmit information across multicellular tissue to coordinate their cells. Although *Trichoplax* has no nervous system, it exhibits behavioral responses to environmental stimuli, and sensitivity to the neuropeptide RFamide has been reported (Schuchert 1993). Similarly, whereas no clear neuronal cells have been identified in sponges, investigations of many species suggest that these animals are capable of sensing information and transmitting it to coordinate the actions of the entire organism (Renard et al. 2009).

Genomically, *Trichoplax* and *Amphimedon*, encode a slew of neuronal proteins (Fig. 4b) (Srivastava et al. 2008, 2010). While both genomes are missing a few genes, homologs of most pre-synaptic and post-synaptic assemblages in bilaterians are present (Sakarya et al. 2007). Components of neurotransmitter biosynthesis and vesicle transport systems, as well as a putative neuroendocrine-like secretory apparatus are also found in these genomes. Putative neurotransmitter and neuropeptide receptors, ion channels as well as abundant seven transmembrane G-protein coupled receptors are also present. Several neural components appear to be unique to placozoans, cnidarians, and bilaterians, i.e., they are missing in sponges. For example, whereas the placozoan genome contains homologs of ionotropic glutamate receptors as well as genes associated with neural migration and axon guidance (Slit, Netrin, and neural cell adhesion molecules), the *Amphimedon* genome appears to be missing counterparts for these proteins.

The *Capsaspora* and *Monosiga* genomes encode the majority of synaptic vescicle trafficking and endocytosis proteins that are known to be broadly distributed among eukaryotes, but lack the majority of scaffolding proteins (Suga et al. 2013). The *Capsaspora* genome has also revealed some neurosecretory and many presynaptic and postsynaptic proteins to be holozoan innovations. For example, the K+ voltage-gated channel of the shaker family, which is missing in the *Amphimedon* genome, is present in both the *Capsaspora* and *Monosiga* genomes. Thus, many neural genes in bilaterians first appeared in the genomes of unicellular ancestors.

What are the functions of neural genes in animals without neurons? It is possible that many neural proteins evolved first for basic cell biological functions (such as vesicle trafficking) that resulted in this apparent sequential assembly of the modern bilaterian synaptic machinery. Voltage-gated ion channels are known to function in non-neural conduction such as in polyspermy and in non-electrical processes such as cell signaling and adhesion. If this is the case for sponge and placozoan neural genes, the absence of neurons in these species could represent the condition of the metazoan ancestor's nervous system. These genes may have been co-opted for neural functions in the cnidarian-bilaterian ancestor. Given the evidence for placozoan and sponge responsiveness to neurotransmitters, it is possible that some of these neural genes endow the two species with the ability to coordinate and communicate information, though not with cells that morphologically resemble bilaterian neurons. In *Trichoplax*, junctions of fiber cells where the cytoplasm of two cells are separated by a disk-like structure with vacuoles on either side have been thought of as 'proto' synapses (Grell 1991). It is also possible that sponges and placozoans have lost neuronal cells that were present in the last metazoan ancestor, but have maintained neural genes for other functions. The recent debate about the phylogenetic placement of ctenophores weighs-in on this question and is discussed in the next section.

Nervous System Evolution and the Phylogenetic Position of Ctenophores

The recent publication of the *M. leidyi* genome sequence, the first representative ctenophore genome, has had tremendous impact on inferences about animal evolution (Ryan et al. 2013). In particular, a striking new hypothesis about the origin of animal nervous systems has emerged as a result of phylogenetic analyses of these sequence data.

Morphology-based phylogenetic analyses placed ctenophores as a sister-lineage to cnidarians, a view that has been supported by some recent phylogenomic analyses based on EST sequences (Dunn et al. 2008; Philippe et al. 2009; Nosenko et al. 2013; Fig. 5). This traditional phylogenetic scenario implies a single origin of the nervous system because ctenophores, cnidarians, and bilaterians share a common ancestor that did not give rise to the two phyla, sponges and placozoans, without neurons (Fig. 5). The most parsimonious interpretation of this scenario suggests that animals evolved the nervous system once, and neural genes in the *Amphimedon* and *Trichoplax* genomes represent "pre-adaptations" in the metazoan ancestor that were co-opted for functioning in cnidarian, ctenophore, and bilaterian nervous systems (Sakarya et al. 2007).

Some recent analyses of EST and genome sequences from two ctenophores, *M. leidyi* and *P. bachei,* had suggested that comb jellies represent the earliest-diverging animal lineage, but some analyses had also recovered support for the traditional view of their phylogenetic position as a sister group to cnidarians (Fig. 5; Dunn et al. 2008; Philippe et al. 2009; Nosenko et al. 2013). Phylogenetic analyses with the whole genome sequence data from *Mnemiopsis* failed to recover support for ctenophores as a sister group to cnidarians (Ryan et al. 2013). In the newly proposed view of

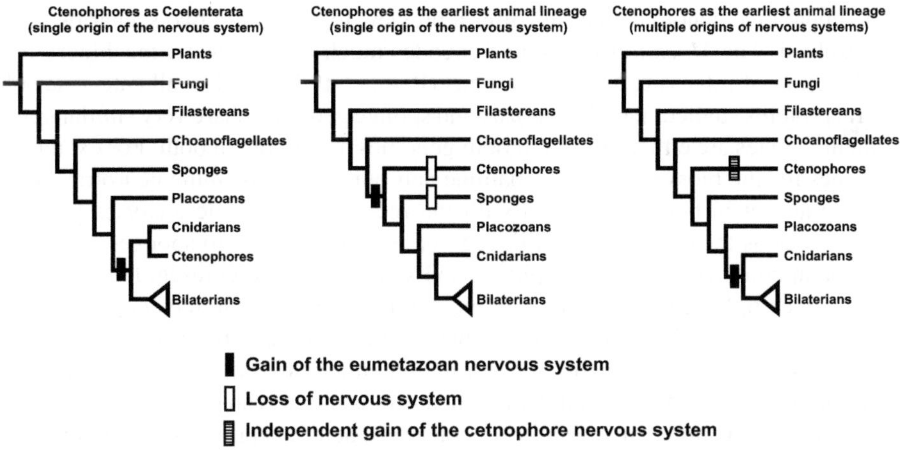

Fig. 5 The phylogenetic position of ctenophores and the evolution of the nervous system. In the traditional view of the phylogenetic placement of ctenophores, comb jellies are closely related to cnidarians and placed within the Coelenterata. Given the absence of neurons in sponges and placozoans, the parsimonious interpretation implies that all animal neurons are homologous and shared a common origin in the common ancestor of the Eumetazoa. The new view, of ctenophores as sister lineage to all other animals, two scenarios explain nervous system evolution. In one, if the same genes that are required for bilaterian and cnidarian nervous systems are found to operate in ctenophore neurons, animal nervous systems likely originated once, with independent loss of the nervous system in sponges and placozoans. And the second scenario, ctenophore and cnidarian/bilaterian nervous systems evolved convergently, without implied loss in sponges and placozoans. The latter scenario would be supported if ctenophore neurons bear different molecular signatures from bilaterian neurons. Two fully sequenced ctenophore genomes and additional sponge genomes will facilitate a detailed study of nervous system specification and function in comb jellies and thus allow us to distinguish between the two scenarios

ctenophores as the earliest-diverging animal phylum, the last common ancestor of cnidarians, ctenophores, and bilaterians also gave rise to the two phyla without neurons (sponges and placozoans). If ctenophore, cnidarian, and bilaterian nervous systems are homologous, i.e., orthologous proteins underlie neurons in both clades, it would imply that the nervous system evolved once, at the dawn of the Metazoa (Fig. 5). In this scenario, sponges and placozoans lost neurons independently, and it is unlikely that the presence of neural genes in sponges and placozoans was a "pre-adaptation". However, if ctenophore neurons do not utilize the orthologs of genes associated with bilaterian and cnidarian neurons, we would infer that the nervous system was invented convergently in two independent lineages, and synaptic genes in sponges and placozoans would then still be pre-adaptations for cnidarian and bilaterian neurons. This scenario would be consistent with several reports of unique features of ctenophore nervous systems (*e.g.*, lack of serotonin immunoreactivity, specialized presynaptic morphology in ultrastructural studies) compared to cnidarian/bilaterian nervous systems.

Comparative genomic analyses offer a preview to how these two scenarios may be resolved. The *Mnemiopsis* genome encodes much of the genetic machinery for the function of bilaterian nervous systems (Ryan et al. 2013). For all categories of neural proteins studied, the *Mnemiopsis* and sponge genomes share very similar gene content—genes are either present in both genomes or missing in both. For example, enzymes for the biosynthesis of dopamine and other catecholamine neurotransmitters are absent in both. Combined with the recent phylogenetic analyses placing ctenophores as the earliest animal lineage or as a sister group to sponges, the neural protein complements of ctenophores and sponges would represent the ancestral metazoan complement. Given the presence of a nervous system in ctenophores, the parsimonious scenario would be one where the animal nervous system originated ancestrally, sponges and placozoans lost it, and cnidarians and bilaterians acquired new components such as catecholamine neurotransmitter synthesis enzymes. However, detailed functional investigation of conserved neuronal proteins in ctenophores, sponge, and cnidarians will be required to establish the precise origins of animal nervous systems. During the production of this book, the genome of another ctenophore, *Pleurobrachia bachei*, was published. This work provides evidence for independent evolution of ctenophore and bilaterian nervous systems (Moroz et al. 2014).

Genomic Correlates of Morphological Diversity in Animals

In addition to identifying genomic events associated with the appearance of animal multicellularity, one of the goals of sequencing non-bilaterian animal genomes was to investigate genomic correlates of morphological diversity in animal body plans. In particular, it was expected that genomic comparisons would identify the genetic basis of how bilaterians gained complex morphological features relative to their earlier-diverged cousins (*e.g.*, non-bilaterians appear to have fewer cell types and organs and thus appear less "complex" by this measure). However, the finding of shared genomic features (gene structure, genome organization, and gene content) between non-bilaterians and bilaterians confounds the search for genomic correlates of morphological complexity.

It is important to note, however, that despite the tremendous conservation described above, there are likely many genomic differences between sponges and eumetazoans, or between cnidarians and bilaterians—it is simply that the significance of these differences to morphology needs to be established. For example, the *Amphimedon* genome lacks a true Hox-like gene (and appears to have few members of all major transcription factor families), whereas *Trichoplax* has one Hox/ParaHox-like gene, and cnidarians have a handful of Hox-related genes (and have several members of the major transcription factor families relative to sponges) (Larroux et al. 2008; Schierwater et al. 2008). Do these differences in Hox genes or transcription factor numbers explain why sponges appear "simpler" than cnidarians (e.g., sponges lack muscles and neurons) or why cnidarians appear simpler than bilaterians?

Gene family expansions that correlate with change in morphological complexity in animals have been identified at a genome-wide scale (Srivastava et al. 2010). For example, though they are found in all animal genomes, homeobox transcription factors, nuclear hormone receptors, and gap junctions proteins are highly enriched in number among invertebrate bilaterians relative to non-bilaterian animal lineages. Do these gene families that expanded in the bilaterian ancestor explain the increase in numbers of cell types in extant bilaterians? The answer lies in understanding the functions of these genes in sponges, placozoans and cnidarians to find the links between genomic complexity and morphology in varied animal lineages.

The non-bilaterian animals selected for sequencing were chosen as representatives of their phyletic lineages because they are also tractable experimental model systems. Thus, the sequencing of these genomes has the additional benefit of enhancing experimental studies of the biology of these animals. Many studies attempting to address the functions of metazoan pathways in non-bilaterian animal phyla are underway. For example, in situ hybridization studies revealed that neural transcription factors and other markers of neural function are expressed in specific subsets of neurons in *Nematostella*. This molecular heterogeneity in neurons offers a view different from the traditional view of the cnidarian nervous system as an undifferentiated nerve net (Marlow et al. 2009). It is possible that, though functionally differentiated neurons are not organized into obvious structures with homology to bilaterian organs (e.g., a brain) in *Nematostella*, there is a diverse array of functionally specialized neurons that may be specified via complex mechanisms in cnidarians. Thus, the body plans of "simple" organisms may be patterned in a complex, albeit cryptic, manner.

In addition to gene content differences (novel genes or gene family expansions), other genomic features may be significantly different between different metazoan clades. *Cis*-regulatory networks are important in bilaterian embryonic development and studies of regulatory linkages between ancestral genes in non-bilaterians may reveal how alterations of *cis*-regulatory networks may have contributed to morphological diversity in animals. Non-coding RNAs, including microRNAs and long non-coding RNAs, have been shown to have important roles in gene regulation and may also be important for understanding morphological differences between non-bilaterian animal lineages and bilaterians. Studies of micro RNAs (miRNAs) in non-bilaterian animal phyla suggest that there may have been a significant increase in the repertoire of regulatory miRNAs before the divergence of Bilateria as miRNAs are found in smaller numbers in cnidarians and sponges and are absent from placozoan and ctenophore genomes (Grimson et al. 2008; Ryan et al. 2013). Together with the whole genome shotgun assemblies described here, new sequencing technologies can enhance the study of these genomic features in non-bilaterians. For example, ChIP-seq experiments can be used to discover regulatory linkages in sponges, placozoans, and cnidarians by aligning reads to the reference assembly.

It has also been argued that origin of bilaterian signaling pathways and patterning mechanisms in the ancestral metazoan is unsurprising (Marshall and Valentine 2010). The survival of single-celled organisms relies on responding to environmental cues; therefore it is reasonable that signal transduction mechanisms that could allow cells to respond to other cells were already in place and used by our unicellular ancestors. In this view, the great diversity of bilaterian body plans is a result of permissive environmental conditions (possibly an increase in atmospheric oxygen levels) that allowed diverse body plans to arise from the raw-material of the large repertoire of signaling and patterning genes ("pre-adaptations") inherited from the metazoan ancestor (Marshall and Valentine 2010; see Chapter "Fossil and Transcriptomic Perspectives on the Origins and Success of Metazoan Multicellularity").

Summary

1. Comparative analyses of the genomes of non-metazoan holozoans, non-bilaterian animals, and bilaterians allowed us to infer that many features of animal genomes have been conserved over the great evolutionary distances that separate these species. Given the shared intron-exon structures of orthologous genes in modern cnidarian, placozoan, sponge, and bilaterian genomes, we infer that the ancestral animal genome was intron-rich. Genes that were putatively linked on contiguous chromosomal segments in the animal ancestor have maintained close proximity in our genome as well as in the genomes of *Nematostella*, *Trichoplax* and *Amphimedon*, despite considerable scrambling in gene order. In contrast to anecdotal examples of conservation of gene order (*e.g.*, the Hox genes in the bilaterian ancestor) provided by early studies, whole genome data now allow us to annotate the animal tree with a large set of genomic characteristics inferred for the metazoan ancestor. Though fruit flies and nematodes have proved to be exquisite model systems for dissecting the genetic underpinnings of metazoan development and physiology, their genomes are relatively poor models for the ancestral bilaterian genome, having lost introns, genes, and gene linkages.
2. The last common ancestor of metazoans likely had a "toolkit" of genes for cell signaling, adhesion, apoptosis, developmental patterning, and for the function of specialized cells types such as epithelia and neurons. In some processes, such as cell cycle and growth, ancient eukaryotic pathways acquired a few new modulators unique to animals, whereas in other cases, such as programmed cell death, vast suites of genes appeared during early animal evolution (Fig. 3). Certain gene families expanded in correspondence with the apparent increase in morphological complexity from non-bilaterians to bilaterians. How did these genes function in the metazoan ancestor, and how important were they for the appearance of multicellularity and the subsequent diversification of animal body plans? Unfortunately, we cannot read from the genome the nature of its gene- and protein-regulatory interactions and networks or how they impact the phenotype. But, non-bilaterian animal genomes provide a platform for testing hypotheses about the functions of ancestral metazoan pathways and interactions. For example, temporal studies of changes in gene expression can be used to infer genetic regulatory networks (Conaco et al. 2012). Ongoing molecular studies in these species have revealed complex biology, arguing for a reevaluation of the superficial "simplicity" of non-bilaterian animals.

3. The largest shifts in our understanding of animal evolution will perhaps emerge from further expanding our studies to include more understudied taxa. Analyses of the filasterean and choanoflagellate genomes showed that the genetic machinery for cell adhesion and several transcription factor families essential for animal biology evolved early in holozoan evolution. In addition, further analyses of ctenophore genomes may result in a revision of our understanding of nervous system evolution. Therefore, sequencing and experimental studies of more animal and non-animal species will be required to develop a nuanced, and more complete picture of how animals evolved.
4. Though we can now better define the genome of the metazoan ancestor, the morphology of this ancestor remains elusive. Of the many forms that may have existed along the lineage leading to animals (after the divergence of choanoflagellates), one was the first organism that was recognizably as an animal—it may have had all the molecular machinery for developmental signaling and other features that we define as uniquely metazoan. However, since none of these intermediate forms survived, or left fossil evidence (to our knowledge), the questions of what the first animal looked like or of the precise sequence of genomic events that generated the ancestral animal genome may remain unanswered (Marshall and Valentine 2010). Nevertheless, the pursuit of defining this ancestral animal at both a molecular and a morphological level promises to yield critical insights about animal multicellularity and the early diversification of animal lineages.

Note added in production During the production of this book, the genome of another ctenophore, *Pleurobrachia bachei*, was published. This work provides evidence for independent evolution of ctenophore and bilaterian nervous systems.

References

Abedin M, King N (2008) The premetazoan ancestry of cadherins. Science 319(5865):946–948. doi:10.1126/science.1151084

Adamska M, Degnan SM, Green KM, Adamski M, Craigie A, Larroux C, Degnan BM (2007) Wnt and TGF-beta Expression in the sponge *Amphimedon queenslandica* and the origin of metazoan embryonic patterning. PLoS One 2(10):e1031

Akam M (1995) Hox genes and the evolution of diverse body plans. Philos Trans R Soc Lond B Biol Sci 349(1329):313–319. doi:10.1098/rstb.1995.0119

Boute N, Exposito JY, Boury-Esnault N, Vacelet J, Noro N, Miyazaki K, Yoshizato K, Garrone R (1996) Type IV collagen in sponges, the missing link in basement membrane ubiquity. Biol Cell 88(1–2):37–44

Brusca RC, Brusca GJ (2002) Invertebrates. Sinauer Associates, Sunderland

Butschli O (1884) Bemerkungen zur Gastraeatheorie. Morph Jb 9:(415–427)

Chapman JA, Kirkness EF, Simakov O, Hampson SE, Mitros T, Weinmaier T, Rattei T, Balasubramanian PG, Borman J, Busam D, Disbennett K, Pfannkoch C, Sumin N, Sutton GG, Viswanathan LD, Walenz B, Goodstein DM, Hellsten U, Kawashima T, Prochnik SE, Putnam NH, Shu S, Blumberg B, Dana CE, Gee L, Kibler DF, Law L, Lindgens D, Martinez DE, Peng J, Wigge PA, Bertulat B, Guder C, Nakamura Y, Ozbek S, Watanabe H, Khalturin K, Hemmrich G, Franke

A, Augustin R, Fraune S, Hayakawa E, Hayakawa S, Hirose M, Hwang JS, Ikeo K, Nishimiya-Fujisawa C, Ogura A, Takahashi T, Steinmetz PR, Zhang X, Aufschnaiter R, Eder MK, Gorny AK, Salvenmoser W, Heimberg AM, Wheeler BM, Peterson KJ, Bottger A, Tischler P, Wolf A, Gojobori T, Remington KA, Strausberg RL, Venter JC, Technau U, Hobmayer B, Bosch TC, Holstein TW, Fujisawa T, Bode HR, David CN, Rokhsar DS, Steele RE (2010) The dynamic genome of Hydra. Nature 464(7288):592–596. doi:10.1038/nature08830

Chourrout D, Delsuc F, Chourrout P, Edvardsen RB, Rentzsch F, Renfer E, Jensen MF, Zhu B, de Jong P, Steele RE, Technau U (2006) Minimal ProtoHox cluster inferred from bilaterian and cnidarian Hox complements. Nature 442(7103):684–687. doi:10.1038/nature04863

Conaco C, Bassett DS, Zhou H, Arcila ML, Degnan SM, Degnan BM, Kosik KS (2012) Functionalization of a protosynaptic gene expression network. Proc Natl Acad Sci USA 109(Suppl 1):10612–10618. doi:10.1073/pnas.1201890109

De Robertis EM, Sasai Y (1996) A common plan for dorsoventral patterning in Bilateria. Nature 380(6569):37–40

Dunn CW, Hejnol A, Matus DQ, Pang K, Browne WE, Smith SA, Seaver E, Rouse GW, Obst M, Edgecombe GD, Sorensen MV, Haddock SH, Schmidt-Rhaesa A, Okusu A, Kristensen RM, Wheeler WC, Martindale MQ, Giribet G (2008) Broad phylogenomic sampling improves resolution of the animal tree of life. Nature 452(7188):745–749. doi:0.1038/nature06614

Fahey B, Degnan BM (2010) Origin of animal epithelia: insights from the sponge genome. Evol Dev 12(6):601–617. doi:10.1111/j.1525-142X.2010.00445.x

Fairclough SR, Chen Z, Kramer E, Zeng Q, Young S, Robertson HM, Begovic E, Richter DJ, Russ C, Westbrook MJ, Manning G, Lang BF, Haas B, Nusbaum C, King N (2013) Premetazoan genome evolution and the regulation of cell differentiation in the choanoflagellate *Salpingoeca rosetta*. Genome Biol 14(2):R15. doi:10.1186/gb-2013-14-2-r15

Genikhovich G, Technau U (2009) The starlet sea anemone *Nematostella vectensis*: an anthozoan model organism for studies in comparative genomics and functional evolutionary developmental biology. Cold Spring Harb Protoc 2009(9):pdb emo129. doi:10.1101/pdb.emo129

Grell KG (1971a) Embryonalentwicklung bei *Trichoplax adhaerens* F. E.Schulze. Naturwiss 58:570

Grell KG, Benwitz G (1971b) Die Ultrastruktur von *Trichoplax adhaerens* F.E. Schulze. Cytobiol 4:216–240

Grell KG, Ruthmann A (1991) Placozoa. In: Harrison FW, Westfall JA (eds) Placozoa, porifera, cnidaria and ctenophora, vol 2. Microscopic Anatomy of the Invertebrates. Wiley-Liss, New York, pp 13–27

Grimson A, Srivastava M, Fahey B, Woodcroft BJ, Chiang HR, King N, Degnan BM, Rokhsar DS, Bartel DP (2008) Early origins and evolution of microRNAs and piwi-interacting RNAs in animals. Nature 455(7217):1193–1197. doi:10.1038/nature07415

Hejnol A, Obst M, Stamatakis A, Ott M, Rouse GW, Edgecombe GD, Martinez P, Baguna J, Bailly X, Jondelius U, Wiens M, Muller WE, Seaver E, Wheeler WC, Martindale MQ, Giribet G, Dunn CW (2009) Assessing the root of bilaterian animals with scalable phylogenomic methods. Proc Biol Sci/Roy Soc 276(1677):4261–4270. doi:10.1098/rspb.2009.0896

Hyman LH (1940) The invertebrates: protozoa through Ctenophora, vol 1, 1st edn. McGraw-Hill, New York

Jaillon O, Aury JM, Brunet F, Petit JL, Stange-Thomann N, Mauceli E, Bouneau L, Fischer C, Ozouf-Costaz C, Bernot A, Nicaud S, Jaffe D, Fisher S, Lutfalla G, Dossat C, Segurens B, Dasilva C, Salanoubat M, Levy M, Boudet N, Castellano S, Anthouard V, Jubin C, Castelli V, Katinka M, Vacherie B, Biemont C, Skalli Z, Cattolico L, Poulain J, De Berardinis V, Cruaud C, Duprat S, Brottier P, Coutanceau JP, Gouzy J, Parra G, Lardier G, Chapple C, McKernan KJ, McEwan P, Bosak S, Kellis M, Volff JN, Guigo R, Zody MC, Mesirov J, Lindblad-Toh K, Birren B, Nusbaum C, Kahn D, Robinson-Rechavi M, Laudet V, Schachter V, Quetier F, Saurin W, Scarpelli C, Wincker P, Lander ES, Weissenbach J, Roest Crollius H (2004) Genome duplication in the teleost fish *Tetraodon nigroviridis* reveals the early vertebrate proto-karyotype. Nature 431(7011):946–957

King N, Westbrook MJ, Young SL, Kuo A, Abedin M, Chapman J, Fairclough S, Hellsten U, Isogai Y, Letunic I, Marr M, Pincus D, Putnam N, Rokas A, Wright KJ, Zuzow R, Dirks W, Good M, Goodstein D, Lemons D, Li W, Lyons JB, Morris A, Nichols S, Richter DJ, Salamov A, Sequencing JG, Bork P, Lim WA, Manning G, Miller WT, McGinnis W, Shapiro H, Tjian R, Grigoriev IV, Rokhsar D (2008) The genome of the choanoflagellate *Monosiga brevicollis* and the origin of metazoans. Nature 451 (7180):783–788. doi:10.1038/nature06617

Koonin EV (2005) Orthologs, paralogs, and evolutionary genomics. Annu Rev Genet 39:309–338. doi:10.1146/annurev.genet.39.073003.114725

Kusserow A, Pang K, Sturm C, Hrouda M, Lentfer J, Schmidt HA, Technau U, von Haeseler A, Hobmayer B, Martindale MQ, Holstein TW (2005) Unexpected complexity of the Wnt gene family in a sea anemone. Nature 433(7022):156–160

Lang BF, O'Kelly C, Nerad T, Gray MW, Burger G (2002) The closest unicellular relatives of animals. Curr Biol 12(20):1773–1778

Larroux C, Luke GN, Koopman P, Rokhsar DS, Shimeld SM, Degnan BM (2008) Genesis and expansion of metazoan transcription factor gene classes. Mol Biol Evol 25(5):980–996. doi:10.1093/molbev/msn047

Lasi M, David CN, Bottger A (2010a) Apoptosis in pre-Bilaterians: hydra as a model. Apoptosis 15(3):269–278. doi:10.1007/s10495-009-0442-7

Lasi M, Pauly B, Schmidt N, Cikala M, Stiening B, Kasbauer T, Zenner G, Popp T, Wagner A, Knapp RT, Huber AH, Grunert M, Soding J, David CN, Bottger A (2010b) The molecular cell death machinery in the simple cnidarian hydra includes an expanded caspase family and pro- and anti-apoptotic Bcl-2 proteins. Cell Res 20(7):812–825. doi:10.1038/cr.2010.66

Ledger PW (1975) Septate junctions in the calcareous sponge *Sycon ciliatum*. Tissue cell 7(1):13–18

Leys SP, Nichols SA, Adams ED (2009) Epithelia and integration in sponges. Integr Comp Biol 49(2):167–177. doi:10.1093/icb/icp038

Marlow HQ, Srivastava M, Matus DQ, Rokhsar D, Martindale MQ (2009) Anatomy and development of the nervous system of *Nematostella vectensis*, an anthozoan cnidarian. Dev Neurobiol 69(4):235–254. doi:10.1002/dneu.20698

Marshall CR, Valentine JW (2010) The importance of preadapted genomes in the origin of the animal bodyplans and the Cambrian explosion. Evolution (Int J Org Evol) 64(5):1189–1201. doi:10.1111/j.1558-5646.2009.00908.x

Moroz LL, Kocot KM, Citarella MR, Dosung S, Norekian TP, Povolotskaya IS, Grigorenko AP, Dailey C, Berezikov E, Buckley KM, Ptitsyn A, Reshetov D, Mukherjee K, Moroz TP, Bobkova Y, Yu F, Kapitonov VV, Jurka J, Bobkov YV, Swore JJ, Girardo DO, Fodor A, Gusev F, Sanford R, Bruders R, Kittler E, Mills CE, Rast JP, Derelle R, Solovyev VV, Kondrashov FA, Swalla BJ, Sweedler JV, Rogaev EI, Halanych KM, Kohn AB (2014) The ctenophore genome and the evolutionary origins of neural systems. Nature 510(7503):109–114. doi:10.1038/nature13400

Nichols SA, Dirks W, Pearse JS, King N (2006) Early evolution of animal cell signaling and adhesion genes. Proc Natl Acad Sci USA 103(33):12451–12456

Nichols SA, Roberts BW, Richter DJ, Fairclough SR, King N (2012) Origin of metazoan cadherin diversity and the antiquity of the classical cadherin/beta-catenin complex. Proc Natl Acad Sci USA 109(32):13046–13051. doi:10.1073/pnas.1120685109

Nosenko T, Schreiber F, Adamska M, Adamski M, Eitel M, Hammel J, Maldonado M, Muller WE, Nickel M, Schierwater B, Vacelet J, Wiens M, Worheide G (2013) Deep metazoan phylogeny: when different genes tell different stories. Mol Phylogenet Evol 67(1):223–233. doi:10.1016/j.ympev.2013.01.010

Philippe H, Derelle R, Lopez P, Pick K, Borchiellini C, Boury-Esnault N, Vacelet J, Renard E, Houliston E, Queinnec E, Da Silva C, Wincker P, Le Guyader H, Leys S, Jackson DJ, Schreiber F, Erpenbeck D, Morgenstern B, Worheide G, Manuel M (2009) Phylogenomics revives traditional views on deep animal relationships. Curr Biol 19(8):706–712. doi:10.1016/j.cub.2009.02.052

Philippe H, Brinkmann H, Lavrov DV, Littlewood DT, Manuel M, Worheide G, Baurain D (2011) Resolving difficult phylogenetic questions: why more sequences are not enough. PLoS Biol 9(3):e1000602. doi:10.1371/journal.pbio.1000602

Putnam NH, Srivastava M, Hellsten U, Dirks B, Chapman J, Salamov A, Terry A, Shapiro H, Lindquist E, Kapitonov VV, Jurka J, Genikhovich G, Grigoriev IV, Lucas SM, Steele RE, Finnerty JR, Technau U, Martindale MQ, Rokhsar DS (2007) Sea anemone genome reveals ancestral eumetazoan gene repertoire and genomic organization. Science 317(5834):86–94

Putnam NH, Butts T, Ferrier DE, Furlong RF, Hellsten U, Kawashima T, Robinson-Rechavi M, Shoguchi E, Terry A, Yu JK, Benito-Gutierrez EL, Dubchak I, Garcia-Fernandez J, Gibson-Brown JJ, Grigoriev IV, Horton AC, de Jong PJ, Jurka J, Kapitonov VV, Kohara Y, Kuroki Y, Lindquist E, Lucas S, Osoegawa K, Pennacchio LA, Salamov AA, Satou Y, Sauka-Spengler T, Schmutz J, Shin IT, Toyoda A, Bronner-Fraser M, Fujiyama A, Holland LZ, Holland PW, Satoh N, Rokhsar DS (2008) The amphioxus genome and the evolution of the chordate karyotype. Nature 453(7198):1064–1071. doi:10.1038/nature06967

Reddy BV, Irvine KD (2008) The Fat and Warts signaling pathways: new insights into their regulation, mechanism and conservation. Development 135(17):2827–2838. doi:10.1242/dev.020974

Renard E, Vacelet J, Gazave E, Lapebie P, Borchiellini C, Ereskovsky AV (2009) Origin of the neuro-sensory system: new and expected insights from sponges. Integr Zool 4(3):294–308. doi:10.1111/j.1749-877.2009.00167.x

Ruiz-Trillo I, Roger AJ, Burger G, Gray MW, Lang BF (2008) A phylogenomic investigation into the origin of metazoa. Mol Biol Evol 25(4):664–672. doi:10.1093/molbev/msn006

Ryan JF, Burton PM, Mazza ME, Kwong GK, Mullikin JC, Finnerty JR (2006) The cnidarian-bilaterian ancestor possessed at least 56 homeoboxes: evidence from the starlet sea anemone, *Nematostella vectensis*. Genome Biol 7(7):R64

Ryan JF, Mazza ME, Pang K, Matus DQ, Baxevanis AD, Martindale MQ, Finnerty JR (2007) Pre-bilaterian origins of the Hox cluster and the Hox code: evidence from the sea anemone, *Nematostella vectensis*. PLoS One 2(1):e153. doi:10.1371/journal.pone.0000153

Ryan JF, Pang K, Schnitzler CE, Nguyen AD, Moreland RT, Simmons DK, Koch BJ, Francis WR, Havlak P, Program NCS, Smith SA, Putnam NH, Haddock SH, Dunn CW, Wolfsberg TG, Mullikin JC, Martindale MQ, Baxevanis AD (2013) The genome of the ctenophore *Mnemiopsis leidyi* and its implications for cell type evolution. Science 342(6164):1242592. doi:10.1126/science.1242592

Sakarya O, Armstrong KA, Adamska M, Adamski M, Wang IF, Tidor B, Degnan BM, Oakley TH, Kosik KS (2007) A post-synaptic scaffold at the origin of the animal kingdom. PLoS One 2(6):e506

Schierwater B (2005) My favorite animal, *Trichoplax adhaerens*. Bioessays 27(12):1294–1302

Schierwater B, Kamm K, Srivastava M, Rokhsar D, Rosengarten RD, Dellaporta SL (2008) The early ANTP gene repertoire: insights from the placozoan genome. PLoS One 3(8):e2457. doi:10.1371/journal.pone.0002457

Schierwater B, Eitel M, Jakob W, Osigus HJ, Hadrys H, Dellaporta SL, Kolokotronis SO, Desalle R (2009) Concatenated analysis sheds light on early metazoan evolution and fuels a modern "urmetazoon" hypothesis. PLoS Biol 7(1):e20. doi:10.1371/journal.pbio.1000020

Schuchert P (1993) *Trichoplax adhaerens* (phylum Placozoa) has cells that react with antibodies against the neuropeptide RFamide. Acta Zool 74(2):115–117

Sebe-Pedros A, de Mendoza A, Lang BF, Degnan BM, Ruiz-Trillo I (2011) Unexpected repertoire of metazoan transcription factors in the unicellular holozoan *Capsaspora owczarzaki*. Mol Biol Evol 28(3):1241–1254. doi:10.1093/molbev/msq309

Sebe-Pedros A, Zheng Y, Ruiz-Trillo I, Pan D (2012) Premetazoan origin of the hippo signaling pathway. Cell Rep 1(1):13–20. doi:10.1016/j.celrep.2011.11.004

Shalchian-Tabrizi K, Minge MA, Espelund M, Orr R, Ruden T, Jakobsen KS, Cavalier-Smith T (2008) Multigene phylogeny of choanozoa and the origin of animals. PLoS One 3(5):e2098. doi:10.1371/journal.pone.0002098

Shinzato C, Shoguchi E, Kawashima T, Hamada M, Hisata K, Tanaka M, Fujie M, Fujiwara M, Koyanagi R, Ikuta T, Fujiyama A, Miller DJ, Satoh N (2011) Using the *Acropora digitifera* genome to understand coral responses to environmental change. Nature 476(7360):320–323. doi:10.1038/nature10249

Simionato E, Ledent V, Richards G, Thomas-Chollier M, Kerner P, Coornaert D, Degnan BM, Vervoort M (2007) Origin and diversification of the basic helix-loop-helix gene family in metazoans: insights from comparative genomics. BMC Evol Biol 7:33. doi:10.1186/147-2148-7-33

Sperling EA, Pisani D, Peterson KJ (2007) Poriferan paraphyly and its implications for Precambrian palaeobiology. Geol Soc, Lond. Special Publ 286:355–368

Srivastava M, Begovic E, Chapman J, Putnam NH, Hellsten U, Kawashima T, Kuo A, Mitros T, Salamov A, Carpenter ML, Signorovitch AY, Moreno MA, Kamm K, Grimwood J, Schmutz J, Shapiro H, Grigoriev IV, Buss LW, Schierwater B, Dellaporta SL, Rokhsar DS (2008) The Trichoplax genome and the nature of placozoans. Nature 454(7207):955–960. doi:10.1038/nature07191

Srivastava M, Simakov O, Chapman J, Fahey B, Gauthier ME, Mitros T, Richards GS, Conaco C, Dacre M, Hellsten U, Larroux C, Putnam NH, Stanke M, Adamska M, Darling A, Degnan SM, Oakley TH, Plachetzki DC, Zhai Y, Adamski M, Calcino A, Cummins SF, Goodstein DM, Harris C, Jackson DJ, Leys SP, Shu S, Woodcroft BJ, Vervoort M, Kosik KS, Manning G, Degnan BM, Rokhsar DS (2010) The *Amphimedon queenslandica* genome and the evolution of animal complexity. Nature 466(7307):720–726. doi:10.1038/nature09201

Suga H, Chen Z, de Mendoza A, Sebe-Pedros A, Brown MW, Kramer E, Carr M, Kerner P, Vervoort M, Sanchez-Pons N, Torruella G, Derelle R, Manning G, Lang BF, Russ C, Haas BJ, Roger AJ, Nusbaum C, Ruiz-Trillo I (2013) The Capsaspora genome reveals a complex unicellular prehistory of animals. Nat Commun 4:2325. doi:10.1038/ncomms3325

Sullivan JC, Ryan JF, Watson JA, Webb J, Mullikin JC, Rokhsar D, Finnerty JR (2006) StellaBase: the Nematostella vectensis genomics database. Nucleic Acids Res 34(Database issue):D495–D499

Yandell M, Ence D (2012) A beginner's guide to eukaryotic genome annotation. Nat Rev Genet 13(5):329–342. doi:10.1038/nrg3174

Young SL, Diolaiti D, Conacci-Sorrell M, Ruiz-Trillo I, Eisenman RN, King N (2011) Premetazoan ancestry of the Myc-Max network. Mol Biol Evol 28(10):2961–2971. doi:10.1093/molbev/msr132

The Evolution of Transcriptional Regulation in the Viridiplantae and its Correlation with Morphological Complexity

Daniel Lang and Stefan A. Rensing

Abstract Since its origin from inorganic matter, biological life undoubtedly has gained complexity. Evidence of this can be found in the lineage of the Viridiplantae or Chlorobionta ("green plants"), represented by the extant diversity of green algae and land plants. The land plants, together with the multicellular animals, arguably represent the two most complex groups of organisms on earth. For both groups, a correlation between the observable morphological complexity and the regulatory networks principally controlling it has been hypothesized. Both groups of organisms not only independently evolved multicellularity, but also underwent ancestral whole genome duplication events that presumably acted as evolutionary playgrounds for the expansion of regulatory and morphological complexity. Within animals, multicellularity evolved once and most genome duplications occurred hundreds of millions of years ago. However, an entirely different scenario unfolds among the Viridiplantae: multicellularity evolved several times independently within the green lineage, and genome duplication is the rule rather than the exception and continues to be utilized. The most successful flavor of green multicellularity evolved within the last common ancestor of extant land plants and their sister group, the charophyte algae. In this chapter, we will review common complexity concepts, introduce and compare means to quantify them, and discuss how the evolution of morphological complexity, as measured by gene regulatory complexity, distinctively affected terrestrial plants and the predominantly aquatic green, red and brown algae.

Keywords Viridiplantae · Land plants · Gene regulatory networks · Transcriptional regulation · Whole genome duplication · Morphological complexity · Multicellularity · Gene retention · Algae

S. A. Rensing (✉)
Faculty of Biology, University of Marburg, Karl-von-Frisch-Str. 8, 35043 Marburg, Germany
e-mail: stefan.rensing@biologie.uni-marburg.de

D. Lang
Plant-Biotechnology, Faculty of Biology, University of Freiburg, Schänzlestr. 1, 79104 Freiburg, Germany
e-mail: daniel.lang@biologie.uni-freiburg.de

I. Ruiz-Trillo, A. M. Nedelcu (eds.), *Evolutionary Transitions to Multicellular Life*, Advances in Marine Genomics 2, DOI 10.1007/978-94-017-9642-2_15

Introduction

Life is Complex

The concept of complexity is unarguably at the very core of how humans perceive, describe and classify themselves and their environment. Even "evolutionarily enlightened", modern biologists are prone to the perception of mankind as the crown of evolution, i.e. the most complex life form on Earth. While the anthropological ramifications of this notion clearly fall into the realms of philosophy and psychology, the concept of complexity has been a long-standing topic of the natural sciences (also see Chapter "Volvocine Algae: From Simple to Complex Multicellularity"). Most scientific disciplines have established one or several complexity concepts. In biology, especially the concepts of biological, organismic or organismal, phenotypic or morphological, or structural complexity and the existence of possible evolutionary/adaptive trends have been intensively debated and studied subjects for a long time.

The term complexity itself is highly ambiguous. Although we intuitively categorize objects as simple or complex, there appears to be no generalizable definition of complexity. In this chapter we will briefly review some complexity concepts which are useful to our understanding of biological complexity and the evolution of complex plants.

Since life originated from inorganic matter, undoubtedly there has been an increase in the complexity of organisms. Nevertheless, if we think about the age of single-celled organisms, the predominant pattern for most lineages seems to be long-term stasis of morphological complexity (Lynch 2007). Payne and collaborators (Payne et al. 2009), on the other hand, identified a two-phased increase in morphological complexity in the animal fossil record that is correlated with the increase of atmospheric oxygen. However, the question of whether there is a general adaptive trend towards complexity in evolution is a highly debated one (Szathmary et al. 2001; Carroll 2001; Lynch 2007; Bell and Mooers 1997; Dawkins and Gould 1997; Gould 1996; McShea 1996, 2005; Pennell et al. 2014), and a detailed discussion would be beyond the scope of this chapter. What is generally undisputed, however, is that some lineages, e.g. animals and plants, did increase in structural or morphological complexity during evolution.

What were the molecular driving forces behind these innovations? How can we measure complexity and the traits shaping it? In the following we will address these questions and review the current state of research that aims to answer them. While most introduced concepts can be applied to the evolution of animals as well, our major focus will be on the evolution of complexity in photosynthetic eukaryotes, i.e. plants and algae.

Morphological complexity of animals has been hypothesized to rely to a large part on the evolutionary expansion of the gene set that regulates transcription (Levine and Tjian 2003). In other words, the increase in complexity of gene regulatory networks by duplication of the underlying genes drives the evolution of animal complexity. Developmental transcription factors were gained and expanded during metazoan evolution, as evidenced by comparative genomics (e.g. (Srivastava et al. 2010; Chapter

"Transcription Factors and the Origin of Animal Multicellularity"). The morphologically complex land plants and multicellular animals share highly complex patterns of proteins involved in gene regulation (de Mendoza et al. 2013).

Genes can be duplicated by a variety of small scale mechanisms, such as tandem duplication, resulting in paralogs (i.e., gene copies derived from a duplication event). However, through mechanisms like auto- or allopolyploidization, whole genome duplications (WGD) can occur as well. In both cases, the majority of paralogs vanish quickly by accumulation of deleterious mutations over a few million years, a process termed gene death (Lynch and Conery 2000). Those paralogs that are retained either acquire a new function (neofunctionalization) or the two paralogs might share the original function, yet become divergent with regard to e.g. expression domains (subfunctionalization).

In animals, two WGD events common to the vertebrate lineage occurred more than 500 Ma (million years) ago. They have been argued to be causally linked to the establishment and success of vertebrates ((Van de Peer et al. 2009) and references therein). Strikingly, however, WGD events in vertebrates are scarce since then—a situation very different from land plants. All seed plants seem to share a WGD event that occurred around 300 Ma ago, while all angiosperms again share such an event that happened around 200 Ma ago (Jiao et al. 2011). All or most eudicotyledonous plants apparently share a genome triplication event that occurred around 120 million years ago (Jiao et al. 2012). Since then, several additional WGD events have occurred in different plant lineages and hence polyploidizations are considered a driving force of plant evolution (Soltis and Soltis 2009; Fawcett et al. 2009) (*cf.* Table 1, column "number of genome duplication events"). For an in-depth review of the impact of gene duplication on plant developmental evolution we refer the reader to more specialized reviews, e.g. (Rensing 2014).

In this chapter we elaborate on how diversification of gene regulatory networks is linked to the evolution of morphological complexity in plants, and compare them to other lineages in that regard. Given the observable patterns in the evolution of complexity, the aforementioned anthropocentric bias could lead to the misconception of equating this gain of complexity with evolutionary success.

Main Text

Complexity as a Concept in Biology

The concept of complexity is used to categorize objects and processes or more general, systems. The search for a common definition of complexity has been pursued in most scientific disciplines (reviewed e.g. in (Adami 2002)). So far no framework has come up with a generalizable definition that is both operational (i.e. defining how to measure complexity in real systems) and universal (i.e. that it can be applied to all systems; (McShea 1996)).

Table 1 Some statistics of selected organisms

species	abbreviation	taxonomic group	is unicellular ?	ploidy level [n]	number of cell types (literature)	number of cell types (predicted by PLS) using all lineages	number of cell types (predicted by PLS) using plastid bearers only	number of cell types (predicted by correlation)	genome size [Mbp]	number of genome duplication events	TAPs	TFs	TRs
Arabidopsis thaliana	arath	Spermatophyta	no	2	27	35	33	n.a.	157	5	2262	1797	355
Zea mays	zeama	Spermatophyta	no	2	100	25	106	n.a.	2671	5	2627	2142	366
Oryza sativa	orysa	Spermatophyta	no	2	44	74	28	n.a.	490	4	2381	1838	439
Physcomitrella patens	phypa	Bryophyta	no	1	20	9	6	n.a.	511	1	1606	1227	310
Volvox carteri	volca	Chlorophyta	no	1	3	2	2	n.a.	120	0	349	217	128
Chlamydomonas reinhardtii	chlre	Chlorophyta	yes	1	2	2	2	n.a.	120	0	421	221	195
Cyanidioschyzon merolae	cyame	Rhodophyta	yes	1	1	2	2	n.a.	16	0	160	98	58
Chondrus crispus	chocr	Rhodophyta	no	1	9	2	2	1	105	0	193	117	72
Ectocarpus siliculosus	ectsi	Phaeophyta	no	1	9	1	2	3	214	0	399	233	158
Homo sapiens	homsa	Metazoa	no	2	411	796	n.a.	n.d.	3000	2	2320	1841	475
Monosiga brevicollis	monbr	Choanoflagellida	yes	1	2	3	n.a.	n.d.	42	0	267	180	84

Species with their five letter code abbreviation (also used in Fig. 2) are shown in the first two columns, followed by the taxonomic group to which the species belongs, by whether it is uni- or multicellular and by the ploidy level. The first colored column shows the literature data for number of individual cell types. The next two columns show the number of cell types as predicted by a PLS analysis (former column if using all species for the PLS, latter if using only those that harbour plastids); *cf.* chapter "(How) can we find the TAP families important for multicellularity?" and chapter "Plant morphological complexity is correlated with the gene set encoding transcriptional regulation" for correlation coefficients. The next column depicts the number of cell types predicted by correlation with the TAP complement, for two multicellular algae that were not part of the analysis in (Lang et al. 2010); significant correlation of number of cell types with TAPs has been demonstrated in that publication for all organisms marked as *n.a.* not available; *n.d.* not determined. The next two columns list the genome size of the species as well as the number of detected WGD events. The tailing three columns contain the total number of TAPs, TFs and TRs for the species in question. Coloring shows low values in red, intermediate values in yellow and high values in green

McShea (1996) established four frequently cited types of complexity that are applicable to describe biological complexity via the description of the complexity of objects and processes, which can be either hierarchical or non-hierarchical.

Object complexity is measured by the number of physical parts of a system, and process complexity describes the number of interactions among them. Systems theory further adds the concept of emergence in complex systems, where the emergent features of a system go beyond the sum of traits of the individual parts. Biological objects can e.g. be a genome, where the genes are the parts encoding molecular functions acting in specific biological processes (e.g. developmental or physiological processes). Biological objects usually appear to be part of a hierarchy (e.g. organism > organ > tissue > cell > organelle > complex > protein).

Hierarchical complexity is defined by the number of levels in a causal sequence of objects or processes. In this regard, developmental processes, like the gene regulatory networks and signaling cascades in embryonic development, do harbor a specific degree of hierarchical process complexity that can be measured by the number of interacting hierarchical layers. Non-hierarchical complexity is measured by the number of parts or interactions at a given temporal or spatial scale. The number of cell types e.g. is a non-hierarchical object complexity.

Building on the above structural (object) and functional (process) complexity concepts, in his essay "What is complexity?" Christoph Adami, one of the pioneers of digital life simulation, introduces the concept of physical complexity as a form of sequence complexity that measures "the amount of information that is coded in the genomes of an adapting population, about the environment to which it is adapting" (Adami 2002). In this respect, "the total complexity of an ecosystem would have to be defined as the mutual entropy of all organisms, about each other and the world they live in". But he has to admit that "This is an information-theoretic formula that is not difficult to write down, but the associated quantity promises to be much more difficult to measure".

Which brings us to an important point: While an increasing number of complexity concepts and definitions have been put forth over the decades, evolutionary biologists had and have a hard time validating the underlying hypotheses and the utility to capture biological complexity, since we simply cannot measure at the required level of detail (e.g. taxonomic resolution). As an example, it would have been impossible to analyze the complexity of gene regulatory networks across all kingdoms of life in the 1990s. Even today it would be a tremendous task to accomplish.

Thus, research has relied on proxies, approximations, projections or in the worst case just assumptions and over-simplifications. When eventually put to the test, some of them already were shown to be not generalizable. For example, while there is a correlation of body size and morphological complexity (as measured by number of cell types) that is independent of phylogenetic dependency in animals and green plants, this trend cannot be extended convincingly to red and brown algae (Bell and Mooers 1997). Another example is genome size, which does not correlate well with structural complexity. For example, some protozoa, amphibians and many plants have much larger genomes than humans. Accompanied by the existence of "*junk DNA*", this phenomenon, also coined the *C-value enigma*, has been and still is fascinating not only to evolutionary biologists (Gregory 2005). Several other proxies have been tested but failed to hold, while the number of cell types or tissues seems to be a reliable, but excruciatingly hard to come by proxy for morphological = structural = non-hierarchical object complexity (McShea 1996). Tenaillon et al. (2007) have proposed phenotypic complexity as the number of genetically uncorrelated traits contributing to an organism's fitness as a proxy for biological complexity. They showed the applicability of this proxy in a study of viral complexity, leaving it open how feasible this abstract metric can be derived for more complex organisms.

Some of the above mentioned authors suggest genetic complexity as a possible proxy that was generally hard to come by at the time of writing the respective articles. Due to the C-value paradox, it is clear that genome size, but also total gene content, cannot be used. This leaves the question whether there is a certain class of genes that is directly or indirectly affecting structural or functional complexity of organisms in a way that it can serve as a proxy. Considering the importance at least for animal complexity evolution, the complexity of gene regulatory networks is a prime suspect to be able to function as proxy (Szathmary et al. 2001). Epigenetic, transcriptional, post-transcriptional, translational and post-translational regulators are key components of gene regulatory networks, which could be used to measure regulatory complexity. So, how can we measure the complements of gene regulators across wide taxonomic ranges and what are the most suitable approaches to compare them? We will tackle this question in the following sections.

As we are moving forward in the post-genomic area, analyzing genomes and transcriptomes even of non-model organisms, covering more and more taxonomic grounds, we are employing next generation sequencing techniques that enable the genome-wide study of entire populations, utilizing ever-growing annotation and sequence databases that rely on ontologies for automated knowledge retrieval and reasoning and an extensive repertoire of comparative and phylogenetic methods. Thus, we can be confident that our means certainly have improved enough to reassess the existing hypotheses and put forth and to test novel theories about the evolution of morphological complexity.

The Evolution of Plants

In the widest possible sense, plants are the photosynthetic eukaryotes, i.e., organisms that possess a plastid that is usually (except for parasites) able to harvest light energy. This definition also includes lineages that have acquired their plastid through symbiosis between two eukaryotes—also referred to as secondary (plastid) endosymbiosis (Reyes-Prieto et al. 2007).

The Archaeplastida, on the other hand, are those organisms that share the synapomorphic establishment of their plastid by primary endosymbiosis (Fig. 1). This event occurred only once (Price et al. 2012) and subsequently gave rise to the glaucophyte algae, the rhodophyte (red) algae, the chlorophyte and prasinophyte (green) algae, as well as to the charophyte algae and their sister group, the land plants (Embryophyta).

As a subclade of the Archaeplastida, the Viridiplantae (green plants) comprise the diverse green algae and the land plants, while the Streptophyta comprise the Charophyta and Embryophyta (Fig. 1). The Embryophyta are the land plants *sensu stricto*. They evolved from charophytes (possessing a dominant haploid gametophytic generation), by the intercalation of mitoses in the diploid zygote, thus developing a multicellular sporophytic (diploid) generation and hence a diplobiontic life cycle (Niklas and Kutschera 2009).

It should be noted that among the Chlorophyta and the Charophyta, multicellularity as well as adaptation to aerial/terrestrial habitats evolved several times independently (Fig. 1) from the lineage giving rise to the land plants (Niklas and Kutschera 2009). The Ulvophyceae (Chlorophyta) also independently evolved the diplobiontic life cycle. It has been hypothesized that the independent gain of multicellularity in the green lineage has evolved via similar stages, namely unicellular to colonial or siphonous to filamentous to pseudoparenchymatous to parenchymatous (Niklas and Newman 2013).

Here, we will first talk about Viridiplantae and will then compare them to other plants *sensu lato*.

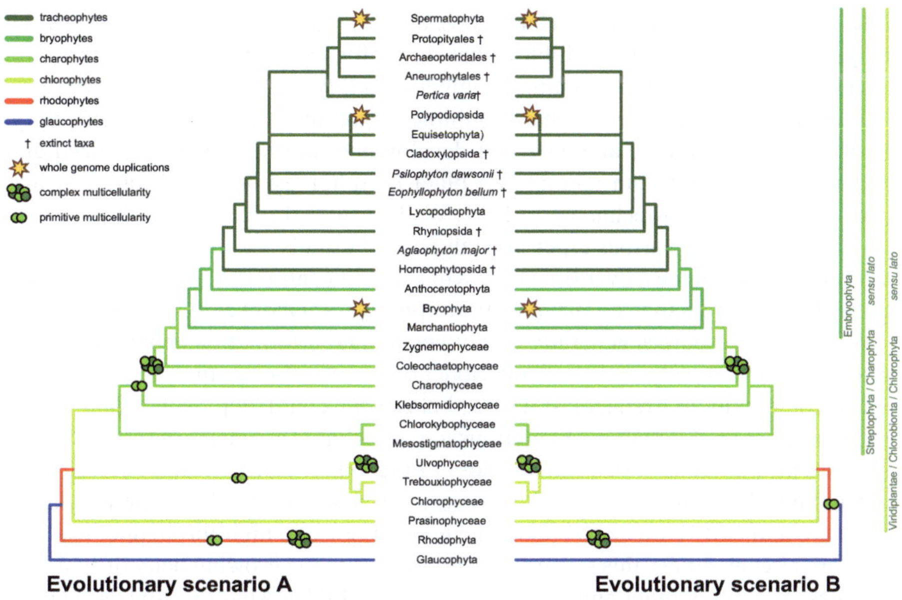

Fig. 1 Two possible evolutionary scenarios for the multiple origins of multicellularity in Archaeplastida. The Archaeplastida comprise all organisms that share plastids derived from the primary endosymbiotic event in which a cyanobacterial-like cell was established as an endosymbiont. The figure displays the phylogenetic relationships of the major lineages of the Archaeplastida. Pictograms symbolize existing evidence for (one or multiple) whole genome duplication events (*stars*) and putative origins of primitive and complex multicellularity (*cell clusters*). Phylogenetic relationships based on cladogram in (Lang et al. 2008) and literature cited within. The two evolutionary scenarios (A/B) and their implications are developed and discussed throughout the chapter

Measuring Morphological and Gene Regulatory Complexity

As discussed earlier, morphological complexity is typically defined as the number of individual cell types that make up an organism. However, such data are notoriously hard to come by and quite diverse estimates can be found in the literature. Part of the problem can be explained by the difficulty to define what a cell type is. Definitions of cell types vary depending on the resolution of the methods used to study them. Commonly, histology and other microscopy-based methods have been used, but as methodology is improving single-cell transcriptomics to derive cell-type specific expression profiles are increasing the resolution and thus also the number of cell types (Shapiro et al. 2013). It is obvious that not all species are studied at the same granularity as e.g. human, resulting in substantial variance in cell type estimates.

The other part of the problem is that the total number of cell types of an organism usually is not the focus of research. Thus, in order to come to realistic estimates of total numbers, authors have to carry out labor-intensive literature studies, which still can result in substantial variance in estimates. For example, the number of cell type estimates for human varies quite extensively, between e.g. 120 (Hedges et al.

2004) and 411 (Vickaryous and Hall 2006), and is predicted to be nearly 800 by the analysis mentioned below (*cf.* Table 1 that shows for selected species literature data for cell type numbers in the column "number of cell types").

There is an alternative view that is gaining importance because of its versatility. Ontologies of cell types, anatomy and development are increasingly utilized and expanded to describe both animal and plant morphology in high-throughput and small-scale analyses (Meehan et al. 2011; Walls et al. 2012; Cooper et al. 2013). As they are employed in various types of analyses and usually maintained and extended as community resources, over time they will serve as central knowledge repositories that will allow cross-species analysis of morphology (Dahdul et al. 2010).

While these resources generally do not yet contain information about non-model organisms, they already cover several model organisms. In our analysis of morphological complexity in Archaeplastida (see below), we could employ the Plant Ontology (Cooper et al. 2013) term annotations to gain reliable estimates of the number of cell types in *Arabidopsis*, rice and the moss *Physcomitrella patens* (Lang et al. 2010).

Although there are gold-standard lists sets of genes involved in gene regulation which are manually curated to reduce the level of noise in high-throughput analysis (Vaquerizas et al. 2009), this surely is not applicable to the growing body of available genomes. Protein domain annotations in form of Hidden Markov Models (HMM) are a reliable and highly sensitive means to identify and define protein domains and families (Wilson et al. 2009; Hunter et al. 2012; Punta et al. 2011). Thus, most studies targeting the relationship of genetic/gene regulatory and morphological complexity did utilize databases of HMM profiles for classification (Lang et al. 2010; de Mendoza et al. 2013; Vogel and Chothia 2006; Zmasek and Godzik 2011). As transcription factors and transcriptional regulators often require protein-protein interactions to act in gene regulation, and protein domains are re-used across gene family boundaries and can be quite promiscuous, the domain architecture of gene regulators can be quite complex, requiring sophisticated rule sets defining which domains are important to be part of a particular family and which are not (de Mendoza et al. 2013; Lang et al. 2010).

Due to this complexity of domain arrangements, the definition of regulatory gene sets based on HMMs has its limitations. HMMs of protein domains usually capture rather broad phylogenetic relationships usually representing superfamilies of multidomain proteins harbouring the same domain in distinct topologies, rather than gene families *sensu strictu*. The latter requires more time and computationally intensive phylogenomics approaches where gene family definition depends on phylogenetic inference on genome-scale. So far these approaches have not been utilized to study morphological complexity.

Comparative vs. Phylogenetic Comparative Approaches

After reliable estimates of morphological and gene regulatory complexity have been obtained, most authors employ a direct comparative (genomics) approach to contrast traits of extant species in order to infer evolutionary hypotheses. Most studies use some form of statistical inference, e.g. regression analysis, to model the evolutionary inter-dependency of regulatory and morphological complexity.

Although easy to apply and interpret, these approaches are conceptually flawed, because they violate an important constraint of conventional statistics: independence of observations. Direct comparison of traits from extant species assumes a star-like topology of the underlying phylogenetic tree where all species are equally (un)related (Fig. 2a). This is clearly not the case if we are comparing traits among species. Extant species descend in hierarchical fashion from common ancestors and are product of an evolutionary process that can be traced by phylogenetic inference. If we want to draw conclusions about this evolutionary process, we need to consider the ancestral nodes and the phylogenetic dependency of the species we look at. Figure 2 illustrates this concept: Fig. 2a shows selected species from our study of Archaeplastida (Lang et al. 2010) as an unrooted cladogram where all branches are connected in a basal polytomy with undefined evolutionary history. In contrast, Fig. 2c depicts the evolutionary history of the same species with internal nodes and branch lengths inferred from the phylogenetic analysis of 14 nuclear-encoded single-copy orthologs.

To overcome this problem, phylogenetic comparative (PC) methods have been developed, which use an underlying phylogenetic tree to correct measured traits for the phylogenetic dependency resulting from the evolutionary process (Felsenstein 1985; Pagel 1994; Garland and Ives 2000; Martins 2000).

Felsenstein's phylogenetically independent contrasts (PIC; Felsenstein 1985) are computationally easy to derive and represent the most commonly employed approach. In order to illustrate the influence of phylogenetic dependency on comparative analysis, we have reanalyzed our previously established data set of Archaeplastida with reliable cell type estimates using principal component analysis (PCA), using direct comparative data (Fig. 2b) and phylogenetically independent contrasts (Fig. 2d). We will discuss the evolutionary implications in more detail in the following paragraph. For now it is important to note that while Fig. 2b depicts the first two components of extant taxa, the phylogenetic comparative PCA in Fig. 2d displays the two principal components of the internal nodes, i.e. ancestral states of these taxa. Knowledge of the latter certainly is crucial to model evolutionary trends or relationships of specific traits.

Morphological Complexity Is Correlated with the Complexity of Gene Regulatory Networks in the Green Lineage

Increasing complexity of gene regulatory networks during plant evolution has been noted in many instances. For example, increasing complexity of circadian network evolution has been argued to be necessary to control circadian expression of target genes in a higher number of tissues evolving in land plants (Farre and Liu 2013). Based on an ancestral network of homodimer forming proteins, the MIKC-type MADS transcription factors have expanded by paralog retention and adaptation of the regulatory network with each round of genome duplication (Veron et al. 2007).

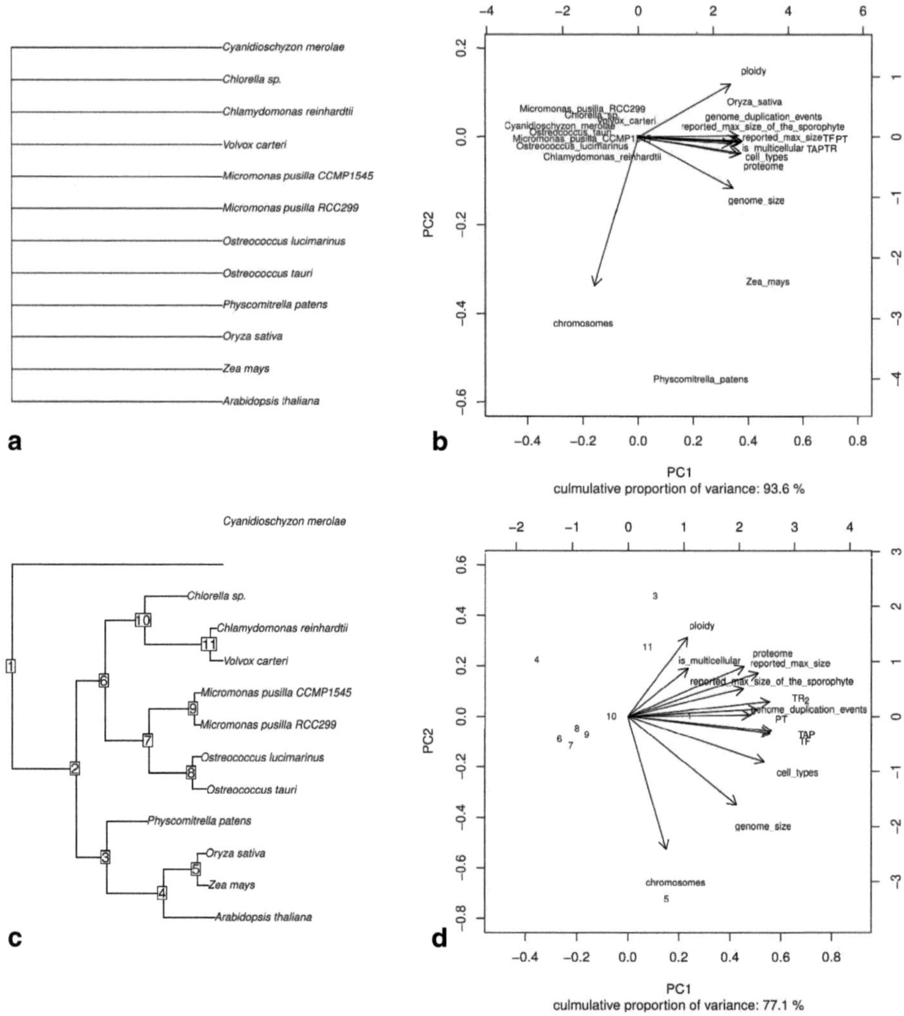

Fig. 2 Comparative vs. phylogenetic comparative (PC) principal component analysis (PCA) of regulatory and structural complexity. Comparison of comparative (a/b) and phylogenetic comparative (PC; c/d) principal component analysis (PCA) of 12 Archaeplastida species with reliable cell type estimates from Lang et al. (2010). The PCA (b, d) biplots depict the two first components inferred from a PCA of 13 life history and genomic traits (is multicellular [1|0], ploidy [n], reported max size [μm], reported max size of the sporophyte [μm], cell types [number], chromosomes [number], genome size [Mbp], genome duplication events [number], TAP [number of transcription associated proteins], TF [number of transcription factors], TR [number of transcriptional regulators], PT [number of putative TAPs], proteome [size]). In the comparative approach, extant species are assumed to be phylogenetically independent, i.e. phylogenetic relationships are as depicted in (**a**), an unrooted, multifurcating/star-like topology. Thus in the comparative PCA (**b**), the common evolutionary history of the 12 Archaeplastida species is ignored and extant taxa are treated as unrelated. Here, the phylogenetic signal of two more closely related species contributes to and biases the variance observed between them. In the PC analysis, knowledge about the phylogenetic relationships of the extant taxa is required. (**c**), depicts the phylogenetic tree of the 12 Arachaeplastida species

The class IV homeodomain leucine zipper family evidently evolved from a single ancestral algal gene, diversifying via independent duplications and mirroring the increasing complexity of plant epidermal characters (Zalewski et al. 2013).

In a general comparative genomics approach (Vogel and Chothia 2006) demonstrated that the global expansion patterns of protein domains involved in gene regulation and extracellular processes are significantly correlated with morphological complexity as measured by number of cell types in 38 eukaryotes. While the results and conclusions drawn in this study are well in line with the generally assumed connection of regulatory and morphological complexity, the methodology does not account for phylogenetic dependence and the taxonomic sampling of the green lineage (three species: *Arabidopsis*, rice and *Chlamydomonas*) is scarce. Thus, in order to generalize conclusions for Viridiplantae we need to consider more species and employ PC methodology.

By 2010, the available Archaeplastida genomes had reached 20. Given the reasonable taxonomic coverage of the Viridiplantae lineages and one red alga as an outgroup we thus set out to assess the relationship of regulatory and morphological complexity (Lang et al. 2010). Utilizing the above mentioned ontology approach, we derived estimates for numbers of individual cell types for eleven Viridiplantae plus outgroup (Table 1), collected 15 additional life history traits for all 20 species and performed phylogenetic comparative (PC) analyses using a chronographic tree (described above; Fig. 2). In addition, we carried out genome-wide classification of all genes encoding transcription associated proteins (TAPs)—comprising transcription factors (TF) that bind in sequence specific fashion to *cis*-regulatory elements, and transcriptional regulators (TR) that act by protein-protein/RNA interaction or chromatin modification.

In a first iteration, the analysis was carried out on the subset of organisms with a cell type estimate (Fig. 2). In the PCA the first component (Fig. 2b and 2d) clearly captures the morphological complexity of the included taxa. If we now ask, which of the life history and genomic traits contribute most to this component, the results from the comparative (Fig. 2b) and PC analyses (Fig. 2d) differ significantly. The first indication of this is found in the cumulative proportion of the total variance that is covered by the first two components. While the first two components basically capture most of the variance in the data (93 %) in the comparative analysis, this value is significantly reduced in the PC analysis (77 %). Thus, there are additional

Fig. 2 (continued) (Lang et al. 2010). In the employed PC method, the inferred branch lengths and internal nodes of the tree are used to infer phylogenetically independent contrasts of the values observed for the extant taxa (Felsenstein 1985). The resulting contrasts are dimensionless whereas the branch lengths and evolutionary history (number of internal nodes in the connecting path) are used to infer phylogenetically independent, weighted contrasts (PIC) for each of the internal nodes of the phylogenetic tree (numbered white boxes 1–11). Thus in the PC-PCA (**d**) we do not use the values of the extant species, but the PIC values at the internal nodes (ancestors: 1–11). Thus, we can infer possible relationships of the studied traits without the bias of the phylogenetic signal that is present in the raw data observed in the extant species. The evolutionary implications of the analyses are discussed in the main text

Table 2 Contribution of the genomic and life history traits to the respective principle components in the comparative (C) and phylogentic comparative (PC) PCAs

principal component	**1. principal component**		**2. principal component**		**3. principal component**		**4. principal component**		**5. principal component**	
cumulative percentage of variance	85.8	62.0	93.6	77.1	96.6	87.5	98.3	93.1	99.0	96.8
type of analysis	**C**	**PC**	**C**	**PC**	**C**	**PC**	**C**	**PC**	**C**	**PC**
is_multicellular	7.1	2.1	0.2	5.5	25.0	49.8	40.9	0.8	5.3	9.1
ploidy	7.0	2.0	9.7	14.9	11.4	22.7	13.0	15.7	34.2	21.6
reported_max_size	8.9	9.7	0.1	4.5	0.3	7.1	0.4	0.5	1.6	3.3
reported_max_size_of_the_sporophyte	8.3	7.6	0.0	1.8	11.0	7.1	2.8	28.1	7.7	0.1
cell_types	8.7	10.6	0.9	4.9	1.4	1.8	0.3	1.5	1.7	0.2
chromosomes	1.6	0.8	78.5	42.1	5.4	0.4	0.0	2.7	4.3	2.7
genome_size	7.4	6.7	9.3	18.6	10.1	0.3	10.6	0.4	1.7	12.6
genome_duplication_events	8.4	9.6	0.1	0.1	7.7	0.6	7.6	0.0	0.9	42.3
TAP	8.8	11.7	0.1	0.5	1.7	0.2	0.0	3.5	4.5	0.1
TF	8.8	11.5	0.1	0.7	2.0	0.1	0.0	4.6	2.9	0.6
TR	8.6	11.5	0.1	0.5	0.0	0.1	2.7	0.0	20.7	1.0
PT	8.0	8.6	0.0	0.0	20.0	9.3	13.7	17.0	0.0	6.0
proteome	8.3	7.7	1.0	5.9	4.0	0.6	7.7	25.3	14.4	0.4
	100.0	100.0	100.0	100.0	100.0	100.0	100.0	100.0	100.0	100.0

Contribution [%] of the 13 used life history and genomic traits to each of five inferred principal components for both PCA analyses presented in Fig. 2. The second row contains the cumulative percentage of the total variance contained in the principal components including the respective component. Main contributors as discussed in the text are shown in green font.

components "hidden" in the data which are covered by the phylogenetic dependency in the comparative analysis. This becomes more clear if we look at the contributions of the individual traits to the respective components (Table 2). In the comparative analysis all variables (except the number of chromosomes) contribute equally to the first component. This is substantially changed in the PC analysis, where the TAP complements (TAP, TF, TR) and the morphological traits (cell types, body size) as well as the number of genome duplication events contribute most to the first component. Only in the PC analysis we can detect a significant second and third component, reflecting genomic complexity (2nd component) and cellularity (3rd component), respectively. Thus, PC analysis allows us to dissect morphological and regulatory complexity from additional inherent components affected by genome size and cellularity. The dissection of cellularity into the 1st and 3rd components could be indicative of the genetic existence of two distinct forms of multicellularity: primitive (3rd component; few cells to filamentous tip-growth found in green algae; internal node 11) and complex (1st component; complex 3D-growth observed in land plants). Interestingly, some putative TAPs (PT) seem to be significantly correlated with primitive multicellularity and ploidy, as indicated by the 3rd component of the PC analysis. We will come back to this in later paragraphs of this chapter.

In the next step, we selected the major contributing traits of the complexity component in the reduced set which were available for all of the 20 species (TF, TR, TAP) to again infer the first component as a proxy for morphological complexity. The proxy showed perfect correlation with the number of cell types and thus could be used to study evolution of morphological complexity in Viridiplantae based on broader taxonomic sampling. Employing this complexity proxy we now can ask which traits are significantly correlated with the evolution of morphological complexity in the green lineage. The results clearly support previous and subsequent findings from comparative genomics approaches (Vogel and Chothia 2006; de Mendoza et al. 2013; Zmasek and Godzik 2011), demonstrating the importance of TFs and TRs for the evolution of complex plants.

All but 18 TAP gene families show evidence of correlated evolution with organismal complexity (10 TFs, 8 TRs). Additionally, our analysis provided novel insights, implicating the importance of miRNAs and genome duplication events for the evolution of structural complexity. Using the established phylogenetic framework we also tested for additional patterns of correlated evolution among traits and could confirm known functional relationships (e.g. between the TAP Aux/IAA and ARF and between miR390 and sporophyte size).

Hence, we now have the phylogenetic basis to hypothesize that—as in multicellular animals (de Mendoza et al. 2013)—the expansion of gene regulatory networks drove the evolution of plant morphological complexity.

Correlation with Genome Duplications

Co-occurrence of expansion of TAP families, adaptive radiation bursts of land plants and WGD events have been revealed based on gene family evolution data and have led to the hypothesis that WGD events, via retention of TAP paralogs, are a driving force of land plant adaptive radiation (Fawcett et al. 2009; Van de Peer et al. 2009; Soltis et al. 2008; Soltis and Burleigh 2009; Soltis and Soltis 2009; Blanc and Wolfe 2004; Seoighe and Gehring 2004; Jiao et al. 2011; Rensing 2014). Using PC methods, it could be confirmed that indeed WGD are significantly correlated with TAP expansion (Lang et al. 2010). Ancestral state reconstruction reveals two bursts of TAP emergence and expansion, one at the time of the water-to-land-transition, the other concurrent with the angiosperm radiation. Interestingly, the double evidence of (i) an ancestral angiosperm WGD event (Jiao et al. 2011), and (ii) of the above-mentioned TAP complexity burst, are correlated with Darwin's "abominable mystery", i.e. the quick radiation of the angiosperms via evolution of morphological novelties, and might help to explain this enigma. The retention of TAPs after the ancestral angiosperm WGD probably led to (i) subfunctionalization of expression domains enabling new cell types, (ii) to novel TAP families specific to plants and (iii) altered and novel gene regulatory networks via sub- and neofunctionalization of TAPs (see (Rensing 2014) for a recent review on this topic).

Is there an Overall Trend Suggesting Correlated Evolution of Regulatory and Structural Complexity in Eukaryotes?

In 2013 de Mendoza and colleagues carried out a comprehensive comparative genomics analysis of transcriptional regulators across all major lineages, comprising 77 species from 22 eukaryotic divisions. While being more comprehensive in terms of taxonomic coverage, their analysis largely confirmed the results of previous studies (Vogel and Chothia 2006; Lang et al. 2010), again demonstrating the importance of TFs for the observable increase of morphological complexity in animals and land plants. In addition, their data allow the generalization of an evolutionary pattern which could be observed previously just considering Viridiplantae (Lang et al. 2010): a significant proportion of the TAP families found in complex multicellular eukaryotes originated or were expanded already in their (most probably) unicellular ancestors (de Mendoza et al. 2013).

However, the study of de Mendoza et al. (2013) for the first time suggested that not all lineages that acquired multicellularity relied on the expansion of TAPs in similar manner. The studied multicellular fungi, the brown alga *Ectocarpus siliculosus* and the red alga *Chondrus crispus* have significantly smaller TAP complements than animals and land plants. de Mendoza and colleagues interpret this phenomenon as a possible consequence of the lack of embryonic development in these lineages. Do these classes represent two distinct types of multicellularity? Clearly, land plants and animals have evolved more complex morphological structures and developmental programs, including embryonic development. Does this represent an adaptation to multicellular life on land with increased abiotic and biotic stress factors?

In order to further elucidate this observation and to provide the most current, comprehensive genomic coverage, for this chapter we have reassessed the relation of regulatory and morphological complexity based on comparative genomics analysis of 83 eukaryotic genomes (Table 3). Using maximum cell type estimates from the literature and classification of TAP complements we can re-evaluate some of the above mentioned aspects on a more comprehensive scale.

Using a partial least squares (PLS) analysis, the covariance of the literature data for number of individual cell types was found to be correlated with the TAP complement for 83 eukaryotes. The global correlation is good ($R^2 = 0.87$), but if one uses the PLS data for prediction of number of cell types, both over- and under-predictions occur (Table 1, columns "number of cell types (predicted by PLS)"). Interestingly, the predictive quality of the PLS becomes much better ($R^2 = 0.97$) for land plants if one uses only those organisms for the analysis that bear plastids—here, the predicted numbers are generally closer to the literature values than for the prediction using a broader taxon set (Table 1).

The direct comparative PCA of the extant taxa (Fig. 3) reveals a distinct trend for the relationship of regulatory and structural complexity in the first component (contribution of TAPs = 88 %; cell types = 12 %) and second component (contribution of TAPs = 20 %; cell types = 80 %). We wonder whether this is another indicator for the existence of two distinct stages of multicellularity which can be distinguished based on their dependence on the expansion of gene regulatory networks.

Missing Links

The data mentioned above have to be interpreted with caution, since we currently lack the genomes of multicellular streptophytic/charophytic algae and ferns, and gymnosperms have not been incorporated. It is highly probable that some of the TAP families detected as novelties or expansions can be traced back to earlier ancestors once such data become available. Indeed, transcriptomic data of charophytes (Timme et al. 2012) suggest that at least 12 out of 30 TAP family gains previously attributed

Table 3 Regulatory and structural complexity traits for 83 sequnced eukaryotes

Species abbrevi-ation	Estimated number of cell types (mainly from lit-erature)	Transcription factors (TF)	Transcriptional regulators (TR)	Taxonomic group	Taxonomy			Cellularity	Former plas-tid	Plastid origin
Nithy	2	119	69	Alga	Charophyta	Charophyceae		Multi	Yes	Primary
Chlat	1	130	86	Alga	Charophyta	Chlorokybophyceae		Uni	Yes	Primary
Chagl	2	87	62	Alga	Charophyta	Coleochaetophyceae		Multi	Yes	Primary
Color	2	90	59	Alga	Charophyta	Coleochaetophyceae		Multi	Yes	Primary
Klefl	2	145	82	Alga	Charophyta	Klebsormidiophyceae		Multi	Yes	Primary
penma	1	165	68	Alga	Charophyta	Zygnemophyceae		Uni	Yes	Primary
Spipr	2	88	19	Alga	Charophyta	Zygnemophyceae		Multi	Yes	Primary
Bigna	1	324	203	Alga	Chlorarachniophyta			Uni	Yes	Secondary
Chlre	2	221	195	Alga	Chlorophyta	Chlamydomonadales		Uni	Yes	Primary
Volca	2	217	128	Alga	Chlorophyta	Chlamydomonadales		Multi	Yes	Primary
Chlsp	2	190	117	Alga	Chlorophyta	Chlorellales		Uni	Yes	Primary
Guith	1	519	182	Alga	Cryptophyta			Uni	Yes	Secondary
Guith_nm	1	3	3	Alga	Cryptophyta			Uni	Yes	Secondary
Cyapa	1	135	43	Alga	Glaucophyta	Cyanophoraceae		Uni	Yes	Primary
Emihu	1	532	351	Alga	Haptophyta			Uni	Yes	Secondary
Fracy	1	181	164	Alga	Heterokonts	Bacillariophyta		Uni	Yes	Secondary
Phatr	1	167	126	Alga	Heterokonts	Bacillariophyta		Uni	Yes	Secondary

Table 3 (continued)

Species abbrevi-ation	Estimated number of cell types (mainly from lit-erature)	Transcription factors (TF)	Transcriptional regulators (TR)	Taxonomic group	Taxonomy			Cellularity	Former plas-tid	Plastid origin
Thaps	1	200	125	Alga	Heterokonts	Bacillariophyta		Uni	Yes	Secondary
Ectsi	9	233	158	Alga	Heterokonts	Phaeophyta		Multi	Yes	Secondary
Auran	1	139	105	Alga	Heterokonts			Uni	Yes	Secondary
Micp1	1	156	120	Alga	Prasinophyta	Mamiellales		Uni	Yes	Primary
Micp2	1	191	130	Alga	Prasinophyta	Mamiellales		Uni	Yes	Primary
Ostlu	1	134	96	Alga	Prasinophyta	Mamiellales		Uni	Yes	Primary
Ostta	1	121	86	Alga	Prasinophyta	Mamiellales		Uni	Yes	Primary
Cyame	1	98	58	Alga	Rhodophyta	Bangiophyceae		Uni	Yes	Primary
Galsu	1	142	66	Alga	Rhodophyta	Bangiophyceae		Uni	Yes	Primary
Chocr	9	117	72	Alga	Rhodophyta	Florideophyceae		Multi	Yes	Primary
Drome	64	1039	446	Animal	Arthropoda	Insecta		Multi	No	
Nemve	20	656	271	Animal	Cnidaria	Anthozoa		Multi	No	
Hydma	15	449	218	Animal	Cnidaria	Hydrozoa		Multi	No	
Ampqu	10	286	196	Animal	Porifera	Demospongiae		Multi	No	
Caeel	25	624	278	Animal	Pseudocoelomata	Nematoda		Multi	No	
Xentr	130	1207	311	Animal	Vertebrata	Amphibia		Multi	No	
Homsa	411	1841	475	Animal	Vertebrata	Mammalia		Multi	No	

Table 3 (continued)

Species abbreviation	Estimated number of cell types (mainly from literature)	Transcription factors (TF)	Transcriptional regulators (TR)	Taxonomic group	Taxonomy			Cellularity	Former plastid	Plastid origin
Fugru	120	960	334	Animal	Vertebrata	Teleostei		Multi	No	
Botci	1	261	110	Fungus	Fungi	Ascomycota		Uni	No	
Picst	1	207	71	Fungus	Fungi	Ascomycota		Uni	No	
Sacce	3	195	73	Fungus	Fungi	Ascomycota		Uni	No	
Schpo	3	121	94	Fungus	Fungi	Ascomycota		Uni	No	
Copci	4	216	138	Fungus	Fungi	Basidiomycota		Multi	No	
Lacbi	4	254	122	Fungus	Fungi	Basidiomycota		Multi	No	
Phach	4	173	98	Fungus	Fungi	Basidiomycota		Multi	No	
Batde	1	157	84	Fungus	Fungi	Chytridiomycota		Uni	No	
Gloin	3	168	118	Fungus	Fungi	Glomeromycota		Multi	No	
Phybl	1	482	161	Fungus	Fungi	Zygomycota		Uni	No	
Rhior	1	446	163	Fungus	Fungi	Zygomycota		Uni	No	
Phypa	20	1227	310	Non-seed	Bryopsida	Funariales	Funariaceae	Multi	Yes	Primary
Selmo	23	700	212	Non-seed	Isoetopsida	Selaginellales	Selaginellaceae	Multi	Yes	Primary
Pteaq	25	474	167	Non-seed	Polypodiopsida	Polypodiales	Dennstaedtiaceae	Multi	Yes	Primary
Phyra	1	237	135	Oomycote	Heterokonts	Oomycota		Uni	No	
Physo	1	262	149	Oomycote	Heterokonts	Oomycota		Uni	No	

Table 3 (continued)

Species abbrevi-ation	Estimated number of cell types (mainly from literature)	Transcription factors (TF)	Transcriptional regulators (TR)	Taxonomic group	Taxonomy			Cellularity	Former plas-tid	Plastid origin
Plafa	1	68	38	Protist	Alveolata	Apicomplexa		Uni	Yes	Secondary
Enthi	1	109	54	Protist	Amoebozoa	Entamoebida		Uni	No	
Dicdi	4	202	127	Protist	Amoebozoa	Mycetozoa		Multi	No	
Dicpu	4	155	96	Protist	Amoebozoa	Mycetozoa		Multi	No	
Monbr	1	180	84	Protist	Choanoflagellida			Uni	No	
Naegr	1	212	185	Protist	Discicristates	Heterolobosea		Uni	No	
Trycr	1	71	20	Protist	Discicristates	Kinetoplastida		Uni	Yes	Secondary
Giala	1	46	25	Protist	Excavates	Diplomonadida		Uni	No	
Enccu	1	48	25	Protist	Microsporidia			Uni	No	
Mimgu	50	1806	370	Seed plant	Asterids	Lamiales	Phyrmaceae	Multi	Yes	Primary
Orysa	44	1838	439	Seed plant	Liliopsida	Poales	Ehrhartoideae	Multi	Yes	Primary
Setit	40	2091	512	Seed plant	Liliopsida	Poales	Panicoideae	Multi	Yes	Primary
Sorbi	40	1847	441	Seed plant	Liliopsida	Poales	Panicoideae	Multi	Yes	Primary
Zeama	100	2142	366	Seed plant	Liliopsida	Poales	Panicoideae	Multi	Yes	Primary
Bradi	40	1480	352	Seed plant	Liliopsida	Poales	Pooideae	Multi	Yes	Primary
Araly	27	1846	386	Seed plant	Rosids	Brassicales	Brassicaceae	Multi	Yes	Primary
Arath	27	1797	355	Seed plant	Rosids	Brassicales	Brassicaceae	Multi	Yes	Primary

Table 3 (continued)

Species abbreviation	Estimated number of cell types (mainly from literature)	Transcription factors (TF)	Transcriptional regulators (TR)	Taxonomic group	Taxonomy			Cellularity	Former plastid	Plastid origin
Carpa	50	1230	232	Seed plant	Rosids	Brassicales	Caricaceae	Multi	Yes	Primary
Cucsa	50	1487	264	Seed plant	Rosids	Cucurbitales	Cucurbitaceae	Multi	Yes	Primary
Glyma	50	3602	605	Seed plant	Rosids	Fabales	Fabaceae	Multi	Yes	Primary
Medtr	50	1499	308	Seed plant	Rosids	Fabales	Fabaceae	Multi	Yes	Primary
Manes	50	2366	425	Seed plant	Rosids	Malpighiales	Euphorbiaceae	Multi	Yes	Primary
Ricco	50	1350	281	Seed plant	Rosids	Malpighiales	Euphorbiaceae	Multi	Yes	Primary
Poptr	50	2479	443	Seed plant	Rosids	Malpighiales	Salicaceae	Multi	Yes	Primary
Eucgr	50	2524	486	Seed plant	Rosids	Myrtales	Myrtaceae	Multi	Yes	Primary
Frave	50	1514	312	Seed plant	Rosids	Rosales	Rosoideae	Multi	Yes	Primary
Maldo	50	3098	492	Seed plant	Rosids	Rosales	Spiraeoideae	Multi	Yes	Primary
Prupe	50	1543	333	Seed plant	Rosids	Rosales	Spiraeoideae	Multi	Yes	Primary
Citcl	50	2160	481	Seed plant	Rosids	Sapindales	Rutaceae	Multi	Yes	Primary
Citsi	50	2442	665	Seed plant	Rosids	Sapindales	Rutaceae	Multi	Yes	Primary
Vitvi	50	1344	279	Seed plant	Rosids	Vitales	Vitaceae	Multi	Yes	Primary
Aquco	25	1386	302	Seed plant	Stem eudicotyledons	Ranunculales	Ranunculaceae	Multi	Yes	Primary

Number of cell types, TFs, TRs in 83 sequenced eukaryotes (Fig. 3). TAP numbers where inferred based on HMM profile searches and the classification rule set developed earlier (Lang et al 2010). Cell type estimates represent maximal values obtained from the literature or by educated guessing. For brevity, species are provided with their five-letter codes (Genus species). In addition to the traits used in Fig. 3, the table also contains three columns describing the precise taxonomic classification of the respective lineage as well as information about plastid evolution

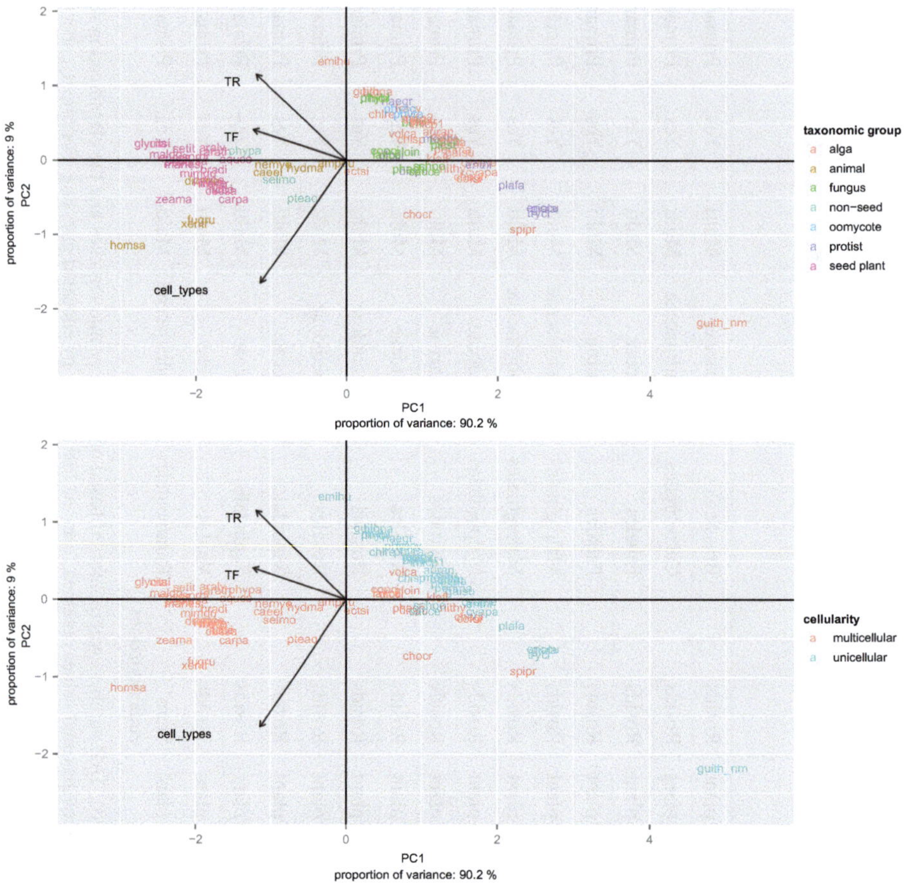

Fig. 3 Comparative PCA of 83 eukaryotes**.** Comparative PCA of the number of cell types, TFs and TRs in 83 sequenced eukaryotes (Table 3). TAP numbers were inferred based on HMM profile searches and the classification rule set developed earlier (Lang et al 2010). Cell type estimates represent maximal values obtained from the literature or by educated guessing. The PCA biplots depict the two first components of the same PCA with different color-coding of species. In the upper panel, individual species are color-coded according to the respective taxonomic group. In the lower panel color-coding is according to cellularity

to land plants (Lang et al. 2010) already occurred in charophytes. Moreover, it has recently been shown that components of phytohormone action already arose in charophytes, namely auxin (Viaene et al. 2013) and strigolactone (Delaux et al. 2012). It is therefore expected that genome availability of charophyte algae will greatly inform our understanding of land plant evolution. It will also resolve the question whether those charophytes that are sister to land plants feature WGD events, and which TAP families already evolved and were expanded in the multicellular algal relatives of land plants.

Multicellularity in Non-Streptophyte Algae

PC analyses demonstrated a significant correlation of the size of the TAP complement with morphological complexity for the Viridiplantae, and most individual TAP families follow this trend (Lang et al. 2010). The sequencing of several genomes of multicellular algae provides us with the opportunity to analyze whether these genomes follow the same trend evident for land plants, and whether particular TAP families can be detected that drive morphological complexity.

Green Algae: Volvox *vs.* Chlamydomonas

As mentioned above, multicellularity probably has evolved more than once within the green lineage (Fig. 1). Besides complex multicellular land plants and streptophyte algae, several chlorophyte algal lineages harbor primitive multicellular forms involving 2D thalli and filamentous growth. One particularly well-suited example for comparative genomics can be found within the Volvocales: *Chlamydomonas reinhardtii*, the unicellular algal model organism, can be compared to its relative *Volvox carteri* that features different cell types with specialized functions (see Chapter "Volvocine Algae: From Simple to Complex Multicellularity"). However, analysis of the TAP complement of the two genomes (Prochnik et al. 2010) reveals no striking differences between the two genomes, and the global TAP complement is even larger in the unicellular alga than in its multicellular relative (Table 1). However, genes encoding components of the extracellular matrix and for regulation of cell division are expanded in *V. carteri*, most probably enabling its particular flavor of multicellularity. It should be noted that evidence for a WGD can be found in neither genome.

Evolutionary conclusions regarding the evolution of multicellularity based on the comparative analysis of the two Chlorophyceae need to be considered with caution. There is a growing body of evidence ((Zimmer et al. 2013) and references therein) suggesting that the comparatively small chlorophyte genomes in fact might by secondarily reduced. Exemplary strong indications for this are the secondarily acquired introns of volvocine algae as well as the strange evolutionary pattern of the pathways involved in miRNA biogenesis. In our previous analysis of Viridiplantae (Lang et al. 2010) we found further indicators in the reconstructed ancestral states for genome and proteome sizes as well as number of cell types. Figure 4 depicts the reconstructed ancestral states for genome size across the Archaeplastida. While the absence of streptophyte algae clearly does account for some error term in the reconstructed ancestral genome sizes (especially the LCA of chlorophyte algae and Streptophyta *sensu lato*), there is a clear indication for secondary reduction and subsequent expansion of genome size in the lineages leading to the two volvocine algae. A possible evolutionary bottleneck along the lineage of chlorophyte algae (e.g. by the competition from their streptophyte cousins) might have resulted in a secondary reduction of genome size and possible loss of genes and families. This was also implicated for animal evolution by the Dollo parsimony analysis of domain

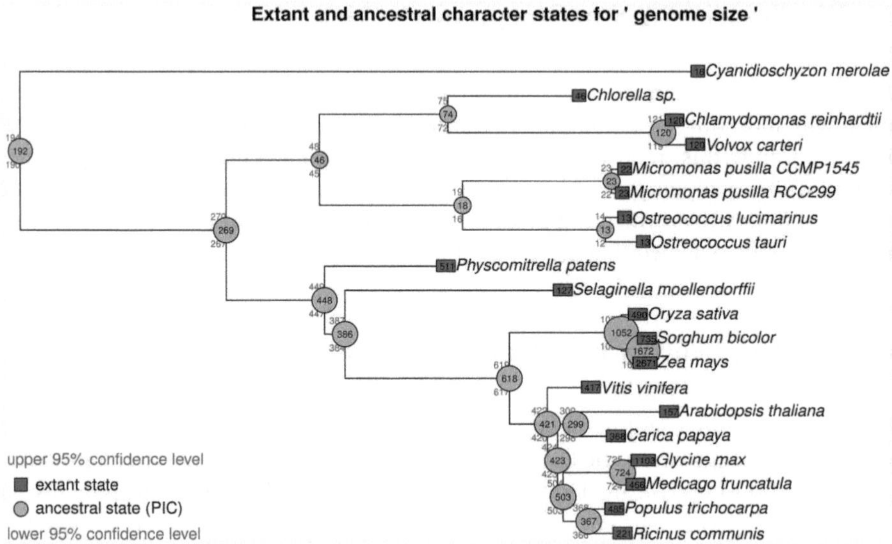

Fig. 4 Ancestral genome sizes. Phylogenetic tree and putative ancestral and extant genome sizes of all Archaeplastida used in Lang et al. (2010). Ancestral states were derived using the PIC method (Felsenstein 1985). Estimates (*light grey circles*) are depicted with 95 % confidence intervals (*light grey* text above and below the circles). The estimates could be confirmed by two additional, independent methods (Garland and Ives 2000; Martins 2000; Lang et al. 2010)

evolution carried out by Zmasek and Godzik (2011). Subsequently, volvocine (and maybe other green algae) might have expanded their genomes again by secondary gain of introns due to transposon activity.

Red Algae

Within the red algae, multicellularity possibly evolved twice (Niklas and Newman 2013). The first genome of a multicellular red alga to be sequenced was that of *Chondrus crispus* (Collen et al. 2013). The TAP complement of *Chondrus* is surprisingly small (193 genes), leading to under-estimation of its number of cell types by PLS (Table 1). In comparison with unicellular red algae, several TAP families are even absent from its genome, and only two are expanded. One TF family significantly contributing to the covariance (with number of cell types) of the PLS analysis is the C2H2 zinc finger family, that is expanded to 34 members (as compared to e.g. eight members in the unicellular red alga *Cyanidioschyzon merolae* and five in *Galdieria sulphuraria*). Possibly, expansion of this TF family is one of the forces behind multicellularity in this lineage.

Intriguing are also Dicer and Argonaute, involved in small RNA processing. Argonaute seems to be generally absent from unicellular red algae and prasinophytes, Dicer is not present in many red/green/heterokont algae or other protists (with the notable exception of the Chlamydomonadales; (Molnar et al. 2007; Casas-Mollano et al. 2008)). *Chondrus* harbours both (and three copies of Argonaute), suggesting a more complex regulation of/by miRNAs than in many unicellular organisms, comparable to multicellular plants and animals. Indeed, ancestral duplications of Dicer and Argonaute in the plant and animal lineages have been noted to coincide with the rise of multicellularity (Mukherjee et al. 2013). In this regard it is interesting to note that the complexity of miRNA families, often targeting TAPs, is significantly correlated with morphological complexity in the Plantae (Lang et al. 2010). Again, there is no WGD evident in this algal genome.

Brown Algae

The first genome of a multicellular brown alga to become available was that of *Ectocarpus siliculosus* (Cock et al. 2010), belonging to the Ectocarpales that are a sister lineage to the Laminariales, harbouring e.g. the giant kelps that form underwater "forests". Brown algae belong to the stramenopiles or heterokonts, a diverse group of organisms also comprising e.g. diatoms or oomycetes. Unlike the Archaeplastida, the plastid-bearing organisms in this group have acquired their plastid by secondary endosymbiosis. The *Ectocarpus* genome harbours about twice as many TAP genes (399) as *Chondrus*, but PLS would still predict a much lower number of cell types than expected (Table 1). Several TAP families are present in *Ectocarpus* and oomycete genomes but absent from unicellular diatoms. While principal component analysis (PCA) on the number of TAPs per family is well able to resolve broad taxonomic groups, it cannot distinguish between uni- and multicellular algae ((Cock et al. 2010) and Fig. 3). While the amount of TAPs encoded by *Ectocarpus* is not striking, the expanded kinase network found in this organism has been argued to be involved in the more complex regulation required by multicellularity (Cock et al. 2010). As in *Chondrus* and *Volvox*, there is no evidence for a WGD event in the brown algal genome. One possible uniting feature of land plants and multicellular animals is their peculiar embryo development that might require an appropriately complex TAP equipment—the less complex modular growth of *Ectocarpus* coincides with its less complex TAP complement (de Mendoza et al. 2013). It shall be interesting to compare the TAP complement of more complex brown algae to those of other morphologically complex organisms. In that regard, the sequencing of e.g. the *Macrocystis* (kelp) genome is eagerly expected.

No Clearcut Correlation of TAPs with Complexity in Non-Green Algae

Non-linear regression between the number of literature-derived cell types and the number of TAPs (data for the 12 Archaeplastida from (Lang et al. 2010)) yield a $R^2 = 0.98$. However, if this regression is used to predict the number of cell types based on the number of TAPs for *Chondrus* and *Ectocarpus*, the values are one and three, thus severely under-predicting the actual number of cell types, nine (Table 1). With regard to red algae, it has been argued that they have passed through an evolutionary bottleneck that resulted in a dramatic reduction of genome size (Collen et al. 2013), rendering their genome quite different from those in the green lineage. This might explain the extremely small TAP complement of *Chondrus*.

Comparable to unicellular green algae ((Lang et al. 2010), Table 3; 213–421 TAPs per genome), the TAP complement of unicellular red (e.g. *Galdieria sulphuraria*; (Schonknecht et al. 2013); 212 TAPs), and glaucophyte algae (*Cyanophora paradoxa*; (Price et al. 2012); 178 TAPs) is in the low range (*cf.* Table 1 and 3; *Cyanidioschyzon merolae* and *Chlamydomonas reinhardtii* represent the lower and higher boundary of the TAP range of archaeplastidal algae, respectively). In the heterokonts, unicellular diatoms and oomycetes are within that same range. However, the nucleomorph-bearing cryptophyte *Guillardia theta* and chlorarachniophyte *Bigelowiella natans* exceed that range with 709 and 534 TAPs, respectively (Curtis et al. 2012), and the haptophyte *Emiliana huxleyi* harbours 893 TAPs (Read et al. 2013)—more than double the amount found in *Ectocarpus* and *Volvox* and four times as many as in *Chondrus* (Table 3).

Clearly, the correlation observed in the green lineage does not hold true outside of that taxonomic group—there are unicellular algae with comparatively large TAP complements and multicellular algae with small ones. In other words, the total size of the TAP complement apparently is a proxy for cell type number/morphological complexity only in the Viridiplantae and Metazoa.

Another Abominable Mystery

If the quick radiation of the angiosperms presented Darwin with a puzzle, the pattern of eukaryotic complexity certainly does the same to us now. Figure 5 displays on the eukaryotic tree of life the presence of plastids (regardless of primary or secondary acquisition), the occurrence of WGD events, and the evolution of multicellularity. It becomes clear that most major groups have acquired and usually kept plastids at some point during their evolutionary history. As outlined above, the presence of a plastid obviously does not lead to unifying principles with regard to the evolution of complexity. Multicellularity apparently evolved several times, possibly three times alone in the green lineage (Scenario A in Fig. 1). However, only two of these events led to the evolution of highly complex plants and animals.

The above mentioned low correlation of cell type estimates and TAP complement sizes in fungi, red and brown algae, mirrored as well in the low correlation of body size and structural complexity for the respective lineages in the PC analysis carried out by (Bell and Mooers 1997), is puzzling. Do these patterns reflect the existence of two distinct phases in the evolution of multicellularity in photosynthetic eukaryotes, mirroring the observed two phases of gain of complexity evident in the animal fossil record (Payne et al. 2009)? This argument is supported by the principle components from both the PC PCA analysis of Archaeplastida (Fig. 2; (Lang et al. 2010)) and the

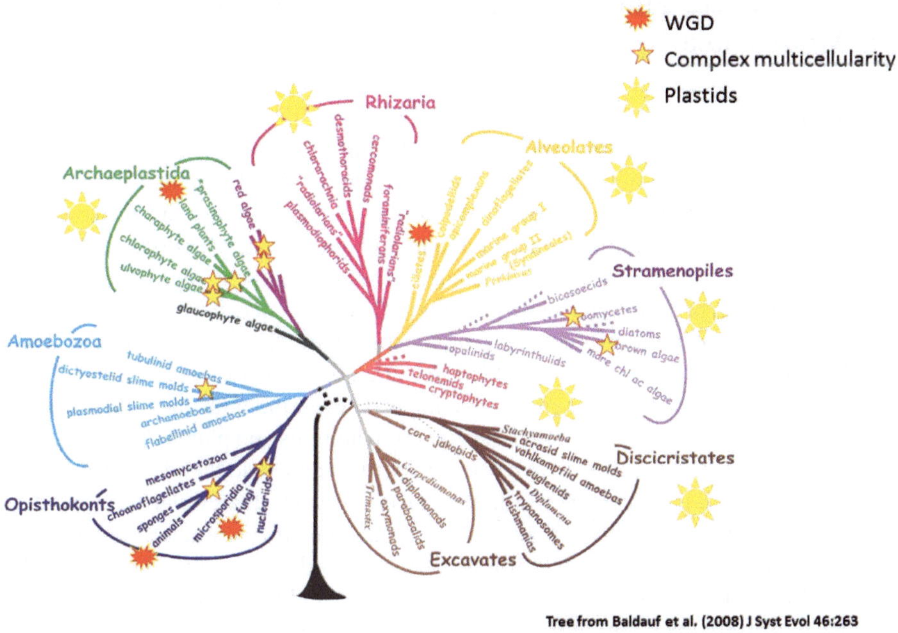

Fig. 5 Another abominable mystery. The eukaryotic tree of life (from (Baldauf 2008)), overlaid with pictograms symbolizing taxonomic groups that harbour plastids (sun), occurrence of ancient whole genome duplication events (explosion) and independent evolution of complex multicellularity (*star*). The mystery to be explained is the apparently uncorrelated pattern of these traits. It should be noted that for many lineages we do not know yet whether WGD occurred, due to lack of data. Also, more primitive multicellularity (attachment of cells) has evolved in several additional lineages (not depicted here)

comparative PCA of 83 eukaryotes (Fig. 3), which both indicated the existence of two cellularity components. These two phases of multicellularity resulting in primitive and more complex multicellular photosynthetic eukaryotes are also included in the two evolutionary scenarios presented in Fig. 1. While scenario A suggest the commonly assumed convergent evolution of multicellularity in Archaeplastida, B suggests an alternative scenario where primitive multicellularity evolved once in the green lineage, leading to the subsequent evolution of more complex forms of multicellularity based on the ancestral toolkits already present in the LCA of the red and green lineage.

Given the similar patterns in the evolution of structural complexity in animals and plants it is tempting to speculate that plant and animal structural complexity co-evolved and might be linked via the levels of atmospheric oxygen and carbon dioxide.

The "use" of WGD as an evolutionary mechanism to enhance genetic and thus morphological complexity is known from only a few groups—the most complex organisms among them. If we consider the expansion of gene regulatory networks a driving force of the rise of cell types and thus morphological complexity, how can we determine which regulators are important?

(How) Can We Find the TAP Families Important for Multicellularity?

In both land plants and multicellular animals, expansion of the TAP complement apparently drives increasing complexity via enabling novel cell types (that might differ by their expression profile, resp. competence for developmental progression, only—and not by their morphology). While the TAP equipment did differentially arise (lineage-specific families) and expand in both groups (de Mendoza et al. 2013), some underlying principles in both lineages seem apparent, e.g. signalling by basic helix-loop-helix (bHLH) heterodimers to control cell fate decisions via downstream target genes has been described in both organism groups (Iso et al. 2003; Macalister and Bergmann 2011; Feller et al. 2011; Pires et al. 2013; Busch et al. 2013). These TFs are also known to combine environmental and developmental cues and to control gene regulation via cascading of bHLH proteins in animals and plants (Yi et al. 2010). Also, TAP expansion occurred in two waves in both lineages (de Mendoza et al. 2013), once in the unicellular ancestors and again with the rise of multicellularity, coinciding with embryo development. Such data suggest that some differentiation processes in complex animals and plants might follow particular underlying signalling principles enacted by (partly) the same TF families. In order to understand these principles we will have to study gene regulatory network evolution in both lineages.

For a start, can we say something about the families putatively involved in complexity regulation? Let's perform three analyses in that regard and see what they tell us: PCA (explaining variance of TAP families), family size bias, and PLS (explaining covariance with literature-derived cell type data). Underlying data for these analyses are shown in Table 1 and 3, the results in Table 4.

I. PCA. If we not only take the total amount of TAPs into account, but perform a PCA on all TAP family sizes encoded by a genome, we can separate organisms into their taxonomic groups (Fig. 3). However, as mentioned above for the example *Ectocarpus*, multicellular organisms group within their taxon rather than forming their own cluster (Fig. 3). Yet, the first component explains 90.2 % of the variance and seems quite capable to separate the green organisms by complexity along its trajectory (algae right, non-seed plants middle, flowering plants left; Fig. 3). If we take a look at the top 20 TAP families (Table 4, PCA) contributing to this trajectory (component 1), we find that all these families were described (Lang et al. 2010) to either originate in land plants (eight) or to have been expanded in land plants (12; Table 4). Also, nine families were secondarily expanded in angiosperms (Table 4). There is a clear bias towards TFs (14 families) in this set.

II. Size bias. As an alternative, we can compare TAP family size between multicellular and unicellular organisms. This analysis identifies 29 families to be consistently larger in multicellular organisms, 17 of which encode TFs. When land plants are compared with algae, 27 families are found to be consistently larger in the former (18 of which encode TFs). When we compare this latter set of families with the families that are larger in multicellular organisms, and remove the 22 families that occur in both sets (to eliminate the influence of the transition to the terrestrial environment in the plant lineage), only seven families remain (Table 4, size bias).

Table 4 Some candidate TAP families potentially involved in the complex multicellularity of land plants

TAP family	TAP type	PCA	PLS	size bias	origin in land plants	expansion in land plants	expansion in angiosperms	monomer	homodimer	heterodimer	multimer
ABI3/VP1	TF		x			x			x	x	
AN1/A20 type zinc finger	TF			x		x			x	x	
AP2/EREBP	TF		x			x			x		
ARF	TF	x			x				x	x	
AS2/LOB	TF	x	x		x				x	x	
Aux/IAA	TR	x			x		x		x	x	
bHLH	TF	x	x			x			x	x	
bZIP	TF	x	x			x			x	x	
C2C2_CO-like	TF	x				x				x	
C2C2_Dof	TF	x				x			x	x	
C2H2	TF		x			x			x		
C3H	TF		x			x					
CCAAT_HAP3	TF	x				x	x				x
DUF246 domain containing	PT	x	x		x						
DUF296 domain containing	PT	x				x					
DUF547 domain containing	PT	x				x					
DUF632 domain containing	PT	x			x		x				
GARP_G2-like	TF	x	x			x			x	x	
GNAT	TF			x				x	x	x	x
GRAS	TF		x		x				x		
GRF	TF	x			x					x	
HB	TF	x	x			x	x			x	
HD-Zip	TF	x			x		x		x		
LIM	TR			x		x	x	x	x		x
MADS	TF		x	x		x	x		x	x	
mTERF	TF		x			x	x	x			
MYB	TF	x	x			x	x	x	x	x	
MYB-related	TF		x			x	x	x	x		
NAC	TF	x	x		x		x		x	x	
PHD	TR		x			x		x	x	x	
SET	TR		x			x			x		
SWI/SNF_BAF60b	TR	x				x	x	x			
SWI/SNF_SNF2	TF			x					x		
SWI/SNF_SWI3	TF			x		x			x	x	
TRAF	TR		x	x		x			x	x	
WRKY	TF	x	x			x	x		x	x	
	performed on:	algae and plants	broad set of eukaryotes	algae and land plants							
	separates:	green lineage by complexity	by number of cell types	uni- from multicellular							

The first column lists the name of the TAP family, the second whether it is a transcription factor (TF), transcriptional regulator (TR) or putative TAP (PT) family; *sensu* (Lang et al. **?**) and using the same colour code

The column PCA marks the top 20 TAP families with an X that contribute to the first component (Eigenrow 1) of a PCA performed on algae and land plants, which is able to separate the green lineage by complexity similar to Fig. 2. The column PLS shows the top 20 TAP families that contribute most to the covariance of literature cell type data with numbers of TAP per family, it was performed on a broad set of organisms (*cf.* chapter "(How) can we find the TAP families important for multicellularity?"). The next column, size bias, marks families that are consistently larger in multicellular than in unicellular algae/land plants (*cf.* chapter "(How) can we find the TAP families important for multicellularity?") for details. Overlap between two of the three columns is shown by grey shading

The next three columns list whether the family in question originated in land plants, was expanded in land plants or expanded in angiosperms, *sensu* (Lang et al. 2010). Overlap between two of the three columns is shown by mauve shading

The tailing four columns summarize literature data on the TAP families with regard to their quarternary structure. Namely, column monomer marks those families that are known to act as monomers, while the next columns mark those that act as homo- or heterodimer, respectively higher order multimers. Overlap between these four columns is shown by light red shading

These are good candidates for TAP genes that play a role in the transition to multicellularity. They include two TR families (TRAF and LIM) and five TF families (GNAT, SWI/SNF_SNF2, SWI/SNF_SWI3, MADS and AN1/A20 type zinc finger (Table 4).

III. PLS. When carrying out a PLS analysis with the cell type literature data (as mentioned above; *cf.* Table 1/3) using a broad set of eukaryotes (including not only plants and algae but also fungi and animals) the first component explains 85.2 % of the covariance. The top 20 families (Table 4, PLS) comprise two of the five families found in the gene family size analysis, namely the TF MADS and the TR TRAF (Table 4). The twenty families determined by PLS also overlap quite substantially with the top 20 families detected by PCA (eight TF families, bHLH, GARP G2-like, AS2/LOB, NAC, WRKY, HB, MYB, bZIP and the putative TAP (PT) family that is characterized by the DUF246 domain; Table 4).

In total, these three analyses identify 37 out of 111 TAP families (Table 4), of which 26 (70 %) are TFs. All but two of these either originate in the plant lineage or were expanded during plant evolution, in line with the more pronounced expansion of TFs over TRs previously observed for plants (Lang et al. 2010). If we omit from the complete set of 37 those nine families that are specific for plants (Table 4, origin in plants), 28 putative candidate gene families remain that might be underlying multicellularity in a broader range of organisms. Among those are some of the candidates pointed out above, e.g. bHLH, C2H2, MADS or TRAF.

As pointed out above for the example of bHLH TFs, TAPs often act as multimers. Literature search was used to annotate TAP families with multimerization data (Table 4) and revealed that TF families as compared to TR families are enriched for multimers, while TRs typically act as monomers (Fisher's exaxt test, fdr corrected p-value 0.05). Interestingly, the top 20 TAP families from the PLS analysis are enriched for families acting as homodimers (fdr corrected p-value 0.03). Out of the 28 candidate families mentioned in the last paragraph, 13 are able to form homo- and heterodimers (four act only as homodimer, two only as heterodimer), while only two act as monomers and one as multimer (Table 4). Among those known to act as homodimers are again e.g. bHLH, C2H2, MADS and TRAF. Homodimers are potentially more stable with regard to gene duplications since their homodimeric interaction is not affected by dosage sensitivity of one of the two partner proteins. If, on the other hand, such TAPs are also able to act as heterodimers (which seems true for nearly half of the 28 candidate families) their combinatorial complexity increases and thus makes them candidates for alteration of gene regulatory networks by duplication/retention and subsequent sub- or neofunctionalization of genes encoding the potential dimerization partners—this is in line with data e.g. on the evolution of the plant MADS family (Veron et al. 2007).

Future research will need to reveal which of the independent but analogous evolutionary expansions of TAP families and the gene regulatory networks enacted by them follow common principles, and, more importantly, what these principles are.

Where are We?

We know today that in the land plants there is a strong correlation of morphological complexity (typically measured as number of individual cell types) with the molecular complexity of the TAPs encoded by the respective genome. Interestingly, plants share this correlation with morphologically complex Metazoa.

There is good evidence that WGD are a driving force of plant TAP and complexity evolution. Making use of genome duplications to increase the number of available paralogs for sub- and neofunctionalization seems to be a hallmark of land plant evolution.

Future research must aim to reveal mechanistic detail of plant genome evolution and how it enables the increase of morphological complexity.

Summary

1. Morphological complexity of land plants is correlated with the expansion of gene families involved in transcriptional regulation (TAPs). The TAP complement of a given genome is a proxy for the organismal complexity of the plant encoded by it.
2. The same is true for multicellular animals, but not for multicellular algae that arose independent from the streptophyte lineage.
3. Similar to animals, photosynthetic eukaryotes may have evolved in two phases coinciding with the level of atmospheric gases and the pattern (expansion) of TAP evolution.
4. The occurrence of ancient WGD events coincides with morphologically complex plants and animals (vertebrates), and with their gene regulatory network complexity.
5. In land plants, many more WGD events can be observed than in vertebrates. To date, no WGD event could be observed in multicellular algae.
6. Common principles of regulatory networks controlling complexity seem evident in plants and animals and are a promising line of future research.

Acknowledgements The authors would like to thank Sebastian Hanke, Erika Lang, Nico van Gessel and Andreas Zimmer for comments on the manuscript

References

Adami C (2002) What is complexity? Bioessays 24(12):1085–1094. doi:10.1002/bies.10192

Baldauf SL (2008) An overview of the phylogeny and diversity of eukaryotes. J Syst Evol 46:263

Bell G, Mooers AO (1997) Size and complexity among multicellular organisms. Biol J Linn Soc 60(3):345–363. doi:10.1111/j.1095-8312.1997.tb01500.x

Blanc G, Wolfe KH (2004) Functional divergence of duplicated genes formed by polyploidy during *Arabidopsis* evolution. Plant Cell 16:1679–1691

Busch H, Boerries M, Bao J, Hanke ST, Hiss M, Tiko T, Rensing SA (2013) Network theory inspired analysis of time-resolved expression data reveals key players guiding P. patens stem cell development. PLoS ONE 8:e60494

Carroll SB (2001) Chance and necessity: the evolution of morphological complexity and diversity. Nature 409(6823):1102–1109. doi:10.1038/35059227

Casas-Mollano JA, Rohr J, Kim EJ, Balassa E, van Dijk K, Cerutti H (2008) Diversification of the core RNA interference machinery in Chlamydomonas reinhardtii and the role of DCL1 in transposon silencing. Genetics 179:69–81

Cock JM, Sterck L, Rouze P, Scornet D, Allen AE, Amoutzias G, Anthouard V, Artiguenave F, Aury JM, Badger JH, Beszteri B, Billiau K, Bonnet E, Bothwell JH, Bowler C, Boyen C, Brownlee C, Carrano CJ, Charrier B, Cho GY, Coelho SM, Collen J, Corre E, Da Silva C, Delage L, Delaroque N, Dittami SM, Doulbeau S, Elias M, Farnham G, Gachon CM, Gschloessl B, Heesch S, Jabbari K, Jubin C, Kawai H, Kimura K, Kloareg B, Kupper FC, Lang D, Le Bail A, Leblanc C, Lerouge P, Lohr M, Lopez PJ, Martens C, Maumus F, Michel G, Miranda-Saavedra D, Morales J, Moreau H, Motomura T, Nagasato C, Napoli CA, Nelson DR, Nyvall-Collen P, Peters AF, Pommier C, Potin P, Poulain J, Quesneville H, Read B, Rensing SA, Ritter A, Rousvoal S, Samanta M, Samson G, Schroeder DC, Segurens B, Strittmatter M, Tonon T, Tregear JW, Valentin K, von Dassow P, Yamagishi T, Van de Peer Y, Wincker P (2010) The Ectocarpus genome and the independent evolution of multicellularity in brown algae. Nature 465:617–621

Collen J, Porcel B, Carre W, Ball SG, Chaparro C, Tonon T, Barbeyron T, Michel G, Noel B, Valentin K, Elias M, Artiguenave F, Arun A, Aury JM, Barbosa-Neto JF, Bothwell JH, Bouget FY, Brillet L, Cabello-Hurtado F, Capella-Gutierrez S, Charrier B, Cladiere L, Cock JM, Coelho SM, Colleoni C, Czjzek M, Da Silva C, Delage L, Denoeud F, Deschamps P, Dittami SM, Gabaldon T, Gachon CM, Groisillier A, Herve C, Jabbari K, Katinka M, Kloareg B, Kowalczyk N, Labadie K, Leblanc C, Lopez PJ, McLachlan DH, Meslet-Cladiere L, Moustafa A, Nehr Z, Nyvall Collen P, Panaud O, Partensky F, Poulain J, Rensing SA, Rousvoal S, Samson G, Symeonidi A, Weissenbach J, Zambounis A, Wincker P, Boyen C (2013) Genome structure and metabolic features in the red seaweed Chondrus crispus shed light on evolution of the Archaeplastida. Proc Natl Acad Sci U S A 110:5247–5252

Cooper L, Walls RL, Elser J, Gandolfo MA, Stevenson DW, Smith B, Preece J, Athreya B, Mungall CJ, Rensing S, Hiss M, Lang D, Reski R, Berardini TZ, Li D, Huala E, Schaeffer M, Menda N, Arnaud E, Shrestha R, Yamazaki Y, Jaiswal P (2013) The plant ontology as a tool for comparative plant anatomy and genomic analyses. Plant Cell Physiol 54(2):e1. doi:10.1093/pcp/pcs163

Curtis BA, Tanifuji G, Burki F, Gruber A, Irimia M, Maruyama S, Arias MC, Ball SG, Gile GH, Hirakawa Y, Hopkins JF, Kuo A, Rensing SA, Schmutz J, Symeonidi A, Elias M, Eveleigh RJ, Herman EK, Klute MJ, Nakayama T, Obornik M, Reyes-Prieto A, Armbrust EV, Aves SJ, Beiko RG, Coutinho P, Dacks JB, Durnford DG, Fast NM, Green BR, Grisdale CJ, Hempel F, Henrissat B, Hoppner MP, Ishida K, Kim E, Koreny L, Kroth PG, Liu Y, Malik SB, Maier UG, McRose D, Mock T, Neilson JA, Onodera NT, Poole AM, Pritham EJ, Richards TA, Rocap G, Roy SW, Sarai C, Schaack S, Shirato S, Slamovits CH, Spencer DF, Suzuki S, Worden AZ, Zauner S, Barry K, Bell C, Bharti AK, Crow JA, Grimwood J, Kramer R, Lindquist E, Lucas S, Salamov A, McFadden GI, Lane CE, Keeling PJ, Gray MW, Grigoriev IV, Archibald JM (2012) Algal genomes reveal evolutionary mosaicism and the fate of nucleomorphs. Nature 492:59–65

Dahdul WM, Balhoff JP, Engeman J, Grande T, Hilton EJ, Kothari C, Lapp H, Lundberg JG, Midford PE, Vision TJ, Westerfield M, Mabee PM (2010) Evolutionary characters, phenotypes and ontologies: curating data from the systematic biology literature. PLoS ONE 5(5):e10708. doi:10.1371/journal.pone.0010708

Dawkins R, Gould SJ (1997) Human chauvinism. Evolution Int J org Evolution 51(3):1015–1015. doi:10.2307/2411179

de Mendoza A, Sebe-Pedros A, Sestak MS, Matejcic M, Torruella G, Domazet-Loso T, Ruiz-Trillo I (2013) Transcription factor evolution in eukaryotes and the assembly of the regulatory toolkit in multicellular lineages. Proc Natl Acad Sci U S A 110(50):E4858–E4866. doi:10.1073/pnas.1311818110

Delaux PM, Xie X, Timme RE, Puech-Pages V, Dunand C, Lecompte E, Delwiche CF, Yoneyama K, Becard G, Sejalon-Delmas N (2012) Origin of strigolactones in the green lineage. New Phytol 195:857–871

Farre EM, Liu T (2013) The PRR family of transcriptional regulators reflects the complexity and evolution of plant circadian clocks. Curr Opin Plant Biol 16(5):621–629. doi:10.1016/j.pbi.2013.06.015

Fawcett JA, Maere S, Van de Peer Y (2009) Plants with double genomes might have had a better chance to survive the Cretaceous-Tertiary extinction event. Proc Natl Acad Sci U S A 106:5737–5742

Feller A, Machemer K, Braun EL, Grotewold E (2011) Evolutionary and comparative analysis of MYB and bHLH plant transcription factors. Plant J 66:94–116

Felsenstein J (1985) Phylogenies and the comparative method. Am Nat 126:1

Garland Jr., Ives AR (2000) Using the past to predict the present: confidence intervals for regression equations in phylogenetic comparative methods. Am Nat 155(3):346–364. doi:10.1086/303327

Gould SJ (1996) Full house: the spread of excellence from Plato to Darwin. Harmony Books, New York, pp 135–230

Gregory TR (2005) The C-value enigma in plants and animals: a review of parallels and an appeal for partnership. Ann Bot (Lond) 95(1):133–146. doi:10.1093/aob/mci009

Hedges SB, Blair JE, Venturi ML, Shoe JL (2004) A molecular timescale of eukaryote evolution and the rise of complex multicellular life. BMC Evol Biol 4:2

Hunter S, Jones P, Mitchell A, Apweiler R, Attwood TK, Bateman A, Bernard T, Binns D, Bork P, Burge S, de Castro E, Coggill P, Corbett M, Das U, Daugherty L, Duquenne L, Finn RD, Fraser M, Gough J, Haft D, Hulo N, Kahn D, Kelly E, Letunic I, Lonsdale D, Lopez R, Madera M, Maslen J, McAnulla C, McDowall J, McMenamin C, Mi H, Mutowo-Muellenet P, Mulder N, Natale D, Orengo C, Pesseat S, Punta M, Quinn AF, Rivoire C, Sangrador-Vegas A, Selengut JD, Sigrist CJA, Scheremetjew M, Tate J, Thimmajanarthanan M, Thomas PD, Wu CH, Yeats C, Yong S-Y (2012) InterPro in 2011: new developments in the family and domain prediction database. Nucleic Acids Res 40(Database issue):D306–D312. doi:10.1093/nar/gkr948

Iso T, Kedes L, Hamamori Y (2003) HES and HERP families: multiple effectors of the Notch signaling pathway. J Cell Physiol 194:237–255

Jiao Y, Wickett NJ, Ayyampalayam S, Chanderbali AS, Landherr L, Ralph PE, Tomsho LP, Hu Y, Liang H, Soltis PS, Soltis DE, Clifton SW, Schlarbaum SE, Schuster SC, Ma H, Leebens-Mack J, dePamphilis CW (2011) Ancestral polyploidy in seed plants and angiosperms. Nature 473(7345):97–100. doi:10.1038/nature09916

Jiao Y, Leebens-Mack J, Ayyampalayam S, Bowers JE, McKain MR, McNeal J, Rolf M, Ruzicka DR, Wafula E, Wickett NJ, Wu X, Zhang Y, Wang J, Carpenter EJ, Deyholos MK, Kutchan TM, Chanderbali AS, Soltis PS, Stevenson DW, McCombie R, Pires CJ, Wong GK, Soltis DE, Depamphilis CW (2012) A genome triplication associated with early diversification of the core eudicots. Genome Biol 13(1):R3. doi:10.1186/gb-2012-13-1-r3

Lang D, Zimmer AD, Rensing SA, Reski R (2008) Exploring plant biodiversity: the Physcomitrella genome and beyond. Trends Plant Sci 13(10):542–549

Lang D, Weiche B, Timmerhaus G, Richardt S, Riano-Pachon DM, Correa LG, Reski R, Mueller-Roeber B, Rensing SA (2010) Genome-wide phylogenetic comparative analysis of plant transcriptional regulation: a timeline of loss, gain, expansion, and correlation with complexity. Genome Biol Evol 2:488–503

Levine M, Tjian R (2003) Transcription regulation and animal diversity. Nature 424(6945):147–151

Lynch M (2007) The frailty of adaptive hypotheses for the origins of organismal complexity. Proc Natl Acad Sci U S A 104(Suppl 1):8597–8604. doi:10.1073/pnas.0702207104

Lynch M, Conery JS (2000) The evolutionary fate and consequences of duplicate genes. Science 290:1151–1155

Macalister CA, Bergmann DC (2011) Sequence and function of basic helix-loop-helix proteins required for stomatal development in Arabidopsis are deeply conserved in land plants. Evol Dev 13:182–192

Martins E (2000) Adaptation and the comparative method. Trends Ecol Evol 15(7):296–299

McShea DW (1996) Metazoan complexity and evolution: is there a trend? Evolution 50(2):477–492

McShea DW (2005) The evolution of complexity without natural selection, a possible large-scale trend of the fourth kind. Paleobiology 31(sp5):146–156. doi:10.1666/0094-8373(031[0146:teocwn]2.0.co;2)

Meehan TF, Masci AM, Abdulla A, Cowell LG, Blake JA, Mungall CJ, Diehl AD (2011) Logical development of the cell ontology. BMC Bioinformatics 12(1):6. doi:10.1186/1471-2105-12-6

Molnar A, Schwach F, Studholme DJ, Thuenemann EC, Baulcombe DC (2007) miRNAs control gene expression in the single-cell alga Chlamydomonas reinhardtii. Nature 447:1126–1129

Mukherjee K, Campos H, Kolaczkowski B (2013) Evolution of animal and plant dicers: early parallel duplications and recurrent adaptation of antiviral RNA binding in plants. Mol Biol Evol 30:627–641

Niklas KJ, Kutschera U (2009) The evolution of the land plant life cycle. New Phytol 23:23

Niklas KJ, Newman SA (2013) The origins of multicellular organisms. Evol Dev 15(1):41–52. doi:10.1111/ede.12013

Pagel M (1994) Detecting correlated evolution on phylogenies: a general method for the comparative analysis of discrete characters. Pro R Soc B: Biol Sci 255(1342):37–45. doi:10.1098/rspb.1994.0006

Payne JL, Boyer AG, Brown JH, Finnegan S, Kowalewski M, Krause RA, Lyons SK, McClain CR, McShea DW, Novack-Gottshall PM, Smith FA, Stempien JA, Wang SC (2009) Two-phase increase in the maximum size of life over 3.5 billion years reflects biological innovation and environmental opportunity. Proc Natl Acad Sci U S A 106(1):24–27. doi:10.1073/pnas.0806314106

Read BA, Kegel J, Klute MJ, Kuo A, Lefebvre SC, Maumus F, Mayer C, Miller J, Monier A, Salamov A, Young J, Aguilar M, Claverie JM, Frickenhaus S, Gonzalez K, Herman EK, Lin YC, Napier J, Ogata H, Sarno AF, Shmutz J, Schroeder D, de Vargas C, Verret F, von Dassow P, Valentin K, Van de Peer Y, Wheeler G, Dacks JB, Delwiche CF, Dyhrman ST, Glockner G, John U, Richards T, Worden AZ, Zhang X, Grigoriev IV (2013) Pan genome of the phytoplankton Emiliania underpins its global distribution. Nature 499:209–213

Pennell MW, Harmon LJ, Uyeda JC (2014) Is there room for punctuated equilibrium in macroevolution? Trends Ecol Evol 29(1):23–32. doi:10.1016/j.tree.2013.07.004

Price DC, Chan CX, Yoon HS, Yang EC, Qiu H, Weber AP, Schwacke R, Gross J, Blouin NA, Lane C, Reyes-Prieto A, Durnford DG, Neilson JA, Lang BF, Burger G, Steiner JM, Loffelhardt W, Meuser JE, Posewitz MC, Ball S, Arias MC, Henrissat B, Coutinho PM, Rensing SA, Symeonidi A, Doddapaneni H, Green BR, Rajah VD, Boore J, Bhattacharya D (2012) Cyanophora paradoxa genome elucidates origin of photosynthesis in algae and plants. Science 335:843–847

Pires ND, Yi K, Breuninger H, Catarino B, Menand B, Dolan L (2013) Recruitment and remodeling of an ancient gene regulatory network during land plant evolution. Proc Natl Acad Sci U S A 110:9571–9576

Prochnik SE, Umen J, Nedelcu AM, Hallmann A, Miller SM, Nishii I, Ferris P, Kuo A, Mitros T, Fritz-Laylin LK, Hellsten U, Chapman J, Simakov O, Rensing SA, Terry A, Pangilinan J, Kapitonov V, Jurka J, Salamov A, Shapiro H, Schmutz J, Grimwood J, Lindquist E, Lucas S, Grigoriev IV, Schmitt R, Kirk D, Rokhsar DS (2010) Genomic analysis of organismal complexity in the multicellular green alga Volvox carteri. Science 329:223–226

Punta M, Coggill PC, Eberhardt RY, Mistry J, Tate J, Boursnell C, Pang N, Forslund K, Ceric G, Clements J, Heger A, Holm L, Sonnhammer ELL, Eddy SR, Bateman A, Finn RD (2011) The Pfam protein families database. Nucleic Acids Res 40(D1):D290-D301. doi:10.1093/nar/gkr1065

Rensing SA (2014) Gene duplication as a driver of plant morphogenetic evolution. Curr Opin Plant Biol 17C:43–48. doi:10.1016/j.pbi.2013.11.002

Reyes-Prieto A, Weber AP, Bhattacharya D (2007) The origin and establishment of the plastid in algae and plants. Annu Rev Genet 41:147–168

Schonknecht G, Chen WH, Ternes CM, Barbier GG, Shrestha RP, Stanke M, Brautigam A, Baker BJ, Banfield JF, Garavito RM, Carr K, Wilkerson C, Rensing SA, Gagneul D, Dickenson NE, Oesterhelt C, Lercher MJ, Weber AP (2013) Gene transfer from bacteria and archaea facilitated evolution of an extremophilic eukaryote. Science 339:1207–1210

Seoighe C, Gehring C (2004) Genome duplication led to highly selective expansion of the *Arabidopsis thaliana* genome. Trends Genet 20:461–464

Shapiro E, Biezuner T, Linnarsson S (2013) Single-cell sequencing-based technologies will revolutionize whole-organism science. Nat Rev Genet 14(9):618–630. doi:10.1038/nrg3542

Soltis DE, Burleigh JG (2009) Surviving the K-T mass extinction: new perspectives of polyploidization in angiosperms. Proc Natl Acad Sci U S A 106:5455–5456

Soltis PS, Soltis DE (2009) The role of hybridization in plant speciation. Annu Rev Plant Biol 60:561–588

Soltis DE, Bell CD, Kim S, Soltis PS (2008) Origin and early evolution of angiosuperms. Ann N Y Acad Sci 1133:3–25

Srivastava M, Simakov O, Chapman J, Fahey B, Gauthier ME, Mitros T, Richards GS, Conaco C, Dacre M, Hellsten U, Larroux C, Putnam NH, Stanke M, Adamska M, Darling A, Degnan SM, Oakley TH, Plachetzki DC, Zhai Y, Adamski M, Calcino A, Cummins SF, Goodstein DM, Harris C, Jackson DJ, Leys SP, Shu S, Woodcroft BJ, Vervoort M, Kosik KS, Manning G, Degnan BM, Rokhsar DS (2010) The *Amphimedon queenslandica* genome and the evolution of animal complexity. Nature 466(7307):720–726. doi:10.1038/nature09201

Szathmary E, Jordan F, Pal C (2001) Molecular biology and evolution. Can genes explain biological complexity? Science 292(5520):1315–1316

Tenaillon O, Silander OK, Uzan JP, Chao L (2007) Quantifying organismal complexity using a population genetic approach. PLoS One 2(2):e217. doi:10.1371/journal.pone.0000217

Timme RE, Bachvaroff TR, Delwiche CF (2012) Broad phylogenomic sampling and the sister lineage of land plants. PLoS ONE 7:e29696

Van de Peer Y, Maere S, Meyer A (2009) The evolutionary significance of ancient genome duplications. Nat Rev Genet 10(10):725–732

Vaquerizas JM, Kummerfeld SK, Teichmann SA, Luscombe NM (2009) A census of human transcription factors: function, expression and evolution. Nat Rev Genet 10(4):252–263. doi:10.1038/nrg2538

Veron AS, Kaufmann K, Bornberg-Bauer E (2007) Evidence of interaction network evolution by whole-genome duplications: a case study in MADS-box proteins. Mol Biol Evol 24(3):670–678

Viaene T, Delwiche CF, Rensing SA, Friml J (2013) Origin and evolution of PIN auxin transporters in the green lineage. Trends Plant Sci 18:5–10

Vickaryous MK, Hall BK (2006) Human cell type diversity, evolution, development, and classification with special reference to cells derived from the neural crest. Biological reviews of the Cambridge Philos Soc 81:425–455

Vogel C, Chothia C (2006) Protein family expansions and biological complexity. PLoS Comput Biol 2(5):e48. doi:10.1371/journal.pcbi.0020048

Walls RL, Athreya B, Cooper L, Elser J, Gandolfo MA, Jaiswal P, Mungall CJ, Preece J, Rensing S, Smith B, Stevenson DW (2012) Ontologies as integrative tools for plant science. Am J Bot 99(8):1263–1275. doi:10.3732/ajb.1200222

Wilson D, Pethica R, Zhou Y, Talbot C, Vogel C, Madera M, Chothia C, Gough J (2009) SUPERFAMILY–sophisticated comparative genomics, data mining, visualization and phylogeny. Nucleic Acids Res 37 (Database issue):D380–D386. doi:10.1093/nar/gkn762

Yi K, Menand B, Bell E, Dolan L (2010) A basic helix-loop-helix transcription factor controls cell growth and size in root hairs. Nat Genet 42(3):264–U108. doi:10.1038/ng.529

Zalewski CS, Floyd SK, Furumizu C, Sakakibara K, Stevenson DW, Bowman JL (2013) Evolution of the class IV HD-zip gene family in streptophytes. Mol Biol Evol 30(10):2347–2365. doi:10.1093/molbev/mst132

Zimmer AD, Lang D, Buchta K, Rombauts S, Nishiyama T, Hasebe M, Van de Peer Y, Rensing SA, Reski R (2013) Reannotation and extended community resources for the genome of the non-seed plant Physcomitrella patens provide insights into the evolution of plant gene structures and functions. BMC Genomics 14:498. doi:10.1186/1471-2164-14-498

Zmasek CM, Godzik A (2011) Strong functional patterns in the evolution of eukaryotic genomes revealed by the reconstruction of ancestral protein domain repertoires. Genome Biol 12(1):R4–R4. doi:10.1186/gb-2011-12-1-r4

Independent Emergence of Complex Multicellularity in the Brown and Red Algae

J. Mark Cock and Jonas Collén

Abstract Brown and red macroalgae represent two of only five eukaryotic groups that have independently evolved complex multicellularity. Compared with animals and land plants, very little is known about the molecular mechanisms underlying multicellular development in the two macroalgal groups, but the recent publication of complete genome sequences for both of these lineages has been a first step towards changing this situation. Comparisons of these genomes with those of other multicellular and unicellular organisms have identified a number of features that may be related to the transitions to complex multicellularity in these macroalgal lineages. One particularly striking feature of the brown algae, for example, is the emergence of a family of membrane-spanning receptor kinases, a class of molecules that is also thought to have been important for the transition to complex multicellularity in animals and land plants. Surprisingly, the genomes of the brown alga *Ectocarpus* and the red alga *Chondrus* are remarkably different at the structural level, despite the fact that both organisms represent lineages that have evolved multicellularity as sedentary organisms in the seashore environment. Current efforts to identify and characterise developmental regulators in macroalgae are expected to enrich comparisons with other complex multicellular lineages.

Keywords *Chondrus crispus* · Complex multicellularity · Developmental regulators · *Ectocarpus* · Genome sequence · Genome structure · Macroalgae · Phylogeny · Receptor kinase · Small RNAs

---

J. M. Cock (✉) · J. Collén
UMR 8227, Integrative Biology of Marine Models, Station Biologique de Roscoff, Sorbonne Universités, UPMC Univ Paris 06, CS 90074, 29688
Roscoff cedex, France
e-mail: cock@sb-roscoff.fr

J. Collén · J. M. Cock
CNRS, UMR 8227, Integrative Biology of Marine Models, Station Biologique de Roscoff, CS 90074, 29688 Roscoff cedex, France
e-mail: collen@sb-roscoff.fr

I. Ruiz-Trillo, A. M. Nedelcu (eds.), *Evolutionary Transitions to Multicellular Life*, Advances in Marine Genomics 2, DOI 10.1007/978-94-017-9642-2_16

Introduction

This chapter focuses on recent insights into the evolution of complex multicellularity in the brown and red macroalgae obtained through the application of genomic approaches, in particular the analysis of the first complete genome sequences for multicellular members of each of these two groups (Cock et al. 2010; Collén et al. 2013; Nakamura et al. 2013). The major findings from the analysis of brown and red algal genomic data will be described in turn, followed by a comparison of the two lineages that will highlight both the similarities and some striking differences.

Defining Complex Multicellularity

Multicellularity evolved many times during the evolutionary history of the eukaryotes (Grosberg and Strathmann 2007); Chapter "Timing the Origins of Multicellular Eukaryotes Through Phylogenomics and Relaxed Molecular Clock Analyses"). In most cases, the emergence of multicellularity produced relatively simple life forms, but in a small number of lineages the transition to multicellularity was followed by an elaborate series of events that led to the evolution of what can be considered to be complex multicellular organisms (Knoll 2011; Niklas and Newman 2013). It is important to distinguish this latter process from the initial transition from a unicellular habit to multicellularity because the second transition occurred very rarely and, indeed, it has been argued that only five of the major eukaryotic groups have produced organisms that can be considered to exhibit complex multicellularity (Cock et al. 2010). These five groups are the animals, green algae/plants, fungi and red and brown algae, and each is thought to have evolved complex multicellularity independently. The argument for selecting these five groups, and excluding other multicellular organisms, is that the former all include macroscopic organisms with defined, recognisable morphologies. Moreover, these organisms are all constructed by developmental programs that involve organised processes combining cell proliferation and differentiation. Despite these definitions, the distinction between simple and complex multicellularity remains somewhat arbitrary and attempts have been made to define complex multicellularity on a more quantitative basis (Rokas 2008; Bell and Mooers 1997). Bell and Mooers, for example, focused on estimations of the number of different cell types present in a particular organism and gave maximum figures for the number of different cell types of 122, 44, 9, 14 and 14 for the metazoan, green, fungal, red algal and brown algal lineages, respectively. Maximal numbers for the Acrasiomycota and Ciliata were three and four, respectively. Using the number of cell types to define complexity does not provide a definitive reply because cell types themselves are not always easy to define (indeed Bell and Mooers state that "We note that cell type number remains a very crude estimate of complexity, and hope that better measures, preferably genome-based, will be developed."). It may, nonetheless be possible to propose an approximate number of cell types (eight?) above which a lineage can be considered to exhibit complex multicellularity, but this should be considered as a rule of thumb rather than an absolute boundary. Another way of looking at this question is ask whether it would be an improvement to be more, or less, inclusive. Outside of the five groups proposed, slime moulds probably exhibit the highest degree of morphological complexity. However, these organisms are constructed by a fundamentally different process, cell aggregation, and are significantly less complex than the most evolved members of the five groups. Similarly, some authors have proposed to define only the metazoa and green plants as complex multicellular organisms on the grounds that these groups exhibit significantly more complexity than any other eukaryote group. This is true but we would argue that the three other groups, the fungi and the red and brown algae, nonetheless exhibit sufficient complexity to be included. In conclusion, even if the transition point between simple and complex multicellularity remains difficult to define precisely, these terms clearly provide a useful framework for investigating the evolution of multicellular development.

Complex Multicellular Development in the Brown Algae

Brown Algal Developmental Complexity

Of the three macroalgal groups (brown, red and green) the brown algae exhibit the highest level of developmental complexity (Charrier et al. 2012; Dawes 1981). The bodyplans of the more complex members of this group, such as the kelps, are divided into a holdfast, a stipe and a blade and may include additional specialised structures such as buoys. The bodyplan includes several different cell types, with epidermal, medulla and cortical cells in the blade for example, together with additional specialised cells such as mucilage secreting cells, physode-packed cells and trumpet cells. The latter transport photosynthate in a manner analogous to the land plant vascular system (Buggeln 1983).

Very little is currently known about the molecular processes that underlie the construction of the multicellular bodyplans of the brown algae. Progress in this domain has been hampered by the absence of a model system adapted for the dissection of developmental pathways. This situation has recently changed with the proposition and emergence of *Ectocarpus* as a genetic and genomic model system for the brown algae (Peters et al. 2004; Charrier et al. 2008; Cock et al. 2011; Coelho et al. 2012; Cock et al. 2012; see Chapter "Emergence of Ectocarpus as a Model System to Study the Evolution of Complex Multicellularity in the Brown Algae "). One of the drawback of *Ectocarpus,* as far as the study of multicellular development is concerned, is that it is a small, filamentous brown alga and therefore lacks many of the more complex developmental features observed in larger brown algae. However, the group to which *Ectocarpus* belongs, the Ectocarpales, is a sister group to the Laminariales (Silberfeld et al. 2010), which includes the large kelps, and the regulatory pathways that control fundamental developmental processes are expected to be shared to a large extent between the two groups. *Ectocarpus* therefore appears to be a relevant model to explore the emergence of complex multicellularity in the brown algae. For the present, however, *Ectocarpus* developmental biology is in its infancy and, although several interesting developmental mutants have been described (Coelho et al. 2011; Le Bail et al. 2011; Peters et al. 2008), it will probably be some time before the molecular mechanisms underlying multicellular development are described and experimentally validated in any detail in this species.

An alternative approach that has been used to identify molecular features that may have been associated with the evolution of complex multicellularity in the brown algae has been to compare the *Ectocarpus* genome sequence with those of well-studied multicellular organisms in other eukaryotic groups. The results that have been obtained using this comparative approach are described in the following sections. Note that, whilst this approach has afforded some interesting insights, the hypotheses that have been generated will need to be tested experimentally, where possible.

General Structural Aspects of the Ectocarpus Genome

The *Ectocarpus* genome exhibits several general features that are typically associated with genomes of complex multicellular organisms (Cock et al. 2010). It is a reasonably large genome (214 Mbp) with a significant content of repeated sequences (22.7 %). Most of the genes have a complex structure, being split into multiple, short exons (242 bp on average) separated by long introns (703 bp and seven per gene on average), and with long 3' untranslated regions (855 bp on average). The *Ectocarpus* genome is predicted to encode 16,256 proteins and this proteome also exhibits features characteristic of proteomes from other multicellular organisms. For example, metazoan proteomes include a large proportion of proteins with small folds, mostly associated with intercellular signalling and regulation (Gerstein and Levitt 1997). Several small folds including ankyrin (IPR002110), tetratricopeptide repeat (IPR011717), WW (IPR001202), notch (IPR000800), NZF (SSF90213) and FNIP (IPR008615) domains were found to be particularly abundant in the *Ectocarpus* proteome compared to the proteomes of 16 diverse eukaryote species (Cock et al. 2010).

A comprehensive analysis of the protein domains present in the predicted proteome of *Ectocarpus* and of the 16 other eukaryotes, including members of the green lineage, animals and fungi, identified eight domains that were consistently more abundant in the proteomes of the multicellular organisms, compared with those of the unicellular organisms (Cock et al. 2010). These domains, classified as endonuclease/exonuclease/phosphatase (IPR005135), AMP-dependent synthetase and ligase (IPR000873), acetyl-CoA synthetase-like (SSF56801), carbohydrate kinase, FGGY (IPR000577), eukaryotic translation initiation factor 4 gamma (PTHR23253), lipoxygenase, C-terminal (IPR013819), UDP-glucuronosyl/UDP-glucosyltransferase (IPR002213) and peptidase T2, asparaginase 2 (IPR000246), included a majority of catalytic motifs. Six protein domains were more abundant in *Ectocarpus* than in any of the other 16 species analysed. These six domains included the ankyrin and notch domains mentioned above. The notch domain proteins are particularly interesting because, in animals, this domain occurs in proteins with key roles in intercellular communication, such as the notch receptor (Guruharsha et al. 2012) and because many of the *Ectocarpus* notch proteins are predicted to be secreted or anchored in the cell membrane.

It was proposed some time ago that multicellular organisms possess a proportionally greater number of extracellular and membrane-localised proteins and it was suggested that these proteins might be important for extracellular signalling between cells in these organisms (Hazkani-Covo et al. 2004). A more global analysis, using 37 complete genomes and different versions of the Hectar program (Gschloessl et al. 2008), confirmed this trend (Cock et al. 2010): on average 14.5 % of the proteins of unicellular organisms were predicted to be processed by the secretory pathway, whereas the figure was 18.6 % for multicellular organisms ($P = 0.038$, Student's t test). However, considerable variation was observed, particularly for multicellular organisms. With *Ectocarpus*, for example, only 12.4 % of proteins were predicted to be processed by the secretory pathway. The observed variation could be due to a number of factors. In particular, depending on the extracellular environment, many unicellular organisms may require a complex set of extracellular proteins to interact with their environment.

Gene Family Evolution

Complex multicellularity is thought to have evolved independently in each of the five lineages listed earlier (animals, green plants/algae, fungi and red and brown algae). The molecular processes underlying multicellular development in these different lineages are therefore expected to be very different in detail, although they may share similar general characteristics and these similarities could provide important insights into the general principles involved.

One approach that has been used to investigate the molecular processes that underlie multicellularity has been to search for gene families that are predicted to have evolved at approximately the same time as a transition to multicellularity was occurring in a particular lineage. Cock et al. (2010; http://bioinformatics.psb.ugent.be/dollo_analysis) compared the complete predicted proteomes of 17 species from across the eukaryotic tree and used Dollo logic to predict when individual gene families emerged (and in some cases were subsequently lost) during eukaryote evolution (briefly, when two contemporary genomes shared a gene family, that family was considered to have been present in their common ancestor). Overall, this analysis indicated a general pattern in which the major eukaryotic groups have retained distinct but overlapping sets of genes since their evolution from a common ancestor, with new gene families evolving independently in each lineage. Comparison of multicellular and unicellular organisms indicated that the former had lost fewer gene families (1518 compared with 2131) and had evolved more new gene families (4069 compared with 2436), on average, since their divergence from the common ancestor. By mapping gene family gains and losses onto a phylogenetic tree, it was possible to identify gene families that were predicted to have evolved at approximately the same time as the transitions to multicellularity in the animal, fungal, green plant/algal and brown algal lineages (no genome sequence was available for a multicellular red alga at the time that this study was carried out; Cock et al. 2010). However, comparison of these sets of gene families between multicellular lineages did not identify any obvious shared features. This suggests that the evolutionary events that lead to the emergence of multicellularity in these four lineages may have been very different. This sort of analysis, across very large evolutionary distances, is complicated because many factors contribute to the divergence between lineages and it is difficult to isolate the effects of one single factor, such as the transition to multicellularity, from the effects of other factors. The individual datasets are nonetheless of interest as starting points for the investigation of the molecular basis of the unicellular to multicellular transitions in each individual lineage.

In the stramenopiles, multicellularity is thought to have evolved since the divergence of the brown algal and diatom lineages because the diatoms are unicellular, as are most of the intermediate groups, such as pelagophytes and raphidophytes (Lin et al. 2012; Guillou et al. 1999; Niklas and Newman 2013). Table 1 shows some examples of genes from families that are predicted to have evolved since the brown algal and diatom lineages diverged. Interestingly, a number of these genes are predicted to encode regulatory proteins and an analysis of gene ontology (GO) categories for the gene families identified a significant enrichment with GO labels related to protein kinase activity (http://bioinformatics.psb.ugent.be/dollo_analysis; Cock et al. 2010). Both these kinases and the other regulatory proteins could have played an important role in the evolution of complex bodyplans.

As far as the kinases are concerned, one of the gene families that is predicted to have evolved since divergence from the diatoms is particularly interesting because it encodes a family of receptor kinases. Located on the cell membrane, these molecules are ideally placed not only to sense changes in the extracellular environment but also to perceive signals emitted by other cells, a process that is essential for a multicellular organism. Both the animal and green plant lineages have independently evolved large families of membrane-located receptor kinases (Cock et al. 2002; Shiu and Bleecker 2001) and these molecules are known to play key roles in developmental processes such as differentiation and cellular patterning (De Smet et al. 2009). Moreover, there is evidence that the acquisition of receptor kinases predates the evolution of complex multicellularity in both lineages (Diévart et al. 2011; Suga et al. 2012; Lehti-Shiu et al. 2009; King and Carroll 2001), suggesting that the emergence of these gene families may have played a key role in the transition to multicellularity in both cases. Membrane-localised receptor kinases typically consist of an extracellular receptor domain separated from an intracellular kinase domain by a single membrane-spanning helix. The extracellular part of the protein can include several different types of protein domain, including for example leucine-rich repeats (LRRs), EGF, Cys-rich, cadherin, lectin, S domain and WAK domains, whereas the intracellular part has a "eukaryotic"-type kinase domain of either the serine/threonine (plants and

Table 1 Examples of genes from families (constructed with MCLBlastline) that are predicted to have evolved in the brown algal lineage since it split from the diatom lineage based on Dollo analysis (Cock et al. 2010). Note that the MCLBlastline "families" generally correspond to subfamilies (e.g. a subfamily of MYB transcription factor). For further details: http://bioinformatics.psb.ugent.be/dollo_analysis

Dollo family	LocusID	Gene description	Functional group
11758	Esi0087_0091	MYB DNA binding protein/transcription factor-like protein	Transcription
12892	Esi0095_0079	Heat shock transcription factor	
10959	Esi0024_0153	AraC family transcriptional regulator	
6464	Esi0252_0001	Kinase	Signal transduction
17846	Esi0038_0021	MAP kinase kinase	
3690	Esi0066_0034	Putative receptor kinase	
9344	Esi0150_0064	Membrane-associated cdc2-inhibitory kinase	
12353	Esi0183_0030	SH2 domain containing protein	
27633	Esi0517_0002	Integrin-related protein	
860	Esi0011_0207	LRR-GTPase of the ROCO family	
9518	Esi0100_0017	RNA-dependent RNA polymerase 1	RNA interference
5396	Esi0306_0033	Histone H1	Chromatin
5694	Esi0303_0006	SUMO deconjugating cysteine peptidase	Protein modification
19444	Esi0032_0082	Small conductance mechanosensitive ion channel	Ion channel
3485	Esi0056_0075	Transient receptor potential channel	
6042	Esi0019_0117	Ankyrin repeat transient receptor Potential Channel	
17628	Esi0050_0067	ABC transporter G family protein	Membrane
12766	Esi0111_0095	Cytochrome P450	Metabolism
17691	Esi0205_0025	Lipase	
17472	Esi0020_0181	Mannitol 1-phosphate dehydrogenase	
1039	Esi0882_0001	Mannuronan C-5-epimerase	

animals) or the tyrosine (animals) class (Lemmon and Schlessinger 2010; Shiu and Bleecker 2001).

The receptor kinases encoded by the *Ectocarpus* genome resemble several animal and plant receptors both in terms of their general structure and in that their extracellular domains include a series of LRRs. However, the Dollo analysis described above indicates that the brown algal family of receptor kinases arose independently, after the divergence of the brown algal and diatom lineages. This conclusion was supported by a more detailed phylogenetic analysis, which showed that the brown algal receptor kinases form a clade that is distinct from those that contain the animal and green plant receptor kinases (Cock et al. 2010). This is an important observation because it suggests that, for three of the five eukaryotic groups that have evolved complex multicellularity, this major evolutionary transition was associated with the acquisition of membrane-localised receptor kinases. Note here that the innovation in each case involved the association of an extracellular receptor domain with an intracellular kinase domain and not the evolution of the domains themselves, which are much more ancient, and were probably present before the crown radiation of the eukaryotes.

The *Ectocarpus* receptor kinases are part of a family that includes 24 members, based on kinase domain homology, of which ten can be defined as bona fide receptor kinases based on the predicted presence of a signal peptide, LRR domains, a membrane-spanning helix and a kinase domain (Fig. 1). Analysis of the conserved catalytic residues of the *Ectocarpus* receptor kinases indicates that these proteins are likely to phosphorylate their downstream targets on serine and threonine residues.

A number of other membrane-localised proteins may also have played a role in the transition to complex multicellularity in the brown algae. For example, the *Ectocarpus* genome also encodes a small number of G-protein coupled receptors (three genes). There does not appear to have been a significant expansion of this class of gene, associated with the transition to multicellularity, but there are indications that downstream signalling from these receptors, mediated by heterotrimeric G-proteins, may be more complex than in unicellular stramenopiles. For example, the *Ectocarpus* genome encodes six Gα subunits, compared to one or two in oomycetes and diatoms (Cock et al. 2010).

The evolution of cell adhesion systems is thought to have been important factor in the transition to multicellularity in metazoans. Interestingly, *Ectocarpus* possesses three genes that are related to integrins but no homologues of these genes were found in diatom and oomycete genomes (Cock et al. 2010). Integrins constitute an important cell adhesion system in metazoans and core components of this system date to before the divergence of the Opisthokonta (Sebé-Pedrós et al. 2010). The function of the integrin-related proteins in *Ectocarpus* is unknown but it is likely to be different to that of the animal integrin system because the cell wall prevents direct cell-to-cell contact in the brown algae. Moreover, the similarity between animal integrins and the integrin-like proteins encoded by the *Ectocarpus* genome is limited to the N-terminal domain, again indicating that the two classes of protein are unlikely to function in exactly the same manner.

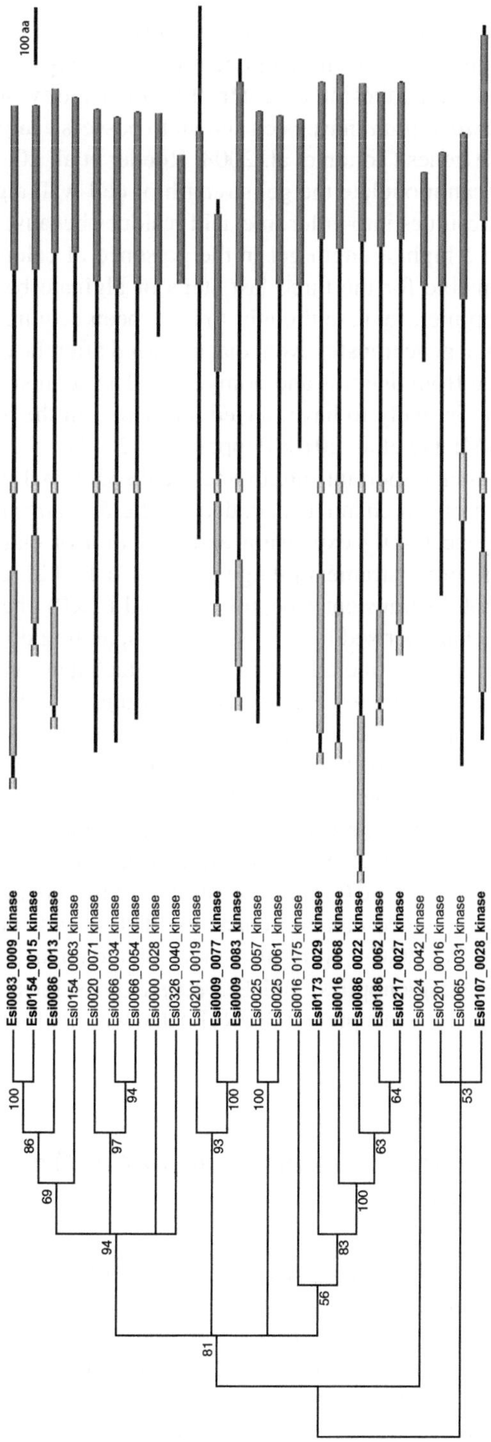

Fig. 1 Neighbour-joining tree showing the phylogenetic relationships between the 24 members of the *Ectocarpus* receptor kinase family. Branches with bootstrap support of less than 50 have been collapsed. The primary protein structure of each family member, drawn to scale, is indicated on the *right*; *blue boxes*, signal peptides and transmembrane domains (note that Esi0065_0031 is predicted to contain a cluster of four membrane-spanning helixes in the marked region); *green boxes*, leucine-rich repeats; *red boxes*, kinase domains

In brown algae, as in green plants, the cell wall plays a key role in cell-cell adhesion. Alginate, which is a major cell-wall component in brown algae, is thought to regulate the maintenance of cell adhesion by controlling the rigidity of the wall. Both *Ectocarpus* and *Laminaria digitata* have been shown to possess a large number of mannuronan C5-epimerase genes (Tonon et al. 2008; Roeder et al. 2005; Michel et al. 2010). These enzymes can modulate the gel strength of cell wall alginates by converting mannuronic acid residues into guluronic acid residues because blocks of guluronic acid residues form a high-strength gel in the presence of calcium. Interestingly, the enzymes responsible for the three last steps of alginate biosynthesis (including mannuronan C5-epimerases) are thought to have been acquired from an actinobacterium by a horizontal gene transfer event that occurred after the divergence of the ancestor of brown algae from diatoms and oomycetes. The acquisition of this alginate biosynthetic pathway is likely to have played a key role in the acquisition of complex multicellularity (Michel et al. 2010; Popper et al. 2011).

Ion channels provide another important membrane-localised signalling system and analysis of the collection of ion channels encoded by the *Ectocarpus* genome revealed instances both of gene family expansion and retention of channels that appear to have been lost from other stramenopile genomes (Cock et al. 2010). *Ectocarpus* has a large family of 18 transient receptor potential (TRP) channels, for example. These proteins are characterised by six membrane-spanning domains and generally act as channels for cations, including Ca^{2+}(Venkatachalam and Montell 2007). TRP channels have been shown to be activated by a range of environmental stimuli (Venkatachalam and Montell 2007), so it is possible that the roles of the *Ectocarpus* proteins are more related to environmental sensing than directly in multicellular development. Nonetheless, the functions carried out by these proteins may have been an important requisite for the evolution of complex multicellularity in this group (see Section "Role of the Cell Wall and Other Adaptations to a Sedentary Lifestyle Involving Long-living Life Cycle Generations" of this chapter). Genes encoding both two-pore voltage gated calcium channels (TPCs) and four-domain voltage gated calcium channels (VGCs) were also more abundant in *Ectocarpus* than in the genomes of other stramenopiles, and *Ectocarpus* possesses an inositol triphosphate (InsP3)/ryanodine type receptor (IP3R/RyR) that was not found in other stramenopiles.

Gene Families that may have been Important for the Transition to Complex Multicellularity

Ectocarpus has a total of 410 transcription-associated proteins (TAPs, which include both transcription factors and transcription regulators). These proteins are interesting because the emergence of multicellularity has been linked with expansion of transcription factor families in both green plants and animals (Rokas 2008; Carroll 2005; Gutiérrez et al. 2004; Hsia and McGinnis 2003; Levine and Tjian 2003; Richardt et al. 2007). However, a detailed analysis of TAPs in 19 genomes, including that of *Ectocarpus*, did not identify any general trends that could be clearly associated with multicellularity (Cock et al. 2010). For example, the genomes of the multicellular organisms included in this study were not predicted to encode a larger proportion of TAPs than the genomes of the unicellular organisms and no individual family was consistently larger in all the multicellular species analysed. This does not mean, however, that that the TAPs did not play an important role in the transition to multicellularity, as it is likely that different TAP families were important for the transition to multicellularity in different lineages. Moreover, trends in TAP family size related to the transition to multicellularity may have been obscured by trends associated with other evolutionary events. Finally, it should also be borne in mind that individual eukaryotic groups might possess lineage-specific transcription factors that still remain to be characterised (e.g. Lohse et al. 2013) and a complete picture will only be possible when these new families have been discovered and characterised. Poorly explored groups such as the brown and red macroalgae are particularly likely to possess such families of uncharacterised transcription factors.

Cellular trafficking plays an important role in many processes in multicellular organisms. Therefore, it is not surprising that, in animals, expansion of the Rab GTPase family, which determines the specificity of vesicle trafficking to different cellular compartments, has been correlated with an increase in complexity (Klöpper et al. 2012). The *Ectocarpus* Rab GTPase family, however, is of a comparable size to those of unicellular stramenopiles such as diatoms and *Aureococcus anophagefferens*, indicating that, as in land plants (Elias et al. 2012), there was no marked expansion of this family during the transition to multicellularity (Cock et al. 2010). *Ectocarpus* does, however, appear to have a highly active intracellular trafficking system, based on microscopic analysis of cellular ultrastructure (Bouck 1965; Oliveira and Bisalputra 1973), and possesses large families of SNARE and coat protein complex proteins (Cock et al. 2010).

Another small G-protein family that has been associated with the evolution of multicellular complexity in the metozoans is the Ras family, which has important roles in cell signalling (Reuther and Der 2000). However, like terrestrial plants, *Ectocarpus* appears to lack this family, underlining the conclusion that complex multicellularity can evolve in the absence of Ras signalling (Cock et al. 2010).

Most studies aimed at identifying genes that were important for the evolution of multicellularity in different eukaryotic lineages have concentrated on the emergence of novel gene families in lineages that have made the transition to multicellularity. Another factor that may have been important, however, is the retention during evolution of key genes, without which the transition to multicellularity would not have been possible. For example, it has been noted that multicellular organisms tend to have lost fewer of the known members of the Rad51 family than have unicellular organisms (Rad51 family proteins are involved in DNA repair and have key roles during meiosis; Lin et al. 2006). It is interesting, therefore, that the Rad51 family of *Ectocarpus* is more complete than that of other stramenopiles for which we have a sequenced genome (Cock et al. 2010). This also appears to be the case for several other gene families, including GTPases and ion channels, for example.

Several authors have suggested that RNA-based regulatory systems such as microRNAs (miRNAs) may have played a key role in the evolution of complex multicellularity (Peterson et al. 2009; Mattick 2004) and, in animals at least, there appears to be a positive correlation between phenotypic complexity and the number of miRNA families present in the genome (Sempere et al. 2006). Analysis of data from deep sequencing of small RNAs from *Ectocarpus* suggests that this brown alga employs this type of molecule as a regulator (Cock et al. 2010). For example, small RNAs mapped preferentially to repeated elements in the genome, suggesting a role in the suppression of transposon activity. Based on the application of a stringent set of criteria, evidence was also found for the presence of several microRNAs (Cock et al. 2010). Evidence for longer non-coding RNAs was obtained using a whole genome tiling array approach, which detected more than eight thousand transcribed regions longer than 200 bp that were not localised to the predicted protein-coding genes (Cock et al. 2010). The vast majority of these regions are not conserved in the genome of the diatom *Thalassiosira pseudonana,* and they are therefore of interest as potential components of processes associated with multicellular development. Taken together, these analyses indicate that non-coding RNAs (both small RNAs and long non-coding RNAs) may be part of the cell regulatory network in brown algae. Future work is expected to provide more information about the exact roles of these molecules and their modes of action.

Role of the Cell Wall and Other Adaptations to a Sedentary Lifestyle Involving Long-living Life Cycle Generations

The emergence of multicellularity in the brown algae was only possible because this group acquired a number of characteristics that are essential for large, long-lived organisms to survive in the harsh environments of coastal ecosystems. As sedentary organisms, brown algae have to be able to resist the multiple abiotic and biotic aggressions of this hostile environment long enough to grow from an initial cell to reproductive maturity. They do not have the option of moving to a more clement niche. The abiotic stresses encountered by brown algae include not only wave action but also large variations in light levels, temperature and salinity, whilst the biotic stresses include many types of pathogens and grazers. Bearing these constraints in mind, we can define a number of features that were probably as important for the evolution of muticellularity in the brown algae as the developmental processes that coordinate the construction of a multicellular bodyplan. These included the emergence of a novel cell wall structure that provided both mechanical strength and flexibility, a highly adaptable photosynthetic system and various defence mechanisms against both biotic and abiotic threats.

The *Ectocarpus* genome exhibits features that can be associated with all of these processes. For example, the protein domains that were highly abundant in *Ectocarpus*, compared to other eukaryotes, included a putative carbohydrate-binding domain (the WSC domain, IPR002889) that may be important in cell wall integrity sensing and a chlorophyll-binding domain (IPR001344) whose diversity in *Ectocarpus* may play an important role in adapting the photosynthetic system to fluctuations in light conditions (Cock et al. 2010). Moreover, detailed annotation of cell metabolism, photosynthesis and both biotic and abiotic defence genes in *Ectocarpus* have revealed the complexity of all these systems in this alga.

Experimental Investigation of the Molecular Mechanisms Underlying Development in Brown Algae

The above discussion illustrates how comparisons of the *Ectocarpus* genome with those of other organisms across the eukaryotic tree has allowed inferences to be made about the molecular processes underlying multicellular development in the brown algae. In the future, however, it will be important to confirm these inferences by experimental dissection of developmental processes in this brown algal model at the molecular level. At present, nothing is known about the molecular machinery that constructs the alga's multicellular bodyplan and no developmental genes have been validated experimentally. The tools necessary to carry out this sort of analysis are being developed however (see Chapter "Emergence of Ectocarpus as a Model System to Study the Evolution of Complex Multicellularity in the Brown Algae" of this book) and genetic approaches have been initiated to identify key developmental regulators.

The life cycle of *Ectocarpus* involves an alternation between two multicellular organisms, the sporophyte and the gametophyte (Kornmann 1956; Müller 1964), each of which represents an independent developmental program. These two generations of the life cycle are of similar size and both consist of a branched thallus of uniseriate filaments. There are however differences between them, both in terms of their early development and with respect to their growth habits (Peters et al. 2008). These differences have been exploited to develop screens for mutant strains affected in life cycle progression. Two mutants of this type have been described to date. In the first mutant, *immediate upright (imm)*, the sporophyte generation exhibits several characteristics of the gametophyte generation during early development, including an asymmetrical initial cell division and the production of a rhizoid and an upright filament rather than the symmetrical initial cell division and basal filament observed in wild type sporophytes (Peters et al. 2008). However, despite exhibiting several characteristics of the gametophyte generation the sporophyte generation remains a sporophyte in functional terms (it produces spores and not gametes). Microarray analysis confirmed that there was a partial switch to the gametophyte program in the mutant, with the induction of genes typically expressed during the gametophyte generation and a repression of a subset of sporophyte generation genes. Moreover, the mutation affected the abundances of transcripts for a large number of genes, suggesting that *IMMEDIATE UPRIGHT* represents a major developmental regulator.

The second life cycle mutant that has been described, *ouroboros (oro)*, has an even more interesting phenotype (Coelho et al. 2011). In this mutant, the sporophyte generation is converted into a fully functional gametophyte, with the consequence that the life cycle becomes simply a continual reiteration of the gametophyte generation. Again, microarray analysis has shown that transcript abundances are modified for a large number of genes in the presence of this mutation.

Finally, a morphogenetic mutant, *étoile (etl)*, has also been described (Le Bail et al. 2011). This mutation effects both cellular differentiation and growth habit (branching). As has also been shown for the *imm* and *oro* mutations, *etl* correspond to single, recessive, Mendelian locus.

Positional cloning projects are underway for all three of these genes, with the *oro* project being the most advanced. Over the coming years, the identification of these genes, together with complementary approaches such as transcriptomic analyses and screens for additional mutants, should start to provide some clues about how developmental processes are regulated in *Ectocarpus,* in particular, and in brown algae in general. With this information in hand, it will start to become possible to compare the molecular machineries that underlie multicellular development in the brown algae with the equivalent processes in other major eukaryotic lineages such as animals and green plants. These analyses, based on functional information rather than sequence similarity, should provide a deeper understanding of the similarities and differences between the various lineages that have independently evolved complex multicellularity. It will also be interesting to compare the results of these genetic analyses, which do not start with any assumptions about the genes involved, with the insights obtained from searching the genome sequence for genes that were potentially important for the transition to multicellularity.

Complex Multicellularity in the Red Algae.

The Red Algae, a Major Group of Photosynthetic Eukaryotes

The red algae constitute one of the three major lineages within the supergroup Archaeplastida, the other two lineages being the chlorophytes (which include the green algae and terrestrial plants) and the glaucophytes. The common ancestor of the Archaeplastida was derived from an endosymbiosis between a eukaryote and a cyanobacterium, which created what may have been the first photosynthetic eukaryote more than 1550 million years ago (Yoon et al. 2004). The red algae represents a well defined monophyletic group of organisms that are characterised by the presence of phycobiliproteins in their photosynthetic apparatus, having non-stacked thylakoids, the utilisation of floridean starch as storage polysaccharide and by the total absence of flagella and centrioles in all life cycle stages (Woelkerling 1990). The rhodophytes are a relatively old group, based on molecular data the split between red and green algae is predicted to have occurred about 1500 million years ago with the divergence between glaucophytes and the ancestor of green and red algae occurring even earlier (Yoon et al. 2004).

The red algae represent a key group in eukaryotic evolution, particularly as far as the photosynthetic taxa are concerned. In addition to being the sister group of the green lineage, which today dominates terrestrial photosynthesis, they have also provided photosynthetic capacity to many other groups, including diatoms, brown algae, haptophytes, cryptophytes and some dinoflagellates, through secondary and

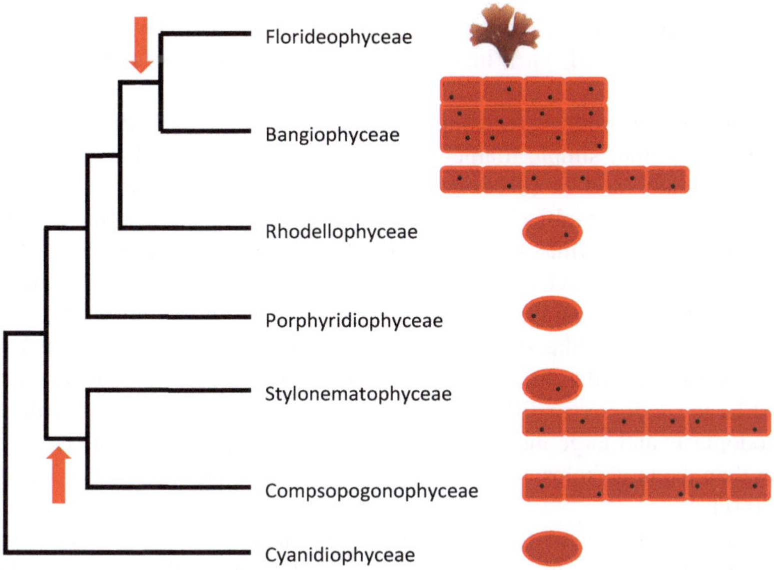

Fig. 2 The phylogeny of red algae, simplified and redrawn after Scott et al. (2011) with a schematic representation of typical morphologies for each group. The *red arrows* show tentatively the acquisition of multicellularity. The image representing the florideophytes shows *Chondrus crispus*

tertiary endosymbiotic events (Keeling 2013). These latter groups today dominate aquatic photosynthesis. The red macroalgae are ecologically important members of the intertidal and subtidal flora on rocky shores, where they can dominate the vegetation. Red seaweeds are also important economically as a source of thickeners used in the food industry, such as agars and carrageenan, provided by for example the genera *Kappaphycus* and *Gracilaria*, and are also consumed as sea vegetables, for example as nori (the genera *Porphyra* and *Pyropia*).

The red algae have historically been divided into two groups: the florideophytes, multicellular seaweeds with more advanced morphology, and the bangiophytes, which include both unicellular species and multicellular species with simple organisation. Recent studies have shown that the bangiophytes do not represent a monophyletic group and this group has therefore been divided into several taxa (Yoon et al. 2006). A representation of the current view of red algal phylogeny is shown in Fig. 2.

Emergence of Multicellularity in the Red Algae

The red algae are thought to have been the first group to have acquired advanced multicellularity: fossils attributed to the red algal species *Bangiomorpha pubescence* have been found in 1200 million year old strata (Butterfield 2000). The fossils show organisms morphologically similar to members of the present day genus *Bangia* and possessing different cell types including reproductive structures indicating sexual reproduction.

As a group, the red algae exhibit a range of complexity. Red algae can be unicellular, colonial, simple filaments and foliose (within the Bangiophyceae and Florideophyceae), and more complex forms can be found within the Florideophyceae. Most of the multicellular red algae form pseudoparenchyma (interwoven filaments) but parenchyma are also known, for example in *Bangia spp.*, *Porphyra* spp. and some Delesseriaceae (Bold and Wynne 1978). Red macroalgae usually have relatively simple morphologies, but some species have stems, holdfasts, bladders, vesicle cells and bladelike structures (e.g. the genera *Botryocladia, Antithamnion* and *Delesseria*). Cellular differentiation is limited within the red algae, but the thalli of most large rhodophytes differentiate to produce small cortical cells with numerous rhodoplasts and large medullar cells with few plastids. Red algae are typically smaller than the large kelps, but some species can reach lengths of up to 3 m (Simon et al. 2001).

In order to understand the evolution of multicellularity in the red algae we compared red algal morphology across the phylogeny of the group. Within the cyanidiophytes, porphyriodeophytes, stylonematophytes and rhodellophytes only unicellular species are found. The stylonematales include unicellular, pseudofilamentous and filamentous forms and the compsopogonophytes include uniseriate, branched filaments with diffuse growth (Adl et al. 2012). The two lineages with the most advanced multicellularity are the florideophytes and bangiophytes. Assuming that the phylogeny in Fig. 2 is correct, multicellularity appears to have evolved at least twice within the red algae: at the base of the clade that includes the Stylonematophyaceae and Compsopogonophyceae and in the ancestor of the Florideophyceae and Bangiophyceae. On the other hand, the phylogeny of Yoon et al. (2010) would suggest at least three independent transitions to multicellularity. Note that in the red algae, complex multicellularity (using the rough definition of at least eight different cell types proposed in the introduction to this chapter) has only emerged once, within the florideophytes. In the other red algal groups only branched or unbranched filaments and sheet-like structures are found. This reinforces the idea that complex multicellularity is an evolutionary rarity.

The red algal group that exhibits complex multicellularity, the florideophytes, is also arguably the most evolutionary successful taxa. The florideophytes are the largest group within the red algae, representing 95 % of the ~6000 extant species (www.algaebase.org) and most of the seaweeds. It is difficult to estimate complexity and number of cell types in red algae and few examples exist in the literature; however, Bell and Mooer (1997) estimated the number of cell types within different florideophytes to between 6 and 14, which is low compared to terrestrial plants and animals, but comparable to brown algae and Fungi. The radiation of the florideophytes from within the bangiophytes has been dated to around 600 million years ago using molecular methods (Yoon et al. 2004) and 600 million year old fossils have

Table 2 Characteristics of red algal genomes. *Cyanidioschyzon merolae* and *Galdieria sulphuraria* are unicellular species while *Pyropia yezoensis* and *Chondrus crispus* are multicellular

Red algal species	Genome size (Mbp)	Gene number	Introns per gene (mean)	Mean intron length (bp)	Reference
Cyanidioschyzon merolae	16.5	5331	0.005	248	(Matsuzaki et al. 2004) (Nozaki et al. 2007)
Pyropia yezoensis	43	10327	0.3–0.7	304	(Nakamura et al. 2013)
Chondrus crispus	105	9606	0.32	123	(Collén et al. 2013)
Galdieria sulphuraria	13.7	6623	2.1	56	(Schönknecht et al. 2013)

been attributed to the florideophytes (Xiao et al. 2004). These fossils also indicate that advanced multicellularity already existed within the red algae at that time.

Red Algal Genome Projects

Until recently very little was known about the genes and processes involved in multicellular development in red algae. One of the reasons for this was the lack of genomic information from unicellular and multicellular red algae. This situation has changed, however, with the recent descriptions of four red algal genome sequences (Table 2): the unicellular extremophiles *Cyanidioschyzon merolae* and *Galdieria sulphuraria,* which belong to the Cyanidiales, the bangiophyte *Pyropia yezoensis* (nori, previously known as *Porphyra yezoensis)* and the florideophyte *Chondrus crispus* (Irish moss). Both *Pyropia* and *Chondrus* are multicellular but *Pyropia* has a relatively simple sheet-like appearance while *Chondrus* is more complex morphologically.

Genome sizes in red algae vary between 10 and 1400 Mbp (Kapraun 2005). The genome sizes of *Pyropia* and *Chondrus* are thus relatively small compared to many red algal species and no obvious correlation exists between genome size and morphological diversity within the multicellular species. The genomes of *Pyropia* and *Chondrus* are also small compared to the genomes of most multicellular organisms, with some notable exceptions such as the fungus *Laccaria bicolour* (64.9 Mbp, Martin et al. 2008) and the land plant *Arabidopsis thaliana* (135 Mbp).

The genes in red algal genomes share several common characteristics, including a low number of introns and small sized introns. These traits are usually associated with less complex organisms, but in the red algae they were also observed for the more morphologically advanced species *Chondrus*. The only exception is *Galdieria* where introns are more abundant although still small in size (Schönknecht et al. 2013). The low number of introns in most red algal genes may have been the result of a genetic bottleneck that caused a reduction in genome size early in the history of red algae (Collén et al. 2013). This bottleneck could potentially have occurred after the split between the Cyanidiales and other red algae, in which case the low intron frequency may not be a feature that was linked with the emergence of complex multicellularity.

The multicellular red algae have more predicted genes than their unicellular counterparts but this correlation between gene number and complexity does not hold in the comparison between the two multicellular species. More genes were predicted in *Pyropia* compared to *Chondrus*, even though *Chondrus* is morphologically more complex. This discrepancy between complexity and gene number could be due to the fact that the *Pyropia* life cycle involves an alternation of life cycles stages with marked changes in morphology from a sheet-like appearance in the gametophyte generation to a microscopic filamentous form in the sporophyte generation (Drew 1949). In *Chondrus* the sporophytes and gametophytes have similar morphologies during their vegetative phases, but differ in their content of carrageenan and in their morphology when reproductive. However, it should be noted that in *Chondrus* a red-algal-specific life cycle stage is also present, the carposporophyte. The carposporophyte is a diploid structure that forms on the female after fertilisation of a female gamete, located on the female gametophyte, by a male gamete.

Gene Families Potentially Involved in Multicellularity in Red Algae

It is presently difficult to draw conclusions about genes and gene families important for complex multicellularity red algae. The main reason for this is the large evolutionary distance between the sequenced unicellular and multicellular red algae. The Cyanidiales diverged from the other red algal lineages more than 1200 million years ago. Moreover, both *Cyanidioschyzon* and *Galdieria* are extremophiles living in hot acids springs whereas *Pyropia* and *Chondrus* are mesophiles that live in marine environments.

Nonetheless, several *Chondrus* gene families have been analysed in the context of the evolution of multicellularity. One example is the class of transcription-associated proteins (including not only transcription factors but associated complexes such as the RNA-induced silencing complex), which consists of 193 proteins. Comparison of the *Chondrus* transcription factors with those of the unicellular red alga *Cyanidioschyzon* identified some gene families that may possibly be involved in processes related to multicellularity (Collén et al. 2013). For example, the C2H2 zinc finger and Sin3 transcription factor families are larger in *Chondrus* than in *Cyanidioschyzon* and cross species partial least squares analysis (Lang et al. 2010) suggested an involvement of the C2H2 zinc finger family in multicellularity (Collén et al. 2013). Genes encoding Argonaute, Dicer and C2C2-CO-like transcription-associated proteins were found in the *Chondrus* and *Pyropia* genomes, but absent from those of *Cyanidioschyzon* and *Galdieria*. This initially suggested that there may have been a correlation between multicellularity and the presence of the RNA-induced silencing complex, but EST data (Chan et al. 2011) indicates that these genes are also present in a unicellular red alga, *Porphyridium cruentum*.

Another group often associated with multicellularity is the eukaryotic protein kinases (ePKs). The *Chondrus* genome encodes 209 ePKs. There is therefore a large kinase family in *Chondrus*, but the size is consistent with what has been observed in the genomes of other eukaryotes, both unicellular and multicellular, where they normally constitute about 1.5–2.5 % of the genes in the genome (Collén et al. 2013). However, given that gene families tend to be exceptionally small in *Chondrus*, the relatively large size of the protein kinase family could indicate a link with multicellularity. Atypical kinases, such as APHs, ABC1 and the bromodomain representatives, are present in relatively large numbers in *Chondrus* compared to *Oryza sativa*, *Phytophthora infestans*, *Thalassiosira pseudonana*, *Saccharomyces cerevisiae* and *Dictyostelium discoideum*. Interestingly, *Chondrus* possesses quite a high number of tyrosine kinases (12 genes). This group of kinases has been correlated with multicellularity in Metazoa (Miller 2012).

Six Rab GTPases were found in *Chondrus* (Rab1, 2, 6, 7, 11a, and 18). All of these families were also identified in *Ectocarpus,* but *Ectocarpus* has seven additional families (Rab5, 8, 22, 28, 32A, 50 and Titan). The Rab family, which contain regulators of vesicle trafficking (Brighouse et al. 2010), is very small in *Chondrus* (Collén et al. 2013). The Rab family in the Cyanidiales appears to be even smaller, as *Cyanidioschyzon* has only one Rab11 paralogue, and *Galdieria* lacks Rab18. The last common eukaryotic ancestor is thought to have possessed about 23 Rab GTPases and extant unicellular organisms contain 10–20 with larger numbers in multicellular organisms (Elias et al. 2012). The Rab family thus seems to have been drastically reduced in red algae in general, and to an even larger extent in the unicellular species. As was observed for the kinases, therefore, the larger size of the Rab GTPase families in the multicellular red algae suggests that the this family may be have been implicated in the transition to multicellularity in the red algae.

Our capacity to identify genes that may have been associated with the transition to multicellularity in the red algae will increase as more red algal genome information become available. This process will not only involve adding new candidates but also the elimination of existing candidates that have been identified based on the currently limited dataset. A previous suggestion that NADPH oxidases occur only in multicellular organisms (Lalucque and Silar 2003) is a good example of this process. A study of the *Chondrus* NADPH oxidase (Hervé et al. 2006) allowed the identification of an NADPH oxidase in the unicellular red alga *Cyanidioschyzon*, thus refuting the initial hypothesis.

Conclusions

Comparative Analysis of the Evolution of Multicellularity in the Brown and Red Macroalgae

The two recently published red macroalgal genome sequences (for *Chondrus crispus* and *Pyropia yezoensis)* not only provide insights into an additional group that has evolved complex multicellularity, but they now also provide the means to compare the evolution of multicellularity in two macroalgal lineages: the red and the brown algae. Comparisons between the red and brown macroalgae are particularly interesting because, whilst the two groups are extremely divergent in evolutionary terms (with their most recent common ancestor dating back to the crown radiation of the major eukaryotic groups), both have evolved as sedentary primary producers in the same general ecological environment; the intertidal and subtidal regions of the rocky shore. Because of the secondary endosymbiosis between a red algae and the ancestor to brown algae, the brown algae also share many of the red algal genes despite their large evolutionary distance. To some extent they started with similar genomic "tool boxes" as the starting point for constructing the gene networks necessary for multicellular development. We might therefore expect that the two lineages have independently evolved comparable solutions to similar environmental challenges. This type of analysis, where multicellularity evolved independently in response to similar environments, is likely to be more informative than comparisons involving groups that have had very different evolutionary strategies, such as land plants and animals for example (sedentary autrophism as opposed to mobility and heterotrophism).

Given the above, one of the most surprising results of the analysis of the red macroalgal genomes was that the red and brown algal genes and genomes have very different structures. For example, introns are rare in red macroalgal genes but brown algal genes tend to be split into many exons. At a larger scale, in the *Chondrus* genome genes tend to be found concentrated in "islands" surrounded by extensive regions that are composed principally of transposon sequences (Collén et al. 2013). This is not the case for the *Ectocarpus* genome, where genes are more evenly spread across the genome (Cock et al. 2010). These differences between structure features of brown and red macroalgal genomes indicate that, although genome expansion appears to be associated with the transition to multicellularity in general, genomes in different multicellular lineages may evolve in quite different ways, presumably as a result of additional (selective and non-selective) factors. Indeed, it has been suggested that the unusual structure of the *Chondrus* genome is a consequence of a reduction in genome size that occurred early in the evolutionary history of the lineage, before the emergence of multicellularity (Collén et al. 2013).

The red macroalgal genome sequences have only recently become available and analysis of these genomes for features that may be associated with the transition to complex multicellularity is ongoing. However, as discussed above, several interesting features have already been described and these can be compared with the features previously identified in the *Ectocarpus* genome. Not surprisingly, transcription factors and signalling proteins, particularly kinases, are strong candidates for genes that may have played key roles in the transition to multicellularity. Another common feature is a link with the size and completeness of the small GTPase family.

Comparative Analysis of Complex Multicellular Lineages Across the Eukaryotic Tree

In this chapter we have highlighted the importance of comparing all five of the eukaryotic lineages that have evolved complex multicellularity in order to understand the key molecular innovations that underlie this important transition. Each of the five lineages evolved complex multicellularity in a different manner, but comparative analysis should allow key evolutionary innovations to be distinguished from events that were contingent on the evolutionary history of each lineage. Work on red and brown macroalgae is likely to make important contributions to this general picture in the future because these two groups are still relatively poorly characterised, compared to land plants and animals for example. Moreover, the two groups represent two independent instances of multicellular organisms evolving to be adapted to the rocky shore environment.

Comparative analysis can provide important insights into key molecular innovations. The discovery of independently evolved membrane-localised receptor kinase families in three of the five complex multicellular lineages, including the brown algae, underlined the importance of cell-to-cell communication. In contrast, pathways that play a central role in one multicellular group, such as the ras pathway in animals, have been shown to be absent from other groups, suggesting that these pathways are not indispensible for the emergence of complex multicellularity.

The role of non-coding RNAs in the transition to complex multicellularity remains an important question (Lozada-Chávez et al. 2011; Mattick 2004; Peterson et al. 2009), which again needs to be addressed by studies that cover the complete range of complex multicellular lineages. It is currently believed that miRNAs evolved independently in both land plants and animals from an ancient small interfering RNA (siRNA) system that dates back to the early origins of the eukaryotes (Ghildiyal and Zamore 2009; Cerutti and Casas-Mollano 2006). Moreover, despite numerous publications reporting miRNAs in diverse eukaryote species, a recent study found that, outside the animal and land plant groups, only candidate sequences from *Chlamydomonas* and *Ectocarpus* met the criteria set for bona fide miRNAs (Tarver et al. 2012). The existence of strong candidate miRNAs for *Ectocarpus* (Cock et al. 2010), suggests that the brown algae may also have evolved their own miRNA system. Similarly the presence of dicer and argonaute in *Chondrus* and *Pyropia* indicates the presence of small RNA-based regulation in red macroalgae. If the existence of an miRNA system in seaweeds can be confirmed experimentally, this would represent another very interesting example of the multiple convergent emergence of a specific molecular system being strongly correlated with the transition to complex multicellularity.

Future work should not only address outstanding questions such as the existence of a microRNA system in macroalgae but also needs to be extended to additional novel regulatory molecules, such as long non-coding RNAs (lncRNAs), for example. Another important goal will be to obtain denser genome sampling of phylogenetic groups that diverged at around the time that multicellular development was emerging. For the brown algal lineage, for example, it would be of interest to sample additional unicellular lineages closely related to the Phaeophyceae in order to date more precisely the emergence of key molecular components such as the brown algal receptor kinases. It would also be very interesting to have access to a genome from a morphologically complex brown alga, such as a kelp or an wrack. For the red lineages a unicellular species closer to the florideophytes would be an important asset.

In the longer term, a deep understanding of the evolutionary processes that lead to the emergence of complex multicellularity will only be possible when the molecular systems that mediate multicellular development are better understood in currently poorly studied groups such as the fungi and red and brown macroalgae. This represents a challenging goal for the coming years.

Summary

1. Brown algae are considered to be one of only five eukaryotic groups that have evolved complex multicellularity, the other four being animals, green algae/plants, fungi and red algae.
2. The brown algae are the most developmentally complex group of algae, with some species growing to produce large organisms with multiple organs, tissues and cell types. At present, however, very little is known about the molecular processes that regulate development in these organisms.
3. Several gene families in the *Ectocarpus* genome have been linked with the evolution of multicellularity in the brown algae, including a family of receptor kinase molecules that is predicted to have evolved after divergence from the lineage that lead to the unicellular diatoms. Similar receptor kinase families evolved independently in the animal and land plant lineages.
4. Additional evolutionary innovations that allowed the emergence of a long-living, sedentary organism adapted to the seashore environment were probably also important for the emergence of complex multicellularity in the brown algae.
5. The red algae, another major group of photosynthetic eukaryotes, are also considered to have evolved complex multicellularity. They may in fact be the oldest group of complex multicellular organisms, with fossil evidence for multicellular forms dating back 1200 million years.
6. The recent publication of multiple red algal genomes provides a means to access the genetic innovations associated with the emergence of multicellularity in the red algae. At the structural level, the genome of the multicellular red algal model

Chondrus is remarkably different from the *Ectocarpus* genome, with many monoexonic genes clustered in islands between regions rich in transposable elements.
7. To fully understand the mechanisms underlying the evolution of complex multicellularity, it is important that comparisons are as broad as possible, including all five groups that have attained this level of developmental complexity. The macroalgal groups are particularly interesting because they have evolved complex multicellularity in similar, aquatic environments and share behavioural and physiological similarities such as sedentary, photosynthetic lifestyles.

Acknowledgments We would like to thank Cecile Hervé and Gurvan Michel (UMR8227, Station Biologique de Rosoff) for adding information about links between the cell wall and the evolution of complex multicellularity. Work at the Station Biologique de Roscoff was supported by the Centre National de Recherche Scientifique, the University Pierre and Marie Curie, the Groupement d'Interet Scientifique Génomique Marine, the Interreg program France (Channel)-England (project Marinexus), the Agence Nationale de la Recherche (project Bi-cycle), IDEALG Grants ANR-10-BTBR-04-02 and 04-04 "Investissements d'avenir, Biotechnologies-Bioressources" and the Groupement de Recherche International "Diversity, Evolution and Biotechnology of Marine Algae" (GDRI N° 0803).

References

Adl SM, Simpson AG, Lane CE, Lukeš J, Bass D, Bowser SS, Brown MW, Burki F, Dunthorn M, Hampl V, Heiss A, Hoppenrath M, Lara E, Le Gall L, Lynn DH, McManus H, Mitchell EA, Mozley-Stanridge SE, Parfrey LW, Pawlowski J, Rueckert S, Shadwick L, Schoch CL, Smirnov A, Spiegel FW (2012) The revised classification of eukaryotes. J Eukaryot Microbiol 59(5):429–493

Bell G, Mooers AO (1997) Size and complexity among multicellular organisms. Biol J Linn Soc 60:345–363

Bold HC, Wynne MJ (1978) Introduction to the algae. Prentice-Hall, Englewood Cliffs

Bouck G (1965) Fine structure and organelle association in brown algae. J Cell Biol 26(2):523–537

Brighouse A, Dacks JB, Field MC (2010) Rab protein evolution and the history of the eukaryotic endomembrane system. Cell Mol Life Sci 67(20):3449–3465

Buggeln RG (1983) Photoassimilate translocation in brown algae. Progr Phycol Res 2:283–332

Butterfield NJ (2000) *Bangiomorpha pubescens* n. gen., n. sp.: implications for the evolution of sex, multicellularity, and the mesoproterozoic/neoproterozoic radiation of eukaryotes. Paleobiology 26:386–404

Carroll SB (2005) Evolution at two levels: on genes and form. PLoS Biol 3(7):e245

Cerutti H, Casas-Mollano JA (2006) On the origin and functions of RNA-mediated silencing: from protists to man. Curr Genet 50(2):81–99

Chan CX, Yang EC, Banerjee T, Yoon HS, Martone PT, Estevez JM, Bhattacharya D (2011) Red and green algal monophyly and extensive gene sharing found in a rich repertoire of red algal genes. Curr Biol 21(4):328–333

Charrier B, Coelho S, Le Bail A, Tonon T, Michel G, Potin P, Kloareg B, Boyen C, Peters A, Cock J (2008) Development and physiology of the brown alga *Ectocarpus siliculosus*: two centuries of research. New Phytol 177(2):319–332

Charrier B, Le Bail A, de Reviers B (2012) Plant Proteus: brown algal morphological plasticity and underlying developmental mechanisms. Trends Plant Sci 17(8):468–477

Cock J, Vanoosthuyse V, Gaude T (2002) Receptor kinase signalling in plants and animals: distinct molecular systems with mechanistic similarities. Curr Opin Cell Biol 14(2):230–236

Cock JM, Sterck L, Rouzé P, Scornet D, Allen AE, Amoutzias G, Anthouard V, Artiguenave F, Aury J, Badger J, Beszteri B, Billiau K, Bonnet E, Bothwell J, Bowler C, Boyen C, Brownlee C, Carrano C, Charrier B, Cho G, Coelho S, Collén J, Corre E, Da Silva C, Delage L, Delaroque N, Dittami S, Doulbeau S, Elias M, Farnham G, Gachon C, Gschloessl B, Heesch S, Jabbari K, Jubin C, Kawai H, Kimura K, Kloareg B, Küpper F, Lang D, Le Bail A, Leblanc C, Lerouge P, Lohr M, Lopez P, Martens C, Maumus F, Michel G, Miranda-Saavedra D, Morales J, Moreau H, Motomura T, Nagasato C, Napoli C, Nelson D, Nyvall-Collén P, Peters A, Pommier C, Potin P, Poulain J, Quesneville H, Read B, Rensing S, Ritter A, Rousvoal S, Samanta M, Samson G, Schroeder D, Ségurens B, Strittmatter M, Tonon T, Tregear J, Valentin K, von Dassow P, Yamagishi T, Van de Peer Y, Wincker P (2010) The *Ectocarpus* genome and the independent evolution of multicellularity in brown algae. Nature 465(7298):617–621

Cock JM, Peters AF, Coelho SM (2011) Brown algae. Curr Biol 21(15):R573–R575

Cock JM, Arun A, Godfroy O, Macaisne N, Strittmatter M, Peters AF, Coelho SM (2012) Genomics of brown algae: current advances and future prospects. Genes Genomics 34:1–5

Coelho SM, Scornet D, Rousvoal S, Peters N, Dartevelle L, Peters AF, Cock JM (2012) *Ectocarpus*: a model system for the brown algae. Cold Spring Harb Protoc 2012(2):193–198

Coelho SM, Godfroy O, Arun A, Le Corguillé G, Peters AF, Cock JM (2011) *Ouroboros* is a master regulator of the gametophyte to sporophyte life cycle transition in the brown alga *Ectocarpus*. Proc Natl Acad Sci U S A 108:11518–11523

Collén J, Porcel B, Carré W, Ball SG, Chaparro C, Tonon T, Barbeyron T, Michel G, Noel B, Valentin K, Elias M, Artiguenave F, Arun A, Aury JM, Barbosa-Neto JF, Bothwell JH, Bouget FY, Brillet L, Cabello-Hurtado F, Capella-Gutiérrez S, Charrier B, Cladière L, Cock JM, Coelho SM, Colleoni C, Czjzek M, Da Silva C, Delage L, Denoeud F, Deschamps P, Dittami SM, Gabaldón T, Gachon CM, Groisillier A, Hervé C, Jabbari K, Katinka M, Kloareg B, Kowalczyk N, Labadie K, Leblanc C, Lopez PJ, McLachlan DH, Meslet-Cladiere L, Moustafa A, Nehr Z, Nyvall Collén P, Panaud O, Partensky F, Poulain J, Rensing SA, Rousvoal S, Samson G, Symeonidi A, Weissenbach J, Zambounis A, Wincker P, Boyen C (2013) Genome structure and metabolic features in the red seaweed *Chondrus crispus* shed light on evolution of the Archaeplastida. Proc Natl Acad Sci U S A 110(13):5247–5252

Dawes CJ (1981) Marine botany. Wiley, New York

De Smet I, Voss U, Jürgens G, Beeckman T (2009) Receptor-like kinases shape the plant. Nat Cell Biol 11(10):1166–1173

Diévart A, Gilbert N, Droc G, Attard A, Gourgues M, Guiderdoni E, Périn C (2011) Leucine-rich repeat receptor kinases are sporadically distributed in eukaryotic genomes. BMC Evol Biol 11:367

Drew KM (1949) Conchocelis-phase in the life-history of *Porphyra umbilicalis* (L.) Kütz. Nature 164:748–749

Elias M, Brighouse A, Gabernet-Castello C, Field MC, Dacks JB (2012) Sculpting the endomembrane system in deep time: high resolution phylogenetics of Rab GTPases. J Cell Sci 125:2500–2508

Gerstein M, Levitt M (1997) A structural census of the current population of protein sequences. Proc Natl Acad Sci U S A 94(22):11911–11916

Ghildiyal M, Zamore PD (2009) Small silencing RNAs: an expanding universe. Nat Rev Genet 10(2):94–108

Grosberg RK, Strathmann RR (2007) The evolution of multicellularity: a minor major transition? Annu Rev Ecol Evol Syst 38:621–654

Gschloessl B, Guermeur Y, Cock J (2008) Hectar: a method to predict subcellular targeting in heterokonts. BMC Bioinform 9:393

Guillou L, Chrétiennot-Dinet M-J, Medlin LK, Claustre H, Goër SL-d, Vaulot D (1999) *Bolidomonas*: a new genus with two species belonging to a new algal class, the Bolidophyceae (Heterokonta). J Phycol 35:368–381

Guruharsha KG, Kankel MW, Artavanis-Tsakonas S (2012) The Notch signalling system: recent insights into the complexity of a conserved pathway. Nat Rev Genet 13(9):654–666

Gutiérrez RA, Green PJ, Keegstra K, Ohlrogge JB (2004) Phylogenetic profiling of the *Arabidopsis thaliana* proteome: what proteins distinguish plants from other organisms? Genome Biol 5(8):R53
Hazkani-Covo E, Levanon E, Rotman G, Graur D, Novik A (2004) Evolution of multicellularity in Metazoa: comparative analysis of the subcellular localization of proteins in *Saccharomyces, Drosophila* and *Caenorhabditis*. Cell Biol Int 28(3):171–178
Hervé C, Tonon T, Collén J, Corre E, Boyen C (2006) NADPH oxidases in Eukaryotes: red algae provide new hints! Curr Genet 49(3):190–204
Hsia CC, McGinnis W (2003) Evolution of transcription factor function. Curr Opin Genet Dev 13(2):199–206
Kapraun DF (2005) Nuclear DNA content estimates in multicellular green, red and brown algae: phylogenetic considerations. Ann Bot 95:7–44
Keeling PJ (2013) The number, speed, and impact of plastid endosymbioses in eukaryotic evolution. Annu Rev Plant Biol 64:583–607
King N, Carroll SB (2001) A receptor tyrosine kinase from choanoflagellates: molecular insights into early animal evolution. Proc Natl Acad Sci U S A 98(26):15032–15037
Klöpper TH, Kienle N, Fasshauer D, Munro S (2012) Untangling the evolution of Rab G proteins: implications of a comprehensive genomic analysis. BMC Biol 10:71
Knoll AH (2011) The multiple origins of complex multicellularity. Annu Rev Earth Planet Sci 39:217–239
Kornmann P (1956) Über die Entwicklung einer *Ectocarpus confervoides*-Form. Pubblicazioni Della Stazione Zoologica di Napoli 28:32–43
Lalucque H, Silar P (2003) NADPH oxidase: an enzyme for multicellularity? Trends Microbiol 11(1):9–12
Lang D, Weiche B, Timmerhaus G, Richardt S, Riaño-Pachón DM, Corrêa LG, Reski R, Mueller-Roeber B, Rensing SA (2010) Genome-wide phylogenetic comparative analysis of plant transcriptional regulation: a timeline of loss, gain, expansion, and correlation with complexity. Genome Biol Evol 2:488–503
Le Bail A Billoud B Le Panse S Chenivesse S Charrier B (2011) *Etoile* regulates developmental patterning in the filamentous brown alga *Ectocarpus siliculosus*. Plant Cell 23:1666–1678
Lehti-Shiu MD, Zou C, Hanada K, Shiu SH (2009) Evolutionary history and stress regulation of plant receptor-like kinase/pelle genes. Plant Physiol 150(1):12–26
Lemmon MA, Schlessinger J (2010) Cell signaling by receptor tyrosine kinases. Cell 141(7):1117–1134
Levine M, Tjian R (2003) Transcription regulation and animal diversity. Nature 424(6945):147–151
Lin Z, Kong H, Nei M, Ma H (2006) Origins and evolution of the recA/RAD51 gene family: evidence for ancient gene duplication and endosymbiotic gene transfer. Proc Natl Acad Sci U S A 103(27):10328–10333
Lin YC, Campbell T, Chung CC, Gong GC, Chiang KP, Worden AZ (2012) Distribution patterns and phylogeny of marine stramenopiles in the North Pacific Ocean. Appl Environ Microbiol 78(9):3387–3399
Lohse MB, Hernday AD, Fordyce PM, Noiman L, Sorrells TR, Hanson-Smith V, Nobile CJ, DeRisi JL, Johnson AD (2013) Identification and characterization of a previously undescribed family of sequence-specific DNA-binding domains. Proc Natl Acad Sci U S A 110(19):7660–7665
Lozada-Chávez I, Stadler PF, Prohaska SJ (2011) "Hypothesis for the modern RNA world": a pervasive non-coding RNA-based genetic regulation is a prerequisite for the emergence of multicellular complexity. Orig Life Evol Biosph 41(6):587–607
Martin F, Aerts A, Ahrén D, Brun A, Danchin EG, Duchaussoy F, Gibon J, Kohler A, Lindquist E, Pereda V, Salamov A, Shapiro HJ, Wuyts J, Blaudez D, Buée M, Brokstein P, Canbäck B, Cohen D, Courty PE, Coutinho PM, Delaruelle C, Detter JC, Deveau A, DiFazio S, Duplessis S, Fraissinet-Tachet L, Lucic E, Frey-Klett P, Fourrey C, Feussner I, Gay G, Grimwood J, Hoegger PJ, Jain P, Kilaru S, Labbé J, Lin YC, Legué V, Le Tacon F, Marmeisse R, Melayah D, Montanini B, Muratet M, Nehls U, Niculita-Hirzel H, Oudot-Le Secq MP, Peter M, Quesneville

H, Rajashekar B, Reich M, Rouhier N, Schmutz J, Yin T, Chalot M, Henrissat B, Kües U, Lucas S, Van de Peer Y, Podila GK, Polle A, Pukkila PJ, Richardson PM, Rouzé P, Sanders IR, Stajich JE, Tunlid A, Tuskan G, Grigoriev IV (2008) The genome of *Laccaria bicolor* provides insights into mycorrhizal symbiosis. Nature 452(7183):88–92

Matsuzaki M, Misumi O, Shin-I T, Maruyama S, Takahara M, Miyagishima SY, Mori T, Nishida K, Yagisawa F, Yoshida Y, Nishimura Y, Nakao S, Kobayashi T, Momoyama Y, Higashiyama T, Minoda A, Sano M, Nomoto H, Oishi K, Hayashi H, Ohta F, Nishizaka S, Haga S, Miura S, Morishita T, Kabeya Y, Terasawa K, Suzuki Y, Ishii Y, Asakawa S, Takano H, Ohta N, Kuroiwa H, Tanaka K, Shimizu N, Sugano S, Sato N, Nozaki H, Ogasawara N, Kohara Y, Kuroiwa T (2004) Genome sequence of the ultrasmall unicellular red alga *Cyanidioschyzon merolae* 10D. Nature 428(6983):653–657

Mattick JS (2004) RNA regulation: a new genetics? Nat Rev Genet 5(4):316–323

Michel G, Tonon T, Scornet D, Cock JM, Kloareg B (2010) The cell wall polysaccharide metabolism of the brown alga *Ectocarpus siliculosus*. Insights into the evolution of extracellular matrix polysaccharides in eukaryotes. New Phytol 188(1):82–97

Miller WT (2012) Tyrosine kinase signaling and the emergence of multicellularity. Biochim Biophys Acta 1823:1053–1057

Müller DG (1964) Life-cycle of *Ectocarpus siliculosus* from Naples, Italy. Nature 26:1402

Nakamura Y, Sasaki N, Kobayashi M, Ojima N, Yasuike M, Shigenobu Y, Satomi M, Fukuma Y, Shiwaku K, Tsujimoto A, Kobayashi T, Nakayama I, Ito F, Nakajima K, Sano M, Wada T, Kuhara S, Inouye K, Gojobori T, Ikeo K (2013) The first symbiont-free genome sequence of marine red alga, Susabi-nori (*Pyropia yezoensis*). PLoS One 8(3):e57122

Niklas KJ, Newman SA (2013) The origins of multicellular organisms. Evol Dev 15(1):41–52

Nozaki H, Takano H, Misumi O, Terasawa K, Matsuzaki M, Maruyama S, Nishida K, Yagisawa F, Yoshida Y, Fujiwara T, Takio S, Tamura K, Chung SJ, Nakamura S, Kuroiwa H, Tanaka K, Sato N, Kuroiwa T (2007) A 100 %-complete sequence reveals unusually simple genomic features in the hot-spring red alga *Cyanidioschyzon merolae*. BMC Biol 5:28

Oliveira L, Bisalputra T (1973) Studies in the brown alga *Ectocarpus* in culture. J Submicrosc Cytol 5:107–120

Peters AF, Marie D, Scornet D, Kloareg B, Cock JM (2004) Proposal of *Ectocarpus siliculosus* (Ectocarpales, Phaeophyceae) as a model organism for brown algal genetics and genomics. J Phycol 40(6):1079–1088

Peters AF, Scornet D, Ratin M, Charrier B, Monnier A, Merrien Y, Corre E, Coelho SM, Cock JM (2008) Life-cycle-generation-specific developmental processes are modified in the immediate upright mutant of the brown alga *Ectocarpus siliculosus*. Development 135(8):1503–1512

Peterson KJ, Dietrich MR, McPeek MA (2009) MicroRNAs and metazoan macroevolution: insights into canalization, complexity, and the Cambrian explosion. Bioessays 31(7):736–747

Popper ZA, Michel G, Hervé C, Domozych DS, Willats WGT, Tuohy MG, Kloareg B, Stengel DB (2011) Evolution and diversity of plant cell walls: from algae to flowering plants. Annu Rev Plant Biol 62:567–590

Reuther G, Der C (2000) The Ras branch of small GTPases: ras family members don't fall far from the tree. Curr Opin Cell Biol 12(2):157–165

Richardt S, Lang D, Reski R, Frank W, Rensing S (2007) PlanTAPDB, a phylogeny-based resource of plant transcription-associated proteins. Plant Physiol 143(4):1452–1466

Roeder V, Collén J, Rousvoal S, Corre E, Leblanc C, Boyen C (2005) Identification of stress gene transcripts in *Laminaria digitata* (Phaeophyceae) protoplast cultures by expressed sequence tag analysis. J Phycol 41:1227–1235

Rokas A (2008) The molecular origins of multicellular transitions. Curr Opin Genet Dev 18(6):472–478

Schönknecht G, Chen WH, Ternes CM, Barbier GG, Shrestha RP, Stanke M, Bräutigam A, Baker BJ, Banfield JF, Garavito RM, Carr K, Wilkerson C, Rensing SA, Gagneul D, Dickenson NE, Oesterhelt C, Lercher MJ, Weber AP (2013) Gene transfer from bacteria and archaea facilitated evolution of an extremophilic eukaryote. Science 339(6124):1207–1210

Scott J, Yang EC, West JA, Yokoyama A, Kim HJ, Loideaux de Goër S, O'Kelly CJ, Orlova E, Kim SY, Park JK, Yoon HS (2011) On the genus *Rhodella, the* emended orders Dixoniellales and Rhodellales with a new order Glaucosphaerales (Rhodellophyceae, Rhodophyta). Algae 26:277–288

Sebé-Pedrós A, Roger AJ, Lang FB, King N, Ruiz-Trillo I (2010) Ancient origin of the integrin-mediated adhesion and signaling machinery. Proc Natl Acad Sci U S A 107(22):10142–10147

Sempere LF, Cole CN, McPeek MA, Peterson KJ (2006) The phylogenetic distribution of metazoan microRNAs: insights into evolutionary complexity and constraint. J Exp Zool B Mol Dev Evol 306(6):575–588

Shiu SH, Bleecker AB (2001) Receptor-like kinases from Arabi*dopsis* form a monophyletic gene family related to animal receptor kinases. Proc Natl Acad Sci U S A 98(19):10763–10768

Silberfeld T, Leigh JW, Verbruggen H, Cruaud C, de Reviers B, Rousseau F (2010) A multi-locus time-calibrated phylogeny of the brown algae (Heterokonta, Ochrophyta, Phaeophyceae): investigating the evolutionary nature of the "brown algal crown radiation". Mol Phylogenet Evol 56(2):659–674

Simon C, Ar Gall E, Deslandes E (2001) Expansion of the red alga *Grateloupia doryphora* along the coasts of Brittany (France). Hydrobiologia 443:23–29

Suga H, Dacre M, de Mendoza A, Shalchian-Tabrizi K, Manning G, Ruiz-Trillo I (2012) Genomic survey of premetazoans shows deep conservation of cytoplasmic tyrosine kinases and multiple radiations of receptor tyrosine kinases. Sci Signal 5(222):ra35

Tarver JE, Donoghue PC, Peterson KJ (2012) Do miRNAs have a deep evolutionary history? Bioessays 34(10):857–866

Tonon T, Rousvoal S, Roeder V, Boyen C (2008) Expression profiling of the mannuronan C5-epimerase multigenic family in the brown alga *Laminaria digitata* (Phaeophyceae) under biotic stress conditions. J Phycol 44:1250–1256

Venkatachalam K, Montell C (2007) TRP channels. Annu Rev Biochem 76:387–417

Woelkerling WJ (1990) An introduction. In: Cole KM, Sheath RG (eds) Biology of the red algae. Cambridge University Press, Cambridge, pp 1–6

Xiao S, Knoll AH, Yuan X, Pueschel CM (2004) Phosphatized multicellular algae in the neoproterozoic Doushantuo formation, China, and the early evolution of florideophyte red algae. Am J Bot 91(2):214–227

Yoon HS, Hackett JD, Ciniglia C, Pinto G, Bhattacharya D (2004) A molecular timeline for the origin of photosynthetic eukaryotes. Mol Biol Evol 21(5):809–818

Yoon HS, Müller KM, Sheath RG, Ott FD, Bhattacharya D (2006) Defining the major lineages of red algae (Rhodophyta). J Phycol 42:482–492

Yoon HS, Zuccarello GC, Bhattacharya D (2010) Evolutionary history and taxonomy of red algae. In: Seckbach J, Chapman DJ (eds) Red algae in the genomic age. Springer, Dordrecht, pp 25–42

Social Amoebae and Their Genomes: On the Brink to True Multicellularity

Gernot Glöckner

Abstract Amoebozoa are one of the main branches of eukaryotes with an independent evolutionary history of more than 1 billion years. The social amoebae within this clade are capable of organizing several thousand cells during their developmental cycle to form sophisticated multicellular structures. Recently, the genomes of a handful of these social amoebae have been sequenced. This wealth of genomic data enables the in-depth analysis of common and derived genomic features, and allows us to draw conclusions regarding the basic requirements of multicellularity within this clade. The genomes of social amoebae are relatively small, but have gene numbers comparable to those of true multicellular systems. Some factors required for multicellularity are readily identifiable in these genomes, but others remain elusive as genome sequences from closely related outgroups to social amoebae are not yet available. Besides genes, genome and gene organization seem to play a role in establishing multicellularity. In this chapter I summarize the data obtained from all currently known genomes of social amoebae with an emphasis on features related to multicellularity.

Keywords Comparative genomics · Transcriptomics · Global regulation of multicellularity

Introduction

Multicellularity evolved several times in the eukaryote kingdom (Parfrey and Lahr 2013, Chapter "Timing the Origins of Multicellular Eukaryotes Through Phylogenomics and Relaxed Molecular Clock Analyses"). The era of genomics makes it now

G. Glöckner (✉)
Leibniz-Institute of Freshwater Ecology and Inland Fisheries, IGB,
Müggelseedamm 301, 12587 Berlin, Germany
Tel.: +49-221-478-87375

Institute for Biochemistry I Medical Faculty, University of Cologne,
Joseph-Stelzmann-Straße 52, 50931 Köln, Germany
e-mail: gloeckner@igb-berlin.de; gernot.gloeckner@uni-koeln.de

I. Ruiz-Trillo, A. M. Nedelcu (eds.), *Evolutionary Transitions to Multicellular Life*,
Advances in Marine Genomics 2, DOI 10.1007/978-94-017-9642-2_17

possible to reveal the genetic basis for this evolutionary outcome. Comparative genomics of closely related species help discern between clade-specific innovations and species-specific modifications of traits. Social amoebae (see Chapter "The Evolution of Developmental Signalling in Dictyostelia from an Amoebozoan Stress Response") have the ability to organize and coordinate themselves into multicellular structures for a limited amount of time. Multicellularity represents only a short episode in the life of social amoebae; for instance, in the laboratory, the slime mold, *Dictyostelium discoideum* completes one cycle within 24 h. During this phase, individual cells cannot feed on bacteria anymore and are not able to divide more than once, presumably to a already started cell cycle. In contrast to other systems where multicellular organisms have somatic cells and a germline that produces gametes, the multicellular life stage in social amoebae serves only the survival of asexually produced spores. Around 20 % of the cells in a fruiting body are sacrificed to form a dead stalk. This part of the sorocarp is thus comparable to the mortal soma of true multicellular organisms.

Social amoebae are very amenable to the study of multicellularity. They exhibit some, but not all features of true multicellular organisms and therefore might be useful to dissect the different organizational tasks, which are implemented in other, true multicellular organisms.

D. discoideum as a Model System

D. discoideum has a sophisticated life style with a multicellular stage, the analysis of which has attracted researchers for many decades. Since the first isolation and cultivation of this species (Raper and Smith 1939), many mutants with aberrant fruiting body morphology, or that arrest at different stages of the developmental cycle, have been identified and generated, and the genes underlying the defective phenotypes have been characterized. In addition to genes that specifically control initiation and proper propagation through the developmental cycle, genes that impact the cycle when mutated but that are common to all eukaryotes were also identified. Presumably the effect of most of these latter genes is only indirect. For one such component it was shown that down-regulation of a major transcriptional regulator of the initiation of the cycle was causative for the mutant phenotype (Lucas et al. 2009).

Thus, the developmental cycle is based on a complex interplay between gene products dedicated only to the cycle and gene products fulfilling normal tasks in the cell but at the same time being needed for progression through the cycle. This might be due to general cellular functions, which are also needed during the developmental cycle; however, it is also possible that some existing gene products with defined functions were recruited to execute additional indispensable steps in development. This intertwining of basic cellular functions with functions only associated with multicellularity makes it difficult to pinpoint the genes invented or recruited solely for the developmental cycle. Another question that remains so far unanswered is which genes are later lineage- or species-specific additions to the developmental cycle, and therefore might have been dispensable in the last common ancestor (LCA) of all social amoebae.

A prerequisite for the understanding of this complex interplay was the sequencing of the *D. discoideum* genome and its subsequent analysis. Despite difficulties caused by the extreme AT nucleotide bias of nearly 80 %, this goal was achieved in 2005 (Eichinger et al. 2005; Glöckner et al. 2002). At only 34 Mb, this genome is relatively small. Although more than 12,000 genes have been initially predicted (Blanco et al. 2007), later manual annotation resulted in several genes being downgraded to pseudogenes (Fey et al. 2009). For comparison, the yeast genome is only 12 Mb in size and encodes less than 6000 genes (Goffeau et al. 1996). Yeasts are autonomously replicating and mating species with a genetic make-up enabling free-living behavior.

The surplus of genes found in social amoebae can in part be attributed to additional capabilities such as a highly sophisticated cytoskeleton enabling, for instance, amoeboid movement and phagocytosis, but a large part is likely directly involved in the developmental cycle.

Additionally, the genome contains 10 % complex repetitive elements, consisting of several classes of RNA and DNA transposons, but no retroviruses (Glöckner et al. 2001). One class of these repeats is clustered at the tip of each chromosome and was proven to function as a centromere (Dubin et al. 2010). The nucleus also contains the rDNA palindrome, which is an extrachromosomal unit that is transmitted to the next generation like a true chromosome. This unit contains the rRNA genes and is highly amplified, presumably to enable the high expression of rRNAs (Sucgang et al. 2003).

The coding portion of genes is littered with simple repeat structures consisting of codons with only A and T nucleotides. These are translated into long homo-polymer runs, contributing considerably to the overall high AT bias of this genome (Szafranski et al. 2005). Micro-RNAs are also present in *D. discoideum* and seem to play a role in the developmental cycle (Avesson et al. 2011).

Since no fossil records of social amoebae are available, a calibration of the evolutionary clock within the social amoebae is impossible. Under the assumption of mutation rates comparable to those in animals and plants the last common ancestor of social amoebae was estimated to have lived more than 600 million years ago (Heidel et al. 2011; Schaap et al. 2006). However, the mutation rate in *D. discoideum* seems to be low compared to other species as was shown by the sequencing of mutation accumulation lines (Saxer et al. 2012). If this low mutation rate was maintained throughout the evolutionary history of social amoebae, the observed divergence of orthologs between the different groups would lead to an underestimation of the time span since the first occurrence of social amoebae.

The Whole Breadth of Social Amoebae Evolution

Genome analysis of one single genome within a certain evolutionary lineage is a prerequisite to understand the make-up of a specific organism. Comparisons to distantly related species then specify, which differences and innovations are specific to this organism. However, a species is subject to evolutionary constraints and selection pressures leading to further speciation and adaptation to specific environmental conditions. The factors underlying a trait of interest can be obscured by those adaptations. Genome comparisons of several related species showing the same trait can be used to circumvent these difficulties. Furthermore, if species covering the evolutionary breadth of an entire branch are being analyzed, the specific time point where a characteristic trait evolved can be identified. In this way it is possible to discern common, lineage-specific features, from species-specific innovations. Analyses of related genomes can also help identify conserved and derived features within a certain evolutionary branch, while comparisons to distantly related species can reveal features shared by the entire branch and that might have been present in the LCA.

Social amoebae comprise four major and several minor evolutionary branches (Romeralo et al. 2011; Schaap et al. 2006). From all major branches at least one genome is currently available in draft or complete state (Glöckner and Noegel 2013; Heidel et al. 2011; Sucgang et al. 2011). All genomes encode more than 10,000 protein-coding genes (Table 1). Interestingly, complex repetitive elements contribute much to the slightly larger genome size in *D. discoideum* since all other genomes have a smaller share of these genomic features. The functional parts of the chromosomal organization (centromeres and telomeres) are different in all genomes examined: while the ancestral state of social amoebae centromeres could have been clusters of specialized transposons, the so-called DIRS elements, like those observed in *D. discoideum* (group 4) and *D. fasciculatum* (group 1) all other genomes have employed different, but so far unknown sequences to fulfill this task (Glöckner and Noegel 2013; Heidel et al. 2011). Telomeres also underwent changes in the different species, which involved the replacement of normal eukaryote telomere repeats by parts of the rDNA palindrome (Table 1). The rDNA palindromes differ in size between species due to species-specific clusters of repeated sequences.

The *D. lacteum* genome (Schaap, P and Glöckner, G. unpublished; for an annotated version see Felder et al. 2013) is surprisingly small, consisting of only 23 Mb. Despite this, it encodes almost as much genes as the other social amoebae. Part of this size reduction is due to the smaller intergenic regions and fewer introns, but another part is due to the reduced size of gene families. It remains to be seen whether a further genome and coding potential reduction can be possible without losing the capability for multicellular development.

The 600 million (or more) years of independent evolution of social amoebae led to the saturation of the genomes with mutations. This saturation sometimes obscures the true evolutionary relationships among genes making it difficult or even impossible to define orthologous pairs based on sequence similarity alone. Analysis of some gene families however revealed that comparable numbers of functional domains often exist despite high diversification of gene and protein sequences (Heidel et al. 2011). While the conserved genes of the basic cellular machinery are alignable even at the DNA level (if their amino acid sequences are used to guide the alignment), most members of several other gene families (e.g., transcription factors) have no detectable orthologs. Overall, only roughly half of all genes have a detectable ortholog in all species. Since syntenic regions are largely absent, positional information is also not usable for this purpose. Hidden functional similarities however can be detected using information on domain structures. If this hidden functional similarity is taken into account, the

Table 1 Genome properties of social amoebae. Data taken from (Heidel et al. 2011) and extended. Only completed genomes are shown

	DD	DP	DL	PP	DF
Contigs	226	1213	54	52	33
Supercontigs	6	838	54	41	25
Total nucleotides (Mbp)	35	33	23	33	31
Average contig length (kbp)	155	27	433	634	1064
Overall nucleotide frequency (A/T %)	77.6	75.4	70.2	68	66.2
Palindrome arm size (kb)	45	26	14	15	28
Mitochondrial genome size (kb)	55	52	47	48	56
Chromosome numbers	6	nd	8 (or more)	7	6
Repeat content (%)	~ 10	3.4	< 1	< 1	< 1
Telomere repeat structure	Palindrome arm	Palindrome arm?	TAGGG + Palindrome arm	TAAGGG	TTAGGG
Predicted coding sequences (CDS)	13,433	12,410	10,958	12,373	12,173
Average gene length	1579	1689	1596	1552	1672
Gene density (CDS per Mb)	396	376	470	375	392
Nucleotide frequency in CDS (A/T %)	72.6	69.9	67.8	63.8	63.2
Predicted tRNAs	401	375	61	273	198

DD Dictyostelium discoideum, *DP D. purpureum* (both group 4), *DL D. lacteum* (group 3), *PP Polysphondilium pallidum* (group 2), *DF D. fasciculatum* (group 1)

genetic make-up of all social amoebae seems to have been stably inherited from their LCA. This fact and the observation that all genomes of social amoebae examined contain roughly the same number of genes suggests that no novelties arose in the social amoebae clade. Species diversification appears to be the result of mainly species-specific gene family expansions or shrinkages. Although we currently do not have a detailed overview of the genes needed for the multicellular life stage we can conclude that the main players in this cycle have orthologous functions in all social amoebae.

Conservation in the Developmental Cycle

The developmental cycle progresses over well-defined stages towards the final fruiting body supported by a stalk. All social amoebae pass through the same steps in the same order, even if the timing and morphology is different in different species. Thus, the coordination of events in the cycle must be tightly regulated and seemingly allows not much alteration. The advent of techniques that allow the study of transcriptomic changes made it possible to analyse global transcriptional changes during the life cycle.

Regulatory Stability

After the initial definition (using EST analyses) of more than 7300 transcribed genes in *D. discoideum* (Morio et al. 1998), microarrays were developed to follow the expression of these genes over the whole life cycle (Van Driessche et al. 2002). This analysis revealed that at least 25 % of all genes in the genome are affected in their expression levels during development. All morphologically distinct states also have a distinct global expression profile, but the most important transition occurs after 6 h following the induction of the developmental cycle. At that point, normal cellular functions are being shut down and the genes required for multicellular development are switched on. After that, different cell types with specific transcriptional activities emerge. Interestingly, the developmental program does not seem to be influenced by external factors such as nutrition. A further study compared the developmental programs of *D. discoideum* and *D. purpureum*, which are both in group 4 (Parikh et al. 2010). The authors showed that in both species the developmental program is remarkably conserved, only a slight shift in timing could be observed. Furthermore, orthologous genes are generally expressed with the same intensity.

Figure 1 shows selected genes from *D. fasciculatum* with different expression profiles. As in *D. discoideum* and *D. purpureum*, cellular functions are being shut down at the beginning of development (green curves), some genes are only needed for the initiation of the development (red curves), and some are only highly expressed when spores are being formed (dark blue curves). These profiles coincide with those observed for the orthologous genes in the two group 4 species discussed above, indicating a conservation of transcriptional activity over the whole breadth of social amoebae evolution. It remains to be shown how many orthologous genes in all groups have corresponding transcriptional profiles and how many differ in that respect. Initial analyses revealed that species-specific developmentally induced genes also exist (Glöckner, unpublished). It is currently unclear whether these genes modulate the developmental program and its outcome in a species-specific manner. However, it is also conceivable that a fraction of these genes are only hitchhiking on the developmental program and have no function in it.

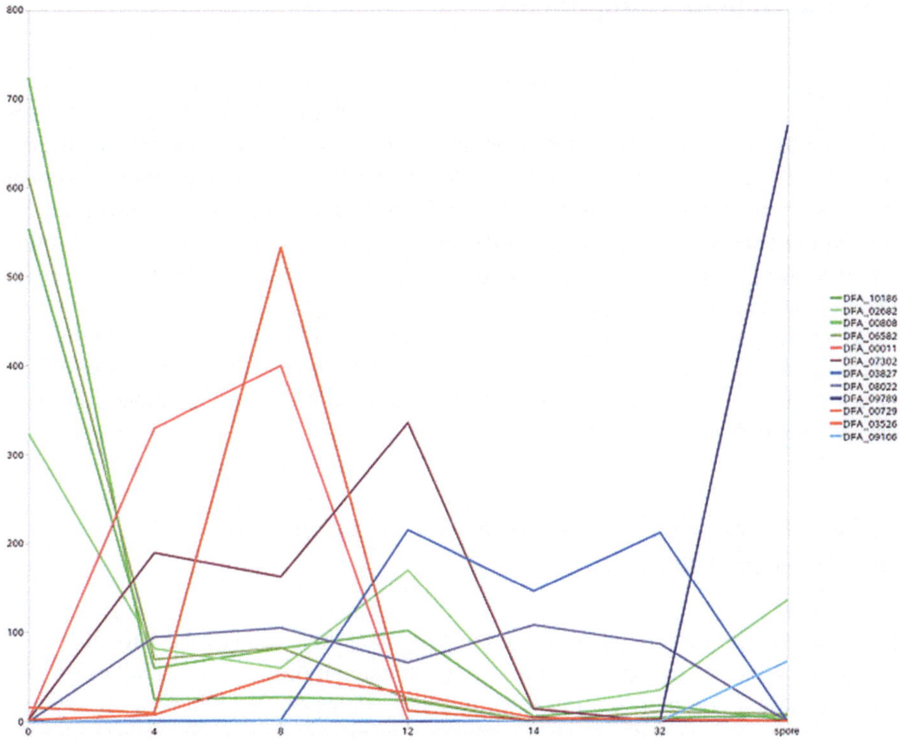

Fig. 1 Transcription patterns during the developmental cycle of *D. fasciculatum*. Data were obtained through GS FLX sequencing. X-axis: time points of harvest from developmental plates. Y-axis: number of reads. Expression profiles depicted in *green* indicate genes that are down-regulated upon initiation of the development. *Red* profiles are from genes that are up-regulated only in an early phase of development, whereas *blue* profiles are from genes that are needed during the whole development or only for spore formation

Constraints on Evolvability

Gene similarities and expression levels of genes at a certain developmental time point can be integrated to yield an estimation of the evolutionary age of developmental phases. Animal and plant development seems to progress in an "hourglass" fashion (Domazet-Loso and Tautz 2010; Quint et al. 2012). The genes needed in early and late stages of development are less conserved in these organisms than genes expressed in between. Thus, the ability to accumulate changes or acquire novel genes exists mainly in these developmental stages, while the middle phase is constrained and contains the deeply rooted developmental genes. A similar analysis was performed by Tian et al. using *D. discoideum* and *D. purpureum*, both from group 4 (Tian et al. 2013). The results suggest that the constraints increase towards the end of the developmental cycle, but no decrease as in the animal and plant kingdom could be observed. The authors propose that a likely explanation for this "half-hourglass" pattern is the lack of modularity in late social amoebae development. The multicellular stage of social amoebae thus would follow the same laws as those of animals and plants with the exception that the latest stages are missing.

Specific Genetic Components

It is not yet totally clear which components should be included in the list of essential constituents for multicellularity. Yet, some functions must be present to enable the coordination of many cells in one organism. Cells must communicate with each other to recognize what actions are planned and to which environmental signals they should respond in a coordinated way. Thus, a signaling system that recognizes its own signals must be present. Cells must also be able to adhere to each other and exclude foreign cells. Differentiation into different tissues, despite a common genetic makeup, is another hallmark of multicellularity.

Since the general mechanism of development in social amoebae likely remained unchanged since the LCA, common functions should be represented by similar genes in all social amoebae groups. Below, some components of these functions are discussed in greater detail.

Adhesion and Self-Recognition as a Prerequisite for Cell–Cell Contacts

During development, social amoebae must regulate their adhesiveness to other cells and to the substrate in order to achieve a coordinated morphogenesis. Adhesion proteins possess a common protein fold, which predates the evolution of eukaryotes. According to the social amoebae comparative genome browser (http://sacgb.fli-leibniz.de; Felder et al. 2013), *D. discoideum* has 39 proteins with EGF/laminin domains (IPR002049) that are known to mediate cell adhesion, growth, migration, and differentiation. Interestingly, the other social amoebae have only between 2 (*D. lacteum*; group 3) to 6 (*D. fasciculatum*; group 1) members of this protein family. This suggests a species-specific adaptation and expansion of this gene family and a prominent role for cell-cell recognition and adhesion in *D. discoideum*.

Another important group of proteins is the family of IPT/TIG domain-containing proteins normally found in cell surface receptors. Since many of these proteins are also predicted to contain transmembrane domains, they are likely located at the cell surface, further underlining their potential involvement in multicellular functions. Indeed, additional analyses revealed that some members (the tiger genes) of this large family are required for the self/self recognition at the beginning of the developmental cycle (Hirose et al. 2011). The tiger genes have a tail to tail organization on chromosome 3 of *D. discoideum*, likely sharing the same promoter (Fey et al. 2009). Presumably, the same family also fulfills this task in the other social amoebae. However, a search for orthologs of the tiger genes in the other species failed to retrieve any results if simple blast-based methods (Altschul et al. 1990) were employed. Since these genes are required for self-recognition, a high degree of sequence divergence is nevertheless expected. Indeed, a search for similarly arranged IPT/TIG domain-containing genes retrieved that at least one such pair exists in each of the completely sequenced social amoebae genomes (Glöckner, unpublished).

D. discoideum also has disintegrins in common with animals. These proteins regulate adhesiveness in Metazoa, and at least one of those proteins (ampA) is involved in the developmental cycle of *D. discoideum*. However, the social amoebae from the other major groups contain proteins only remotely similar to ampA. This apparent lack of orthology suggests that some regulatory functions in the multicellular development are species-specific.

Some components that are associated with adherens junctions (catenin, actinin, formins, VASP and myosin VII) in metazoa also have distant relatives in *D. discoideum* (Grimson et al. 2000). All these components can be found in all currently known social amoebae genomes. The recent finding of a further cell type forming a polarized epithelium emphasizes the similarities between multicellularity in metazoa and social amoebae, though some components like cadherin are missing (Dickinson et al. 2011, 2012).

Signaling Components

The initiation of the developmental cycle demands recognition of species-specific signals from the environment. The signaling system needed for that purpose very likely evolved from the already present chemotaxis system. To be able to sense environmental stimuli a cell needs receptors. In many species this task is performed by G-protein-coupled cell surface receptors (GPCRs), which can sense a variety of different signals such as Ca^{2+}, light, nucleotides, peptides, and secondary metabolites. The GPCRs are subdivided into six families without significant sequence similarity between the families. Analysis of the full gene complement of GPCRs in all groups showed that this family underwent species-specific amplification, and a clear orthology relationship is not detectable for most of the family members (Heidel et al. 2011).

Social amoebae have adapted ABC transporters to control various developmental signaling events. In *D. discoideum*, several ABC transporters (TagA, TagB and TagC) have been shown to be used for peptide-based signaling, similar to that previously observed for mating in *S. cerevisiae* and antigen presentation in human T cells (Anjard and Loomis 2002; Anjard et al. 1998; Asghar et al. 2012; Good et al. 2003; Good and Kuspa 2000). Interestingly, the Tag proteins have a unique domain architecture, where a serine protease domain is adjacent to a single transporter domain, which likely adds functionality to these proteins. Additional common signaling systems are involved in the developmental cycle at various stages.

Social amoebae also possess a wealth of polyketide synthases (PKS). Our current knowledge regarding which secondary metabolites are generated by these enzymes is scarce. The lack of orthology between members of this heavily expanded family from different species suggests a highly species-specific repertoire of signaling molecules (Heidel et al. 2011). Most of these are likely used for defense or other response to environmental clues, but at least one PKS is involved in the developmental cycle (Ghosh et al. 2008; Zucko et al. 2007). Additional secondary metabolites produced likely by PKSs were also proven to act in the developmental cycle (Saito et al. 2006).

A hallmark of all signaling systems in social amoebae that were analysed thus far is that some family members seem to be involved in the developmental cycle, while others fulfill different purposes not related to development. The divergent amplification of the families, the lack of orthologs for most members, and the supposed partly overlapping functions of the signaling components render their analysis complicated. Only if a single ortholog of a developmentally important gene is present for a certain family member in each species, conservation of function and therefore its membership in the core gene set of the developmental cycle can be taken as granted.

DNA Organization is Alike in Multicellular Systems

A eukaryote genome is maintained in the nucleus as DNA strands wound around histones. The expression level of genes can depend on the mode of packaging, since nucleosomes can make transcription factor binding sites inaccessible (Moreira and Holmberg 1998). Next generation sequencing made it possible to analyse the patterns in which these histones organize the DNA. DNA sequences associated with nucleosomes can be identified by treating chromatin with micrococcal nucleases. These enzymes digest DNA only if it is freely accessible, and therefore sequences of nucleosomal DNA remain intact. In several model organisms the nucleosome positions were already determined showing that protein coding genes are covered with arrays of nucleosomes, whereas the promoter regions are mainly nucleosome free (Lee et al. 2004; Schones and Zhao 2008). Unicellular organisms like yeasts place the first nucleosome over the transcription start site (TSS). This mode of placement requires that before transcription the contact between histones and DNA be lost so that the transcription factors have free access to their binding sites. In contrast, multicellular species like *Drosophila* place their first genic nucleosome downstream of the TSS, which leaves transcription factor binding sites accessible. When the polymerase encounters the first nucleosome, this causes pausing of the polymerase which, at least in animals, may contribute to timing of the developmental programs (Levine 2012). The *D. discoideum* pattern of nucleosome occupancy is surprisingly alike to that of multicellular species in this respect. While currently not many uni- and multicellular systems have been analysed with respect to nucleosome positioning, this finding points to a deep involvement of chromatin states in the regulation of multicellularity in such diverse organisms like animals, plants, and social amoebae. Interestingly, the global pattern of occupancy shows moderate change upon entering the developmental cycle (Chang et al. 2012). The canonical spacing of 162 bases in the vegetative phase shifts towards 169 bases upon onset of starvation, which is indicative of transcriptional quiescence. This does not exclude that the expression of specific genes could be increased by changes in nucleosome positioning during the developmental cycle, since the chosen resolution allows only a global analysis.

Other Transiently Multicellular Systems

The social behavior of individual cells during a part of the life cycle is not unique to social amoebae within the Amoebozoa. Rather, this type of multicellularity evolved several times in diverse eukaryote lineages. The protist, *Copromyxa protea*, is not a slime mold but belongs to the Tubulinea. Despite residing in the Amoebozoa lineage, this species very likely evolved this behavior independently of the true slime molds (Brown et al. 2011). Such an independent evolution points to a simple and common basic toolbox, of which Amoebozoa species made use to evolve this life style. A complete genome analysis of such species outside the social amoebae clade is needed to confirm this conclusion. Further transiently multicellular systems can be found in Alveolata (Ciliates) (Lasek-Nesselquist and Katz 2001; Sugimoto and Endoh 2006), Opisthokonta (Brown et al. 2009), Excavata (Brown et al. 2012b), and Rhizaria (Brown et al. 2012a). All these sorocarp forming species are evolutionary distant and reside in different major branches of the eukaryote tree. The requirement for sorocarp formation (signalling to other cells, coordinated movements, and synchronized timing of developmental stages) however is the same. Due to the vast evolutionary distance between these species they likely have only the basic eukaryote layout in common, which stems from the LCA of all eukaryotes. This form of multicellularity is thus either intrinsically built in the make-up of all eukaryotes and can be easily evolved under various selective pressures, or each lineage has to acquire a handful of additional genes to make use of the genetic material already present. It will be of utmost interest to compare the genomes of all these divers species to elucidate how different the strategies are to achieve multicellularity.

Summary

1. Social amoebae genomes are remarkably uniform in size and coding capacity.
2. Orthology detection is hampered by vast evolutionary distances.
3. The core gene set of the developmental program is unchanged since the establishment of the LCA of social amoebae.
4. Expression throughout the cycle is tightly regulated and similar in all social amoebae.
5. Conservation of genes and their expression in the developmental cycle is highest towards the end of the cycle (spore formation).
6. Social amoebae share some features of multicellular development with animal and plants, especially signaling systems.
7. Self recognition is mediated by compatible pairs of adhesion molecules.
8. Nucleosome patterns are alike those in other multicellular systems.
9. Nucleosome spacing shifts upon entering the developmental cycle presumably to silence global gene expression.

References

Altschul SF, Gish W, Miller W, Myers EW, Lipman DJ (1990) Basic local alignment search tool. J Mol Biol 215(3):403–410. doi:10.1016/S0022-2836(05)80360-2

Anjard C, Loomis WF (2002) Evolutionary analyses of ABC transporters of *Dictyostelium discoideum*. Eukaryot Cell 1(4):643–652

Anjard C, Zeng C, Loomis WF, Nellen W (1998) Signal transduction pathways leading to spore differentiation in *Dictyostelium discoideum*. Dev Biol 193(2):146–155. doi:10.1006/dbio.1997.8804

Asghar A, Groth M, Siol O, Gaube F, Enzensperger C, Glöckner G, Winckler T (2012) Developmental gene regulation by an ancient intercellular communication system in social amoebae. Protist 163(1):25–37. doi:10.1016/j.protis.2010.12.002

Avesson L, Schumacher HT, Fechter P, Romby P, Hellman U, Soderbom F (2011) Abundant class of non-coding RNA regulates development in the social amoeba Dictyostelium discoideum. RNA Biol 8(6):1094–1104

Blanco E, Parra G, Guigo R (2007) Using geneid to identify genes. In: Baxevanis AD et al (ed) Current protocols in bioinformatics, Chapter 4, Unit 4 3

Brown MW, Spiegel FW, Silberman JD (2009) Phylogeny of the "forgotten" cellular slime mold, Fonticula alba, reveals a key evolutionary branch within Opisthokonta. Mol Biol Evol 26(12):2699–2709

Brown MW, Silberman JD, Spiegel FW (2011) "Slime molds" among the Tubulinea (Amoebozoa): molecular systematics and taxonomy of Copromyxa. Protist 162(2):277–287

Brown MW, Kolisko M, Silberman JD, Roger AJ (2012a) Aggregative multicellularity evolved independently in the eukaryotic supergroup Rhizaria. Curr Biol 22(12):1123–1127

Brown MW, Silberman JD, Spiegel FW (2012b) A contemporary evaluation of the acrasids (Acrasidae, Heterolobosea, Excavata). Eur J Protistol 48(2):103–123

Chang GS, Noegel AA, Mavrich TN, Muller R, Tomsho L, Ward E, Felder M, Jiang C, Eichinger L, Glöckner G, Schuster SC, Pugh BF (2012) Unusual combinatorial involvement of poly-A/T tracts in organizing genes and chromatin in Dictyostelium. Genome Res 22(6):1098–1106

Dickinson DJ, Nelson WJ, Weis WI (2011) A polarized epithelium organized by beta- and alpha-catenin predates cadherin and metazoan origins. Science 331(6022):1336–1339. doi:10.1126/science.1199633 (New York, NY)

Dickinson DJ, Nelson WJ, Weis WI (2012) An epithelial tissue in Dictyostelium challenges the traditional origin of metazoan multicellularity. Bioessays 34(10):833–840

Domazet-Loso T, Tautz D (2010) A phylogenetically based transcriptome age index mirrors ontogenetic divergence patterns. Nature 468(7325):815–818. doi:10.1038/nature09632

Dubin M, Fuchs J, Graf R, Schubert I, Nellen W (2010) Dynamics of a novel centromeric histone variant CenH3 reveals the evolutionary ancestral timing of centromere biogenesis. Nucleic Acids Res 38(21):7526–7537

Eichinger L, Pachebat JA, Glöckner G, Rajandream MA, Sucgang R, Berriman M, Song J, Olsen R, Szafranski K, Xu Q, Tunggal B, Kummerfeld S, Madera M, Konfortov BA, Rivero F, Bankier AT, Lehmann R, Hamlin N, Davies R, Gaudet P, Fey P, Pilcher K, Chen G, Saunders D, Sodergren E, Davis P, Kerhornou A, Nie X, Hall N, Anjard C, Hemphill L, Bason N, Farbrother P, Desany B, Just E, Morio T, Rost R, Churcher C, Cooper J, Haydock S, van Driessche N, Cronin A, Goodhead I, Muzny D, Mourier T, Pain A, Lu M, Harper D, Lindsay R, Hauser H, James K, Quiles M, Madan Babu M, Saito T, Buchrieser C, Wardroper A, Felder M, Thangavelu M, Johnson D, Knights A, Loulseged H, Mungall K, Oliver K, Price C, Quail MA, Urushihara H, Hernandez J, Rabbinowitsch E, Steffen D, Sanders M, Ma J, Kohara Y, Sharp S, Simmonds M, Spiegler S, Tivey A, Sugano S, White B, Walker D, Woodward J, Winckler T, Tanaka Y, Shaulsky G, Schleicher M, Weinstock G, Rosenthal A, Cox EC, Chisholm RL, Gibbs R, Loomis WF, Platzer M, Kay RR, Williams J, Dear PH, Noegel AA, Barrell B, Kuspa A (2005) The genome of the social amoeba: *Dictyostelium discoideum*. Nature 435(7038):43–57

Felder M, Romualdi A, Petzold A, Platzer M, Sühnel J, Glöckner G (2013) GenColors-based comparative genome databases for small eukaryotic genomes. Nucleic Acids Res 41(Database issue):D692–D699. doi:10.1093/nar/gks1176

Fey P, Gaudet P, Curk T, Zupan B, Just EM, Basu S, Merchant SN, Bushmanova YA, Shaulsky G, Kibbe WA, Chisholm RL (2009) dictyBase–a *Dictyostelium* bioinformatics resource update. Nucleic Acids Res 37(Database issue):D515–D519

Ghosh R, Chhabra A, Phatale PA, Samrat SK, Sharma J, Gosain A, Mohanty D, Saran S, Gokhale RS (2008) Dissecting the functional role of polyketide synthases in *Dictyostelium discoideum*: biosynthesis of the differentiation regulating factor 4-methyl-5-pentylbenzene-1,3-diol. J Biol Chem 283(17):11348–11354. (Epub 5 Feb 2008)

Glöckner G, Noegel AA (2013) Comparative genomics in the Amoebozoa clade. Biol Rev Camb Philos Soc 88(1):215–225

Glöckner G, Szafranski K, Winckler T, Dingermann T, Quail MA, Cox E, Eichinger L, Noegel AA, Rosenthal A (2001) The complex repeats of *Dictyostelium discoideum*. Genome Res 11(4):585–594

Glöckner G, Eichinger L, Szafranski K, Pachebat JA, Bankier AT, Dear PH, Lehmann R, Baumgart C, Parra G, Abril JF, Guigo R, Kumpf K, Tunggal B, Cox E, Quail MA, Platzer M, Rosenthal A, Noegel AA (2002) Sequence and analysis of chromosome 2 of *Dictyostelium discoideum*. Nature 418(6893):79–85

Goffeau A, Barrell BG, Bussey H, Davis RW, Dujon B, Feldmann H, Galibert F, Hoheisel JD, Jacq C, Johnston M, Louis EJ, Mewes HW, Murakami Y, Philippsen P, Tettelin H, Oliver SG (1996) Life with 6000 genes. Science 274(5287):546, 563–567

Good JR, Kuspa A (2000) Evidence that a cell-type-specific efflux pump regulates cell differentiation in *Dictyostelium*. Dev Biol 220(1):53–61. doi:10.1006/dbio.2000.9611

Good JR, Cabral M, Sharma S, Yang J, Van Driessche N, Shaw CA, Shaulsky G, Kuspa A (2003) TagA, a putative serine protease/ABC transporter of *Dictyostelium* that is required for cell fate determination at the onset of development. Development 130(13):2953–2965

Grimson MJ, Coates JC, Reynolds JP, Shipman M, Blanton RL, Harwood AJ (2000) Adherens junctions and beta-catenin-mediated cell signalling in a non-metazoan organism. Nature 408(6813):727–731. doi:10.1038/35047099

Heidel AJ, Lawal HM, Felder M, Schilde C, Helps NR, Tunggal B, Rivero F, John U, Schleicher M, Eichinger L, Platzer M, Noegel AA, Schaap P, Glöckner G (2011) Phylogeny-wide analysis of social amoeba genomes highlights ancient origins for complex intercellular communication. Genome Res 21(11):1882–1891

Hirose S, Benabentos R, Ho H-I, Kuspa A, Shaulsky G (2011) Self-recognition in social amoebae is mediated by allelic pairs of tiger genes. Science 333(6041):467–470. doi:10.1126/science.1203903 (New York, NY)

Lasek-Nesselquist E, Katz LA (2001) Phylogenetic position of Sorogena stoianovitchae and relationships within the class Colpodea (Ciliophora) based on SSU rDNA sequences. J Eukaryot Microbiol 48(5):604–607

Lee CK, Shibata Y, Rao B, Strahl BD, Lieb JD (2004) Evidence for nucleosome depletion at active regulatory regions genome-wide. Nat Genet 36(8):900–905. doi:10.1038/ng1400

Levine M (2012) Paused RNA polymerase II as a developmental checkpoint. Cell 145(4):502–511

Lucas J, Bilzer A, Moll L, Zündorf I, Dingermann T, Eichinger L, Siol O, Winckler T (2009) The carboxy-terminal domain of *Dictyostelium* C-module-binding factor is an independent gene regulatory entity. PLoS One 4(4):e5012. doi:10.1371/journal.pone.0005012

Moreira JM, Holmberg S (1998) Nucleosome structure of the yeast CHA1 promoter: analysis of activation-dependent chromatin remodeling of an RNA-polymerase-II-transcribed gene in TBP and RNA pol II mutants defective in vivo in response to acidic activators. EMBO J 17(20):6028–6038. doi:10.1093/emboj/17.20.6028

Morio T, Urushihara H, Saito T, Ugawa Y, Mizuno H, Yoshida M, Yoshino R, Mitra BN, Pi M, Sato T, Takemoto K, Yasukawa H, Williams J, Maeda M, Takeuchi I, Ochiai H, Tanaka Y (1998) The

Dictyostelium developmental cDNA project: generation and analysis of expressed sequence tags from the first-finger stage of development. DNA Res 5(6):335–340

Parfrey LW, Lahr DJ (2013) Multicellularity arose several times in the evolution of eukaryotes. Bioessays 35(4):339–347 (Response to doi:10.1002/bies.201100187)

Parikh A, Miranda ER, Katoh-Kurasawa M, Fuller D, Rot G, Zagar L, Curk T, Sucgang R, Chen R, Zupan B, Loomis WF, Kuspa A, Shaulsky G (2010) Conserved developmental transcriptomes in evolutionarily divergent species. Genome Biol 11(3):R35

Quint M, Drost HG, Gabel A, Ullrich KK, Bonn M, Grosse I (2012) A transcriptomic hourglass in plant embryogenesis. Nature 490(7418):98–101

Raper KB, Smith NR (1939) The growth of *Dictyostelium discoideum* upon pathogenic bacteria. J Bacteriol 38(4):431–445

Romeralo M, Cavender JC, Landolt JC, Stephenson SL, Baldauf SL (2011) An expanded phylogeny of social amoebas (Dictyostelia) shows increasing diversity and new morphological patterns. BMC Evol Biol 11:84

Saito T, Taylor GW, Yang JC, Neuhaus D, Stetsenko D, Kato A, Kay RR (2006) Identification of new differentiation inducing factors from *Dictyostelium discoideum*. Biochim Biophys Acta 1760(5):754–761

Saxer G, Havlak P, Fox SA, Quance MA, Gupta S, Fofanov Y, Strassmann JE, Queller DC (2012) Whole genome sequencing of mutation accumulation lines reveals a low mutation rate in the social amoeba *Dictyostelium discoideum*. PLoS One 7(10):e46759

Schaap P, Winckler T, Nelson M, Alvarez-Curto E, Elgie B, Hagiwara H, Cavender J, Milano-Curto A, Rozen DE, Dingermann T, Mutzel R, Baldauf SL (2006) Molecular phylogeny and evolution of morphology in the social amoebas. Science 314(5799):661–663

Schones DE, Zhao K (2008) Genome-wide approaches to studying chromatin modifications. Nat Rev Genet 9(3):179–191. doi:10.1038/nrg2270

Sucgang R, Chen G, Liu W, Lindsay R, Lu J, Muzny D, Shaulsky G, Loomis W, Gibbs R, Kuspa A (2003) Sequence and structure of the extrachromosomal palindrome encoding the ribosomal RNA genes in *Dictyostelium*. Nucleic Acids Res 31(9):2361–2368

Sucgang R, Kuo A, Tian X, Salerno W, Parikh A, Feasley CL, Dalin E, Tu H, Huang E, Barry K, Lindquist E, Shapiro H, Bruce D, Schmutz J, Salamov A, Fey P, Gaudet P, Anjard C, Babu MM, Basu S, Bushmanova Y, van der Wel H, Katoh-Kurasawa M, Dinh C, Coutinho PM, Saito T, Elias M, Schaap P, Kay RR, Henrissat B, Eichinger L, Rivero F, Putnam NH, West CM, Loomis WF, Chisholm RL, Shaulsky G, Strassmann JE, Queller DC, Kuspa A, Grigoriev IV (2011) Comparative genomics of the social amoebae *Dictyostelium discoideum* and *Dictyostelium purpureum*. Genome Biol 12 (2):R20

Sugimoto H, Endoh H (2006) Analysis of fruiting body development in the aggregative ciliate Sorogena stoianovitchae (Ciliophora, Colpodea). J Eukaryot Microbiol 53(2):96–102

Szafranski K, Lehmann R, Parra G, Guigo R, Glöckner G (2005) Gene organization features in A/T-rich organisms. J Mol Evol 60(1):90–98

Tian X, Strassmann JE, Queller DC (2013) Dictyostelium development shows a novel pattern of evolutionary conservation. Mol Biol Evol 30(4):977–984

Van Driessche N Shaw C Katoh M Morio T Sucgang R Ibarra M Kuwayama H Saito T Urushihara H Maeda M Takeuchi I Ochiai H Eaton W Tollett J Halter J Kuspa A Tanaka Y Shaulsky G (2002) A transcriptional profile of multicellular development in *Dictyostelium discoideum*. Development 129(7):1543–1552

Zucko J, Skunca N, Curk T, Zupan B, Long PF, Cullum J, Kessin RH, Hranueli D (2007) Polyketide synthase genes and the natural products potential of *Dictyostelium discoideum*. Bioinformatics 23(19):2543–2549

Part V
Molecular Mechanisms

Transcription Factors and the Origin of Animal Multicellularity

Arnau Sebé-Pedrós and Alex de Mendoza

Abstract The transition from a unicellular life-style, with temporal cell differentiation to a multicellular life-style, with both temporal and spatial cell differentiation, required the expansion of the regulatory capabilities of ancient animals. In this chapter, we describe how this change occurred from the perspective of transcription factor (TF) evolution. First, we revise TF diversification throughout eukaryotes. We trace the evolutionary origins of major TF classes and describe general patterns of TF content and how they correlate with multicellular life-styles in eukaryotes. We then focus on the animal TF developmental toolkit at the gene family level. Many of the metazoan developmental TFs originated in a unicellular context; yet there are also many TFs that evolved at the onset of Metazoa. Finally, we describe the changes that led to the establishment of gene regulatory networks that control animal multicellularity and review different case-examples that have provided illuminating insights into this question.

Keywords *Cis* evolution · *Trans* evolution · Co-option · Gene regulatory networks · Developmental toolkit · Choanoflagellates · *Capsaspora owczarzaki* · Porifera

Introduction

Animal development involves the orchestrated deployment of gene batteries to control spatial and temporal cell differentiation. Conversely, unicellular life cycles mainly require temporal changes, in order to regulate the transitions from one life stage to another or to regulate metabolic activities and responses to environmental cues. Moreover, control over cell proliferation (in order to minimize the emergence of non-cooperating cells) is a critical requirement in multicellular lineages (Grosberg and Strathmann 2007). Transcription factors (TFs) are key players in these processes, as they bind DNA in a sequence-specific manner and enhance or repress gene expression. Indeed, it has long been hypothesized that the complexity of the transcription

A. de Mendoza (✉) · A. S.-Pedrós
Institut de Biologia Evolutiva, CSIC-Universitat Pompeu Fabra, Passeig Marítim de la Barceloneta 37–49, 08003 Barcelona, Spain
e-mail: alexmendozasoler@gmail.com

I. Ruiz-Trillo, A. M. Nedelcu (eds.), *Evolutionary Transitions to Multicellular Life*, Advances in Marine Genomics 2, DOI 10.1007/978-94-017-9642-2_18

regulation system is correlated with organismal complexity (Levine and Tjian 2003). TFs can be viewed as regulatory hubs, where the information processed by the cell must be translated into biogenesis. For this reason, TFs are often key downstream targets of the major metazoan developmental signaling pathways (Pires-daSilva and Sommer 2003). Taking all this into account, it has been hypothesized that the onset of animal multicellularity was accompanied by an increase in transcription factor diversity (Rokas 2008).

Despite the striking diversity of body plans found in metazoans, there is a transcription factor toolkit largely shared by all metazoans (Vaquerizas et al. 2009; Degnan et al. 2009). This toolkit is known to be involved in embryonic development in both bilaterian and non-bilaterian animal lineages (Technau and Steele 2011; Adamska et al. 2011, Chapter "The Evolution of Developmental Signalling in Dictyostelia from an Amoebozoan Stress Response"). Therefore, animal TFs represent a shared patterning language that facilitate multicellularity and govern development. This fact brings up several questions. Since multicellularity evolved more than once in the history of eukaryotes, can we identify similarities or differences between animal multicellularity and other multicellularities in terms of transcription factor toolkits? When did the metazoan TF types emerge and how did they evolve? Finally, if metazoan TFs were present in the unicellular ancestors of Metazoa, how were they adapted for use in the context of a multicellular developmental program?

A Natural History of Eukaryotic Transcription Factors

The acquisition of multicellularity by eukaryotes is a story of evolutionary convergence, since at least 26 independent transitions to multicellularity have been reported (Grosberg and Strathmann 2007; Parfrey and Lahr 2013, Chapter "Timing the Origins of Multicellular Eukaryotes Through Phylogenomics and Relaxed Molecular Clock Analyses"). If we restrict our definition of multicellularity to complex multicellularity, which involves extensive spatial cellular differentiation (see Chapter "Independent Emergence of Complex Multicellularity in the Brown and Red Algae" for a discussion of complex multicellularity), we can identify at least seven independent groups that have acquired complex multicellularity, including animals, plants, fungi, and brown and red algae (Knoll 2011; Niklas and Newman 2013). Thus, it is possible to analyze the sequenced genomes of species belonging to complex multicellular lineages and compare their transcription factor repertoires with those of the closely related unicellular species. Indeed, eukaryotes are known to exhibit great diversity in terms of TF abundance and TF type composition (Weirauch and Hughes 2011), which we define as the TFome, and by looking at patterns of abundance and diversity we can infer trends associated with multicellularity.

We recently performed such analyses and identified some general trends in the total abundance of TFs (Fig. 1; de Mendoza et al. 2013). For example, taxa with complex multicellular development taxa, i.e. Embryophyta and Metazoa, present a dramatic increase in TF numbers compared to other eukaryotes. Moreover, the morphologically simpler forms within these groups (sponges, mosses, etc.) have fewer

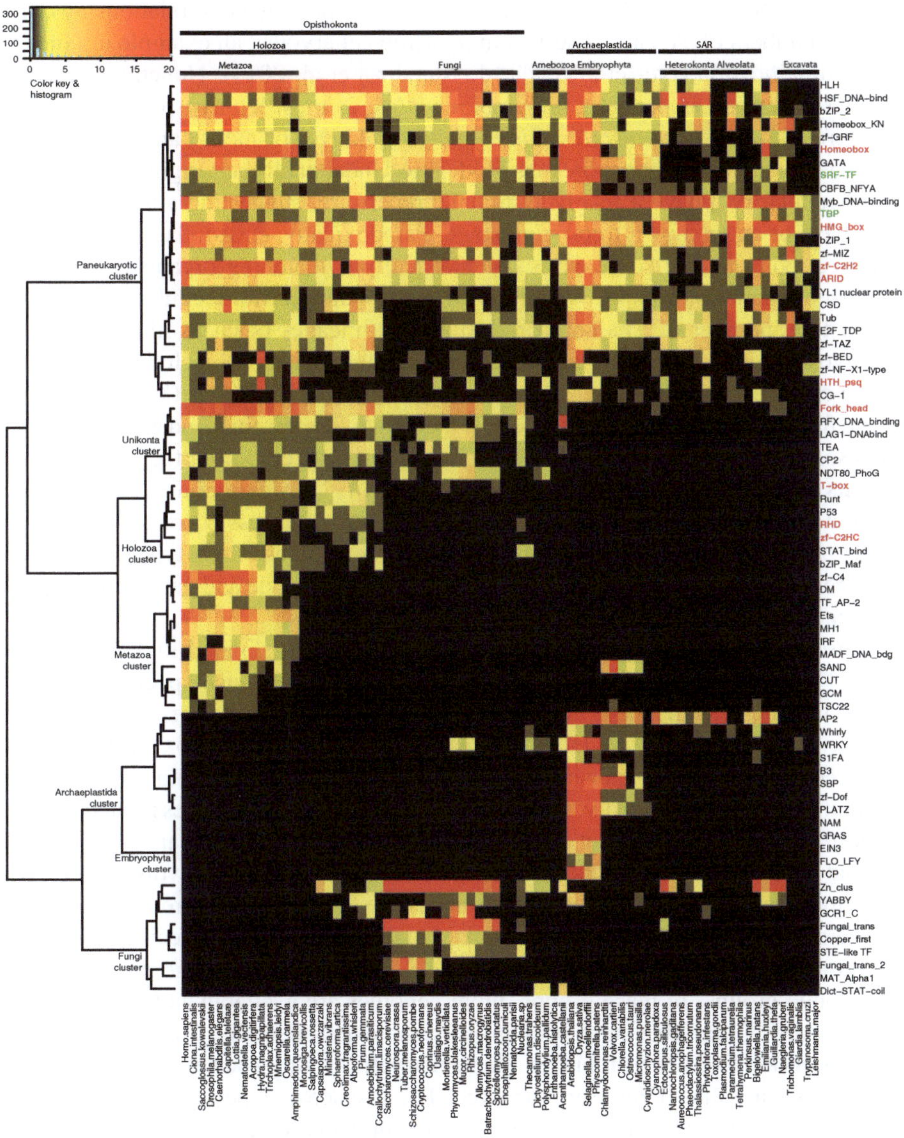

Fig. 1 Presence and abundance of transcription factors (TFs) in eukaryotes. The heat map depicts absolute TF counts according to the colour scale. TFs were identified using the HMM profiles of their respective DNA-binding domains. TF types (*rows*) are clustered according to abundance and distribution, and species (*columns*) are grouped according to phylogenetic affinity. Major eukaryotic lineages are indicated (*Top*). (Adapted from de Mendoza et al. 2013)

TFs (de Mendoza et al. 2013). Irrespective of their phylogenetic relationships, the lowest numbers of TFs are observed in parasitic eukaryotes, an example of convergent simplification (Iyer et al. 2008; de Mendoza et al. 2013). Nonetheless, some species described as parasitic or symbiotic, such as most ichthyosporeans and the filasterean *Capsaspora owczarzaki* (both close relatives of metazoans; see Chapter "Filastereans and Ichthyosporeans: Models to Understand the Origin of Metazoan Multicellularity") have a relatively rich TF repertoire, suggesting a more complex life cycle or an unknown free-living stage. Paradoxically, choanoflagellates, which are free-living and in some cases colonial and prey-catching organisms (Chapter "Choanoflagellates: Perspective on the Origin of Animal Multicellularity"), have a lower total number of TFs compared to related lineages (such as filastereans and ichthyosporeans). Another factor that can explain particularly rich TF repertoires is the occurrence in some groups of Whole Genome Duplications (WGD) (such as those found in vertebrates, zygomycetes, the ciliate *Paramecium* and embryophytes; also see Chapter "The Evolution of Transcriptional Regulation in the Viridiplantae and its Correlation with Morphological Complexity") (Maere et al. 2005; Van De Peer et al. 2009). WGD tends to lead to the deletion of excess copies of duplicated genes; however, the TFs are one of the most resilient genes to this type of loss (Van De Peer et al. 2009; De Smet et al. 2013). Therefore life-style and genome dynamics influence the total number of TFs in eukaryotic genomes.

Based on their phylogenetic distribution, TFs can be divided into two groups (Fig. 1): those that are paneukaryotic and those that are lineage-specific. The first group includes TFs that have undergone independent expansions in several lineages, such as Myb_DNA-binding, HLH, GATA, SRF-TF, bZIP, Homeobox, HMGbox and zf-C2H2 (Fig. 1). The expansion of ancient TF families was an evolutionary source of innovation that allowed diversification of the TFome, in some cases related to multicellularity (de Mendoza et al. 2013). The second group includes plant-, fungal- and animal-specific clusters of TFs, indicating that some lineages have evolved their own repertoires of TFs by diversifying their ancestral TFome. This correlation between clusters of lineage-specific TFs and eukaryotic groups with multicellular lineages might also be due to sampling bias as those lineages represent the most-sampled taxonomic groups. In these groups, more genomes are available and, more importantly, many TF families have been investigated experimentally. Thus, new specific TF clusters are likely to be found in other eukaryotic groups, such as stramenopiles or rhizarians, once their TFome are surveyed. The absence of experimental studies may also explain some exceptionally TF-poor taxa in which unknown TF families may have an important role. The case of the IBD family unique to *Trichomonas vaginalis* provides a good example of this (Iyer et al. 2008). It seems clear that ancestral TF types evolved dynamically through gene family expansions in some lineages, while new TFs were added to the ancestral repertoire in the lineages that led to fungi, plants and animals.

As we discussed above, there are strong phylogenetic patterns of TF diversity, in terms of both lineage-specificity and abundance. In addition, there are patterns in the relative contribution of each TF type in the TFome of each species, measured as the number of genes of each TF class as a percentage of the total number of TFs in that genome (Fig. 2). For example, animal genomes are dominated by Homeobox, zf-C2H2 and bHLH TFs (accounting for more than 50 % of the TFome), whereas unicellular holozoans have a distinct TFome profile, indicating that the transition to multicellularity involved a system-level change in TF type proportions (de Mendoza et al. 2013). The TF types that became predominant in animal genomes are mainly those involved in patterning and differentiation in animals, a structural requirement for an organism with diverse cell types (Degnan et al. 2009; Sebé-Pedrós et al. 2011). Conversely, p53-like TFs (e.g. T-box, Runx, p53 and others) or bZIP seem to be proportionally more important in non-bilaterian metazoans, with less cell types and simpler patterning. In contrast, they represent a small percentage in bilaterian lineages. The higher proportion of p53-like or bZip TFs in non-bilaterians and unicellular holozoan TFomes could reflect their potential role in regulating house-keeping functions, such as metabolic processes, proliferation or immunity response, although T-box is an interesting exception (Hammonds et al. 2013; Sebé-Pedrós et al. 2013).

Finally, the simplicity of the TFome of other complex multicellular eukaryote lineages, mainly red algae and brown algae (phaeophytes) (discussed in Chapter "Independent Emergence of Complex Multicellularity in the Brown and Red Algae"), is an intriguing question. Indeed, there is no expansion of total number of TFs when comparing the multicellular brown algae *Ectocarpus silicosus* and its unicellular relative *Nanochloropsis gaditana* (Cock et al. 2010; Radakovits et al. 2012). Judging from the large number of unique TFs in other complex multicellular lineages (plants and animals), there are likely to be undiscovered stramenopile- or brown-algae-specific TFs, but there is a dearth of functional studies in this group (Peters et al. 2008; Coelho et al. 2011). It is worth mentioning that *E. siliculosus* does not have a typical embryonic development, but rather a modular growth strategy, although this is not the case for other multicellular brown algae, such as *Fucus spiralis* (Bouget et al. 1998). We hypothesize that a richer TF repertoire will be found in brown algae with embryonic development. The case of red algae is similar, since the TFomes of both *Pyropia yezoensis* and *Chondrus crispus* are surprisingly simple compared to the unicellular red algae *Cyanidioschyzon merolae* (Nakamura et al. 2013; Collen et al. 2013).

The simplicity of the TF repertoires of some multicellular eukaryotes may be explained by their modes of development, although the identification of new lineage-specific TF types remains a critical issue to be resolved through future research in these groups. We can conclude that complex multicellularity is associated with enrichment of the TF toolkit (both in terms of abundance and innovation) in lineages with complex embryonic development: plants and animals. However, this toolkit is greatly influenced by the TF repertoires of their respective unicellular ancestors (de Mendoza et al. 2013). In the next section, we will focus on the early evolution of animal TF families in order to gain further insights into which genes evolved in a unicellular context.

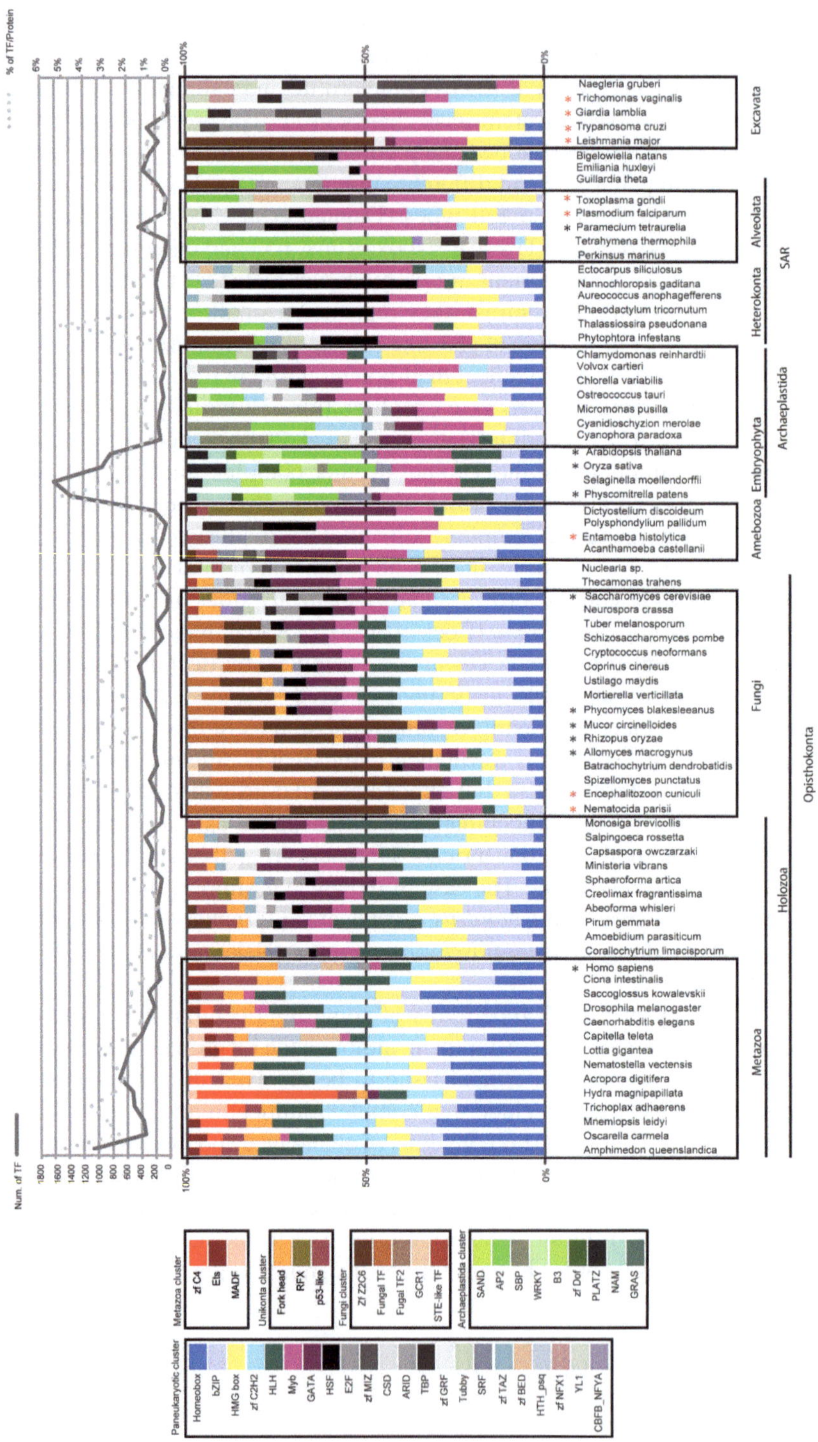
% of TF/Protein
6%
5%
4%
3%
2%
1%
0%
Num. of TF
1800
1600
1400
1200
1000
800
600
400
200
0
100%
50%
0%
Naegleria gruberi
Trichomonas vaginalis
Giardia lamblia
Trypanosoma cruzi
Leishmania major
Excavata
Bigelowiella natans
Emiliania huxleyi
Guillardia theta
Toxoplasma gondii
Plasmodium falciparum
Paramecium tetraurelia
Tetrahymena thermophila
Perkinsus marinus
Alveolata
Ectocarpus siliculosus
Nannochloropsis gaditana
Aureococcus anophagefferens
Phaeodactylum tricornutum
Thalassiosira pseudonana
Phytophtora infestans
Heterokonta
SAR
Chlamydomonas reinhardtii
Volvox cartieri
Chlorella variabilis
Ostreococcus tauri
Micromonas pusilla
Cyanidioschyzion merolae
Cyanophora paradoxa
Arabidopsis thaliana
Oryza sativa
Selaginella moellendorffii
Physcomitrella patens
Embryophyta
Archaeplastida
Dictyostelium discoideum
Polysphondylium pallidum
Entamoeba histolytica
Acanthamoeba castellanii
Amebozoa
Nuclearia sp.
Thecamonas trahens
Saccharomyces cerevisiae
Neurospora crassa
Tuber melanosporum
Schizosaccharomyces pombe
Cryptococcus neoformans
Coprinus cinereus
Ustilago maydis
Mortierella verticillata
Phycomyces blakesleeanus
Mucor circinelloides
Rhizopus oryzae
Allomyces macrogynus
Batrachochytrium dendrobatidis
Spizellomyces punctatus
Encephalitozoon cuniculi
Nematocida parisii
Fungi
Monosiga brevicollis
Salpingoeca rossetta
Capsaspora owczarzaki
Ministeria vibrans
Sphaeroforma artica
Creolimax fragrantissima
Abeoforma whisleri
Pirum gemmata
Amoebidium parasiticum
Corallochytrium limacisporum
Holozoa
Homo sapiens
Ciona intestinalis
Saccoglossus kowalevskii
Drosophila melanogaster
Caenorhabditis elegans
Capitella teleta
Lottia gigantea
Nematostella vectensis
Acropora digitifera
Hydra magnipapillata
Trichoplax adhaerens
Mnemiopsis leidyi
Oscarella carmela
Amphimedon queenslandica
Metazoa
Opisthokonta
Metazoa cluster
zf C4
Ets
MADF
Unikonta cluster
Fork head
RFX
p53-like
Fungi cluster
Zf Z2C6
Fungal TF
Fugal TF2
GCR1
STE-like TF
Archaeplastida cluster
SAND
AP2
SBP
WRKY
B3
zf Dof
PLATZ
NAM
GRAS
Paneukaryotic cluster
Homeobox
bZIP
HMG box
zf C2H2
HLH
Myb
GATA
HSF
E2F
zf MIZ
CSD
ARID
TBP
zf GRF
Tubby
SRF
zf TAZ
zf BED
HTH_psq
zf NFX1
YL1
CBFB_NFYA

Early Origins of Metazoan Developmental TFs

TFs have very disparate evolutionary origins and patterns. While some TF families seem to have remained quite static throughout evolution, preserving a high degree of similarity between orthologous gene families (Sebé-Pedrós et al. 2011), others such as zf-C2H2 evolved at a faster rate, making it extremely difficult to assign them to specific orthologous gene families over large evolutionary distances (Vaquerizas et al. 2009). Although the field of evolutionary developmental biology has identified many well-known animal developmental TF families in different phyla, these are only a subset of all TFs in metazoan genomes. Even among the members of a particular TF type, usually only some of them are developmental TFs. Therefore the identification of such TFs requires precise class and family phylogeny-based classifications. Here we will review the evolutionary histories of animal multi-gene TF families.

Homeobox is one of the most abundant and diverse developmental TF types in animals, and a milestone in the evolution of animal development (Bürglin 2011). During metazoan evolution, Homeobox genes increased from a rather simple gene complement to a highly diversified toolkit (Larroux et al. 2008). This expansion coincided with an increase in domain combinations (Bürglin 2011), including Pou, Lim, Cut or Paired domains that are found adjacent to the homeodomain. Some of these co-occurring domains are metazoan innovations (Pou, Cut, Paired, Six, Prospero or Iro-box), while others are ancient eukaryotic domains (e.g. LIM or zf-C2H2) fused to Homeobox proteins by domain shuffling. Unicellular relatives of metazoans do not possess a rich Homeobox complement; for example, choanoflagellates have just two Homeobox genes, both belonging to the TALE superclass (King et al. 2008). In fact, TALE Homeoboxes represent one of the two ancient lineages of Homeobox that, together with non-TALE Homeoboxes, have been present since the origin of eukaryotes (Derelle et al. 2007). In animals, the TALE superclass diversified into few families, namely Iroquois, PBX, Meis/PREP and Tgif (Larroux et al. 2008; Bürglin 2011; Fig. 3). In contrast, non-TALE Homeoboxes are extremely diverse. Sponges and ctenophores, which are potentially the two earliest branching metazoans, already possess members of the 5 classes of non-TALE Homeobox (ANTP, PRD-like, POU, LIM and SINE) (Larroux et al. 2008, Ryan et al. 2010). Later in

Fig. 2 Phylogenetic patterns of TFome composition across eukaryotes. For clarity, TF types representing $< 2\,\%$ of the corresponding TFome are not considered, and some are summarized in higher-level categories according to structural similarities. This is the case of (i) the Homeobox supergroup, which comprises Homeobox and Homeobox_KN/TALE; (ii) the bZIP supergroup, which comprise bZIP_1, bZIP_2, and bZIP_Maf; and (iii) the p53-like supergroup, which comprise p53, STAT, Runx, NDT80, LAG1, and RHD. To the *Left*, the total number of TF types present in each taxon and the relative abundance of each DNA Binding Domain (DBD) type in the TFome of every species are depicted using the colour code in the legend of DBDs. In the line graph (*Top*), the *solid line* indicates the total number of TFs in each species, and the *dotted line* indicates this number as a percentage of the total number of proteins. The *black asterisks* indicate species with WGDs. The *red asterisks* indicate strict parasites. (Adapted from de Mendoza et al. 2013)

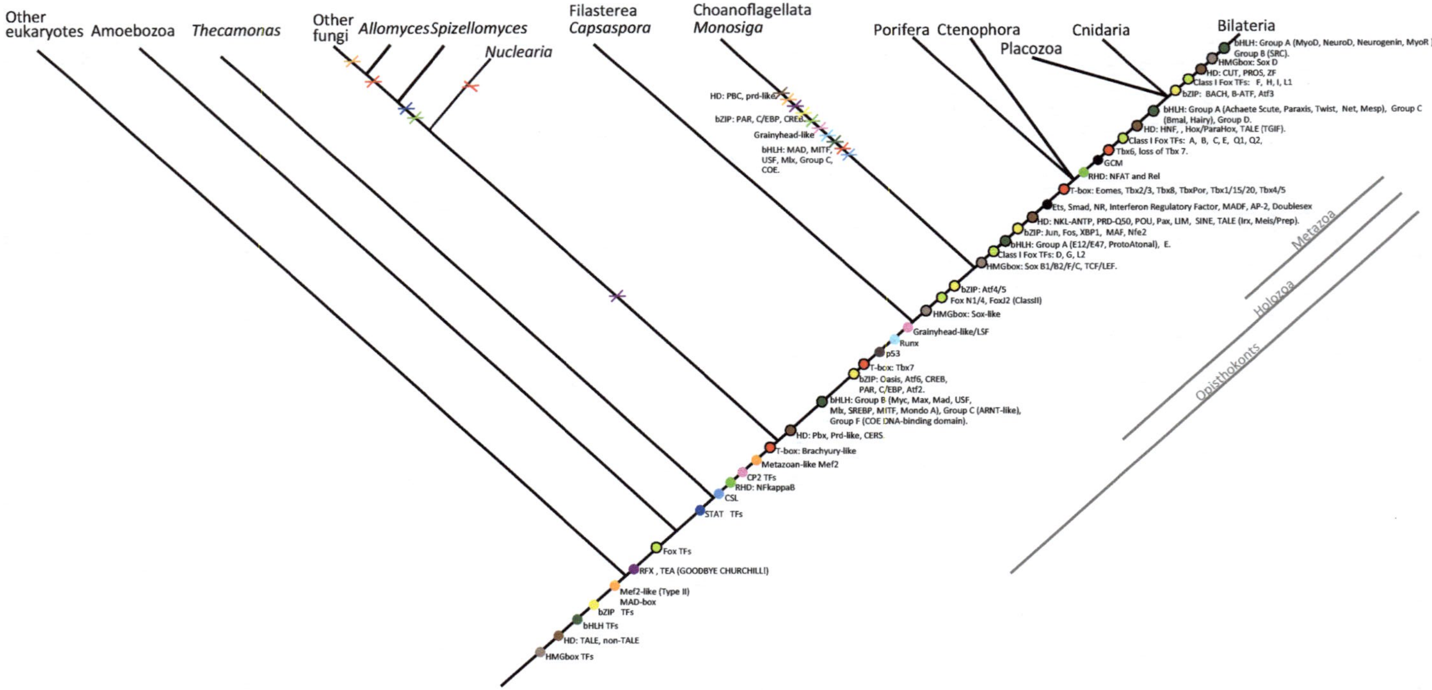

Fig. 3 Cladogram representing metazoan TF evolution. Colours are unique to each domain class. A *coloured dot* indicates the hypothetical origin of the domain. A *black-circled dot* indicates where a specific protein family appears in our taxon sampling. A *cross* means the loss of the domain or specific protein family in a lineage. (Adapted from Sebé-Pedrós et al. 2011)

animal evolution, other classes emerged, including HNF in the last common ancestor of Placozoa, Cnidaria, and Bilateria, and CUT, PROS and ZF in the bilaterian stem (Putnam et al. 2007; Srivastava et al. 2010). This rapid process of duplication and subfunctionalization is characteristic of metazoan Homeobox gene families, which, once they evolve, tend to maintain a conserved domain architecture and clear aminoacidic motifs typical of each gene (Bürglin 2011). In contrast, clear orthologs outside animals are scarce and generally do not bear the key amino acids or concurrent protein domains that characterize animal Homeobox families (Sebé-Pedrós et al. 2011).

T-box genes are also key developmental TFs in animals. They originated in the common ancestor of Opisthokonta and were secondarily lost in Dikarya fungi and in choanoflagellates. Brachyury was the ancestral T-box class; a second class, Tbx7, originated in the common ancestor of Holozoa (Sebé-Pedrós et al. 2013). All other T-box classes (including Tbx1/15/20, Tbx2/3, Tbx4/5, Eomes, TbxPor and Tbx8) originated at the stem of Metazoa, except Tbx6, which appeared in the cnidarian-bilaterian ancestor (Fig. 3; Sebé-Pedrós et al. 2013). Therefore, like Homeobox TFs, T-box TFs underwent rapid radiation and subfunctionalization at the origin of Metazoa, and some classes, like Tbx8 or TbxPor, were later lost in many metazoan lineages. In contrast to Homeoboxes, little innovation occurred in the T-box TF family after the early stages of animal evolution, except for the diversification of Tbx1/15/20 at the stem of the cnidarian-bilaterian ancestor.

Forkhead domain containing genes (Fox) have an ancient eukaryotic origin, and although they are scarce in bikonts, they are quite abundant in opisthokont genomes (Figs. 1, 2). All Fox genes found in non-metazoans are of Class II, including clear homologs of Fox N1/4 and Fox J2 in choanoflagellates (Larroux et al. 2008; Sebé-Pedrós et al. 2011). Class I Fox genes originated at the onset of Metazoa, with Fox D, G and L2 present in non-bilaterian metazoans; Fox A, B, C, E, Q1 and Q2 are found in Placozoa + Cnidaria + Bilateria; and Fox F, H, L1 and I originated in Bilateria (Fig. 3; Larroux et al. 2008; Sebé-Pedrós et al. 2011).

Basic Helix-Loop-Helix (bHLH) transcription factors are one of the main eukaryotic TF types (Figs. 1 and 3) and metazoans have six groups (A, B, C, D, E and F) (Simionato et al. 2007). Most bHLH TFs conserved between metazoans and their unicellular relatives belong to Group B. For example, Myc/Max/Mad, SREBP, USF, Mlx, MondoA, and MITF can be found in both lineages, while Src, AP4, FigAlpha and MNT are metazoan innovations (Fig. 3; Sebé-Pedrós et al. 2011). We find Group C genes in *C. owczarzaki*, which are co-orthologs of metazoan Group C bHLH TFs. Group F, also known as the COE family, includes TFs with a COE DNA-binding domain that also has an HLH domain at the C-terminal part of the protein. This COE domain can be found in the *C. owczarzaki* genome, but it does not have any trace of the HLH domain (also degenerated in some metazoan orthologs) (Suga et al. 2013). Finally, of the six metazoan groups, bHLH A, D and E are unique to metazoans and have diversified into many subfamilies (Group A genes), most of which are heterodimerizing classes (Simionato et al. 2007).

The HMGbox domain encompasses a large family of DNA-binding proteins, some of which have sequence-specific TF activity. Metazoans have a unique set of subtypes of these TFs, with the Sox family being the most diverse. Of the five known groups of human Sox genes, B1, B2, C, E and F are present in the non-bilaterian metazoans, while Group D appeared at the stem of Bilateria (Fig. 3; Fortunato et al. 2012). Tcf/Lef genes are also animal-specific HMGbox TFs, and act as the regulators of the Wnt pathway (Larroux et al. 2008). No Sox or Tcf/Lef genes are found outside Metazoa.

bZIP is another paneukaryotic TF family (Fig. 1), in this case quite homogenously distributed throughout eukaryotes in terms of relative abundance (Fig. 2). Some metazoan bZIP classes, including Oasis, Atf6, CREB, Atf2, C/EBP, PAR and Atf4/5, originated in unicellular holozoans (Fig. 3), constituting homodimerizing classes in most cases. In contrast, the heterodimerizing bZIP classes Jun, Fos, XBP1, MAF and Nfe2 are metazoan innovations, with bZIP BACH, B-ATF and Atf3 evolving later in the metazoan lineage (Sebé-Pedrós et al. 2011).

NFkappaB genes have also been found to have a more ancient origin than previously thought (Sebé-Pedrós et al. 2011; de Mendoza et al. 2013). As for T-box genes, they originated in the last common ancestor of opisthokonts and were lost in both choanoflagellates and fungi (Fig. 3). This ancestral gene had the typical NFkappaB domain structure, with a RHD domain followed by Ankyrin repeats. Later on in metazoans, two other RHD domain-containing TFs without Ankyrin repeats evolved: NFAT in Placozoa + Cnidaria + Bilateria and Rel in Bilateria (Gauthier and Degnan 2008; Sebé-Pedrós et al. 2011).

STAT TFs were present in the last common ancestor of opisthokonts and apusozoans and were lost in fungi. STAT TFs are found in organisms with tyrosine kinase genes, such as choanoflagellates and filastereans. In this context, STAT TFs interact with Y-phosphorylated proteins through their C-terminal SH2 domain, constituting the transcriptional outputs of tyrosine kinase signaling.

Although MADS-box (SRF domain) TFs are paneukaryotic, TFs with specific Mef2 signatures appeared at the stem of the Opisthokonta, are were secondarily lost in Dikarya fungi and choanoflagellates. Finally, p53 and Runx TFs are holozoan innovations.

In summary, metazoan developmental TFs have three main sources: first, *de novo* types that emerged at the onset of Metazoa; second, co-opted genes that originated in their unicellular relatives; and third, duplication of pre-existing TFs, sometimes accompanied by diversification of co-occurring domains. Among the truly metazoan innovations we find Smad (MH1 domain), Ets, DoubleSex, AP-2 and the Nuclear Receptors (zf-C4 domain) (Fig. 3). Key TF types that originated in unicellular Holozoa include NFkappaB, T-box, p53, the Myc/Max network, Grainyhead and LSF, and Runx. Finally, significantly enriched developmental TF types in metazoans include Homeobox, T-box, Fox, Sox and NFkappaB (Fig. 1). All of these TF types were present before the emergence of animal multicellularity, but new specific families involved in development emerged during metazoan evolution.

Establishment of the Metazoan TF Regulatory System

Transcriptional regulation by sequence-specific TFs goes beyond the TFome content itself. TFs act in a cellular context and their activity requires interaction with other proteins and, of course, with specific DNA sites. TFs bind to enhancers and promoters of genes and also interact with cofactors that modulate their activity in various ways (for example, controlling translocation to the nucleus or restricting DNA binding specificity). Moreover, TF binding is influenced by the epigenetic context, such as nucleosome positioning, histone modifications and DNA methylation patterns (Spitz and Furlong 2012). The cross-talk between TFs and all of these elements is what ultimately defines the TF regulatory system.

Currently, little is known about the functions of TFs in the unicellular relatives of animals, so it is not possible at this time to have a clear picture on how TF activities changed during the transition to animal multicellularity. Nevertheless, experimental approaches have provided some insights into this issue. There are two main evolutionary changes that may have accompanied an increase in regulatory interactions in animals: *trans* and *cis* regulatory changes. *Trans* regulatory changes involve the appearance of new physical interactions between TFs and cofactors (which may be either new or ancient) and/or new DNA-binding motif specificity. *Cis* regulatory changes involve the evolution of new DNA binding sites for a particular TF, which results in new downstream targets controlled by a TF and, therefore, the re-wiring of the TF network. The following cases provide examples of such processes.

The evolution of Brachyury, a well-known T-box TF with essential roles in gastrulation and mesoderm specification in animals (Technau 2001), is an example of a *trans* regulatory change. The presence of a clear Brachyury ortholog in *C. owczarzaki* was one of the most striking findings in the first unicellular holozoan TF survey (Sebé-Pedrós et al. 2011), and the exploration of its functional conservation using heterologous expression in *Xenopus laevis* and protein binding microarray (PBM) experiments revealed a suggestive pattern. *C. owczarzaki* Brachyury can rescue artificially inhibited embryo gastrulation in *Xenopus laevis*, although it does so in a rather unspecific way, activating diverse downstream genes that are commonly activated by different T-box gene families in metazoans. In contrast, sponge and ctenophore Brachyury homologs behave very much like endogenous *Xenopus laevis* Brachyury (Yamada et al. 2010; Sebé-Pedrós et al. 2013). Moreover, the results of the PBM experiments showed that the differences between *C. owczarzaki* Brachyury and metazoan Brachyury were not due to changes in the DNA-binding motif specificity of the different Brachyury genes studied (Sebé-Pedrós et al. 2013). Taken together, these two results strongly suggest that Brachyruy target specificity in metazoans arose through interaction with cofactors (Smad and probably others) that were probably established at the onset of Metazoa (Fig. 4a).

Generally, transcription binding site recognition motifs seem to be highly conserved in some TF families, such as bHLH or T-box (Jolma et al. 2013; Sebé-Pedrós et al. 2013). In contrast, other families are more labile in their sequence binding preferences (Nakagawa et al. 2013), such as in Homeoboxes, zinc fingers (C2H2)

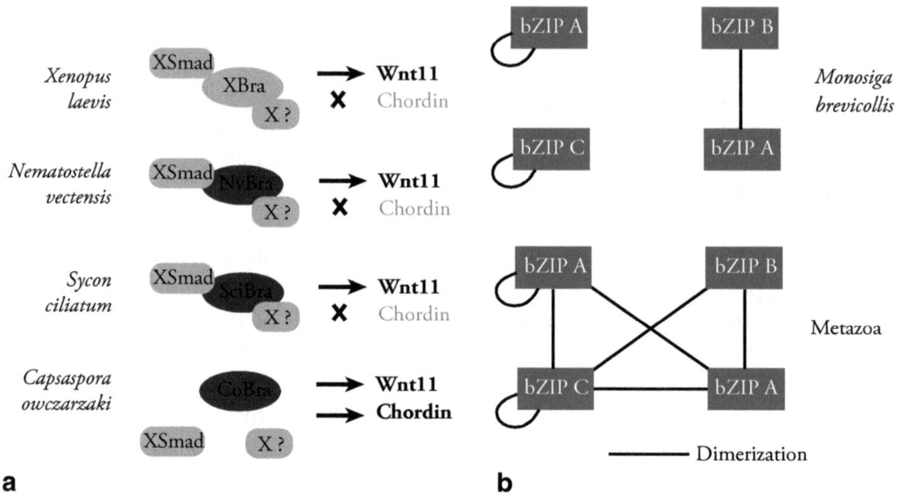

Fig. 4 *Trans* changes during the evolution of metazoan TF networks. **a** Schematic representation of the results from Sebé-Pedrós et al. 2013. Brachyury, an essential TF for animal development, is found in the unicellular *Capsaspora owczarzaki*. In a heterologous expression assay, only the early-metazoan Brachyury orthologs (of the sponge *Sycon ciliatum* and the cnidarian *Nematostella vectensis*) show the same molecular phenotype as the endogenous *Xenopus laevis* Brachyury ortholog (Activation of Wnt11 and non-activation of Chordin). This suggests that the *trans* regulatory interactions between Brachyury and cofactors like Smad (and probably other unknown cofactors, shown as "X?") were established at the onset of Metazoa. **b** Schematic representation of the results of Reinke et al. 2013. Analysis of the *in vitro* interactions of all bZIPs of several metazoans and of the choanoflagellate *Monosiga brevicollis* showed that the proportion of heterodimeric interactions (versus homodimeric interactions) was much lower in a unicellular context. This result suggests an increase in the complexity of the bZIP heterodimerizing network during the transition to animal multicellularity, which probably allowed new regulatory outputs by combining old (and also new) bZIP TFs

and Nuclear receptors (Jolma et al. 2013), all of which are major components of animal TF toolkits (de Mendoza et al. 2013). The plasticity of DNA-binding motif recognition is crucial to avoid cross-activation between TF paralogs. Thus, the transition to animal multicellularity involved the expansion of some of these labile families, which may have facilitated the acquisition of developmental complexity by allowing denser, non-overlapping readers of *cis*-regulatory information.

An open question is how these gene regulatory networks changed, not only quantitatively (more interactions), but also qualitatively. For example, it is conceivable to find that *C. owczarzaki* Brachyury regulates fewer downstream genes, and is more directly connected to batteries of effector genes than to other regulatory TFs. The degree of hierarchy of these interactions (where each level is defined as a master regulatory gene regulating another regulatory gene) would probably be lower in the case of unicellular gene regulatory networks. As a working hypothesis, shallower regulatory networks could have been intercalated into more complex wiring when developmental multicellularity evolved (Davidson and Erwin 2006).

Another example of *trans* evolution is the increase in interactivity found in bZIP TFs when comparing metazoans with unicellular eukaryotes (Reinke et al. 2013). bZIP TFs bind to DNA as dimers, usually homodimers (two bZIPs of the same class) but also as heterodimers, although not all possible heterodimeric interactions occur. Phylogenetically, some heterodimeric interactions can be predicted, but most remain elusive in non-model organisms. Reinke and co-workers examined the *in vitro* dimerization affinities of all bZIPs in several species, including a unicellular relative of metazoans, the choanoflagellate *Monosiga brevicollis*. They found that metazoan bZIPs have a higher proportion of heterodimeric interactions than *M. brevicollis* Therefore, there was an increase in complexity of the bZIP interaction network, which generated new combinatorial binding specificities (Fig. 4b).

bHLHs are another type of heterodimerizing TF. Myc is an animal bHLH TF responsible for cell cycle control, growth and apoptosis, and is a crucial oncogene in many types of cancer. It binds to DNA in association with Max, another bHLH TF. Max has the ability to interact with other bHLH TFs, mainly Mxd/Mad and MNT. Mxd/Mad and MNT proteins antagonize Myc, controlling cell cycle arrest and gene repression. This network was already established in a pre-metazoan context, where we find members of the Myc, Max and Mlx/Mad families, while MNT appeared later on, in the Eumetazoan split (Sebé-Pedrós et al. 2011; Young et al. 2011). It has been shown that *M. brevicollis* Myc and Max orthologs heterodimerize, revealing experimentally that the heterodimerization network was present and functionally conserved (Young et al. 2011). Moreover, a core set of ancient eukaryotic genes involved in ribosome biogenesis seems to be regulated by Myc in Holozoan genomes. The E-box motif (the typical DNA-binding motif of Myc/Max dimers) is found to be enriched in the promoters of *M. brevicollis* and animals that have Myc and Max orthologs, whereas fungi and *C. elegans*, which both lack Myc genes, are depleted of E-boxes in the promoters of those ribosomal genes (Brown et al. 2008). Myc network data from unicellular holozoans tell us that not only were the physical interactions between different TFs already in place, but also that there is some cis-regulatory conservation in the downstream genes involved in basic cell processes.

Despite this rather simple cis-regulatory conservation between unicellular taxa and metazoans, there is evidence of an increase in cis-regulatory interactions in metazoans. The 5′ intergenic regions of TFs and other regulatory genes (where many regulatory proteins bind) are expanded in metazoan genomes (compared to the mean intergenic distance of the genome), which allows for a more complex regulatory landscape (both in terms of the number and combination of regulatory modules, such as enhancers or repressors) and, ultimately, for complex spatiotemporal regulatory states (Nelson et al. 2004; Suga et al. 2013). This complex regulatory landscape explains why some TFs, such as NK or Hox, have evolved in syntenic blocks. Other TFs have retained by-stander genes, forming micro-syntenic blocks. The by-stander gene contains cis-regulatory elements embedded in its gene body that regulate the expression of the TF, preventing genomic recombination that would separate the two neighbours (Irimia et al. 2012). Overall, we can observe how the unique metazoan genomic architecture is greatly influenced by regulatory interactions between enhancers and downstream genes. These interactions allow complex developmental gene regulation by TFs.

In summary, current evidence suggests that not only did the TF repertoire itself changed during the origin of animal multicellularity, but a fundamental change also occurred in the gene regulatory networks in which these genes were embedded. These created more complex patterns of spatiotemporal regulation of gene expression, an essential feature of a complex multicellular entity. Some of these major regulatory changes occurred early in metazoan history, and became virtually frozen, producing what is known as kernel gene regulatory networks; while others were incorporated later in a phylum-, class- or species-specific manner, forming the basis of the morphological and functional diversity of extant metazoans (Davidson and Erwin 2006).

Summary

1. From a broad eukaryotic perspective, it is clear that phylogenetic inertia is an important factor that conditions the TF toolkit of different origins of multicellularity. Therefore, studying the TF repertoire of metazoans' unicellular relatives is essential for understanding the foundation of the metazoan TFome.
2. Various evolutionary forces have shaped the metazoan TFome, including *de novo* gene origin, gene family expansion and gene co-option.
3. The establishment of complex gene regulatory networks accompanied the origin of Metazoa. In the context of these networks many TFs were locked into specific developmental processes.
4. A global rearrangement of both TFome content and *cis* and *trans* interactions facilitated an explosion in the regulatory capabilities of TFs in Metazoa.

References

Adamska M, Degnan BM, Green K, Zwafink C (2011) What sponges can tell us about the evolution of developmental processes. Zoology (Jena) 114:1–10. doi:10.1016/j.zool.2010.10.003

Bouget FY, Berger F, Brownlee C (1998) Position dependent control of cell fate in the Fucus embryo: role of intercellular communication. Development 125:1999–2008

Brown SJ, Cole MD, Erives AJ (2008) Evolution of the holozoan ribosome biogenesis regulon. BMC Genomics 9:442. doi:10.1186/1471-2164-9-442

Bürglin T (2011) Homeodomain subtypes and functional diversity. In: Hughes TR (ed) A handbook of transcription factors, SE–5. Springer, The Netherlands, pp 95–122

Cock JM, Sterck L, Rouzé P et al (2010) The Ectocarpus genome and the independent evolution of multicellularity in brown algae. Nature 465:617–621. doi:10.1038/nature09016

Coelho SM, Godfroy O, Arun A et al (2011) OUROBOROS is a master regulator of the gametophyte to sporophyte life cycle transition in the brown alga Ectocarpus. Proc Natl Acad Sci U S A 108:11518–11523. doi:10.1073/pnas.1102274108/

Collen J, Porcel B, Carre W et al (2013) Genome structure and metabolic features in the red seaweed *Chondrus crispus* shed light on evolution of the Archaeplastida. Proc Natl Acad Sci U S A 110:5247–5252. doi:10.1073/pnas.1221259110

Davidson EH, Erwin DH (2006) Gene regulatory networks and the evolution of animal body plans. Science 311:796–800. doi:10.1126/science.1113832

De Mendoza A, Sebé-Pedrós A, Sestak MS et al (2013) Transcription factor evolution in eukaryotes and the assembly of the regulatory toolkit in multicellular lineages. Proc Natl Acad Sci U S A 110:E4858–4866. doi:10.1073/pnas.1311818110

De Smet R, Adams KL, Vandepoele K et al (2013) Convergent gene loss following gene and genome duplications creates single-copy families in fl owering plants. Proc Natl Acad Sci U S A 110:2898–2903. doi:10.1073/pnas.1300127110/

Degnan BM, Vervoort M, Larroux C, Richards GS (2009) Early evolution of metazoan transcription factors. Curr Opin Genet Dev 19:591–599. doi:10.1016/j.gde.2009.09.008

Derelle R, Lopez P, Le Guyader H et al (2007) Homeodomain proteins belong to the ancestral molecular toolkit of eukaryotes. Evol Dev 9:212–219. doi:10.1111/j.1525-142X.2007.00153.x

Fortunato S, Adamski M, Bergum B et al (2012) Genome-wide analysis of the sox family in the calcareous sponge Sycon ciliatum: multiple genes with unique expression patterns. Evodevo 3:14. doi: 10.1186/2041-9139-3-14

Gauthier M, Degnan BM (2008) The transcription factor NF-kappaB in the demosponge Amphimedon queenslandica: insights on the evolutionary origin of the Rel homology domain. Dev Genes Evol 218:23–32. doi:10.1007/s00427-007-0197-5

Grosberg RK, Strathmann RR (2007) The evolution of multicellularity: a minor major transition? Annu Rev Ecol Evol Syst 38:621–654. doi:10.1146/annurev.ecolsys.36.102403.114735

Hammonds AS, Bristow CA, Fisher WW et al (2013) Spatial expression of transcription factors in Drosophila embryonic organ development. Genome Biol 342:R140. doi:10.1186/gb-2013-14-12-r140

Irimia M, Tena JJ, Alexis MS et al (2012) Extensive conservation of ancient microsynteny across metazoans due to cis-regulatory constraints. Genome Res 22:2356–2367. doi:10.1101/gr.139725.112

Iyer LM, Anantharaman V, Wolf MY, Aravind L (2008) Comparative genomics of transcription factors and chromatin proteins in parasitic protists and other eukaryotes. Int J Parasitol 38:1–31. doi:http://dx.doi.org/10.1016/j.ijpara.2007.07.018

Jolma A, Yan J, Whitington T et al (2013) DNA-binding specificities of human transcription factors. Cell 152:327–339. doi:10.1016/j.cell.2012.12.009

King N, Westbrook M, Young S, Kuo A et al (2008) The genome of the choanoflagellate Monosiga brevicollis and the origin of metazoans. Nature 21:4300–4305. doi:10.1091/mbc.E10-03-0251

Knoll AH (2011) The multiple origins of complex multicellularity. Annu Rev Earth Planet Sci 39:217–239. doi:10.1146/annurev.earth.031208.100209

Larroux C, Luke GN, Koopman P et al (2008) Genesis and expansion of metazoan transcription factor gene classes. Mol Biol Evol 25:980–996. doi:10.1093/molbev/msn047

Levine M, Tjian R (2003) Transcription regulation and animal diversity. Nature 424:147–151. doi:10.1038/nature01763

Maere S, De Bodt S, Raes J et al (2005) Modeling gene and genome duplications in eukaryotes. Proc Natl Acad Sci U S A 102:5454–5459. doi:10.1073/pnas.0501102102

Nakagawa S, Gisselbrecht SS, Rogers JM et al (2013) DNA-binding specificity changes in the evolution of forkhead transcription factors. Proc Natl Acad Sci U S A 110:12349–12354. doi:10.1073/pnas.1310430110

Nakamura Y, Sasaki N, Kobayashi M et al (2013) The first symbiont-free genome sequence of marine red alga, Susabi-nori (Pyropia yezoensis). PLoS One 8:e57122. doi:10.1371/journal.pone.0057122

Nelson CE, Hersh BM, Carroll SB (2004) The regulatory content of intergenic DNA shapes genome architecture. Genome Biol 5:R25.1–R25.15

Niklas KJ, Newman SA (2013) The origins of multicellular organisms. Evol Dev 15:41–52. doi:10.1111/ede.12013
Parfrey LW, Lahr DJG (2013) Multicellularity arose several times in the evolution of eukaryotes. Bioessays 35:339–347. doi:10.1002/bies.201200143
Peters AF, Scornet D, Ratin M et al (2008) Life-cycle-generation-specific developmental processes are modified in the immediate upright mutant of the brown alga Ectocarpus siliculosus. Development 135:1503–1512. doi:10.1242/dev.016303
Pires-daSilva A, Sommer RJ (2003) The evolution of signalling pathways in animal development. Nat Rev Genet 4:39–49. doi:10.1038/nrg977
Putnam NH, Srivastava M, Hellsten U et al (2007) Sea anemone genome reveals ancestral eumetazoan gene repertoire and genomic organization. Science 317:86–94. doi:10.1126/science.1139158
Radakovits R, Jinkerson RE, Fuerstenberg SI et al (2012) Draft genome sequence and genetic transformation of the oleaginous alga Nannochloropis gaditana. Nat Commun 3:686. doi:10.1038/ncomms1688
Reinke AW, Baek J, Ashenberg O, Keating AE (2013) Networks of bZIP protein-protein interactions diversified over a billion years of evolution. Science 340:730–734. doi:10.1126/science.1233465
Rokas A (2008) The origins of multicellularity and the early history of the genetic toolkit for animal development. Annu Rev Genet 42:235–251. doi:10.1146/annurev.genet.42.110807.091513
Ryan JF, Pang K, Mullikin JC et al (2010) The homeodomain complement of the ctenophore Mnemiopsis leidyi suggests that Ctenophora and Porifera diverged prior to the ParaHoxozoa. Evodevo 1:9. doi:10.1186/2041-9139-1-9
Sebé-Pedrós A, de Mendoza A, Lang BF et al (2011) Unexpected repertoire of metazoan transcription factors in the unicellular holozoan Capsaspora owczarzaki. Mol Biol Evol 28:1241–1254. doi:10.1093/molbev/msq309
Sebé-Pedrós A, Ariza-Cosano A, Weirauch MT et al (2013) Early evolution of the T-box transcription factor family. Proc Natl Acad Sci U S A 110:16050–16055. doi:10.1073/pnas.1309748110
Simionato E, Ledent V, Richards G et al (2007) Origin and diversification of the basic helix-loop-helix gene family in metazoans: insights from comparative genomics. BMC Evol Biol 7:1–18. doi:10.1186/1471-2148-7-33
Spitz F, Furlong EEM (2012) Transcription factors: from enhancer binding to developmental control. Nat Rev Genet 13:613–626. doi:10.1038/nrg3207
Srivastava M, Simakov O, Chapman J et al (2010) The Amphimedon queenslandica genome and the evolution of animal complexity. Nature 466:720–726. doi:10.1038/nature09201
Suga H, Chen Z, de Mendoza A et al (2013) The Capsaspora genome reveals a complex unicellular prehistory of animals. Nat Commun 4:1–9. doi:10.1038/ncomms3325
Technau U (2001) Brachyury, the blastopore and the evolution of the mesoderm. Bioessays 23:788–794. doi:10.1002/bies.1114
Technau U, Steele RE (2011) Evolutionary crossroads in developmental biology: Cnidaria. Development 138:1447–1458. doi:10.1242/dev.090472
Van De Peer Y Maere S Meyer A (2009) The evolutionary significance of ancient genome duplications. Nat Rev Genet 10:725–732. doi:10.1038/nrg2600
Vaquerizas JM, Kummerfeld SK, Teichmann SA, Luscombe NM (2009) A census of human transcription factors: function, expression and evolution. Nat Rev Genet 10:252–263. doi:10.1038/nrg2538
Weirauch MT, Hughes TR (2011) A catalogue of eukaryotic transcription factor types, their evolutionary origin, and species distribution. In: A handbook of transcription factors. Springer, The Netherlands. Subcell Biochem 52:25–73. doi:10.1007/978-90-481-9069-0
Yamada A, Martindale MQ, Fukui A, Tochinai S (2010) Highly conserved functions of the Brachyury gene on morphogenetic movements: Insight from the early-diverging phylum Ctenophora. Dev Biol 339:212–222. doi:10.1016/j.ydbio.2009.12.019
Young SL, Diolaiti D, Conacci-Sorrell M et al (2011) Premetazoan ancestry of the Myc–Max network. Mol Biol Evol 28:2961–2971. doi:10.1093/molbev/msr132

How to Build an Allorecognition System: A Guide for Prospective Multicellular Organisms

Laura F. Grice and Bernard M. Degnan

Abstract The multicellular condition cannot be maintained without safeguards protecting the integrity of the individual. Tissue contact and fusion with other conspecific individuals may threaten this integrity, as genetically non-identical cells may shirk their somatic duties and gain disproportionate access to the germ line. Allorecognition capabilities appear to be widespread amongst metazoans. However, although similar functional mechanisms underlie the different metazoan allorecognition systems, all evidence to date rejects the idea of homology and a shared evolutionary history.

This chapter attempts to reconstruct an allorecognition system prototype that could support a successful transition to multicellularity. We begin by discussing the significance and evolutionary origins of self-nonself recognition, before describing the three essential phases of all self-nonself recognition reactions. Within this framework, we then discuss the molecular and genomic requirements of an allorecognition system capable of executing these three phases, using the commonalities of known invertebrate self-nonself recognition systems as supporting evidence. Finally, we explore the possibilities of a "pre-allorecognition" genomic state, and speculate that an early self-nonself recognition system may have been constructed via the coupling of existing molecules and pathways, including the incorporation of cell-cell interaction or adhesion components.

Keywords Allorecognition · Cheating · Evolution of multicellularity · Genomics · Innate immunity · Invertebrates · Histocompatibility · Self-nonself recognition

B. M. Degnan (✉) · L. F. Grice
School of Biological Sciences, University of Queensland, 4072 Brisbane, QLD, Australia
e-mail: b.degnan@uq.edu.au

I. Ruiz-Trillo, A. M. Nedelcu (eds.), *Evolutionary Transitions to Multicellular Life*,
Advances in Marine Genomics 2, DOI 10.1007/978-94-017-9642-2_19

Why: The Importance of Allorecognition for the Multicellular Condition

Simple Multicellularity Offers Direct Benefits

Transition to the multicellular state is a key step in the evolution of organismal complexity and has occurred independently multiple times across life on Earth (Buss 1987; Bonner 1988, 2000; King 2004; Grosberg and Strathmann 2007; Chapter "Timing the Origins of Multicellular Eukaryotes Through Phylogenomics and Relaxed Molecular Clock Analyses"). Multicellularity can arise by the aggregation of individual cells (as occurs for example in the social amoebae *Dictyostelium discoideum*; Raper 1935; Bonner 2000; Dormann et al. 2002; Chapter "The Evolution of Developmental Signalling in Dictyostelia From an Amoebozoan Stress Response"), or via cell division without subsequent daughter cell separation (Bonner 1988; Tarnita et al. 2013). As cell division and adhesion processes are early evolutionary innovations, antedating true multicellularity (Boland et al. 2000; Fairclough et al. 2010; Dayel et al. 2011; Sebe-Pedros et al. 2013), this latter process does not require any new cellular behaviour or machinery, making coloniality and a basic multicellular state mechanistically simple to achieve.

The size increase associated with the transition to multicellularity offers immediate potential selective advantages, such as greater habitat occupation and the ability to 'outsize' predators (Bonner 1966, 1988, 2000; Chapter "The Evolutionary Ecology of Multicellularity: The Volvocine Green Algae as a Case Study"). Another potential benefit of transition to a multicellular state is the new capacity for the division of labour, whereby different cells within an organism become responsible for producing and sharing different key gene products or performing useful functions (Kirk 2005; Rossetti et al. 2010; Gavrilets 2010; Goldsby et al. 2012; Ratcliff et al. 2012; Ispolatov et al. 2012; Chapters "Multicellularity in Bacteria: From Division of Labor to Biofilm Formation and Evolutionary Transitions in Individuality and Recent Models of Multicellularity"). The division of labour allows an organism to increase metabolic efficiency by dividing different cellular tasks between specialised cell types (Goldsby et al. 2012), and by partitioning incompatible cellular processes such as motility and cell division (Buss 1987; Chapter "The Evolutionary Ecology of Multicellularity: The Volvocine Green Algae as a Case Study"), or nitrogen fixation and photosynthesis (Fay 1992; Chapter "Multicellularity in Bacteria: From Division of Labor to Biofilm Formation"). It should be noted, however, that division of labour is not an inevitable consequence of transition to multicellularity (Rueffler et al. 2012) and requires additional innovations, including mechanisms to control cell type-specific gene expression and the spatial patterning of cells (Degnan et al. 2005; Richards and Degnan 2009).

Multicellular Conglomerates Invite Cheating

Successful multicellularity, particularly in organisms with multiple cell types, requires cooperation between and amongst different cells and cell types, with each cell performing its required role and receiving support in return (Buss 1987; Chapter "Evolutionary Transitions in Individuality and Recent Models of Multicellularity"). This cooperation requires individual cells to sacrifice their own autonomy to benefit the fitness of the higher-order organismal unit. A clear example of this requirement can be seen in organisms with distinct somatic and germ cell groups, with somatic cells relinquishing the capacity to contribute their genetic material to subsequent generations (Michod 2007). Mechanisms are therefore required to ensure these cells do not abandon their somatic duties in favour of a more individually-advantageous path, for example by unchecked cell replication or neglect of key cellular roles. Such behaviour is termed cheating, that is, exploitative behaviour that benefits an individual unit (in this case, a cell) at the expense of other members of a usually cooperative group (Strassmann and Queller 2011).

Cell cheating typically takes one of two forms, depending on the source of the cheater—either internal or external cheating. Internal cheaters arise when mutations cause cells to exploit otherwise-genetically identical cells within the multicellular body, as occurs in cancers. Multiple mechanisms exist to aid the control of internal cheating. For example, apoptosis, DNA repair and the arrest of cell division can minimise the expression of somatic mutations (Kastan and Bartek 2004), while sequestration of the germ line and a unicellular bottleneck stage of development both limit the potential for transmission of deleterious cheater mutations to the next generation (Grosberg and Strathmann 2007). External cheating occurs when other individuals threaten organismal integrity, for example by tissue or organismal fusion. This is potentially problematic, because the altruism of somatic cellular cooperation and sacrifice of germ line contribution can only be maintained if genetically identical (or at least, closely related) cells are able to contribute genetic material to the next generation (Eberhard 1975). Unrelated cells, therefore, have no 'motivation' to contribute fairly, and can thus exploit resources provided by the somatic cells, potentially using these resources to increase their own reproductive output at the expense of the host.

Amongst extant animals, control of external cheating has been best documented in the colonial ascidian *Botryllus schlosseri*. In this species, colonies sharing one or more alleles for the highly polymorphic locus *FuHC* are considered self and will undergo vasculature fusion, while those with disparate *FuHC* alleles reject each other (Oka and Watanabe 1957). As large numbers of *FuHC* alleles are present in *B. schlosseri* populations, fusion is effectively limited to closely related colonies. However, fusion between histocompatible individuals has been observed at relatively high rates (Rinkevich et al. 1998); when this does occur, it tends to be followed by a process of resorption, whereby one fusion partner is partially or entirely eliminated, in a competitive and reproducible fashion (Rinkevich and Weissman 1987). Intriguingly, however, the resorptive winner can experience germ or somatic cell parasitism, which, in extreme cases, may lead to total replacement of winner cells with those from the resorptive loser (Stoner and Weissman 1996; Stoner et al. 1999). This parasitism occurs despite the presence of a complex self-nonself recognition system, which emphasises the importance of restricting fusion, and therefore potential germ line control, to self or close kin. Systems allowing the recognition of and discrimination between self and nonself allow successful multicellular organisms to limit wasted resources and potential loss of reproductive output. The rest of this chapter will focus on the requirements and execution of self-nonself recognition systems that allow the distinction between conspecific members of a single species.

Where: Self-Nonself Recognition Occurs Throughout the Tree of Life

Allorecognition is vital to multicellular life forms in order to maintain organismal integrity in the face of external cheaters. It is interesting to note, then, that the capacity to distinguish between ingroups and outgroups (i.e. self-nonself recognition in varying forms) antedates the multiple transitions to multicellularity (Table 1), although it is likely that such systems are, to a large extent, the result of independent evolution. Here we present three diverse case studies from extant organisms that demonstrate that the ability to distinguish between self and nonself can exist in the absence of, and is thus not reliant upon, the continuous multicellular state.

Case Study 1: Group-Specific Quorum Sensing in Staphylococcus aureus Promotes Virulence Factor Production

Quorum sensing is a form of cell-cell communication that allows bacteria to gather information about the population densities and activities of other bacteria in the surrounding environment, and change their own behaviour accordingly (Waters and Bassler 2005). The production of virulence factors by pathogenic bacteria is one process under the control of quorum sensing mechanisms, allowing bacterial populations to reach a sufficiently large size before a full pathogenic attack is mounted against an infected host (Williams et al. 2000). The quorum sensing system of the medically significant bacterium *Staphylococcus aureus* (Firmicutes, *Bacilli*) has been well studied and demonstrates the existence of self-nonself recognition capacities in bacteria.

The *S. aureus agr* (accessory gene regulation) locus is responsible for quorum sensing and is comprised of two adjacent operons, each under the control of separate promoters, P2 and P3 (Janzon and Arvidson 1990; Novick et al. 1993). P2 controls the expression of four genes, *agrA, B, C* and *D* (Novick et al. 1995), whilst P3 drives expression of a regulatory RNA, *RNAIII* (Janzon et al. 1989; Janzon and Arvidson 1990). A small autoactivating peptide (AIP) is produced from *agrD*, which is modified and secreted with the aid of agrB and binds to membrane receptor agrC. Once

Table 1 Examples of self-nonself recognition in non-metazoan groups. (Multicellularity status from Grosberg and Strathmann 2007)

Group	Multicellularity Status	Examples	Well-characterised phenomena	Well-characterised molecules
Bacteria	U	*Escherichia coli* *Staphylococcus aureus*	Biofilm behaviour Interference pathways Quorum sensing Restriction enzymes	agr CRISPR (Marraffini and Sontheimer 2010)
Ciliates	UM	*Tetrahymena thermophila*	Kin-based social aggregation (Chaine et al. 2010) Mating types	Er-1
Green algae	UCM	*Volvox carteri* *Chlamydomonas reinhardtii*	Herbivore defence (Hay et al. 1987) Mating types	mt (Ferris et al. 2002)
Red algae	M	*Cyanidioschyzon merolae*	Herbivore defence	
Flowering plants	M	*Arabidopsis thaliana* *Nicotiana benthamiana*	Herbivore defence (Karban and Shiojiri 2010) R proteins (Glowacki et al. 2010) Root growth—nonself avoidance (Gruntman and Novoplansky 2004)	R proteins (Glowacki et al. 2010) SI proteins (Takayama and Isogai 2005)
Slime moulds	UCM	*Dictyostelium discoideum* *Dictyostelium purpureum*	Kin-based social aggregation Mating types (Erdos et al.1973)	tgrB1/C1
Fungi	UCM	*Neurospora crassa* *Saccharomyces cerevisiae*	Heterokaryon incompatibility (Hall et al. 2010)	het-c, pin-c (Hall et al. 2010)

All example species have a sequenced genome.
U unicellular, *C* colonial, *M* multicellular

extracellular AIP levels reach a certain threshold concentration, agrA is activated, leading to the upregulation of expression from the P2 and P3 promoters. This induces transcription of *RNAIII*, as well as further autoregulation of the agr cycle (George and Muir 2007). *RNAIII* is a key regulatory RNA that is responsible for modulating the expression of a large number of genes, including the upregulation of some virulence factors and other secretory proteins, and the downregulation of some cell surface molecules (Dunman et al. 2001).

Interestingly, the *agr* locus is polymorphic, possessing a hypervariable region that includes *agrD* and portions of *agrB* and *agrC* (Ji et al. 1997; Bonner 1998). Sequence comparisons of this region allow *S. aureus* to be partitioned into four distinct groups (Ji et al. 1997; Jarraud et al. 2000) that may indicate incipient speciation (Wright et al. 2005). Such partitioning is also seen in other *Staphylococcus* species (Van Wamel et al. 1998; Dufour et al. 2002). Intergroup or interspecies AIP molecules are not capable of agr pathway cross-stimulation. While the presence of "self group" AIP stimulates further autoactivation of the agr cycle, "nonself group" AIP inhibits this process (Ji et al. 1997; Otto et al. 1999), thereby blocking the entire pathway. This may allow different *S. aureus* groups to compete for control of a particular infection site (Otto et al. 1999), or may prevent individual bacteria from producing virulence factors, a costly metabolic process, when the population density of nonself bacteria is too high. The *agr* gene system therefore serves to partition *S. aureus* into types or groups, with discrimination between these groups serving as a key element of *S. aureus* pathogenicity.

Case Study 2: Euplotes raikovi Rejects Self-Mating

The unicellular ciliate *Euplotes raikovi* (Alveolata, Ciliophora) provides an interesting example of recognition and passive discrimination between self and nonself. This species is part of a wider group of ciliates that display both sexual and asexual life stages. *E. raikovi* and other *Euplotes* species exhibit a large number of mating types, which act to restrict mating in the sexual stage to pairs of genetically different individuals (Kimball 1939; Luporini et al. 1983; Miceli et al. 1983). Mating type in *E. raikovi* is dictated by a single allelic gene called *Er.* Each *Er* allele produces two different gene products via alternative splicing: a soluble pheromone and a longer, membrane-bound form (Miceli et al. 1992); the two form a ligand-receptor pair (Ortenzi et al. 2000). The membrane-bound protein can bind any *Er* pheromone variant (Ortenzi and Luporini 1995), but is involved in two opposing processes: homotypic binding between a receptor and its matching pheromone causes asexual division and growth, while heterotypic binding of a foreign pheromone triggers the mating process (Ortenzi and Luporini 1995; Vallesi et al. 1995). In this way, *E. raikovi* individuals employ self-nonself recognition to successfully avoid mating unless genetically different individuals are available. As optional sexual life stages are usually associated with periods of stress, limiting sexual mating to unrelated partners increases genetic diversity of offspring and offers a greater chance at surviving stressful environments.

Case Study 3: Dictyostelium discoideum Preferentially Aggregate with Kin

The social amoeba *Dictyostelium discoideum* (Amoebozoa, Mycetozoa) is a third example of a microorganism that displays self-nonself recognition capacities. The *D. discoideum* lifecycle involves three stages: vegetative, sexual and social (Morgan 1903; Harper 1926; Raper 1935, 1940). *D. discoideum* individuals in the vegetative stage are solitary unicellular organisms that feed on bacteria (Raper 1937). If this food source becomes scarce, however, the amoebae enter the social stage (Oehler 1922; Schuckmann 1924; Schuckmann:1925tw; cited by Konijn and Raper 1961), utilising a chemotaxis mechanism to identify and aggregate with other conspecific individuals in the nearby environment. These aggregates then form motile 'slugs' which move towards light and heat (Bonner and Lamont 2005), and later differentiate into a fruiting body, where approximately 85 % of cells form spores and the remaining 15 % die and form a supporting stalk structure (Bonner and Slifkin 1949; Chapter "The Evolution of Developmental Signalling in Dictyostelia From an Amoebozoan Stress Response"). As this fruiting body is formed via the aggregation of genetically-distinct cells, rather than by cell division by a single individual, and requires an altruistic sacrifice by a number of incorporated cells, this strategy seemingly exposes cooperative cells to exploitation by cheating cells. However, aggregating *D. discoideum* cells possess a mechanism that allows them to preferentially aggregate in a strain-specific manner.

Benabentos et al. (2009) and Hirose et al. (2011) have demonstrated that a matched set of alleles for two genes, *tgrB1* and *tgrC1* (formerly *lagB1* and *lagC1*), is necessary and sufficient for strain-specific self-recognition to occur in *D. discoideum*. Both genes encode polymorphic transmembrane proteins (Benabentos et al. 2009) that allow neighbouring cells to interact, such that a tgrC1 variant on one cell binds to a compatible tgrB1 protein on another cell, and vice versa (Hirose et al. 2011). This binding thus promotes the aggregation of only genetically similar, and therefore closely related, cells. Experimental elimination of *tgrB1* and *tgrC1* abolishes self-recognition, while the presence of additional *tgrB1* and *tgrC1* alleles does not affect recognition as long as at least one compatible allele pair is shared between cells (Hirose et al. 2011). With this system, *D. discoideum* possess a molecular mechanism that allows more-closely related individuals to recognize one another and preferentially aggregate, while non-kin individuals segregate away. This reduces the potential deleterious effects of mixed populations of genetically distinct, or cheating and non-cheating (Ho et al. 2013) individuals in supposedly-cooperative groups.

These three case studies demonstrate that the general ability to distinguish between self and nonself is not reliant on the true multicellular condition, but rather exists in a variety of forms in unicellular, colonial and transient or simple multicellular (e.g. *D. discoideum*) organisms. This ability was thus most likely present in those unicellular ancestors that gave rise to multicellular lineages. The examples also highlight that the methods of distinguishing between self and nonself in different taxa are diverse—many different strategies may be employed to achieve a mechanistically similar outcome—suggesting that such systems are the result of independent evolutionary origins; we will return to this theme later in this chapter. However, despite the diversity in these and other self-nonself recognition systems, commonalities do exist in the broad requirements of any self-nonself recognition process, which we discuss in the following section.

What: The Functional and Molecular Requirements of Allorecognition

The Three-Phase Model of Self-Nonself Recognition

All self-nonself recognition reactions occur as a three-phase process. The first phase of the process is *detection*—a particular individual unit (e.g. a cell type, organism, etc.) must detect the presence of another biological entity in its vicinity. Phase two is *recognition*, whereby the first unit must then determine the identity of the detected unit as self or nonself. Different systems may recognise the presence (or absence) of self, of nonself, or be able to directly recognise both self and nonself. The simplest, and thus probably most ancient, of these hypothetical systems is one based on self recognition, whereby cells or molecules lacking some label identifying them as self are rejected (Coombe and Ey 1984; Boehm 2006). The final phase of the self-nonself recognition process is *discrimination*, where some action is taken on the basis of the recognition decision. The outcome of this action varies. For example, self could be favoured (or nonself disfavoured) as is the case in immune reactions, whereas nonself may be favoured (or self disfavoured) in mate selection processes. The mechanisms employed to execute this discrimination also vary, and may be passive or aggressive.

The three phases of self-nonself recognition may not necessarily occur as distinct events. For example, the detection and recognition phases may occur simultaneously in systems where recognition is possible only through the binding of particular homotypic or heterotypic recognition labels, as occurs in the *E. raikovi* mate recognition system as described above. Here, detection and recognition occur as a single step, but the outcome of the recognition decision (self or nonself) determines which of two possible discrimination processes follows—asexual growth or sexual reproduction. Alternatively, all three phases may occur simultaneously, as in *D. discoideum* strain-specific aggregation. In this case, detection can only occur if a homotypic tgrB1/C1 pair is present and bound; passive discrimination is a direct consequence of this binding. Regardless of the precise mechanisms of action, however, all three phases should occur in some capacity in any self-nonself recognition reaction.

The Functional Requirements of Self-Nonself Recognition Systems Predict Their Underlying Molecular Features

All allorecognition systems must possess one or more molecules capable of executing the three phases of self-nonself recognition outlined above. Therefore, consideration of the functional requirements of such allorecognition systems allows the prediction of the expected features of their underlying molecules. Such predictions are of practical value, for example acting as useful criteria when attempting to identify putative allorecognition molecules from a set of newly identified candidate genes (see, for example, Rosa et al. 2010). However, as few allorecognition systems have been thoroughly characterised, these criteria are not likely to apply to all systems.

Detection

The first phase of self-nonself recognition reactions, detection, involves sensing the presence of other individuals in the nearby environment. Such a task must be performed by a molecule capable of mediating intercellular interactions, either via direct cellular contact or the binding of secreted molecules. This predicts the existence of an allorecognition molecule with an extracellular region capable of binding molecules attached to, or secreted by, neighbouring cells—although intracellular receptors are known in other signalling pathways (e.g. in endocytosis, nuclear receptor activation and pathogen detection; Geuze et al. 1984; Baumann et al. 1999; Meylan et al. 2006, respectively) and thus their presence here cannot be excluded. Indeed, the recent identification of a cytosolic gene in the *B. schlosseri FuHC* locus reveals that not all allorecognition factors are on the cell surface or secreted (Voskoboynik et al. 2013).

Proteins fulfilling this requirement are prevalent in the molecular suites of most well-characterised allorecognition systems. These are usually transmembrane or secreted proteins featuring large extracellular regions with tandemly repeated protein domains (Fig. 1a, Table 2). For example, the allodeterminants alr1 (Rosa et al. 2010) and alr2 (Nicotra et al. 2009) from the cnidarian *Hydractinia symbiolongicarpus*, and mFuHC, whose encoding gene resides within the *Botryllus schlosseri* FuHC locus (De Tomaso et al. 2005; Nydam et al. 2013b; Voskoboynik et al. 2013), are all equipped with multiple immunoglobulin-like domains, while the aggregation factor (AF) proteins from the sponges *Amphimedon queenslandica* and *Clathria prolifera (*formerly *Microciona)* are all predicted to possess numerous tandemly-repeated Calx-beta domains (Fernandez-Busquets et al. 1996; Grice, Gauthier and Degnan, unpublished). Such extracellular domains are commonly comprised of β-sheets and related folds such as the β-sandwich structure (Table 2). These folds are structurally robust to amino acid change (Wright et al. 2004), which may be of key importance for the maintenance of molecule functionality despite the high levels of intraspecific sequence diversity required of allorecognition molecules (discussed below).

Recognition

The primary requirement of this phase is a capacity for high-precision recognition decisions, in order to prevent costly self or nonself rejection or acceptance, depending on the circumstance (Tsutsui 2004). Such precision requires an underlying highly polymorphic molecular system, in order to produce unique labels for each individual self unit (Hildemann 1979; Grosberg 1988; Tsutsui 2004). The presence in a population of such levels of polymorphism means that, for recognition reactions between conspecific individuals, there is a strong probability that tags matching an individual's self signature are true representatives of self, rather than random matches due to chance. Mechanistically, this occurs via sequence differences that potentially confer structural changes to allorecognition protein secondary, tertiary and quaternary structure. This in turn affects the binding properties and specificities between mature proteins, allowing self-nonself recognition to occur.

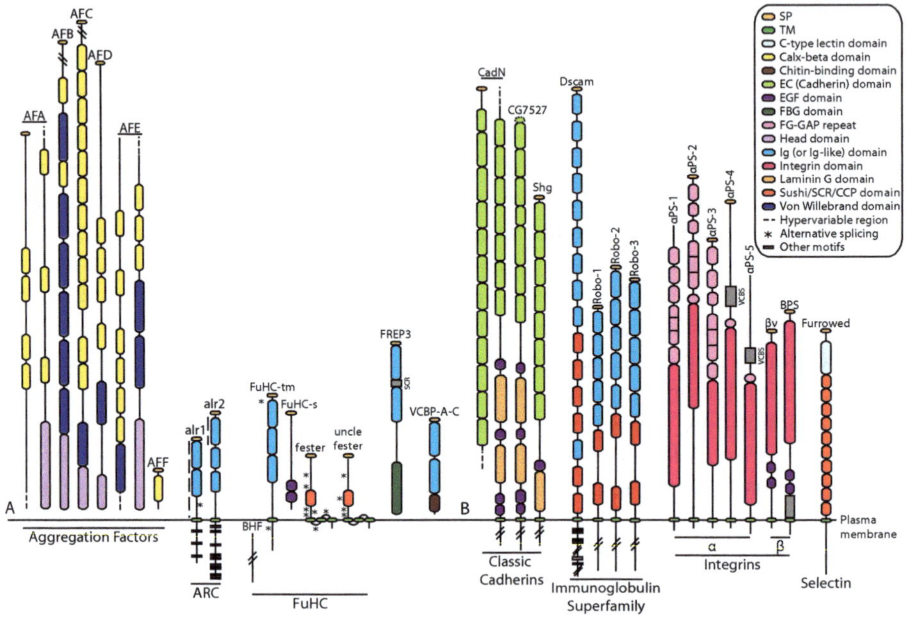

Fig. 1 Invertebrate allorecognition, self-nonself recognition and cell adhesion proteins. The secondary protein structures of **a** selected invertebrate allorecognition and self-nonself recognition associated molecules and **b** *Drosophila melanogaster* cell adhesion molecules. **a** Featured molecules are the aggregation factors AFA—AFF from *Amphimedon queenslandica* (Gauthier 2010), alr1 and alr2 from the *Hydractinia symbiolongicarpus* allorecognition complex (ARC; Nicotra et al. 2009; Rosa et al. 2010), *Botryllus schlosseri* FuHC locus proteins BHF, $FuHC^{tm}$, $FuHC^{s}$, fester and uncle fester (De Tomaso et al 2005; McKitrick et al. 2011; Nyholm et al. 2006; Nydam et al. 2013b; Voskoboynik et al. 2013), FREP3 from the *Biomphalaria glabrata* parasite defense system (Zhang et al. 2001), and a single representative structure of VCBP forms A—C from the anti-pathogen system of the urochordate *Ciona intestinalis* (Dishaw et al. 2011). As FREP3 and the VCBPs are not involved in allorecognition processes, they are here categorised as self-nonself recognition molecules. **b** Members of the key cell adhesion protein families—classic cadherins (Hill et al. 2001), immunoglobulins (Kidd et al. 1998; Schmucker et al. 2000; Simpson et al. 2000), integrins (Narasimha and Brown 2000) and selectins (Leshko-Lindsay and Corces 1997)—from the representative invertebrate species *D. melanogaster* are shown. All identified members for this species of the classic cadherins, integrins and selectins are shown. As the *D. melanogaster* immunoglobulin superfamily is very large, we here show only four members, the axon guidance receptor molecules Dscam and Robo 1–3. Boxes indicate protein domains and other key features; the linear structure of the proteins are shown. The *line* symbolises the plasma membrane, with the region above representing the extracellular space, and below representing the cytoplasm. All structures are drawn to scale except where indicated by *crossed lines*. As AFA, AFB and CadN are very large, these structures have been split in two as represented by *dashed lines*. *SP* signal peptide, *TM* transmembrane domain

Different strategies may be employed to generate the high levels of polymorphism required by allorecognition systems. Allorecognition genes are often richly allelic. For example, fusion-rejection decisions in *H. symbiolongicarpus* are largely under the control of two tightly-linked, highly polymorphic genes, *alr1* and *alr2* (Rosa

Table 2 Structural properties of key cell adhesion and self-nonself recognition domains

Domain	Pfam code	Secondary structure	Additional structural features	Reference
Immunoglobulin	CL0011	β-sandwich	Disulphide bond joins β-strands	Bork et al. 1994; Harpaz and Chotia 1994
EGF	CL0001	Two β-sheets	Three disulphide bridges	Wouters et al. 2005
Calx-beta	PF03160	β-sheet	–	Schwarz and Benzer 1997
Sushi/SCR/CCP	PF00084	β-sandwich	Stabilised by disulphide bridges	Norman et al. 1991
Fibrinogen C	PF00147	α-helices, β-sheets	Two disulphide bridges	Middha and Wang 2008
FG-GAP repeat	PF01839	β-sheet	Seven repeats form β-propeller	Springer 1997
FNIII	PF00041	β-sandwich	–	Leahy et al. 1992
Laminin G	PF00054 PF02210 PF13385	β-sandwich	–	Hohenester et al. 1999
Cadherin	PF00028	β-sandwich	–	Shapiro et al. 1995
C-type lectin	PF00059	Loop-within-a-loop structure with two β-sheets and two α-helices	Two disulphide bridges	Zelensky and Gready 2005
Chitin-binding domain	PF01607	β-sandwich	Three disulphide bridges	Ikegami et al. 2000
Von Willebrand	CL0128	Twisted β-sheet flanked by α-helices	Two disulphide bridges	Edwards and Perkins 1995

et al. 2010); two contacting colonies require at least one shared allele at both *alr1* and *alr2* for recognition as self and subsequent successful fusion. cDNA sequencing has identified around 200 unique *alr2* alleles within a single Connecticut *H. symbiolongicarpus* population (Gloria-Soria et al. 2012). The rich allelic nature of these genes facilitates only low rates of colony fusion—experimental manipulations of *H. symbiolongicarpus* have demonstrated fusion rates at less than 5 % (Rosa et al. 2010). Similarly, fusibility assays in three Israeli populations of *B. schlosseri* estimate the existence of over 300 FuHC alleles per population (Rinkevich et al. 1995). The putative *B. schlosseri* histocompatibility receptor, *fester*, is also richly allelic, with at least 21 alleles observed in one study (Nyholm et al. 2006).

Although the function of allorecognition proteins predicts that they be equipped with polymorphic extracellular regions, known molecules associated with allorecognition processes vary in their precise localisation and distribution of polymorphisms across their lengths (Fig. 1a). Sequence polymorphism in *alr1* and *alr2* is largely restricted to particular hypervariable regions (Nicotra et al. 2009; Rosa et al. 2010) (Fig. 1a). Within the FuHC locus, variation in the new candidate allorecognition gene *BHF* (Voskoboynik et al. 2013) and in *sFuHC* and *mFuHC* (De Tomaso et al. 2005; Nydam et al. 2013b; Voskoboynik et al. 2013) is distributed across each protein's length; in *BHF,* polymorphism is somewhat more prominent within the first 300 nucleotides and is absolutely predictive of fusibility outcomes (Voskoboynik et al. 2013). *fester* polymorphism is restricted to the extracellular region (Nyholm et al. 2006). The recently-characterised *Hsp40-L* also resides within the FuHC locus, and despite being a cytoplasmic protein, is similarly highly polymorphic with diversity localised to the C-terminal region (Nydam et al. 2013a).

In addition to sequence polymorphism, numerous other mechanisms, such as alternative splicing, post-transcriptional modification, recombination and RNA editing, may also be used to create diversity in allorecognition systems, either individually or in combination with one or more other processes (Ghosh et al. 2011).

Discrimination

The final self-nonself recognition phase, discrimination, may proceed in diverse ways, complicating attempts to make generalisations about the molecular components facilitating this stage. System-specific information is required in order to make predictions about the nature of the particular processes occurring therein. For example, systems that utilise differential cell adhesion as a passive discrimination mechanism may be predicted to possess a membrane-bound receptor molecule capable of tethering self cells together. Alternatively, in processes with differential outcomes, where recognition activates or represses a particular cascade or pathway, we can predict the presence of transmembrane receptor proteins with cytoplasmic tails linking to downstream effector molecules or completely internalised cytoplasmic proteins. The precise nature of these receptor and effector molecules will vary depending on their precise mechanisms of action. There is, however, evidence of a degree of conservation in the downstream response to allorecognition challenge in marine invertebrates, with particular binding and catalytic proteins, including heat shock proteins, pattern recognition receptors and immunophilins, being implicated in the responses to allorecognition challenge in both cnidarians and ascidians (Oren et al. 2013).

How: The Genomic Basis of Allorecognition

Self-nonself recognition appears to be a ubiquitous feature of metazoans, however research into the genetic basis of metazoan allorecognition has failed to find preserved evidence of a directly-shared evolutionary history between the 'frontline' allorecognition molecules of different taxa (Table 3). Regardless of the evolutionary origins

Table 3 Selected key self-nonself molecular systems in metazoans.

Group	Example	Molecular system and putative functions	IM	TLR	CMP	AI
Poriferans	*Amphimedon queenslandica* *Clathria prolifera**	AFs—histocompatibility, cell adhesion[a]	+	+	–	–
Cnidarians	*Hydra magnipapillata* *Nematostella vectensis*	ARC (alr1 and alr2)—histocompatibility[a]	+	+	+	–
Crustaceans	*Daphnia pulex* *Penaeus monodon**	Crustins—antimicrobial peptides (Smith et al. 2008) Penaeidins—antimicrobial peptides (Destoumieux et al. 1997)	+	+	+	–
Insects	*Drosophila melanogaster*	Dscam—neuronal patterning (Schmucker et al. 2000), pattern recognition receptor function (Dong et al. 2006)	+	+	+	–
Nematodes	*Caenorhabditis elegans*	Various antimicrobial peptides (Bogaerts et al. 2010)	+	+	+	–
Gastropods	*Biomphalaria glabrata** *Haliotis* spp.* *Lottia gigantea*	FREPs—parasite defense (Zhang et al. 2004)	+	+	+	–

Table 3 (continued)

Group	Example	Molecular system and putative functions	IM	TLR	CMP	AI
Echinoderms	*Strongylocentrotus purpuratus*	Extensive expansion of TLR and NLR families (Hibino et al. 2006) RAG1/2-like molecules—possible gene rearrangement role (Fugmann et al. 2006)	+	+	+	–
Cephalochordates	*Branchiostoma floridae*	VCBPs—host-microbe interactions (Cannon et al. 2002)	+	+	+	–
Ascidians	*Botryllus schlosseri* *Ciona intestinalis*	FuHC locus (s/tmFuHC, fester, uncle fester, BHF -?)—histocompatibility[a] VCBPs—host-microbe interactions (Dishaw et al. 2011)	+	+	+	–
Jawless vertebrates	*Petromyzon marinus* *Eptatretus burgeri**	VLRs—adaptive immunity (Pancer et al. 2004)	+	+	+	–
Jawed vertebrates	*Mus musculus* *Danio rerio* *Homo sapiens*	MHC, TCR, RAG and Ig molecules—adaptive immunity	+	+	+	+

IM innate immunity, *TLR* TLR (Toll-like receptor) pathway, *CMP* complement system, *AI* 'true' adaptive immunity
[a]discussed in text, refer for references
All example species have a sequenced genome except where otherwise indicated (*). Molecules listed are either unique or characteristic of the phylogenetic group, or well-studied therein. The far right of the table indicates the presence (+) or absence (-) of major immune pathways

and initial genetic sources (discussed further in the next section) of these allorecognition systems, these systems have and continue to diverge along different evolutionary lineages via mutation, exon (domain) shuffling and molecular tinkering. In conjunction with the shared molecular features that exist between diverse allorecognition systems discussed earlier, it is becoming increasing clear that allorecognition loci often possess commonalities in various genomic features as well. Here we discuss two trends apparent in genomic loci encoding diverse allorecognition systems that have been identified with existing sequencing data.

Given the essential nature of self-nonself recognition in disparate metazoans, it would appear likely that allorecognition genes, many encoding extracellular or cell surface molecules, probably originated from some common ancestral gene or genes. However, the lack of evidence for orthology amongst known extant allorecognition molecules and recent evidence that some allorecognition molecules are cytoplasmic (Voskoboynik et al. 2013) questions this supposition. Although the existence of taxon-restricted allorecognition systems amongst characterised extant species argues for the independent evolution of these systems, the selective pressures for defense against conspecific nonself invasion is consistent with allorecognition systems being in continuous operation over the course of metazoan evolution. This, in turn, suggests that there may exist deeper homologies between allorecognition systems, beyond specific gene families, or that these systems evolved independently, being co-opted from a common reservoir of structural genes (such as those encoding extracellular or cell surface proteins). In theory, co-option to a role in allorecognition simply requires that the protein possesses a domain that can maintain structural integrity whilst accommodating extensive amino acid change (e.g. beta-sheet containing domains; Table 2).

Here we discuss two trends apparent in the genomic loci encoding diverse allorecognition systems that have already been identified with existing data.

Clustering of Allorecognition Genes

One striking feature of the allorecognition systems characterised to date is that their component genes tend to co-occur in clusters of multiple, usually structurally similar genes (Fig. 2); but see Voskoboynik (2013) for an exception. The large modular structure of the individual genes, coupled with the tandemly repeated nature of the loci, mean that these regions are often large. The *H. symbiolongicarpus alr1* and *alr2* genes have been mapped to a single genomic interval, the allorecognition complex (ARC; Cadavid et al. 2004). A 700 kb sub-complex resides within the ARC, in which *alr1* is clustered amongst an additional ten Ig-like domain-encoding genes; at least four of these genes are polymorphic (Rosa et al. 2010). Although the precise role of these genes is unknown, the variable members remain plausible candidates for other currently unidentified allodeterminants within this species. Similarly, aggregation factors (AFs), putative allorecognition molecules in sponges (Moscona 1968; Humphreys 1970; Muller and Zahn 1973; Henkart et al. 1973; Fernandez-Busquets et al. 1998; Fernandez-Busquets and Burger 1999), are also encoded by a set of clustered genes in the *A. queenslandica* genome. Here, five *AF* genes sit together within an 80 kb cluster of the genome, with a sixth putative *AF* sitting alone elsewhere in the genome (Gauthier 2010; Grice, Gauthier and Degnan, unpublished). Finally, while new evidence suggests that *B. schlosseri* histocompatibility may be encoded by a single gene, *BHF* (Voskoboynik et al. 2013), the FuHC locus also contains other genes that appear to contribute to the allorecognition phenotype (De Tomaso et al. 2005; Nyholm et al. 2006; McKitrick and De Tomaso 2010; Nydam et al. 2013b; discussed in theory by Harada 2013). *sFuHC and mFuHC* genes (De Tomaso et al. 2005; Nydam et al. 2013b), which correlate well with predicted allorecognition properties (De Tomaso et al. 2005; Nydam et al. 2013b) and fusibility outcomes (Voskoboynik et al. 2013), are situated within ~400 kb of other candidate regulators of allorecognition, *fester* and *uncle fester* (Nyholm et al. 2006; McKitrick et al. 2011). Clustered genes have also been reported from the immune or self-nonself recognition systems of other species (Fig. 2) including *Drosophila melanogaster* (Werner et al. 2000), the purple sea urchin *Strongylocentrotus purpuratus* (Miller et al. 2010), chickens and zebra finches (Hellgren and Ekblom 2010) and the fungus *Neurospora crassa* (Micali and Smith 2006).

The clustering of allorecognition genes in part reflects their origins through tandem duplication, but cluster maintenance appears to have occurred via natural selection. Clustering of allorecognition genes facilitates the transfer of sequence information between regions within an immune locus, which may be executed in a number of ways including gene conversion, recombination and unequal crossing over, alternative splicing and gene inversion (Graham 1995; Ghosh et al. 2011). The clustering of related allorecognition genes may increase the efficiency and precision of co-regulated gene expression if required (Blumenthal 1998), as has been observed in suites of non-allorecognition genes from diverse taxa, such as zebrafish (Ng et al. 2009), *C. elegans* (Spieth et al. 1993), *S. cerevisiae* (Zhang and Smith 1998) and *D. melanogaster* (Spellman and Rubin 2002). Clustering can also increase the co-inheritance of particular 'matched set' gene variants (Pál and Hurst 2003), although this hypothesis has not held up in other tests of non-immune ligand-receptor linkage in humans (Hurst and Lercher 2005). Birth and death evolution also can contribute to the maintenance of species-specific features amongst these grouped allorecognition genes (Nei and Rooney 2005). Finally, the primary driving force behind cluster maintenance in allorecognition and other immune systems may be the need to generate high levels of sequence diversity between individuals or species. The mutational divergence of duplicated genes, and the gain or loss of various functional domains, can further increase the rate of diversification within these clusters.

Positive Selection

Allorecognition molecules are expected to display a high level of diversity within species, to produce different molecular signatures of self for distinct individuals. We

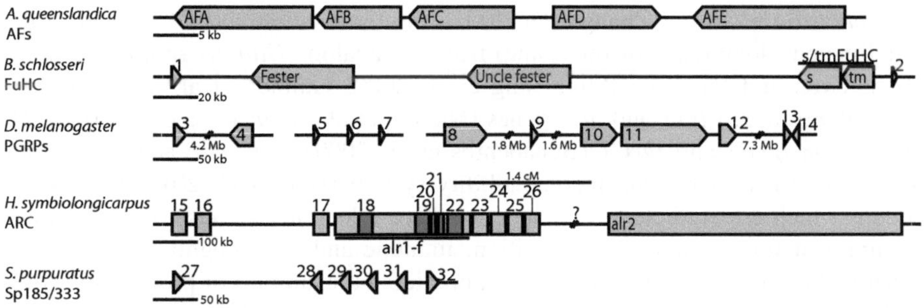

Fig. 2 Genomic clustering of invertebrate self-nonself recognition genes. The genomic organisation of clustered self-nonself recognition and allorecognition genes, from selected invertebrate species. Shown are the *Amphimedon queenslandica* AFs (Gauthier 2010), various reported FuHC locus genes from *Botryllus schlosseri* (De Tomaso et al. 2005; Nyholm et al. 2006; McKitrick et al. 2011), the peptidoglycan recognition proteins (PGRPs) from *Drosophila melanogaster* (Werner et al. 2000), the *Hydractinia symbiolongicarpus* ARC, including the uncharacterised IgSF-like genes present in the region (Nicotra et al. 2009; Rosa et al. 2010), and the *Sp185/333* gene cluster from *Strongylocentrotus purpuratus* (Miller et al. 2010). The *D. melanogaster* PGRP genes sit in three separate genomic regions, corresponding left to right to the X, 2R and 3 L chromosomes, respectively. PGRP genetic coordinates are taken from the *D. melanogaster* genomic assembly hosted by Ensembl. The *H. symbiolongicarpus* ARC has not yet been fully mapped beyond linkage analysis, therefore the precise distance between the *alr1* and *alr2* regions is unknown. Five genes (*IgSF-like-1, -4, -7, -X* and *–Y*) sit within the current limits of the *alr1*-containing interval. In all cases, only known, clustered gene family members are shown. For numbered genes, names and Ensembl accession numbers (in brackets, for PGRP genes) are as follows; 1: *BHF*, 2: *HSP40*, 3: *PGRP-SA* (FBgn0030310), 4: *PGRP-LE* (FBgn0030695), 5: *PGRP-SC1A* (FBgn0043576), 6: *PGRP-SC1B* (FBgn0033327), 7: *PGRP-SC2* (FBgn0043575), 8: *PGRP-LD* (FBgn0260458), 9: *PGRP-SD* (FBgn0035806), 10: *PGRP-LA* (FBgn0035975), 11: *PGRP-LC* (FBgn0035976), 12: *PGRP-LF* (FBgn0035977), 13: *PGRP-SB2* (FBgn0043577), 14: *PGRP-SB1* (FBgn0043578), 15: *IgSF-like-F*, 16: *IgSF-like-G*, 17: *IgSF-like-A*, 18: *IgSF-like-7*, 19: *IgSF-like-4*, 20: *IgSF-like-X*, 21: *IgSF-like-Y*, 22: *IgSF-like-1*, 23: *IgSF-like-B*, 24: *IgSF-like-C*, 25: *IgSF-like-D*, 26: *IgSF-like-E*, 27: *Sp185/333-A2*, 28: *Sp185/333-B8*, 29: *Sp185/333-D1y*, 30: *Sp185/333-D1g*, 31: *Sp185/333-D1b*, 32: *Sp185/333-E2*

have mentioned different methods of gene or transcript rearrangement to facilitate this variation above. However, mutation and nucleotide-level variants also play a large role in the establishment of allorecognition diversity. Within the expectations of Kimura's neutral theory (Kimura 1968), synonymous mutations are predicted to be selectively neutral and therefore be observed at a higher frequency than non-synonymous mutations when comparing allele sequences within a species (Kimura 1977). Various statistical models that compare frequencies of synonymous (d_S) and non-synonymous (d_N) polymorphisms have been developed (Jensen et al. 2007). Examples where non-synonymous differences are observed at a higher frequency than synonymous changes provide evidence that particular sequences or codons may be under positive selection, whereby amino acid change and protein diversification is selectively favoured (Jensen et al. 2007).

A number of examples of positive selection have been observed in characterised self-nonself recognition systems to date. For example, *Sp185/333* from *S. purpuratus* (Terwilliger et al. 2006), the parasite defense gene *FREP3* from the freshwater snail

Biomphalaria glabrata (Zhang et al. 2001), the fertilisation genes *lysin* and *VERL* (vitelline envelope receptor for lysine) from the abalone *Haliotis* spp. (Metz et al. 1998; Lyon and Vacquier 1999; Yang et al. 2000; Galindo et al. 2003), the *H. symbiolongicarpus alr1* and *alr2* genes (Nicotra et al. 2009; Rosa et al. 2010), *D. discoideum tgrB1* and *tgrC1* (Benabentos et al. 2009) and *het-c* and *pin-c* from the *N. crassa* heterokaryon incompatibility system (Hall et al. 2010) all possess codons which are predicted to be under positive selection. Because of the inherent requirement for self-nonself recognition, immune and allorecognition proteins to generate high levels of diversity, we can predict that examples of positive selection will be identified at increasing rates as more genome data become available and alleles from a greater number of individuals are surveyed.

Whence: The Origins of Animal Allorecognition Loci

Self-nonself recognition can be conceptualised as a three-step process involving the detection of nearby conspecifics, the recognition of these as either self or nonself, and a discriminatory outcome dependent on the previous recognition decision. The modular nature of this process suggests a basic mechanism for how a primitive allorecognition system, capable of supporting an early multicellular animal, might first evolve. If this organism could couple together several existing processes to ultimately perform the three phases of allorecognition, a new self-nonself recognition system is born. Addition to or modification of existing genetic elements could produce further novel functionality.

A Relationship Between Allorecognition 'Detection' and Cell Adhesion

Profound similarities exist between the functional requirements of the allorecognition detection phase and of cell adhesion processes; it is likely that animal cell adhesion and allorecognition systems are evolutionarily related (Curtis 1979; Fernandez-Busquets and Burger 1999). Both systems require the presence of compatible ligands and receptors, which interact specifically to facilitate binding and/or communication between their respective cells. Each ligand or receptor may have multiple possible binding partners. The structural features of each class of molecule are also similar, including the frequent inclusion of transmembrane domains and large extracellular regions comprised of tandemly repeated extracellular protein domains (Fig. 1, Table 2). Examples of these repeated structures can be seen in the various members of the cadherin, immunoglobulin, integrin and selectin cell adhesion families (Fig. 1b).

Cell adhesion molecules also play a role in cell recognition and sorting events, for example during tissue development and organogenesis (McNeill 2000). Cell aggregation experiments have demonstrated the key role of cadherins in differential cell adhesion, again demonstrating a clear functional relationship between cell adhesion processes and allorecognition molecules. However, while differential cell interactions in allorecognition are underpinned by highly polymorphic self-nonself recognition molecules, differential cadherin binding is largely mediated by the control of cell surface deployment of invariant molecules (Leckband and Prakasam 2006; Halbleib and Nelson 2006).

Analysis of sponge allorecognition processes further demonstrates the relationship between cell adhesion and allorecognition functionality, in two ways. First, the putative sponge allorecognition molecules, aggregation factors (AFs), mediate the species-specific reaggregation of dissociated sponge cells (Humphreys 1963; Moscona 1968; Humphreys 1970; Muller and Zahn 1973; Henkart et al. 1973), acting as bridging molecules between compatible cells (Jarchow et al. 2000). In addition, however, AFs have also been demonstrated to accumulate at nonself tissue graft interfaces (Fernandez-Busquets et al. 1998), and display molecular and genetic features characteristic of allorecognition molecules, such as a clustered gene organisation (Gauthier 2010), tandemly-repeated extracellular protein domains (Fernandez-Busquets et al. 1996) and a high level of sequence polymorphism (Fernandez-Busquets and Burger 1997). Second, observation of the effects of chimerism in juvenile sponges also demonstrates a functional relationship between cell adhesion and histocompatibility. Experimental fusion of pairs of fluorescently-labeled sponge postlarvae and juveniles leads to an initial period of cellular intermingling followed by a near-complete cell sorting, whereby cells from one individual contribute predominantly to the choanocytes, while the cells of the other individual form the pinacocytes and mesohyl (Gauthier and Degnan 2008). This differential cell sorting process is reminiscent of the cadherin-mediated sorting of cell populations discussed above; however, the intriguing strict separation of cell types by individual further demonstrates a link between cell adhesion and self-nonself recognition.

Metazoans evolved from a unicellular holozoan ancestor that possessed many of the domain types characteristic of the modern metazoan cell adhesion and signalling systems. For example, cadherin (King et al. 2003, 2008; Abedin and King 2008; Nichols et al. 2012; Suga et al. 2013; Fairclough et al. 2013), immunoglobulin (King et al. 2008), integrin (Sebé-Pedrós et al. 2010; Suga et al. 2013; Sebe-Pedros et al. 2013) and tyrosine kinase (King and Carroll 2001; King et al. 2003, 2008; Manning et al. 2008; Pincus et al. 2008; Sebé-Pedrós et al. 2010, 2013; Suga et al. 2012, 2013; Fairclough et al. 2013) components have all been identified in the unicellular holozoan relatives of the Metazoa. Interestingly, however, the domain organisation in these proteins often differs from their metazoan counterparts (King et al. 2008), highlighting the importance of cooption, gene duplication and divergence, and domain shuffling for the formation of the proteins that currently populate the cell surface and extracellular matrix of multicellular animals. As self-nonself recognition systems probably evolved in concert with the first experiments in metazoan multicellularity, and given the similarities between cell adhesion and allorecognition molecule structure and function highlighted above, it is not unreasonable to suppose that self-nonself recognition systems also evolved from the pre-existing holozoan cell surface and extracellular protein set.

A Model for the Origin of Allorecognition Systems

We propose a mechanism by which an allorecognition system could evolve in a pre-metazoan that does not yet possess this capacity (Fig. 3). This scenario describes how such a process is possible in the stem metazoan lineage in light of the known general characteristics of allorecognition systems, and is not intended to serve as a 'true' reconstruction of evolutionary history based on homology in extant organisms. As a minimal requirement, a primordial allorecognition system should be able to maintain a multicellular aggregate of self cells, while excluding unrelated nonself cells from the organismal milieu. Execution of an allorecognition reaction triggered by nonself 'invasion' should proceed via detection of the invader, recognition of this invader as nonself (or the absence of self), and a discriminatory action that prevents the inclusion of the invader into the aggregate.

In this model, the ancestor of this early animal allorecognition system may have been a cell-cell interaction gene or locus (Fig. 3a). As discussed above, metazoan cell adhesion molecules predate the origin of stem metazoans, and are structurally and functionally related to molecules involved in the allorecognition detection and recognition phases. The precursor gene would probably encode a protein that was at least partially extracellular, and that possessed common cell interaction features such as repeated extracellular domains that are evolvable yet structurally robust (for example, immunoglobulin and other beta-sheet containing domains; Table 2). We make no specific predictions regarding when the ancestral locus would need to be originally coopted, in terms of the organisms' evolutionary trajectory from unicellularity to multicellularity. However, as self-nonself recognition requirements differ greatly between unicellular and obligatorily multicellular organisms, any systems coopted prior to the transition to multicellularity would most likely require significant modifications thereafter.

Duplication of the ancestral cell adhesion locus would facilitate its cooption into a self-nonself recognition system, by relaxing the selective pressure acting upon the duplicate. This would leave one copy free to diverge and elaborate upon its existing properties. Multiple duplication events could lead to the formation of a clustered, structurally-related gene family, whose members could then independently diverge to take on novel functions and features. One example of such a diversified cluster can be seen in the AF genes from the sponge *A. queenslandica* (Fig. 1a; Gauthier 2010).

Further family diversification could proceed by the development of novel protein architecture, either through within-gene domain expansion and shuffling, or the introduction of new domain types from other genes via recombination and shuffling (Gilbert 1978; Patthy 2003). Such novel architecture would then be found in one,

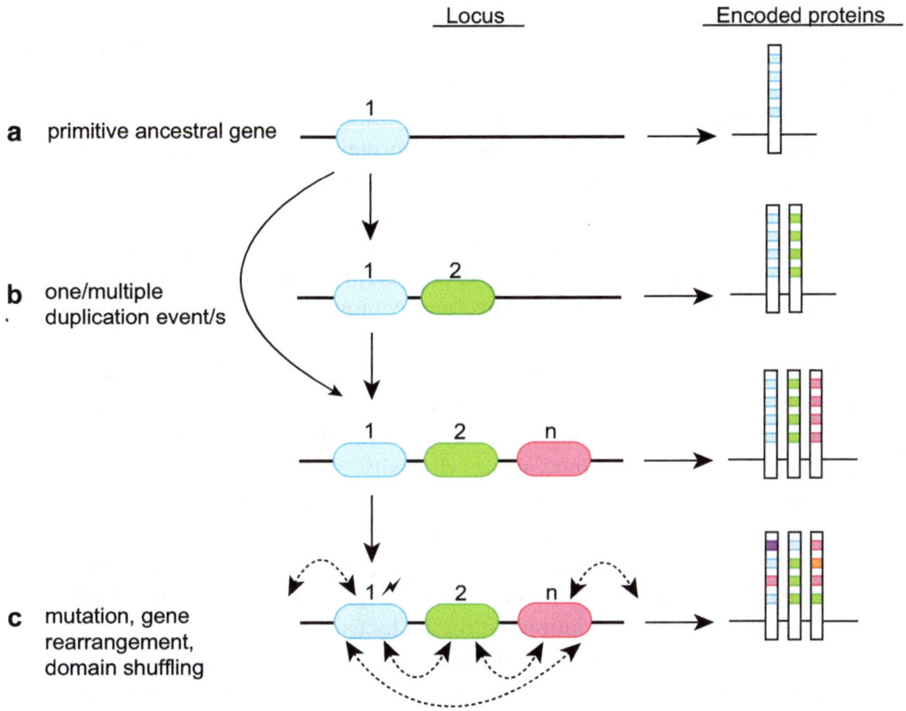

Fig. 3 Schematic model of the origin of a primitive self-nonself recognition system. A possible mechanism for the evolution of an early allorecognition locus, and corresponding proteins, is shown. **a** The original ancestral gene may have been a cell-cell interaction gene encoding a transmembrane protein with a modular extracellular domain. **b** One or more duplication events generate a gene cluster; duplication relaxes the selection pressure acting on the region, allowing divergence within and between genes. **c** Diversity is generated in the locus via mutation and gene rearrangement processes (*lightning bolt* and *dashed arrows*, respectively), leading to protein diversification

some, or all members of the new protein family, depending on the relative orders of gene and domain duplication, incorporation and loss events. Domain shuffling appears to be an ongoing feature of the evolution of cell surface and extracellular proteins underlying cell adhesion, communication and self-nonself recognition (Lee 2009; for example see Patel et al. 1987; Fahey and Degnan 2010, 2012; Hynes 2012). The sponge AFs provide one example of both within- and between-gene domain architecture modifications (Fig. 1a). The AFs from *A. queenslandica* and *C. prolifera* are both comprised of multiple Calx-beta domains (Fernandez-Busquets et al. 1996; Gauthier 2010) and a single Head domain (Grice et al. unpublished). Each AF protein possesses a different number of Calx-beta domains, a signature of multiple gene-specific domain expansion events. Additionally, four of the *A. queenslandica* AFs also contain a Von Willebrand domain, a domain type found in a range of other cell-cell and cell-matrix adhesion proteins (Bork and Rohde 1991), but these domains have not been identified in the AFs of *C. prolifera* or other studied sponge species (Fernandez-Busquets et al. 1996; Gauthier 2010). This demonstrates that the inclusion of the Von Willebrand domain into the *A. queenslandica* AFs may have been a lineage-specific event within the demosponges.

It is of interest to note that extant allorecognition proteins tend to be comprised of domains commonly found in cell adhesion proteins (see Fig. 1a and Table 2 for select examples). The structural and functional similarities between adhesion and recognition processes is consistent with adhesion proteins serving as a source of domain types for allorecognition molecules. In this way, developing allorecognition genes could obtain functionally similar, structurally robust domains from an existing pool of adhesion or interaction domains, and deploy them in new combinations to achieve new specificity or functionality.

Additional variation could potentially also be introduced to the new allorecognition locus by other element sharing processes, for example by recombination, gene conversion or alternative splicing (Fig. 3c). Such methods potentially allow for the generation of intralineage or intraspecies diversity, such that each individual could express individual-specific extracellular tags to flag particular cells or organisms as self or nonself, by way of differential binding. In this way, as long as the proteins maintained their basic structural and binding properties, a new protein family could emerge that is capable of intercellular binding, with novel binding specificities conferred by the sequence variation within the locus.

The final requirement of this hypothetical allorecognition system is the capability for discrimination. This could most simply be achieved by passive discrimination, whereby homophilic (or, depending on the system, heterophilic) binding, where cells recognised as self (or nonself, where applicable) remain bound, while nonself cells are excluded. This method would most likely be employed by a primitive allorecognition system with roots in cell adhesion.

The Evolution of Metazoan Allorecognition

The model proposed above suggests a possible mechanism and source of genetic material for the origin of a simple system capable of executing the three phases of self-nonself recognition. We do not attempt to explain the divergence of this hypothetical primitive system into the wealth of biological complexity present in the extant metazoan allorecognition systems. However, it remains possible that the process suggested here could produce a basic allorecognition system that could then diverge into multiple disparate allorecognition systems with no preserved evidence of homology. Alternatively, and perhaps more likely, it remains possible that metazoan allorecognition systems could emerge in parallel, whilst still taking "source material" from a common pool—that is, the large reservoir of cell surface and extracellular proteins that are comprised of a range of domains that enable protein-protein interactions, that were present in stem metazoans and the last common ancestors to extant animals. In theory, cooption of cell-cell interaction genes to a role in allorecognition simply requires that the protein possess a domain that can maintain structural integrity while accommodating extensive amino acid changes (e.g. beta-sheet containing domains; Table 2). This would allow different lineages to adapt different allorecognition molecules from a conserved pool of cell adhesion molecules, while still explaining the complexity and lack of apparent homology present in the extant allorecognition systems known today.

Conclusions

Effective multicellularity requires the constituent cells of an organism to sacrifice their own autonomy and, for most cells, reproductive contribution; the multicellular state is therefore potentially compromised in incidences of tissue fusion and cell transfer between conspecific individuals. True cooperation can only be maintained by natural selection if all constituent cells of an organism are genetically identical; nonself invaders of a host do not face the same selective pressures for cooperation. Allorecognition systems, which prevent the invasion of an individual by nonself cells, are therefore widespread amongst metazoans.

All allorecognition systems function in the same basic way: they must detect the presence of a cell, determine whether the cell is self or nonself, and take some discriminatory action based upon this decision. For this reason, many allorecognition systems share similar features. Allorecognition loci are often clusters of multiple allorecognition genes, which encode for large, partly extracellular modular proteins that are highly variable within a species. Invertebrate allorecognition systems are not immediately homologous, although it is likely that they represent highly divergent forms of one or more similar, ancient loci whose original role was in cell adhesion.

Summary

1. The successful transition to multicellularity requires a self-nonself recognition system that can maintain individual integrity by excluding foreign cells.
2. Allorecognition antedates the transition to multicellularity, as it has been identified in unicellular eukaryotes and bacteria.
3. All self-nonself recognition reactions occur as a three-phase process, involving the detection of a neighbouring entity, the recognition of this entity as self or nonself, and an appropriate discriminatory action.
4. Metazoan allorecognition systems differ greatly between taxa, but in general share common molecular and genomic features that fulfill the requirements of the three-phase self-nonself recognition process.
5. Extant metazoan allorecognition systems do not appear to be homologous but of those that have been characterised many appear to have evolved from existing cell-cell interaction or adhesion systems; these may have subsequently undergone diversification and pathway recoupling to produce novel allorecognition and innate immunity functionalities.

References

Abedin M, King N (2008) The premetazoan ancestry of cadherins. Science 319:946–948

Baumann CT, Lim CS, Hager GL (1999) Intracellular localization and trafficking of steroid receptors. Cell Biochem Biophys 31:119–127

Benabentos R, Hirose S, Sucgang R et al (2009) Polymorphic members of the lag gene family mediate kin discrimination in *Dictyostelium*. Curr Biol 19:567–572

Blumenthal T (1998) Gene clusters and polycistronic transcription in eukaryotes. Bioessays 20:480–487

Boehm T (2006) Quality control in self/nonself discrimination. Cell 125:845–858

Bogaerts A, Beets I, Schoofs L, Verleyen P (2010) Antimicrobial peptides in *Caenorhabditis elegans*. Invertebrate Surviv J 7:45–52

Boland T, Latour RA, Stutzenberger FJ (2000) Molecular basis of bacterial adhesion. In: An YH, Friedman RJ (eds) Handbook of bacterial adhesion: principles, methods and applications. Springer, Towata, pp. 29–41

Bonner JT (1966) Size and cycle. Princeton University Press, Princeton

Bonner JT (1988) The evolution of complexity. Princeton University Press, Princeton

Bonner JT (1998) The origins of multicellularity. Integr Biol 1:27–36

Bonner JT (2000) First signals: the evolution of multicellular development. Princeton University Press, Princeton

Bonner JT, Lamont DS (2005) Behavior of cellular slime molds in the soil. Mycologia 97:178–184

Bonner JT, Slifkin MK (1949) A study of the control of differentiation: the proportions of stalk and spore cells in the slime mold *Dictyostelium discoideum*. Am J Bot 36:727–734

Bork P, Rohde K (1991) More von Willebrand factor type A domains? Sequence similarities with malaria thrombospondin-related anonymous protein, dihydropyridine-sensitive calcium channel and inter-alpha-trypsin inhibitor. Biochem J 279:908–910

Bork P, Holm L, Sander C (1994) The immunoglobulin fold. J Mol Biol 242:309–320

Buss LW (1987) The Evolution of individuality. Princeton University Press, Princeton

Cadavid LF, Powell AE, Nicotra ML et al (2004) An invertebrate histocompatibility complex. Genetics 167:357–365

Cannon JP, Haire RN, Litman GW (2002) Identification of diversified genes that contain immunoglobulin-like variable regions in a protochordate. Nat Immunol 3:1200–1207

Chaine AS, Schtickzelle N, Polard T et al (2010) Kin-based recognition and social aggregation in a ciliate. Evolution 64:1290–1300

Coombe D, Ey P (1984) Self/non-self recognition in invertebrates. Q Rev Biol 59:231–255

Curtis AS (1979) Histocompatibility systems, recognition and cell positioning. Dev Comp Immunol 3:379–387

Dayel MJ, Alegado RA, Fairclough SR et al (2011) Cell differentiation and morphogenesis in the colony-forming choanoflagellate *Salpingoeca rosetta*. Dev Biol 357:73–82

Degnan BM, Leys SP, Larroux C (2005) Sponge development and antiquity of animal pattern formation. Integr Comp Biol 45:335–341

Destoumieux D, Bulet P, Loew D et al (1997) Penaeidins, a new family of antimicrobial peptides isolated from the shrimp *Penaeus vannamei* (Decapoda). J Biol Chem 272:28398–28406

De Tomaso AW, Nyholm SV, Palmeri KJ et al (2005) Isolation and characterization of a protochordate histocompatibility locus. Nature 438:454–459.

Dishaw LJ, Giacomelli S, Melillo D et al (2011) A role for variable region-containing chitin-binding proteins (VCBPs) in host gut-bacteria interactions. Proc Natl Acad Sci U S A 108:16747–16752

Dong Y, Taylor HE, Dimopoulos G (2006) AgDscam, a hypervariable immunoglobulin domain-containing receptor of the *Anopheles gambiae* innate immune system. PLoS Biol 4:e229

Dormann D, Vasiev B, Weijer CJ (2002) Becoming multicellular by aggregation; the morphogenesis of the social amoebae *Dicyostelium discoideum*. J Biol Phys 28:765–780

Dufour P, Jarraud S, Vandenesch F et al (2002) High genetic variability of the agr locus in *Staphylococcus* species. J Bacteriol 184:1180–1186

Dunman PM, Murphy E, Haney S et al (2001) Transcription profiling-based identification of *Staphylococcus aureus* genes regulated by the *agr* and/or *sarA* loci. J Bacteriol 183:7341–7353

Eberhard M (1975) The evolution of social behavior by kin selection. Q Rev Biol 50:13

Edwards YJ, Perkins SJ (1995) The protein fold of the von Willebrand factor type A domain is predicted to be similar to the open twisted β-sheet flanked by α-helices found in human ras-p21. FEBS Lett 358:283–286

Erdos GW, Raper KB, Vogen LK (1973) Mating types and macrocyst formation in *Dictyostelium discoideum*. Proc Natl Acad Sci U S A 70:1828–1830

Fahey B, Degnan BM (2010) Origin of animal epithelia: insights from the sponge genome. Evol Dev 12:601–617

Fahey B, Degnan BM (2012) Origin and evolution of laminin gene family diversity. Mol Biol Evol 29:1823–1836

Fairclough SR, Dayel MJ, King N (2010) Multicellular development in a choanoflagellate. Curr Biol 20:R875–876

Fairclough SR, Chen Z, Kramer E et al (2013) Premetazoan genome evolution and the regulation of cell differentiation in thechoanoflagellate Salpingoeca rosetta. Genome Biol 14:R15

Fay P (1992) Oxygen relations of nitrogen fixation in cyanobacteria. Microbiol Rev 56:340–373

Fernandez-Busquets X, Burger MM (1997) The main protein of the aggregation factor responsible for species-specific cell adhesion in the marine sponge *Microciona prolifera* is highly polymorphic. J Biol Chem 272:27839–27847

Fernandez-Busquets X, Burger MM (1999) Cell adhesion and histocompatibility in sponges. Microsc Res Tech 44:204–218

Fernandez-Busquets X, Kammerer RA, Burger MM (1996) A 35-kDa protein is the basic unit of the core from the 2 × 10(4)-kDa aggregation factor responsible for species-specific cell adhesion in the marine sponge Microciona prolifera. J Biol Chem 271:23558–23565

Fernandez-Busquets X, Gerosa D, Hess D (1998) Accumulation in marine sponge grafts of the mRNA encoding the main proteins of the cell adhesion system. J Biol Chem 273:29545–29553

Ferris PJ, Armbrust EV, Goodenough UW (2002) Genetic structure of the mating-type locus of *Chlamydomonas reinhardtii*. Genetics 160:181–200

Fugmann SD, Messier C, Novack LA et al (2006) An ancient evolutionary origin of the Rag1/2 gene locus. Proc Natl Acad Sci U S A 103:3728–3733

Galindo BE, Vacquier VD, Swanson WJ (2003) Positive selection in the egg receptor for abalone sperm lysin. Proc Natl Acad Sci U S A 100:4639–4643

Gauthier M (2010) Developing a sense of self. PhD Dissertation, University of Queensland

Gauthier M, Degnan BM (2008) Partitioning of genetically distinct cell populations in chimeric juveniles of the sponge Amphimedon queenslandica. Dev Comp Immunol 32:1270–1280

Gavrilets S (2010) Rapid transition towards the division of labor via evolution of developmental plasticity. PLoS Comput Biol 6:e1000805

George EA, Muir TW (2007) Molecular mechanisms of *agr* quorum sensing in virulent staphylococci. Chembiochem 8:847–855

Geuze HJ, Slot JW, Strous GJAM et al (1984) Intracellular receptor sorting during endocytosis: comparative immunoelectron microscopy of multiple receptors in rat liver. Cell 37:195–204

Ghosh J, Lun CM, Majeske AJ et al (2011) Invertebrate immune diversity. Dev Comp Immunol 35:959–974

Gilbert W (1978) Why genes in pieces? Nature 271:501

Gloria-Soria A, Moreno MA, Yund PO et al (2012) Evolutionary genetics of the hydroid allodeterminant *alr2*. Mol Biol Evoll 29:3921–3932

Glowacki S, Macioszek VK, Kononowicz AK (2010) R proteins as fundamentals of plant innate immunity. Cell Mol Biol Lett 16:1–24

Goldsby HJ, Dornhaus A, Kerr B, Ofria C (2012) Task-switching costs promote the evolution of division of labor and shifts in individuality. Proc Natl Acad Sci U S A 109:13686–13691

Graham GJ (1995) Tandem genes and clustered genes. J Theor Biol 175:71–87

Grosberg RK (1988) The evolution of allorecognition specificity in clonal invertebrates. Q Rev Biol 63:377–412

Grosberg RK, Strathmann RR (2007) The evolution of multicellularity: a minor major transition? Annu Rev Ecol Evol Syst 38:621–654

Gruntman M, Novoplansky A (2004) Physiologically mediated self/non-self discrimination in roots. Proc Natl Acad Sci U S A 101:3863–3867

Halbleib JM, Nelson WJ (2006) Cadherins in development: cell adhesion, sorting, and tissue morphogenesis. Genes Dev 20:3199–3214

Hall C, Welch J, Kowbel DJ, Glass NL (2010) Evolution and diversity of a fungal self/nonself recognition locus. PLoS One 5:e14055

Harada Y (2013) Allorecognition between compound ascidian colonies. Zoolog Sci 30:694–698

Harpaz Y, Chothia C (1994) Many of the immunoglobulin superfamily domains in cell adhesion molecules and surface receptors belong to a new structural set which is close to that containing variable domains. J Mol Biol 238:528–539

Harper RA (1926) Morphogenesis in *Dictyostelium*. Bull Torrey Bot Club 53:229–268

Hay ME, Fenical W, Gustafson K (1987) Chemical defense against diverse coral-reef herbivores. Ecology 68:1581–1591

Hellgren O, Ekblom R (2010) Evolution of a cluster of innate immune genes (β-defensins) along the ancestral lines of chicken and zebra finch. Immunome Res 6:1–15

Henkart P, Humphreys S, Humphreys T (1973) Characterization of sponge aggregation factor. A unique proteoglycan complex. Biochemistry 12:3045–3050

Hibino T, Loza-Coll M, Messier C et al (2006) The immune gene repertoire encoded in the purple sea urchin genome. Dev Biol 300:349–365

Hildemann WH (1979) Immunocompetence and allogeneic polymorphism among invertebrates. Transplantation 27:1–3

Hill E, Broadbent ID, Chothia C, Pettitt J (2001) Cadherin superfamily proteins in *Caenorhabditis elegans* and *Drosophila melanogaster*. J Mol Biol 305:1011–1024

Hirose S, Benabentos R, Ho H-I et al (2011) Self-recognition in social amoebae is mediated by allelic pairs of tiger genes. Science 333:467–470

Ho H-I, Hirose S, Kuspa A, Shaulsky G (2013) kin recognition protects cooperators against cheaters. Curr Biol 23:1590–1595

Hohenester E, Tisi D, Talts JF, Timpl R (1999) The crystal structure of a laminin G-like module reveals the molecular basis of alpha-dystroglycan binding to laminins, perlecan, and agrin. Mol Cell 4:783–792

Humphreys T (1963) Chemical dissolution and in vitro reconstruction of sponge cell adhesions. Dev Biol 8:27–47

Humphreys T (1970) Biochemical analysis of sponge cell aggregation. Symp Zool Soc Lond 25:325–334

Hurst LD, Lercher MJ (2005) Unusual linkage patterns of ligands and their cognate receptors indicate a novel reason for non-random gene order in the human genome. BMC Evol Biol 5:1–12

Hynes RO (2012) The evolution of metazoan extracellular matrix. J Cell Biol 196:671–679

Ikegami T, Okada T, Hashimoto M et al (2000) Solution structure of the chitin-binding domain of *Bacillus circulans* WL-12 chitinase A1. J Biol Chem 275:13654–13661

Ispolatov I, Ackermann M, Doebeli M (2012) Division of labour and the evolution of multicellularity. Philos Trans R Soc Lond B Biol Sci 279:1768–1776

Janzon L, Arvidson S (1990) The role of the delta-lysin gene (*hld*) in the regulation of virulence genes by the accessory gene regulator (*agr*) in *Staphylococcus aureus*. EMBO J 9:1391–1399

Janzon L, Löfdahl S, Arvidson S (1989) Identification and nucleotide sequence of the delta-lysin gene, *hld*, adjacent to the accessory gene regulator (*agr*) of *Staphylococcus aureus*. Mol Gen Genet 219:480–485

Jarchow J, Fritz J, Anselmetti D et al (2000) Supramolecular structure of a new family of circular proteoglycans mediating cell adhesion in sponges. J Struct Biol 132:95–105

Jarraud S, Lyon GJ, Figueiredo AM et al (2000) Exfoliatin-producing strains define a fourth *agr* specificity group in *Staphylococcus aureus*. J Bacteriol 182:6517–6522

Jensen JD, Wong A, Aquadro CF (2007) Approaches for identifying targets of positive selection. Trends Genet 23:568–577
Ji G, Beavis R, Novick RP (1997) Bacterial interference caused by autoinducing peptide variants. Science 276:2027–2030
Karban R, Shiojiri K (2010) Identity recognition and plant behavior. Plant Signal Behav 5:854–855
Kastan MB, Bartek J (2004) Cell-cycle checkpoints and cancer. Nature 432:316–323
Kidd T, Brose K, Mitchell KJ et al (1998) Roundabout controls axon crossing of the CNS midline and defines a novel subfamily of evolutionarily conserved guidance receptors. Cell 92:205–215
Kimball RF (1939) Mating types in *Euplotes*. Am Nat 73:451–456
Kimura M (1968) Evolutionary rate at the molecular level. Nature 217:624
Kimura M (1977) Preponderance of synonymous changes as evidence for the neutral theory of molecular evolution. Nature 267:275–276
King N (2004) The unicellular ancestry of animal development. Dev Cell 7:313–325
King N, Carroll SB (2001) A receptor tyrosine kinase from choanoflagellates: molecular insights into early animal evolution. Proc Natl Acad Sci U S A 98:15032–15037
King N, Hittinger CT, Carroll SB (2003) Evolution of key cell signaling and adhesion protein families predates animal origins. Science 301:361–363
King N, Westbrook MJ, Young SL et al (2008) The genome of the choanoflagellate Monosiga brevicollis and the origin of metazoans. Nature 451:783–788
Kirk DL (2005) A twelve-step program for evolving multicellularity and a division of labor. BioEssays 27:299–310
Konijn TM, Raper KB (1961) Cell aggregation in *Dictyostelium discoideum*. Dev Biol 3:725–756
Leahy DJ, Hendrickson WA, Aukhil I, Erickson HP (1992) Structure of a fibronectin type III domain from tenascin phased by MAD analysis of the selenomethionyl protein. Science 258:987–991.
Leckband D, Prakasam A (2006) Mechanism and dynamics of cadherin adhesion. Annu Rev Biomed Eng 8:259–287
Lee B (2009) Comparison of exon-boundary old and young domains during metazoan evolution. Genomics Inform 7:131–135
Leshko-Lindsay LA, Corces VG (1997) The role of selectins in *Drosophila* eye and bristle development. Development 124:169–180
Luporini P, Miceli C, Ortenzi C (1983) Evidence that the ciliate *Euplotes raikovi* releases mating inducing factors (gamones). J Exp Zool 226:1–9
Lyon JD, Vacquier VD (1999) Interspecies chimeric sperm lysins identify regions mediating species-specific recognition of the abalone egg vitelline envelope. Dev Biol 214:151–159
Manning G, Young SL, Miller WT, Zhai Y (2008) The protist, Monosiga brevicollis, has a tyrosine kinase signaling network more elaborate and diverse than found in any known metazoan. Proc Natl Acad Sci U S A 105:9674–9679.
Marraffini LA, Sontheimer EJ (2010) CRISPR interference: RNA-directed adaptive immunity in bacteria and archaea. Nat Rev Genet 11:181–190
McKitrick TR, De Tomaso AW (2010) Molecular mechanisms of allorecognition in a basal chordate. Semin Immunol 22:34–38
McKitrick TR, Muscat CC, Pierce JD et al (2011) Allorecognition in a basal chordate consists of independent activating and inhibitory pathways. Immunity 34:616–626
McNeill H (2000) Sticking together and sorting things out: adhesion as a force in development. Nat Rev Genet 1:100–108
Metz EC, Robles-Sikisaka R, Vacquier VD (1998) Nonsynonymous substitution in abalone sperm fertilisation genes exceeds substitution in introns and mitochondrial DNA. Proc Natl Acad Sci U S A 95:10676–10681
Meylan E, Tschopp J, Karin M (2006) Intracellular pattern recognition receptors in the host response. Nature 442:39–44
Micali CO, Smith ML (2006) A nonself recognition gene complex in *Neurospora crassa*. Genetics 173:1991–2004
Miceli C, Concetti A, Luporini P (1983) Isolation of the mating-inducing factor of the ciliate *Euplotes*. Exp Cell Res 149:593–598

Miceli C, La Terza A, Bradshaw RA, Luporini P (1992) Identification and structural characterization of a cDNA clone encoding a membrane-bound form of the polypeptide pheromone Er-1 in the ciliate protozoan *Euplotes raikovi*. Proc Natl Acad Sci U S A 89:1988–1992
Michod RE (2007) Evolution of individuality during the transition from unicellular to multicellular life. Proc Natl Acad Sci U S A 104(Suppl 1):8613–8618
Middha S, Wang X (2008) Evolution and potential function of fibrinogen-like domains across twelve *Drosophila* species. BMC Genomics 9:260
Miller CA, Buckley KM, Easley RL, Smith LC (2010) An *Sp185/333* gene cluster from the purple sea urchin and putative microsatellite-mediated gene diversification. BMC Genomics. doi:10.1186/1471-2164-11-575
Morgan AP (1903) *Dictyostelieae* or *Acrasieae*. J Mycol 9:84–86
Moscona AA (1968) Cell aggregation: properties of specific cell-ligands and their role in the formation of multicellular systems. Dev Biol 18:250–277
Muller WEG, Zahn RK (1973) Purification and characterization of a species-specific aggregation factor in sponges. Exp Cell Res 80:95–104
Narasimha M, Brown NH (2000) Integrins and associated proteins in *Drosophila* development. In: Danen E (ed) Integrins and development. Landes Bioscience, Texas
Nei M, Rooney AP (2005) Concerted and birth-and-death evolution of multigene families. Annu Rev Genet 39:121–152
Ng YK, Wu W, Zhang L (2009) Positive correlation between gene coexpression and positional clustering in the zebrafish genome. BMC Genomics. doi:10.1186/1471-2164-10-42
Nichols SA, Roberts BW, Richter DJ et al (2012) Origin of metazoan cadherin diversity and the antiquity of the classical cadherin/β-catenin complex. Proc Natl Acad Sci U S A 109:13046–13051
Nicotra ML, Powell AE, Rosengarten RD et al (2009) A hypervariable invertebrate allodeterminant. Curr Biol 19:583–589
Norman DG, Barlow PN, Baron M et al (1991) Three-dimensional structure of a complement control protein module in solution. J Mol Biol 219:717–725
Novick RP, Ross HF, Projan SJ et al (1993) Synthesis of staphylococcal virulence factors is controlled by a regulatory RNA molecule. EMBO J 12:3967–3975
Novick RP, Projan SJ, Kornblum J et al (1995) The *agr* P2 operon: an autocatalytic sensory transduction system in *Staphylococcus aureus*. Mol Gen Genet 248:446–458
Nydam ML, Hoang TA, Shanley KM, De Tomaso AW (2013a) Developmental and comparative immunology. Dev Comp Immunol 41:128–136
Nydam ML, Netuschil N, Sanders E et al (2013b) The candidate histocompatibility locus of a basal chordate encodes two highly polymorphic proteins. PLoS One 8:e65980
Nyholm SV, Passegue E, Ludington WB et al (2006) Fester, a candidate allorecognition receptor from a primitive chordate. Immunity 25:163–173
Oehler R (1922) *Dictyostelium mucoroides* (Brefeld). Centr Bakteriol Parasitenk 89:155–156
Oka H, Watanabe H (1957) Colony specificity in compound ascidians as tested by fusion experiments. Proc Jpn Acad 33:657–659
Oren M, Paz G, Douek J et al (2013) Marine invertebrates cross phyla comparisons reveal highly conserved immune machinery. Immunobiology 218:484–495
Ortenzi O, Luporini P (1995) Competition among homologous polypeptide pheromones of the ciliate *Euplotes raikovi* for binding to each other's cell receptors. J Eukaryot Microbiol 42:242–248
Ortenzi C, Alimenti C, Vallesi A et al (2000) The autocrine mitogenic loop of the ciliate *Euplotes raikovi*: the pheromone membrane-bound forms are the cell binding sites and potential signaling receptors of soluble pheromones. Mol Biol Cell 11:1445–1455
Otto M, Süssmuth R, Vuong C et al (1999) Inhibition of virulence factor expression in *Staphylococcus aureus* by the *Staphylococcus epidermidis* agr pheromone and derivatives. FEBS Lett 450:257–262
Pál C, Hurst LD (2003) Evidence for co-evolution of gene order and recombination rate. Nat Genet 33:392–395

Pancer Z, Amemiya CT, Ehrhardt GR et al (2004) Somatic diversification of variable lymphocyte receptors in the agnathan sea lamprey. Nature 430:174–180
Patel RS, Odermatt E, Schwarzbauer JE, Hynes RO (1987) Organization of the fibronectin gene provides evidence for exon shuffling during evolution. EMBO J 6:2565–2572
Patthy L (2003) Modular assembly of genes and the evolution of new functions. Genetica 118:217–231
Pincus D, Letunic I, Bork P, Lim WA (2008) Evolution of the phospho-tyrosine signaling machinery in premetazoan lineages. Proc Natl Acad Sci U S A 105:9680–9684
Raper KB (1935) *Dictyostelium discoideum*, a new species of slime mold from decaying forest leaves. J Agric Res 50:135–147
Raper KB (1937) Growth and development of *Dictyostelium discoideum* with different bacterial associates. J Agric Res 55:289–316
Raper KB (1940) The communal nature of the fruiting process in the Acrasieae. Am J Bot 27:436–448
Ratcliff WC, Denison RF, Borrello M, Travisano M (2012) Experimental evolution of multicellularity. Proc Natl Acad Sci U S A 109:1595–1600
Richards GS, Degnan BM (2009) The dawn of developmental signaling in the Metazoa. Cold Spring Harb Symp Quant Biol 74:81–90
Rinkevich B, Weissman IL (1987) A longâŁłterm study on fused subclones in the ascidian *Botryllus schlosseri*: the resorption phenomenon (Protochordata: Tunicata). J Zool 213:717–733
Rinkevich B, Porat R, Goren M (1995) Allorecognition elements on a urochordate histocompatibility locus indicate unprecedented extensive polymorphism. Proc R Soc Lond B 259:319–324
Rinkevich B, Weissman IL, De Tomaso AW (1998) Transplantation of Fu/HC-incompatible zooids in *Botryllus schlosseri* results in chimerism. Biol Bull 195:98–106
Rosa SFP, Powell AE, Rosengarten RD et al (2010) Hydractinia allodeterminant *alr1* resides in an Immunoglobulin superfamily-like gene complex. Curr Biol 20:1122–1127
Rossetti V, Schirrmeister BE, Bernasconi MV, Bagheri HC (2010) The evolutionary path to terminal differentiation and division of labor in cyanobacteria. J Theor Biol 262:23–34
Rueffler C, Hermisson J, Wagner GP (2012) Evolution of functional specialization and division of labor. Proc Natl Acad Sci U S A 109:E326–E335
Schmucker D, Clemens JC, Shu H et al (2000) *Drosophila* Dscam is an axon guidance receptor exhibiting extraordinary molecular diversity. Cell 101:671–684.
Schuckmann WV (1924) Zur biologie von *Dictyostelium mucoroides* Bref. Centr Bakteriol Parasitenk 91:302–309
Schwarz EM, Benzer S (1997) Calx, a Na-Ca exchanger gene of *Drosophila melanogaster*. Proc Natl Acad Sci U S A 94:10249–10254
Sebé-Pedrós A, Roger AJ, Lang FB et al (2010) Ancient origin of the integrin-mediated adhesion and signaling machinery. Proc Natl Acad Sci U S A 107:10142–10147
Sebe-Pedros A, Irimia M, del Campo J et al (2013) Regulated aggregative multicellularity in a close unicellular relative of metazoa. eLife 2:e01287–e01287
Shapiro L, Kwong PD, Fannon AM et al (1995) Considerations on the folding topology and evolutionary origin of cadherin domains. Proc Natl Acad Sci U S A 92:6793–6797
Simpson JH, Kidd T, Bland KS, Goodman CS (2000) Short-range and long-range guidance by slit and its Robo receptors. Robo and Robo2 play distinct roles in midline guidance. Neuron 28:753–766
Smith VJ, Fernandes JMO, Kemp GD, Hauton C (2008) Crustins: enigmatic WAP domain-containing antibacterial proteins from crustaceans. Dev Comp Immunol 32:758–772
Spellman PT, Rubin GM (2002) Evidence for large domains of similarly expressed genes in the *Drosophila* genome. J Biol 1(5):1–8
Spieth J, Brooke G, Kuersten S et al (1993) Operons in *C. elegans*: polycistronic mRNA precursors are processed by trans-splicing of *SL2* to downstream coding regions. Cell 73:521–532
Springer TA (1997) Folding of the N-terminal, ligand-binding region of integrin alpha-subunits into a beta-propeller domain. Proc Natl Acad Sci U S A 94:65–72

Stoner DS, Weissman IL (1996) Somatic and germ cell parasitism in a colonial ascidian: possible role for a highly polymorphic allorecognition system. Proc Natl Acad Sci U S A 93:15254–15259

Stoner DS, Rinkevich B, Weissman IL (1999) Heritable germ and somatic cell lineage competitions in chimeric colonial protochordates. Proc Natl Acad Sci U S A 96:9148–9153

Strassmann JE, Queller DC (2011) Evolution of cooperation and control of cheating in a social microbe. Proc Natl Acad Sci U S A 108(Suppl 2):10855–10862

Suga H, Dacre M, de Mendoza A et al (2012) Genomic survey of premetazoans shows deep conservation of cytoplasmic tyrosine kinases and multiple radiations of receptor tyrosine kinases. Sci Signal 5:ra35

Suga H, Chen Z, de Mendoza A et al (2013) The Capsaspora genome reveals a complex unicellular prehistory of animals. Nat Commun 4:1–9

Takayama S, Isogai A (2005) Self-incompatibility in plants. Annu Rev Plant Biol 56:467–489

Tarnita CE, Taubes CH, Nowak MA (2013) Evolutionary construction by staying together and coming together. J Theor Biol 320:10–22

Terwilliger DP, Buckley KM, Mehta D et al (2006) Unexpected diversity displayed in cDNAs expressed by the immune cells of the purple sea urchin, *Strongylocentrotus purpuratus*. Physiol Genomics 26:134–144

Tsutsui N (2004) Scents of self: the expression component of self/non-self recognition systems. Ann Zool Fennici 41:713–724

Vallesi A, Giuli G, Bradshaw RA, Luporini P (1995) Autocrine mitogenic activity of pheromones produced by the protozoan ciliate *Euplotes raikovi*. Nature 376:522–524

Van Wamel WK, van Rossum G, Verhoef J et al (1998) Cloning and characterization of an accessory gene regulator (*agr*)-like locus from *Staphylococcus epidermidis*. FEMS Microbiol Lett 163:1–9

Voskoboynik A, Newman AM, Corey DM et al (2013) Identification of a colonial chordate histocompatibility gene. Science 341:384–387

Waters CM, Bassler BL (2005) Quorum sensing: cell-to-cell communication in bacteria. Annu Rev Cell Dev Biol 21:319–346

Werner T, Liu G, Kang D et al (2000) A family of peptidoglycan recognition proteins in the fruit fly *Drosophila melanogaster*. Proc Natl Acad Sci U S A 97:13772–13777

Williams P, Camara M, Hardman A et al (2000) Quorum sensing and the population-dependent control of virulence. Philos Trans R Soc Lond B Biol Sci 355:667–680

Wouters MA, Rigoutsos I, Chu CK et al (2005) Evolution of distinct EGF domains with specific functions. Protein Sci 14:1091–1103

Wright CF, Christodoulou J, Dobson CM, Clarke J (2004) The importance of loop length in the folding of an immunoglobulin domain. Protein Eng Des Sel 17:443–453

Wright JS, Traber KE, Corrigan R et al (2005) The agr radiation: an early event in the evolution of staphylococci. J Bacteriol 187:5585–5594

Yang Z, Swanson WJ, Vacquier VD (2000) Maximum-likelihood analysis of molecular adaptation in abalone sperm lysin reveals variable selective pressures among lineages and sites. Mol Biol Evol 17:1446–1455

Zelensky AN, Gready JE (2005) The C-type lectin-like domain superfamily. FEBS J 272:6179–6217

Zhang X, Smith TF (1998) Yeast "operons". Microb Comp Genomics 3:133–140

Zhang SM, Léonard PM, Adema CM, Loker ES (2001) Parasite-responsive IgSF members in the snail *Biomphalaria glabrata*: characterization of novel genes with tandemly arranged IgSF domains and a fibrinogen domain. Immunogenetics 53:684–694

Zhang S-M, Adema CM, Kepler TB, Loker ES (2004) Diversification of Ig superfamily genes in an invertebrate. Science 305:251–254

Developmental Signalling and Emergence of Animal Multicellularity

Maja Adamska

Abstract Five major signalling pathways (Wnt, TGF-beta, Hedgehog, Notch and FGF) orchestrate short and long range cell-to-cell communication during development of cnidarians and bilaterians, and are often involved in homologous processes. Pre-metazoan ancestry of the pathways is evidenced by presence of some components in non-metazoans: Notch and proto-Hedgehog (Hedgling) pathways components are present in choanoflagellates, and some intracellular Wnt pathway components in slime molds. In contrast, long range signalling through diffusible ligands apparently coincided with emergence of animal multicellularity. Conservation of the signalling pathways in earlier branching animal lineages (sponges, ctenophores and placozoans) varies widely. Wnt and TGF-beta pathways display strongest conservation in all lineages. In sponges, the Wnt pathway appears to be involved in patterning of the body axis, as it is in cnidarians and bilaterians. On the other hand, the Hedgehog/Hedgling pathway has been repeatedly lost, as it is absent from ctenophores and placozoans. Thus, the developmental signalling toolkits of extant animal lineages have been shaped by loss and gain of entire pathways and their selected components.

Keywords Developmental signalling · Evolution · Wnt · TGF-beta · Hedgehog · Notch · FGF · Sponges · Ctenophores · Placozoans

Introduction

The key question of developmental biology is how complex multicellular organisms arise from single cells, usually fertilized eggs. This question is intimately linked with equally fundamental evolutionary question: how complex multicellular organisms arose from their single-cell ancestors? Decades of studies demonstrated that during animal development, cells use a handful of signaling pathways, especially Wnt, TGF-beta, Hedgehog, Notch and FGF, to coordinate movement, differentiation, proliferation and death (Pires da Silva and Sommer 2003). In model animals, such as

M. Adamska (✉)
Sars International Centre for Marine Molecular Biology, University of Bergen,
5008 Bergen, Norway
e-mail: Maja.Adamska@sars.uib.no

I. Ruiz-Trillo, A. M. Nedelcu (eds.), *Evolutionary Transitions to Multicellular Life*,
Advances in Marine Genomics 2, DOI 10.1007/978-94-017-9642-2_20

insects and vertebrates, these pathways have been shown to be re-used iteratively in developmental processes beginning with specification of embryonic axes, to fine-tuning of spacing of bristles or feathers on adult organisms. These different processes require intercellular communication to occur over a variety of ranges, from long range—with gradients encompassing many cell diameters, to direct communication of the adjacent cells. In pathways involved in short-range communication, such as Notch, the ligands are transmembrane proteins allowing only adjacent cells to communicate. On the other hand, diffusible proteins able to travel long distances, such as Hedgehog, can form gradients spanning multiple cell diameters when secreted at a localized source.

While the different signaling pathways are not related to each other (although some components of the Wnt and Hedgehog pathways are shared, and some are descendants of common ancestral genes (Nusse 2003), they usually work according to a similar model: An extracellular ligand binds a transmembrane receptor, which transduces the signal through a small set of cytoplasmic components, resulting in activation of a transcription factor– the key effector of the pathway, and finally affecting target gene expression (Fig. 1). Activation of the transcription factor often means switching off its repression function, as in the absence of the signal the effectors usually remain bound to their target sequences in the nucleus, actively repressing target gene expression. Binding of the ligand to the receptor and signal transduction through the cytoplasmic components results in availability of a co-activator to bind the transcription factor and turn repression into activation (Barolo and Posakony 2002). The specificity of the response to pathway activation is additionally strengthened by interactions of the key effector with cell- or tissue specific, cooperatively binding transcription factors. In addition, a range of extracellular antagonists able to bind ligands or the receptors is active in modulation of the signal.

Significantly, the signaling pathways regulating development of bilaterians (animals encompassing flatworms, insects, molluscs and vertebrates) are also found in cnidarians (such as jellyfish, corals and sea anemones) (Technau et al. 2005). This implies that the last common ancestor of cnidarians and bilaterians already possessed all of the key pathways, and utilized them during its development. On the other hand, no complete developmental signaling pathway has been found in the so-far sequenced genomes of single cell (and colonial) relatives of animals, such as choanoflagellates or other protists. Therefore, the phyla that diverged after choanoflagellates, and before cnidarians, hold the key to understanding of evolution of these pathways. Unfortunately, the branching order of these "basal" animal phyla: ctenophores (comb jellies), sponges and placozoans, remains a highly disputed issue, with a plethora of contradictory scenarios (see Nosenko et al. 2013, Chapter "Origin of Metazoan Developmental Toolkits and their Expression in the Fossil Record" and references therein for current discussion of the subject, and Fig. 2 for two examples of tree topologies).

Traditionally, the ctenophores—jellyfish-like marine pelagic predators with sensory organs and nervous system—were considered the sister group to the cnidarians. Presence of striated muscles, and bi-radial rather than radial symmetry suggested even a closer affinity to bilaterians than to cnidarians. Within metazoans (animals), the three clades of ctenophores, cnidarians and bilaterians are collectively referred to

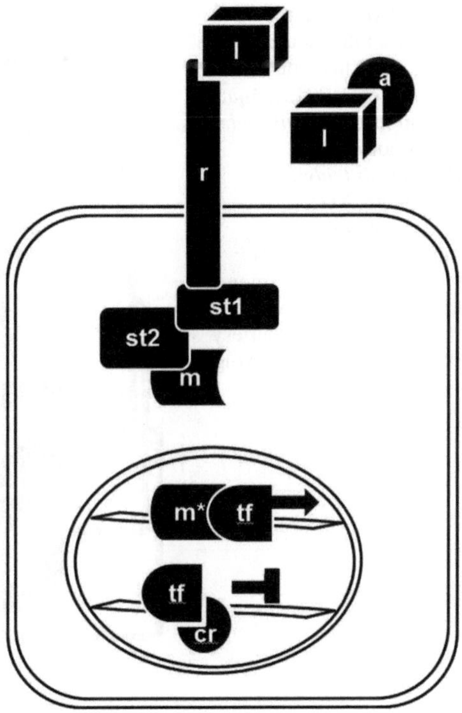

Fig. 1 Basic components of a metazoan developmental signalling pathway: *l* ligand, *r* receptor, *m* mediator (*activated), *st* signal transduction proteins, *tf* transcription factor, *cr* co-repressor, *a* antagonist

as the Eumetazoa, "true animals", to the exclusion of significantly simpler placozoans and sponges, neither of which have nerves nor muscles. Placozoans, represented by a single described species, *Trichoplax adhaerens* (many other morphologically similar, but genetically distinguished species likely exist), are marine benthic animals composed of only four morphologically distinguishable cell types. They have an apical-basal (top-bottom) polarity, but no anterior-posterior axis. Placozoa feed on algae, with the basal epithelium temporarily sealing the food on the substrate, as there is no internal feeding cavity. The simplicity of the placozoan body plan and its benthic lifestyle suggested to early researchers that *Trichoplax* represents the earliest branch of multicellular animals, supporting the placula hypothesis of metazoan evolution, in which the last common ancestor of animals was a flat two-cell-layered animal, facing the substrate with its ciliated layer (which later gave rise to the gut) (reviewed by Schierwater 2005). Another line of classic evolutionary thinking has seen sponges as the first multicellular animals: they are predominantly marine, benthic filter feeders, and their defining cell type—choanocytes—bear striking resemblance to choanoflagellates. While most of adult sponges display little similarity to the eumatazoan body plans, there is a wide array of larval types, with some (e.g. parenchymella of many demosponges) quite similar to larvae of cnidarians (Adamska et al. 2011). On the other hand, observations of calcisponges, and their metamorphosing larvae, suggested and supported the gastrea hypothesis of metazoan evolution, in which the

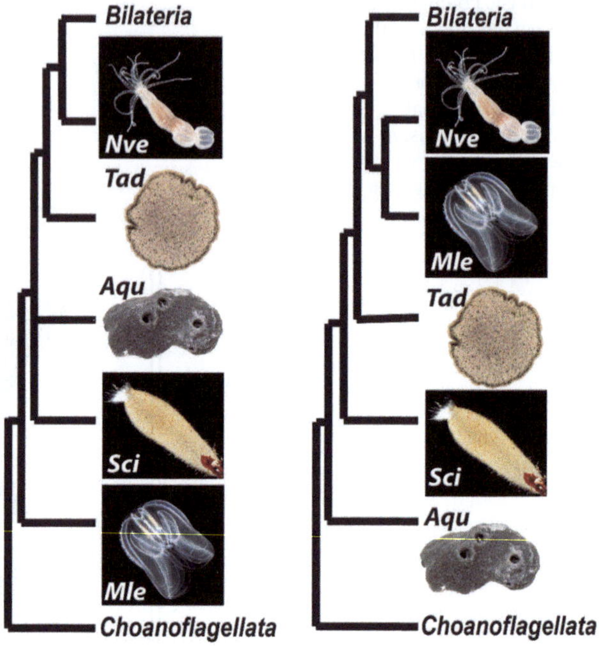

Fig. 2 Two of multiple versions of the metazoan tree of life (modified after Nosenko et al. 2013). *Aqu*: Amohimedon *queenslandica*, Demosponge; *Mle*: *Mnemiopsis leidyi*, ctenophore; *Nve*: *Nematostella vectensis*, anthozoan cnidarian; *Sci*: *Sycon ciliatum*, Calcisponge; *Tad*: *Trichoplax adhaerens*, placozoan. Images courtesy of M. Adamski and O. Voigt

last common ancestor of animals was also a two-cell-layered animal, but with inner ciliated cavity corresponding (and homologous) to the gut of eumetazoans (reviewed by Leys and Eerkes-Medrano 2005).

Instead of solving the problem of whether the placozoans or sponges branched first of the animal tree of life, advent of molecular phylogenies—single gene based at first, then utilizing entire genomes—additionally muddled the picture. In particular, a recent series of papers demonstrated a number of often mutually exclusive scenarios, pointing to virtually each of the four clades as the basal one. For example, some analyses suggests that ctenophores branched earlier than sponges and placozoans, with striking implications of either secondary loss of morphological complexity in sponges and placozoans, or convergent evolution of nerves and muscles in ctenophores (Ryan et al. 2013). Another surprising outcome was a suggestion of sponge paraphyly, with the siliceous sponges branching before the calcisponges and the homoscleromorphs, and an implication that all eumetazoans are in fact derived sponges. It appears that various topologies depend on choice of sequences and outgroups, and the conundrum appears unsolvable at the moment (Nosenko et al. 2013). Ideally, a sound phylogenetic framework should be the basis for mapping of the developmental signaling features. However, while it is still lacking, we can still compare the signaling toolkits of bilaterians, cnidarians, placozoans and sponges, and see if the toolkit complexity can be correlated with morphological complexity, and if it might give us hints as to interrelationships of these phyla.

Wnt Pathway

Wnt Pathway Components in Eumetazoans

Interactions of molecular components of the canonical (= beta-catenin dependent) Wnt pathway have been extensively reviewed, including a recent book (Nusse et al. 2013) and the regularly updated resources at The Wnt Homepage http://www.stanford.edu/group/nusselab/cgi-bin/Wnt (summarized in Table 1 and Fig. 3). The ligands of the pathway, the Wnts, are secreted glycoproteins with highly conserved cysteine residues. Several proteins act as Wnt receptors, with two families considered key for the canonical Wnt pathway: the seven-transmembrane Frizzled proteins and the single-pass Lrp5/6 (Arrow in *Drosophila*). TCF transcription factors are key effectors of the pathway, and (together with co-repressor Groucho) they remain bound to the cis-regulatory elements of target genes repressing their expression. The expression of target genes is promoted when beta-catenin, the key mediator of the pathway, translocates to the nucleus, displaces Groucho and switches TCF's function from repression to activation. In the absence of the signal, the cytoplasmic complex of multi-domain proteins Axin and APC allows CK1 and GSK3 kinases to phosphorylate beta-catenin, marking it for degradation. Binding of the Wnt ligand to Frizzled and Lrp5/6 results in change of the conformation of their intracellular portions, and allows for interaction between Frizzled and Dishevelled as well as Lrp5/6 and Axin. In result, the destruction complex becomes inactivated, and the non-phosphorylated beta-catenin can translocate to the nucleus.

The activity of the pathway is modulated by several antagonists, present in bilaterian and cnidarian genomes (Technau et al. 2005). Secreted Frizzled-Related Proteins (SFRPs), which share the Wnt-binding domain CRD with the Frizzled receptors, as well as unrelated to them WIF (Wnt Inhibitory Factor)-1 and Gremlin/Cerberus, bind the ligands making them unavailable for the receptors. On the other hand, Dickkopf proteins have affinity for Lrp5/6 receptors, and their binding results in degradation of LRP5/6. Thus, while SFRPs, WIF or Gremlin/Cerberus can inhibit all branches of the pathway, Dickkopf selectively affects the canonical Wnt pathway.

Many of the pathway components are encoded by, sometimes large, gene families. In particular, there are 13 subfamilies of the Wnt ligands in cnidarians and bilaterians (WntA, Wnt1 to Wnt11, Wnt16); most of which are shared by the majority of species with minor exceptions (WntA absent in vertebrates, Wnt3 in protostomes, Wnt10 in cnidarians) (Cho et al. 2010). Additional duplications within the subfamilies result in 19 Wnt ligands in mammals and 13 in the sea anemone *Nematostella*. Similarly, there are 5 Frizzled subfamilies shared by cnidarians and bilaterians: Fzd1/2/7, Fzd3, Fzd4, Fzd5/8 and Fzd9/10. Chordates lack Fzd3, but instead have an unrelated subfamily (Fzd3/6) likely derived from Fzd1/2/7. In addition, Smoothened, which is involved in Hedgehog signalling (see below), is closely related to the Frizzleds.

Table 1 Presence and absence of key signalling pathway components in Placozoans, Ctenophores, Sponges (*Amphimedon* and/or *Oscarella*) and non-metazoans (*Monosiga* and/or *Dictyostelium*). *White*: absent, *black*: present, *grey*: not analysed, *y*: yes, *n*: no, *y** indicates presence weakly supported by phylogenetic or domain composition analysis

pathway	role	gene	*Trichoplax*	*Mnemiopsis*	*Amphimedon + Oscarella*	*Monosiga + Dictyostelium*
Wnt	ligand	Wnt	y	y	y	n
	receptor	Frizzled	y	y	y	y
	receptor	LRP	y	y	y	n
	mediator	B-catenin	y	y	y	y
	signal transduction	Axin	y	n	y	n
	signal transduction	APC	y	y	y	n
	signal transduction	Dishevelled	y	y	y	n
also Hh	signal transduction	CK1	y	y	y	y
also Hh	signal transduction	Gsk3	y	y	y	y
	transcription factor	Tcf	y	y	y	n
	co-repressor	Groucho	y	y	y	n
	antagonist	SFRP	y	y	y	n
	antagonist	Wif	n	n	n	n
	antagonist	Dickkopf	n	n	y	n
TGF-beta	ligand	BMP-like	y	y*	n	n
	ligand	TGFB-like	y	y	y	n
	receptor	Type II	y	y	y	n
	receptor	Type I	y	y	y	n
	transcription factor	R-Smads	y	y	y	n
	transcription factor	Smad1/5/8	y	y	y	n
	transcription factor	Smad2/3	y	y	y	n
	transcription factor	I-Smad	y	y	n	n
	signal transduction	SARA	y	n	n	n
	co-repressor	Ski/sno	y	n	n	n
	antagonist	Noggin	y	n	y	n
	antagonist	Chordin	n	n	n	n
	antagonist	Follistatin	y	n	y	n
also Wnt	antagonist	Cerberus	y	n	n	n
Hedgehog	ligand	N-hh	n	n	y	y
	receptor	Patched	n		y	y
	receptor	I-hog	y		y	n
	signal transduction	Smoothened	n		n	n
	signal transduction	Fused	y		y	y
	signal transduction	Sufu	n		y	n
	signal transduction	Cos	y		y	n
	transcription factor	Ci/Gli	n	y	y	n
FGF	ligand	FGF	n	n	n	n
	receptor	FGFR	y		n	n
	signal transduction	Grb2	y		y	y
	signal transduction	Sos	y		y	y
	signal transduction	SHP2	y		y	y
	signal transduction	Ras	y		y	y
	signal transduction	Raf	y		y	n
	signal transduction	MEK	y		y	y
	signal transduction	ERK	y		y	y
Notch	ligand	Delta	y	y*	y	n
	ligand	Jagged	n		n	n
	receptor	Notch	y	y	y	y
	processing	Fringe	n	n	y	n
	processing	O-fut	y	n	y	y
	processing	Furin	y	y*	y	y
	transcription factor	Csl/su(H)	y	y	y	y
	co-activator	Mastermind	n	n	n	n

Wnt Pathway Functions in Eumetazoans

In many bilaterians and in cnidarians, one of the earliest and most profound developmental functions of the canonical Wnt pathway is establishment of the major

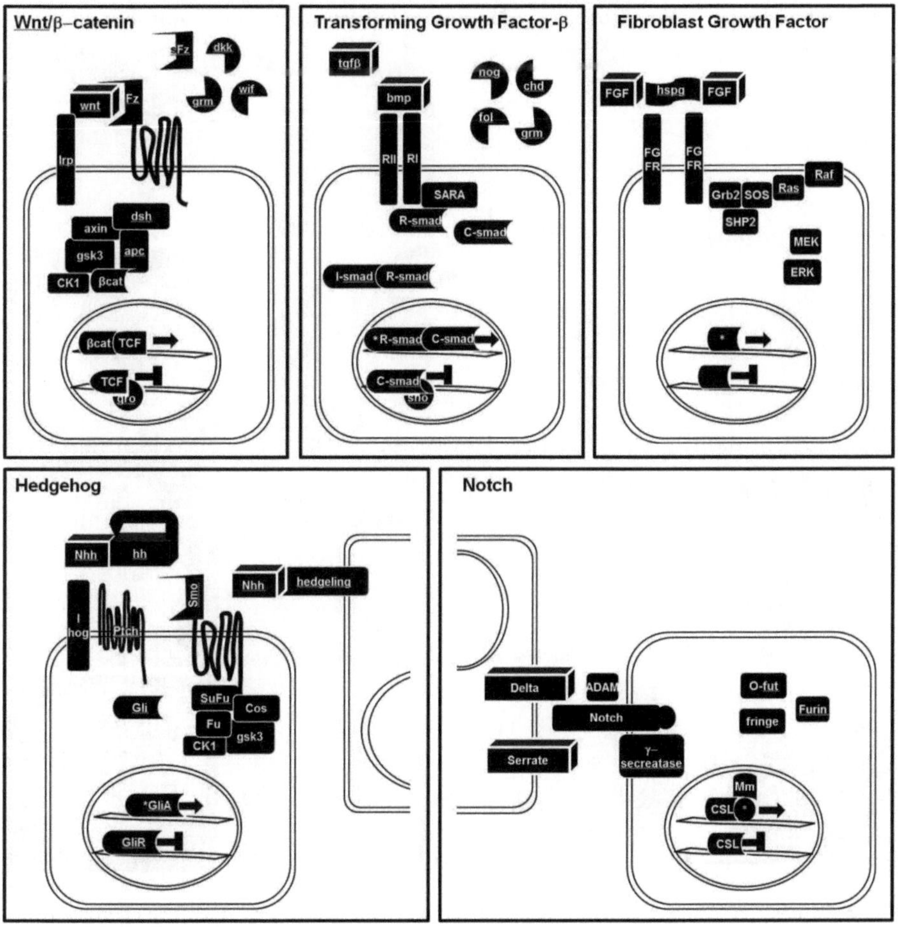

Fig. 3 Schematic representations of the major metazoan developmental signalling pathways, see text for details

embryonic axis, anterior-posterior in bilaterians and oral-aboral in cnidarians. In particular, specific expression of multiple Wnt genes is associated with the posterior pole of a variety of animals, from larvae of sea anemones, through planarians to leeches and vertebrates (Holland et al. 2000; Lee et al. 2006; Fig. 4a, 4b, 4e, 4f). This eumetazoan-wide function of the Wnt pathway can be seen as supporting homology of the cnidarian and bilaterian axes, with the oral end of cnidarians (derived from blastopore at the posterior pole of the larvae, as defined by swimming direction) corresponding to the posterior end of bilaterians (which in many cases, e.g. chordates, is also the position of the blastopore) (e.g. Petersen and Reddien 2009; Holstein 2012). On the other hand, expression of Hox genes in the sea anemone *Nematostella* appears to support an inverse relationship, with the oral end of cnidarians corresponding to the anterior end of bilaterians, although the situation in Hydrozoans

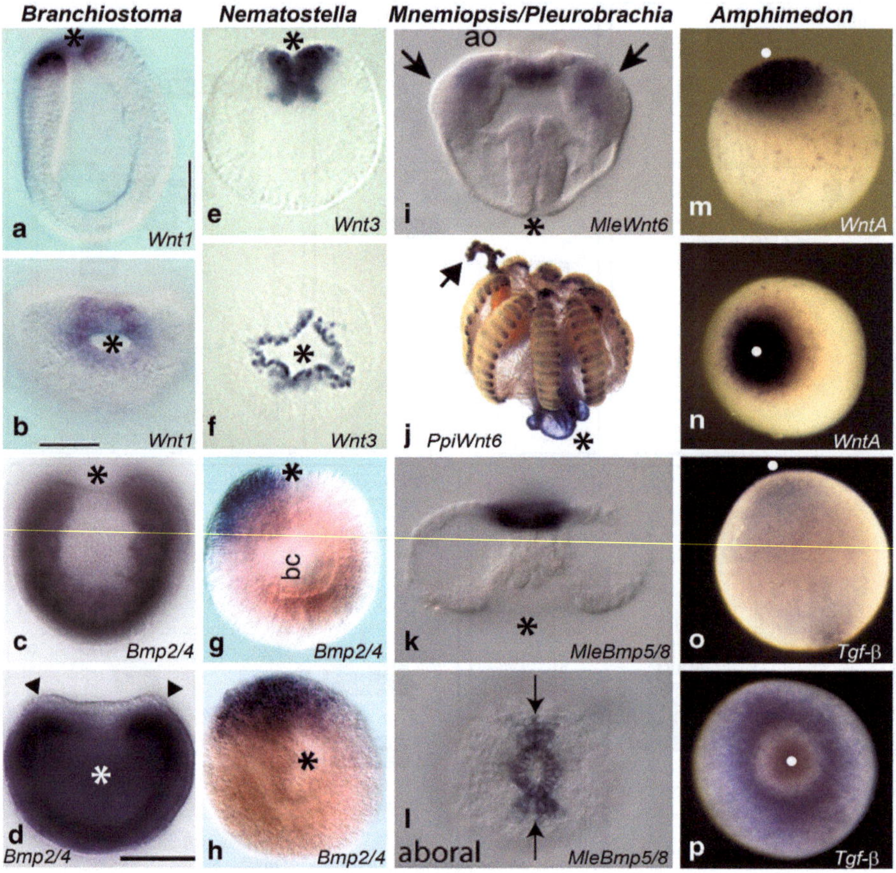

Fig. 4 Expression of selected Wnt and TGF-beta ligands in metazoans. **a–d**, The chordate *Branchiostoma floridae* (*Bilateria*): **a** (side view) and **b** (blastopore/posterior view): Wnt1 is expressed around the blastopore (in the posterior region of the embryo) (Holland et al. 2000). **c** (dorsal view) and **d** (blastopore/posterior view) BMP2/4 is expressed in a broad ventral domain (only the dorsal region, corresponding to the forming neural tube, is devoid of transcripts, as indicated by *arrowheads*) (Yu et al. 2008). **e–h** The sea anemone *Nematostella vectensis* (*Cnidaria*). **e** (side view) and **f** (blastopore/oral view): Wnt3 is expressed around the blastopore (the future oral region) (Lee et al. 2006). **g** (side view) and **h** (blastopore/oral view): BMP2/4 is asymmetrically expressed near the blastopore (Finnerty et al. 2004). **i–l** The comb jellies *Mnemiopsis leidyi* and *Pleurobrachia pileus* (*Ctenophora*). **i** (side view) of a *Mnemiposis* embryo, demonstrating Wnt6 expression in the tentacle bulbs (*arrows*) and the apical organ (*ao*) (Pang et al. 2010). **j** (side view) of adult *Pleurobrachia* demonstrating prominent Wnt6 expression in the oral region in addition to expression in the tentacle (*arrow*), apical organ and comb rows (Jager et al. 2013). **k** (side view) and **l** (aboral view): BMP5/8 is expressed at the aboral pole and along the sagittal plane (*arrows*) in the *Mnemiopsis* embryo (Pang et al. 2011). **m–p** The sponge *Amphimedon queenslandica* (*Demospongia*). **m** (side view) and **n** (posterior view): WntA is expressed in the posterior pole of the embryo. **o** (side view) and **p** (posterior view): TGF-betaA is dynamically expressed along the anterior-posterior axis: first stronger at both poles (**o**), and then excluded from the posterior pole (**p**) (Adamska et al. 2007a). *Asterisk* indicates blastopore, *white dot* indicates posterior region of the sponge embryos

appears to be more complex (Finnerty et al. 2004). With the two key axial patterning systems giving conflicting clues as to homology of the body axes, it is clear that significantly more in-depth knowledge of the genetic mechanism underlying axial patterning, especially in basal metazoans, is needed before valid comparisons can be made and conclusions reached (see Wagner 2007, for discussion on developmental genetics view on homology). Less controversially, beta-catenin plays a central role in the network leading to establishment of the innermost germ layer, the endoderm, in cnidarians and bilaterians (Roettinger et al. 2012).

In subsequent steps of the development, all branches of the Wnt pathway are involved in patterning and subsequent homeostasis of multiple organs, including gut, nervous system, skin and limbs (Nusse et al. 2013). The name "Wnt" itself testifies to multitude of functions of the pathway, as it is derived from a combination of the original independent *Drosophila* and mouse gene names, "wingless" (with the mutation affecting wing formation) and "Int" (Integration-1) resulting in mammary tumours. In fact, genes encoding components of the canonical Wnt pathway are often mutated in human cancers, with one of the cytoplasmatic complex proteins, APC (Adenomatous Polyposis Coli), being particularly often associated with the colorectal cancers. The uniting feature of these cancer-associated mutations is that they result in overactivation of the canonical Wnt pathway. In contrast, mutations leading to underactivation of the Wnt pathway lead to loss of intestinal stem cells (Taipale and Beachy 2001).

Wnt Pathway in Early-branching Metazoans

Wnt Pathway in Placozoans

Three Wnt ligands are present in the genome (Srivastava et al. 2008), but protein predictions of only two of them are long enough for phylogenetic analysis. Both of these proteins associate (with weak support) with Wnt8 subfamily, only one of 13 subfamilies recognized in cnidarians and bilaterians (Pang et al. 2010). Additional pathway components include the receptors (Frizzled), the main mediator (beta-catenin), proteins involved in regulation of its activity (Dishevelled, Axin and GSK3) and the key transcription factor of the pathway, TCF. Among the antagonists of the pathway, Dickkopf is absent, but a single SFRP is present and expressed in somatic tissue (Srivastava et al. 2008). While presence of all core components of the pathway suggests that the pathway is likely to be functional, it is important to keep in mind that no data are available regarding the activity of the pathway, both in terms of protein-protein interactions and role in the life of the animal. Thus, until experimental evidence is available, the assumption that the pathway functions in the same way as in bilaterians has to be treated with caution.

Wnt Pathway in Ctenophores

Components of the Wnt pathway in the complete draft genome of the ctenophore *Mnemiopsis leidyi* have been analysed by Pang et al. (2010). Three of the four Wnt ligands associate (with weak support) with Wnt6, Wnt9 and WntA subfamilies, and the last one does not fall into any recognized cnidarian/bilaterian subfamily. The identified receptors include two Frizzleds and one Lrp5/6/Arrow; as in case of the Wnt ligands it is not clear which cnidarian + bilaterian Frizzled subfamilies are represented in the ctenophores. All of the components of the cytoplasmic complex are also present in single copies, except for Axin, which has not been found. The specific antagonist of the canonical Wnt pathway, Dickkopf, is also apparently missing, but the likely general antagonist, SFRP, is present. Given the uncertain phylogenetic position of the ctenophores, it is unclear whether this simplicity reflects an ancestral condition or is a result of secondary gene loss in the ctenophore lineage, or in this particular species (as sponges clearly contain Axin and Dickkopf, see below). It is also formally possible that these sequences are present in the genome of *Mnemiopsis*, but are highly derived and thus cannot be easily identified by bioinformatics means. Analysis of Wnt pathway components present in the transcriptome dataset of another ctenophore, *Pleurobrachia pileus*, revealed the same picture of absence and presence of components (Jager et al. 2013).

Expression of all of the key components has also been studied during embryonic development of *Mnemiopsis* (Pang et al. 2010). Rather surprisingly, none of the four Wnt ligands is expressed before gastrulation is completed in *Mnemiopsis*, although two of the key downstream components, Dvl and FzdA, are indeed expressed already in the oocytes. Perhaps even more surprisingly, when the Wnt expression starts in *Mnemiopsis*, it is associated with the aboral pole (Fig. 4i), although again all of the downstream components, including beta-catenin, are specifically expressed in the oral region even before Wnt expression is detectable. It has been suggested that some of the components might be maternally loaded as proteins and thus assembly of the functional pathway during development would not be dependent on zygotic expression. Analysis of the localization of the Wnt pathway components by specific antibodies would be then necessary to clarify their function during early ctenophore development (Pang et al. 2010).

Knowledge whether the activity of the Wnt pathway is associated with the oral or aboral pole of the ctenophores appears critical to our understanding of evolution of metazoan body plans. As mentioned above, activity of the canonical Wnt signalling pathway in the oral end of cnidarians can be taken as support to homology of the cnidarian oral with bilaterian posterior poles, and its involvement in gastrulation (happening at the oral region in both cnidarians and ctenophores) appears to be a unifying metazoan feature (e.g. Holstein 2012).

Intriguingly, the recent study of Wnt pathway components expression in another ctenophore species, *Pleurobrachia pileus* (Jager et al. 2013), revealed expression patterns more consistent with the expectations: in addition to multiple expression domains associated with the apical organ, tentacles and comb rows, the most prominent region of co-expression of all components is the oral area (Fig. 4j). What is more, expression domains of the four Wnt genes in adult *Pleurobrachia* (which are orthologous to the *Mnemiopsis* genes) are nested, reminiscent of the nested Wnt domains around the oral pole of the cnidarians (e.g. Kusserow et al. 2005). The expression data from *Pleurobrachia* provide support for the notion that Wnt function in determination of the primary body axis is a universal metazoan feature, although clearly functional studies are needed before we can gain more confidence in that matter.

Wnt Pathway in Sponges

The Wnt pathway components in *Amphimedon* have been analysed in detail, including detection of putative protein-protein interaction motifs based on similarity to bilaterian sequences with known function, as well as analysis of developmental expression of all of the key components (Adamska et al. 2007a, 2010). Three Wnt ligands and two Frizzled receptors have been identified, although none of them can be confidently assigned to a recognized cnidarian + bilaterian subfamily. All of the other key components, including Lrp5/6/arrow, beta-catenin, Tcf, GSK3 and Groucho are present and display bilaterian protein-protein interaction motifs. This is not the case for APC and Axin, in which the expected beta-catenin interaction domains could not be detected, despite careful analysis of the sequences, including molecular cloning of full length open reading frames in addition to consulting of the genomic and transcriptomic assemblies (Adamska et al. 2010). As these domains are also not found in the cnidarian APC and Axin sequences, their ability to interact with each other and beta-catenin in the non-bilaterians has been called to question. While the apparently missing regions constitute significant portion of the APC, in case of Axin, the stretch of aminoacids which is conserved within bilaterians and demonstrated to interact with beta-catenin, is a relatively short sequence in the middle portion of the protein (Xing et al. 2003). It is thus possible that the interaction can take place, but the sequence responsible for the interaction cannot be recognized by bioinformatics analysis only. In line with that, axin from *Oscarella carmela*, a homoscleromorph sponge, was recently recovered as a binding partner of beta-catenin in a yeast two-hybrid screen, despite also missing the obvious beta-catenin interaction motif (Nichols et al. 2012). This result highlights the need of biochemical studies, rather than relying on in-silico sequence comparisons only, to identify true interaction within the components of the pathways.

Among the modulators of the pathway, several SFRPs are present, but in some of them, sequence similarity to the *Amphimedon* FzdA rises a possibility that they might be a result of convergent evolution rather than being homologues of bilaterian/ cnidarian SFRPs. Strikingly, no Dickkopf gene has been identified, while one has been previously found in a homoscleromorph sponge, *Oscarella carmela* (Nichols et al. 2006), demonstrating necessity to study gene content in representatives of several sponge lineages before conclusions regarding absence and presence of components "in sponges" can be made. All of the pathway components are expressed during *Amphimedon* development from the earliest stages, and indeed expression of AmqWntA is the first non-symmetrically distributed marker in the embryos (Adamska et al. 2007a; Fig. 4m, 4n). Significantly, expression of this gene marks the posterior pole of the embryo and forming larva, consistent with (although certainly not proving) homology of the embryonic A-P axis between sponges, cnidarian and bilaterians (e. g. Petersen and Reddien 2009; Holstein 2012).

While no functional data are available for *Amphimedon*, Wnt pathway interference study in another demosponge, *Ephydatia mulleri*, indicates that the Wnt pathway is also involved in specification of the adult body plan (Windsor and Leys 2010). In this study, upregulation of the canonical Wnt signalling by pharmacological inhibition of GSK3 resulted in formation of multiple oscula, reminiscent of multiplication of axis by similar experiments in cnidarians, supporting the view that involvement of the canonical Wnt pathway in determination of the body plan is a universal metazoan feature.

Pre-metazoan Ancestry of the Wnt Pathway

Protein kinases GSK3 and CK1, which in metazoans are not specific to the Wnt pathway, are the only Wnt pathway components in the choanoflagellates (Adamska et al. 2010). Surprisingly, beta-catenin-like and Frizzled related genes, in addition to GSK3, are present in *Dictyostelium discoideum*, a slime mold (reviewed by Harwood 2008). GSK3 and the b-catenin-like gene, Aardvark, which is similar to plant proteins with armadillo repeats, are parts of a signalling network involved in formation of the fruiting body. However, in contrast to the Wnt pathway, GSK3 positively regulates Aardvark, and it is not clear whether this pathway represents an evolutionary precursor of the metazoan Wnt pathway, or is a result of independent assembly of common components. Of the 25 *Dictyostelium* Frizzled proteins, two carry KTXXXW motif, which is specific to Frizzled proteins acting in the canonical Wnt pathway. Their function remains a mystery, and so is their evolutionary origin—have Frizzled genes been lost from multiple lineages, or is their presence in slime molds a result of horizontal gene transfer?

TGF-beta Pathway

TGF-beta Pathway Components in Eumetazoans

The core components of the Transforming Growth Factor beta pathway are extracellular ligands belonging to the TGF-beta superfamily, their transmembrane receptors of type I and II, and transcription factors of the SMAD class. Several auxiliary proteins, such as Smad Anchor for Receptor Activation (SARA), which facilitates Smad-receptor activation, as well as co-repressors Ski/Sno are also involved (Kitisin et al. 2007, Table 1, Fig. 3). The large superfamily of TGF-beta ligands (encompassing over 40 members in humans) can be roughly divided into two major families,

BMP-like and TGF-beta-like, both of which include defined subfamilies, such as BMPs, nodals, TGF-beta *sensu stricto*, Activins and others. Many of the subfamilies are represented in cnidarians, including BMPs and Activins, demonstrating that at least part of the diversification of the family happened before the bilaterian-cnidarian divergence (Technau et al. 2005). All of the ligands are produced as pre-proteins, with a large weakly conserved N-terminal part and a shorter, strongly conserved C-terminal. During maturation, they are cleaved by furin convertase to release the C-terminal signalling fragment, characterized by presence of 7–9 highly conserved cysteine residues. The ligands are received at the cell membrane by type II receptors, which are single pass serine-threonine kinases, displaying specificity for the type of the ligand. In deuterostomes, the type II receptors fall into three categories: BMPRII, ActRII, and TβRII; while only BMPRII, ActRII types are represented in so far studied protostomes and the cnidarian *Nematostella* (Pang et al. 2011). Upon binding of the ligand, the type II receptors associate with and activate type I receptors, which are also single pass serine-threonine kinases divided into three categories (BMPRI, ActRI, and TβRI, also known as ALK1–7 in mammals). All three of these sub-families are also represented in cnidarians (Pang et al. 2011). The activated type I receptors phosphorylate Smads of the R-Smad category (Receptor activated Smads, including two subfamilies: Smad1/5/8 and Smad2/3), which interact with common (or co-mediator) Smad4. Following the binding, the complex R-Smad + Smad4 translocates to the nucleus to activate target gene expression. Another type of Smads, I-Smads (Inhibitory Smads, Smad6 and Smad7) antagonize the pathway by several independent mechanisms. There is specificity of response to the signalling molecule, with Smads2/3 activated by TGF-beta- like signals, and Smads1/5/8 by BMP-like signals. All Smad sub-families are also represented in the cnidarians, as are all of the key antagonists of the pathway: Follistatin, DAN/gremlin/Cerberus, Tolloid and Chordin/SOG (Technau et al. 2005).

TGF-beta Pathway Functions in Eumetazoans

Perhaps the most famous function of the TGF-beta signalling is in establishment of the dorsal-ventral axis in bilaterians (reviewed by De Robertis and Sasai 1996). In embryos of protostomes like the fruit fly, Dpp (a homologue of BMP2/4) is expressed dorsally, while its antagonist, Short Gastrulation (SOG, a homologue of Chordin) is expressed ventrally. In chordates the situation is opposite, with BMPs expressed ventrally (Yu et al. 2008; Fig. 4c, 4d), and chordin dorsally. In both cases, the nervous system develops on the SOG/Chordin side, and the non-neural ectoderm is specified by BMP/Dpp signals. More surprisingly, TGF-beta signalling molecules are also asymmetrically expressed along the axis perpendicular to the larval anterior/posterior (adult aboral/oral) axis in anthozoan cnidarians, suggesting that they might function in specification of the secondary axis (Finnerty et al. 2004, Fig. 4g, 4h). However, in contrast to the bilaterian situation, chordin is expressed at the same side as dpp in *Nematostella*, and the logic of the Bmp-chordin interaction is different between cnidarians and bilaterians (Saina et al. 2009). Whether involvement of the TGF-beta pathway in patterning of the second body axis in bilaterians and cnidarians is a result of convergent evolution, or testifies to deep evolutionary origin of bilateral symmetry (with its subsequent loss in majority of cnidarians) remains a disputable issue. Analyses in a variety of cnidarian models, and functional studies revealing possible role and mechanisms of actin of the TGF-beta signalling in cnidarians are urgently needed. In later development of bilaterians, TGF-beta signalling is involved in a plethora of developmental processes, from germ layer specification to cardiac, bone and kidney development, as well as vasculogenesis and hematopoiesis (reviewed by Kitisin et al. 2007).

TGF-beta Pathway in Early-branching Metazoans

TGF-beta Pathway in Placozoans

There are five TGF-beta superfamily ligands encoded in the *Trichoplax* genome, four of them belong to the BMP-like family, and one to the TGF-like family (Srivastava et al. 2008; Pang et al. 2011). Only four receptors have been identified; one of them is the type II Activin receptor, the others are type I receptors, (one clear BMP receptor and the other two of more labile position, associating with Activin or BMP type I receptors depending on analysis) (Srivastava et al. 2008; Pang et al. 2011; Huminiecki et al. 2009). In contrast, the four *Trichoplax* Smads represent all recognized subfamilies of these transcription factors: R-Smad4, co-Smads Smad2/3 and Smad1/5/8, and the inhibitory Smad6/7 (Huminiecki et al. 2009; Pang et al. 2011). Other components include SARA, which in bilaterians associates with non-phosporylated R-Smads at the membrane, and the transcription co-repressor Ski/Sno. Most of the major antagonists of the pathway are also present, with Noggin, Follistatin and Cerberus/Gremlin identified. A gene similar to Chordin is also present, but lacking the key Chordin domain, so it is unclear whether its function is related to the TGF-beta pathway (Srivastava et al. 2008; Richards and Degnan 2009). As all key components are present, it is likely that the pathway can be functional, although lack of information on expression patterns as well as biochemical and genetic interactions of the components does not allow even a prediction of its possible function in the placozoans.

TGF-beta Pathway in Ctenophores

Expression and function of the TGF-beta pathway components in *Mnemiopsis leidyi* have been analysed (Pang et al. 2011). The ligands encompass two likely BMP-like genes, one TGF-beta *sensu stricto* gene, one additional TGF-beta-like gene and five genes that do belong to the TGF-beta superfamily but cannot be assigned to any cnidarian-bilaterian classes. The identified receptors belong to type I and type

II classes, but cannot be clearly assigned to any sub-class. In contrast, Smads of all classes recognized in cnidarians + bilaterians have been identified. The expression of ligands is identified by in situ hybridization during gastrulation, and while patterns suggest roles in axial patterning (Fig. 4k, 4l), the axes of ctenophores are already established before this expression becomes apparent. Intriguingly, Smad-4 (the co-Smad, which is the key element in TGF-beta signalling in bilaterians) has relatively narrow expression pattern in comparison to other Smads, suggesting that it might not be necessary for transduction of the signal in all contexts. An alternative explanation—that the pathway is active only where Smad4 is expressed—is unlikely, as pharmacological inhibition of the pathway results in defects in comb rows, where Smad4 is not detectable.

TGF-beta Pathway in Sponges

There are eight TGF-beta family ligands in *Amphimedon*, with two belonging to TGF-beta *sensu-stricto* subfamily, and the remaining ones not falling into any recognisable cnidarian + bilaterian category (Srivastava et al. 2010). BMP ligands, which are present in both ctenophore and placozoan genomes, could not be identified in *Amphimedon* or any other sponge for which sequence information is avaialable. As the identity of most of the identified ligands is unclear, and sequence information is available only for a limited number of sponge species, it is impossible to ascertain whether lack of the BMP ligands is due to their ancestral absence or secondary loss or divergence, and if this situation is representative to all sponges. On the other hand, the suite of receptors in *Amphimedon* (but not *Trichoplax* or *Mnemiopsis*) includes TGF-beta *sensu stricto* receptors type II. Three identified type I receptors cannot be confidently assigned to any category (Pang et al. 2011). Smads include R-Smads of Smad 1/5/8 and Smad2/3 categories as well as co-Smad4, but no inhibitory Smads. Cerberus/Gremlin inhibitors were not found in sponges so far, and a Chordin related gene lacks the Chordin domain, as it is also the case in *Trichoplax* (Srivastava et al. 2010). On the other hand, Noggin has been reported in *Suberites domuncula* (Müller et al. 2003) and is also present in *Amphimedon*, and follistatin has been identified in *Oscarella carmela* (Nichols et al. 2006). While no functional data are available, expression of at least one ligand indicates function in development. Similarly to one of the Wnt genes, this TGF-beta gene is expressed in a fraction of micromeres beginning at early cleavage stages. Subsequently, it is dynamically expressed throughout the outer cell layer of the embryo, and in a concentric pattern in connection with the forming pigment ring (the sensory organ of the larva). In contrast to bilaterians and anthozoan cnidarians, where TGF-beta (BMP) expression patterns the axis perpendicular to the one patterned by Wnt signals, *Amphimedon* Wnt and TGF-beta are expressed along one embryonic axis (Adamska et al. 2007a, Fig. 4o, 4p).

Pre-metazoan Ancestry of the TGF-beta Pathway

In contrast to the remaining pathways, the TGF-beta pathway appears to be a metazoan novelty, as no components have been identified elsewhere (Srivastava et al. 2010).

Hedgehog Pathway

Hedgehog Pathway Components in Eumetazoans

The core components of the Hedgehog signalling pathway are the Hedgehog ligand, Patched and I-hog receptors, the mediator Smoothened, a microtubule-associated complex of Cos2/Kif7, Fused (Fu), Suppressor of Fused (Sufu), PKA, GSK3, and CK1, and the key effector Ci/Gli zink finger transcription factor (Robbins et al. 2012, summarized in Table 1 and Fig. 3). All of these components are also present in the cnidarians (Matus et al. 2008). The ligands of the Hedgehog pathway are unique proteins which can auto-catalyse cleavage of the N-terminal signalling fragment from the C-terminal intein domain. The N-terminal undergoes lipid modification, including covalent attachment of cholesterol; this modification is critical for Hedgehog's ability to signal across long distances. A twelve-transmembrane domain protein, Patched, which belongs to a family of sterol-sensing proteins, is the core receptor of the cholesterol-linked Hedgehog molecule, and a single-transmembrane protein I-hog acts as a co-receptor. Formation of Hh-Patched-Ihog complex results in phosphorylation of a cytoplasmic tail of the key mediator of the pathway: a seven-transmembrane protein Smoothened (which is related to the Wnt receptors, Frizzled). Phosphorylated Smoothened translocates from micro-vesicles, where it remained in the absence of Hh signal, to the plasma membrane (in vertebrates, to the plasma membrane of the primary cilium). From there, it promotes dissociation of the microtubule-associated complex of several proteins (including kinases PKA, GSK3 and CK1) which in the absence of signal phosphorylated the transcription factor Ci/Gli. Phosphorylated Ci/Gli undergoes proteolytic cleavage, which removes its C-terminal activator domain and generates a transcriptional repressor form (Ci/GliR). Presence of the Hedgehog signal, through dissociation of the kinases complex, results in accumulation of non-phosphorylated Ci/Gli and its subsequent (but not fully understood) processing resulting in formation of an activator form (Ci/GliA) and allowing strong activation of the target genes.

Hedgehog Pathway Functions in Eumetazoans

Drosophila Hedgehog has been originally discovered and named through its role in regulation of number of bristles on the larval cuticle. Subsequently, the single *Drosophila* gene, and its several homologues in vertebrates (Sonic, Indian and Desert

Hedgehog in the mouse) have been shown to play a wide array of roles in bilaterian development (Jiang and Hui 2008). In many of the developmental contexts, Hedgehog molecules act as classical morphogens—they elicit different response in the cells receiving the ligand depending on concentration; in others the signal is short range, or even contact dependent. The examples of short range signalling include induction of the neural floor plate by the notochord expressing Shh in vertebrate embryos, or Hedgehog action in developing *Drosophila* wing disk. Perhaps the best studied example of long range action of the Hedgehog signalling is the vertebrate limb. The signalling ligand is produced in the aptly named Zone of Polarizing Activity in the posterior part of the limb, and patterns the forming digits in a dose-dependent manner. Different parts of the limb are exposed to different levels and forms of the signal: the posterior digits derive from cells that at some point express Shh; the middle digits depend on the signalling molecule long-range diffusion, while the anterior-most digit formation is independent from Hedgehog signals (McGlinn and Tabin 2006).

Although the function of the Hedgehog pathway has not been studied in the cnidarians, expression of two Hedgehog genes, as well as Patched and Gli, has been analysed in *Nematostella*. Intriguingly, while Nvhh1 is expressed in pharyngeal ectoderm, and Nvhh2 in the endoderm, Patched and Gli are expressed only in the endoderm, suggesting both short and long range signalling is likely to be occurring (Matus et al. 2008).

Hedgehog Pathway in Early-branching Metazoans

Hedgehog Pathway in Placozoans and Ctenophores

No evidence of the ligand (N-terminal part of Hedgehog), the receptor (Patched), the mediator Smoothened or the key transcription factor, Gli have been found in *Trichoplax* (Srivastava et al. 2008). Similarly, the ligand and other key components of the pathway, Smoothened and SUFU (suppressor of fused) were also not detected in the ctenophore *Mnemiopsis leidyi* (Ryan et al. 2013). As at least some components of this pathway are present in pre-metazoans, it appears that the pathway has been secondarily, and independently lost in these two lineages.

Hedgehog Pathway in Sponges

While the genome of *Amphimedon* contains no true Hedgehog gene, both domains of the protein are present and constitute parts of non-related proteins (Adamska et al. 2007b). In particular, the C-terminal intein portion is found in several copies, associated with genes encoding laccases. The N-terminal, signalling Hedgehog domain forms the N-terminal part of a large transmemebrane protein, which also encompasses cadherin repeats, immunoglobulin and EGF-domains. The gene, which we named Hedgling, is developmentally expressed, with rather late embryonic expression associated with formation of the sensory pigment ring at the posterior pole of the larvae. Analysis of other metazoan genomes identified a highly conserved Hedgling gene in *Nematostella vectensis* (which also possess two "canonical" Hedgehogs), but not in any of analysed bilaterians. The cnidarian Hedgling is also developmentally expressed, and in addition to domains identified in *Amphimedon* possesses an SH2 domain, associated with signal transduction. It is not known whether Hedgling proteins undergo modifications and/or cleavage. If not, in contrast to true Hedgehog, which can act as a long-distance morphogen, Hedgling could only signal across adjacent cells, similarly to the Notch signalling. The genome of *Amphimedon* contains several other components of the Hedgehog pathway, including one of the receptors (Ihog), several signal transduction components (Fu, Sufu, Cos) as well as Gli transcription factors (Srivasatava et al. 2010; Richards and Degnan 2009), but no Patched and no Smoothened. It is not clear whether Hedgling signals through a pathway similar to the canonical Hedgehog pathway or through an unidentified mechanism.

Pre-metazoan Ancestry of the Hedgehog Pathway

While true Hedgehog genes have not been identified outside cnidarians and bilaterians, a Hedgling homologue is present in a choanoflagellate, *Monosiga brevicolis* (King et al. 2008). In addition, the receptor Patched, and a protein involved in transport of Hedgehog in bilaterians, Dispatched, are also present, along with the signal transduction protein Fused. On the other hand, the key transcription factor, Gli, is absent, as are other key elements (reviewed by Richards and Degnan 2009). Thus, as in case of sponges, it is unclear whether the components that are shared with cnidarians + bilaterians are already assembled in a pathway, or whether they are involved in different processes.

FGF pathway

FGF Pathway Components in Eumetazoans

Fibroblast Growth Factors are one of families of multiple growth factor types that signal through Receptor Tyrosine Kinases (RTKs), and transduce the signal through several cascades, including Ras/Raf/MEK/ERK (Goldfarb 2001, summarized in Table 1 and Fig. 3). While the growth factor families are not related to each other (so FGFs display no sequence similarity to PDGF, VEGFs or EGFs), all of their receptors carry a related intracellular tyrosine kinase domain. In particular, FGF receptors (FGFRs) belong to immunoglobulin-like RTKs, which also encompass PDGFRs and VEGFs. The signalling is activated upon formation of a complex including two FGFR molecules and two FGF molecules linked by heparin sulfate proteoglycan. This initiates a series of phosphorylation events, which include GRB2, SOS, SHP2,

RAS, RAF, MEK and ERK proteins, and result in phosphorylation of multiple transcription factors, including, but not limited to, those from ETS family. Thus, there is not a single type of dedicated transcription factor of the FGF (or even broadly RTK) pathway, such as Tcf for Wnt, Smad for TGF-beta, or Ci/Gli for the Hedgehog pathway. Importantly, the multiple downstream components of the FGF pathway are largely shared between a multitude of the RTKs, including receptors of EGFs, and are also regulated by non-RTKs signalling events.

FGF Pathway Functions in Eumetazoans

In bilaterians, FGF signalling is indispensable for a variety of developmental processes, from the early stages, when acting in specification of both dorsal/ventral and anterior-posterior axes, through gastrulation movements, induction of germ layers and neural tissue, to organ patterning (Thisse and Thisse 2005). In one of its axial patterning roles, FGF signals convey posterior identity to both neural plate and mesoderm of vertebrate embryos by acting upstream of Hox genes. In the cnidarian *Nematostella*, and likely in bilaterians whose larvae possess apical tuft (a putative larval sensory organ), FGF signalling through the Erk cascade controls development of this structure (Rentzsch et al. 2008).

FGF signalling appears to also play conserved role in boundary formation. In vertebrates, FGF signals are critical for establishment and function of the midbrain-hindbrain boundary. In the cnidarian *Hydra*, FGFR *kringelchen* is expressed in a ring pattern at the boundary between the parental polyp and the forming foot of the bud, and has been shown to be critical for bud detachment (Sudhop et al. 2004). Among the later developmental events, FGFs' role in branching morphogenesis is particularly well studied in both *Drosophila* and vertebrate systems, where it is involved in formation of the tracheal system and the lungs, respectively. Developing vertebrate limb is a key example of function of FGFs in regulating growth of structures, as FGF signals emanating from the Apical Ectodermal Ridge (AER) are responsible for maintenance of proliferative potential of the underlying mesenchymal tissue.

FGF Pathway in Early-branching Metazoans and Pre-metazoans

Investigations of genomes of the three basally branching metazoans (*Trichoplax*, *Mnemiopsis* and *Amphimedon*) as well as the choanoflagellate *Monosiga* revealed absence of FGF-ligands (Srivastava et al. 2008, 2010; Ryan et al. 2013; King et al. 2008). However, proteins with domain similar to the one found in FGFs, and called FGF-like, have recently been identified in a broad range of metazoans and pre-metazoans, including sponges, arthropods, lophotrochozoans, cephalochordates and choanoflagellates, and were suggested to be descendants of evolutionary precursors of the true FGFs (Bertrand et al. 2014). Function of these proteins remains unclear, as the FGF receptors—composed of extracellular immunoglobulin-like loops responsible for interaction with the ligand and the intracellular receptor tyrosine kinase part responsible for signal transduction—are not found outside of cnidarians and bilaterians. The only exception is an FGFR-like gene described in *Trichoplax* (Rebscher et al. 2009), which might be a result of independent evolution via domain shuffling (Bertrand et al. 2014). On the other hand, multiple receptor tyrosine kinases (RTKs) are present—and constitute a huge gene family—in *Monosiga brevicolis*, and some of them display weak similarity to the intracellular portion of the FGFR receptors (Manning et al. 2008), perhaps evidencing pre-FGF ancestry of the receptors.

The downstream signal transduction components, such as Sos, Grb2, Shp2, Ras, Mek and Erk, are all present in *Monosiga*, and some even in other eukaryotes (e.g. Srivastava et al. 2010). This demonstrates a gradual assembly of the pathway, from signal transduction cascade utilized by single cell eukaryotes, through addition of receptors allowing more sophisticated sensing of the environment (or perhaps neighbouring cells in the colony), up to invention of a highly effective morphogen in the ancestor of cnidarians and bilaterians.

Notch Pathway

Notch Pathway Components in Eumetazoans

In contrast to Wnt, TGF-beta, Hedgehog and FGF signalling pathways, which derive their names from the diffusible ligands employed, the Notch pathway is named after its receptor, and the ligands of the pathway—Delta and Serrate/Jagged—are also transmembrane proteins (Bray 2006; summarized in Table 1 and Fig. 3). Thus, the signalling through the Notch pathway can only occur from one adjacent cell to another. The ligands and receptors are large, single pass transmembrane proteins with multiple EGF-like repeats in the extracellular domains. Both are glycosylated by O-fucosyltransferase (O-fut) and Fringe proteins, and the extent of glyosylation affects receptor-ligand binding. Notch undergoes another modification—proteolytic cleavage by Furin—before it is targeted to the plasma membrane. Upon binding of the ligand, Notch is cleaved again by an ADAM metalloprotease, and then subsequently by gamma-secretase complex. The cleavages result in release of the intracellular portion of the protein—termed Notch intracellular domain (Nicd)—which then translocates to the nucleus. In there, Nicd associates with DNA-binding protein CSL (known as Su(H) in *Drosophila*), which acts as a repressor in its absence, and switches CSL's role to activator of transcription. A co-activator Mastermind (Mam) completes the trimeric CSL-Nicd-Mam complex to promote transcription of the target genes.

Notch Pathway Functions in Eumetazoans

Function of the Notch signalling has been extensively studied in development and homeostasis of bilaterians and cnidarians. It is involved in formation of virtually all organs, including brain, heart, pancreas and skin, and is crucial for segmentation in the vertebrates (Andersson et al. 2011). In both bilaterians and cnidarians, it is particularly important in formation of the nervous system (Marlow et al. 2012). The mode of action of the Notch signalling can be assigned to one of three types, with all examples found during *Drosophila* development: (1) lateral inhibition, as in the developing ommatidia, where the cells acquiring neural fate inhibit surrounding cells from adopting the same fate; (2) lineage decisions, as in sensory organ precursors, where daughter cells inherit different levels of Numb; (3) boundary formation, as in the wing primordium, where the boundary between dorsal and ventral compartments forms at the edge of Serrate (a Notch ligand) expression (Bray 2006). In cnidarians, there is direct evidence for function of Notch in boundary formation during budding of *Hydra* (Münder et al. 2010). On the other hand, no direct evidence for lateral inhibition or lineage decisions roles in cnidarians have been reported so far, although Notch pathway clearly affects differentiation of oocytes, nematoblasts and neurons (Käsbauer et al. 2007; Marlow et al. 2012).

Notch Pathway in Early-branching Metazoans

Notch Pathway in Placozoans and Ctenophores

While most of the Notch pathway components are present, including the receptor, one type of ligand (Delta), some of the processing components (Furin, O-fucosyltransferase, ADAM metaloprotease, gamma-secretase complex), Fringe genes were not found in *Trichoplax* (Richards and Degnan 2009). Similarly, Fringe, and also O-fucosyltransferase were not detected in the ctenophore *Mnemiopsis*, and the Delta ligand appears to have unusual domain composition (Ryan et al. 2013). Unfortunately, no data on expression and possible developmental function of the Notch pathway has been published so far in either placozoans or ctenophores, and as complete genome data are available for only one ctenophore species so far, it remains unclear how representative is the situation in *Mnemiopsis* for the complete phylum.

Notch Pathway in Sponges

Notch pathway components and their developmental expression in *Amphimedon* embryos have been analysed in detail (Richards et al. 2008, Richards and Degnan 2012; Gazave et al. 2009). Significant expansion of the gene families encoding the key components is present: five different Delta ligands and six different Fringe genes are reported. Other components of the pathway are also represented, including the receptor Notch, Furin, O-fucosyltransferase, gamma-secretase complex and ADAM proteases, as well as the key transcription factor CSL. Expression of Notch and Delta genes indicates they are involved in specification of putative sensory cells, with Notch expression being broad, and the five Delta genes displaying dynamic, unique patterns. The domain composition of the identified genes, as well as their expression patterns, indicate the pathway operates in a manner similar to the cnidarian + bilaterian situation, and is involved in evolutionarily conserved processes, such as cell-type specification and differentiation.

Pre-metazoan Ancestry of the Notch Pathway

The Notch pathway appears to be the evolutionarily oldest one, with sequences similar to the receptor Notch, its processing proteins Furin and O-fucosyltransferase, the transcription factor CSL, as well as proteases cleaving Notch as a response to signal—ADAM and gamma-secretase complex—present in the choanoflagellate genome (King et al. 2008; Gazave et al. 2009). No Delta or Serrate related ligands have been identified in *Monosiga*, and it is suggested that in choanoflagellates Notch might be involved in sensing environmental stimuli, rather than intercellular signalling (reviewed by Richards and Degnan 2009).

Developmental Signalling Pathways and the Metazoan Tree of Life

It might have been expected that analysis of metazoan developmental toolkits would reveal a straight increase in complexity from pre-metazoan roots, reflected by the choanoflagellates, through "basal" metazoans, to the bilaterians. It turns out however that entire pathways or their components were lost by certain animal lineages while at the same time other components were being added. Unfortunately, lack of resolution of the branching order of placozoans, sponges and ctenophores precludes definite assessment of which of the components present in one lineage have been lost by others, and in which absence represents the original simplicity. The presence/absence of components does not follow a consistent pattern (see Table 1). For example, a Wnt inhibitor, Dickkopf, has been identified in a homoscleromorph sponge, *Oscarella carmela*, but not in demosponges, placozoans or ctenophores. Axin is present in demosponges and placozoans, but not in the ctenophores. The inhibitory Smads6/7 are absent from *Amphimedon*, but present in placozoans and ctenophores. Fringe genes are present in demosponges, but not in placozoans.

On the other hand, in case of the Notch and Hedgehog/Hedgling pathways, components of which are present in the choanoflagellates, it is clear that all of the lineages have sustained some losses: majority of the Hedgehog pathway components is absent from *Trichoplax*, Hedgling is absent from *Mnemiopsis*, and Patched is not found in *Amphimedon*. It is thus clear that the developmental signalling toolkits of extant metazoans were shaped not only by increase in complexity, but also by secondary simplification. Perhaps paradoxically, the pathways that have deep roots in pre-metazoans, such as Hedgehog and Notch, appear to be more dispensable in the metazoans as evidenced by losses of pathway components in multiple lineages. On the other hand, pathways which are clearly metazoan novelties, such as Wnt and TGF-beta, appear indispensable, certifying to their importance in development and/or maintenance of animal body plans.

An important conclusion that can be drawn from these comparisons is that the signalling toolkits of the sponges, ctenophores and placozoans are in fact very similar to each other, accounting for the patchy loss of components. That is, the complexity of the toolkit does not correlate with morphological complexity (for example, if we compare *Trichoplax* to *Mnemiopsis*). Also, none of the toolkits is significantly more similar to the choanoflagellate or the cnidarian one on the other side of the spectrum. Whether these similarities reflect the evolutionary history of the animals—placozoans, ctenophores and sponges diverging rapidly within short time after establishment but before major diversification of the Wnt and TGF-beta pathways in the ancestor of cnidarians and bilaterians—or are a result of convergent simplification of the pathways in the basal metazoans, is unclear at the moment. It remains to be seen whether sequencing of additional species representing previously unsampled lineages of sponges, ctenophores and placozoans will shed more light into evolution of animals and their developmental mechanisms.

Summary

1. Development of cnidarians and bilaterians in orchestrated by five major signalling pathways: Wnt, TGF-beta, Hedgehog, FGF and Notch.
2. Analysis of genomes of "basal metazoans": placozoans, ctenophores and sponges, demonstrates that many of the Wnt, TGF-beta, Hedgehog and Notch pathway components are present, and are likely involved in development.
3. Notch and Hedgehog pathway are evolutionarily oldest, with some components present in the choanoflagellates.
4. A few components of the Wnt pathway are present in a slime mold, *Dictyostelium*, but evolutionary significance of this finding is not clear.
5. The developmental signalling toolkits of basal metazoans (sponges, placozoans and ctenophores) are similar, despite major differences in their morphological complexity, and are not helpful in establishing of their branching order.

Acknowledgments Research in my laboratory is funded by Sars International Centre for Marine Molecular Biology.

References

Adamska M, Degnan SM, Green KM, Adamski M, Craigie A, Larroux C, Degnan BM (2007a) Wnt and TGF-beta expression in the sponge *Amphimedon queenslandica* and the origin of metazoan embryonic patterning. PLoS One 2:e1031

Adamska M, Matus DQ, Adamski M, Green K, Rokhsar DS, Martindale MQ, Degnan BM (2007b) The evolutionary origin of Hedgehog proteins. Curr Biol 17(19):R836–837

Adamska M, Larroux C, Adamski M, Green K, Lovas E, Koop D, Richards GS, Zwafink C, Degnan BM (2010) Structure and expression of conserved Wnt pathway components in the demosponge *Amphimedon queenslandica*. Evol Dev 12(5):494–518

Adamska M, Degnan BM, Green K, Zwafink C (2011) What sponges can tell us about the evolution of developmental processes. Zoology (Jena) 114:1–10

Andersson ER, Sandberg R, Lendahl U (2011) Notch signaling: simplicity in design, versatility in function. Development 138(17):3593–3612

Barolo S, Posakony JW (2002) Three habits of highly effective signaling pathways: principles of transcriptional control by developmental cell signaling. Genes Dev 16(10):1167–1181

Bertrand S, Iwema T, Escriva H (2014) FGF signaling emerged concomitantly with the origin of *eumetazoans*. Mol Biol Evol 31(2):310–318

Bray SJ (2006) Notch signalling: a simple pathway becomes complex. Nat Rev Mol Cell Biol 7(9):678–689

Cho SJ, Vallès Y, Giani VC Jr, Seaver EC, Weisblat DA (2010) Evolutionary dynamics of the Wnt gene family: a lophotrochozoan perspective. Mol Biol Evol 27(7):1645–1658

De Robertis EM Sasai Y (1996) A common plan for dorsoventral patterning in *bilateria*. Nature 380(6569):37–40

Finnerty JR, Pang K, Burton P, Paulson D, Martindale MQ (2004) Origins of bilateral symmetry: hox and dpp expression in a sea anemone. Science 304(5675):1335–1337

Gazave E, Lapébie P, Richards GS, Brunet F, Ereskovsky AV, Degnan BM, Borchiellini C, Vervoort M, Renard E (2009) Origin and evolution of the Notch signalling pathway: an overview from eukaryotic genomes. BMC Evol Biol 9:249

Goldfarb M (2001) Signaling by fibroblast growth factors: the inside story. Sci STKE 106:pe37

Harwood AJ (2008) *Dictyostelium* development: a prototypic Wnt pathway? Methods Mol Biol 469:21–32

Holland LZ, Holland NN, Schubert M (2000) Developmental expression of AmphiWnt1, an amphioxus gene in the Wnt1/wingless subfamily. Dev Genes Evol 210(10):522–524

Holstein TW (2012) The evolution of the Wnt pathway. Cold Spring Harb Perspect Biol 4(7):a007922

Huminiecki L, Goldovsky L, Freilich S, Moustakas A, Ouzounis C, Heldin CH (2009) Emergence, development and diversification of the TGF-beta signalling pathway within the animal kingdom. BMC Evol Biol 9:28

Jager M, Dayraud C, Mialot A, Quéinnec E, le Guyader H, Manuel M (2013) Evidence for involvement of Wnt signalling in body polarities, cell proliferation, and the neuro-sensory system in an adult ctenophore. PLoS One 8(12):e84363

Jiang J, Hui CC (2008) Hedgehog signaling in development and cancer. Dev Cell 15(6):801–812

Käsbauer T, Towb P, Alexandrova O, David CN, Dall'armi E, Staudigl A, Stiening B, Böttger A (2007) The Notch signaling pathway in the cnidarian *Hydra*. Dev Biol 303(1):376–90

King N, Westbrook MJ, Young SL, Kuo A, Abedin M, Chapman J, Fairclough S, Hellsten U, Isogai Y, Letunic I, Marr M, Pincus D, Putnam N, Rokas A, Wright KJ, Zuzow R, Dirks W, Good M, Goodstein D, Lemons D, Li W, Lyons JB, Morris A, Nichols S, Richter DJ, Salamov A, Sequencing JG, Bork P, Lim WA, Manning G, Miller WT, McGinnis W, Shapiro H, Tjian R, Grigoriev IV, Rokhsar D (2008) The genome of the choanoflagellate *monosiga brevicollis* and the origin of metazoans. Nature 451(7180):783–788

Kitisin K, Saha T, Blake T, Golestaneh N, Deng M, Kim C, Tang Y, Shetty K, Mishra B, Mishra L (2007) TGF-beta signaling in development. Sci STKE 399:cm1

Kusserow A, Pang K, Sturm C, Hrouda M, Lentfer J, Schmidt HA, Technau U, von Haeseler A, Hobmayer B, Martindale MQ, Holstein TW (2005) Unexpected complexity of the Wnt gene family in a sea anemone. Nature 433:156–160

Lee PN, Pang K, Matus DQ, Martindale MQ (2006) A WNT of things to come: evolution of Wnt signaling and polarity in cnidarians. Semin Cell Dev Biol 17(2):157–167

Leys SP, Eerkes-Medrano D (2005) Gastrulation in calcareous sponges: in search of Haeckel's gastraea. Integr Comp Biol 45(2):342–351

Manning G, Young SL, Miller WT, Zhai Y (2008) The protist, *monosiga brevicollis*, has a tyrosine kinase signaling network more elaborate and diverse than found in any known metazoan. Proc Natl Acad Sci U S A 105(28):9674–9679

Marlow H, Roettinger E, Boekhout M, Martindale MQ (2012) Functional roles of Notch signaling in the cnidarian *Nematostella vectensis*. Dev Biol 362(2):295–308

Matus DQ, Magie CR, Pang K, Martindale MQ, Thomsen GH (2008) The Hedgehog gene family of the cnidarian, *nematostella vectensis*, and implications for understanding metazoan Hedgehog pathway evolution. Dev Biol 313(2):501–518

McGlinn E, Tabin CJ (2006) Mechanistic insight into how shh patterns the vertebrate limb. Curr Opin Genet Dev 16(4):426–432

Müller WE, Korzhev M, Le Pennec G, Müller IM, Schröder HC (2003) Origin of metazoan stem cell system in sponges: first approach to establish the model (*suberites domuncula*). Biomol Eng 20(4–6):369–379

Münder S, Käsbauer T, Prexl A, Aufschnaiter R, Zhang X, Towb P, Böttger A (2010) Notch signalling defines critical boundary during budding in *hydra*. Dev Biol 344(1):331–345

Nichols SA, Dirks W, Pearse JS, King N (2006) Early evolution of animal cell signaling and adhesion genes. Proc Natl Acad Sci U S A 103(33):12451–12456

Nichols SA, Roberts BW, Richter DJ, Fairclough SR, King N (2012) Origin of metazoan cadherin diversity and the antiquity of the classical cadherin/β-catenin complex. Proc Natl Acad Sci U S A 109(32):13046–13051

Nosenko T, Schreiber F, Adamska M, Adamski M, Eitel M, Hammel J, Maldonado M, Müller WE, Nickel M, Schierwater B, Vacelet J, Wiens M, Wörheide G (2013) Deep metazoan phylogeny: when different genes tell different stories. Mol Phylogenet Evol 67(1):223–233

Nusse R (2003) Wnts and Hedgehogs: lipid-modified proteins and similarities in signaling mechanisms at the cell surface. Development 130(22):5297–5305

Nusse R, He X, van Amerongen R (eds) (2013) Wnt Signaling Cold Spring Harbor Laboratory Press, ISBN-10: 1936113236

Pang K, Ryan JF, NISC Comparative Sequencing Program, Mullikin JC, Baxevanis AD, Martindale MQ (2010) Genomic insights into Wnt signaling in an early diverging metazoan, the ctenophore *mnemiopsis leidyi*. Evodevo 1(1):10

Pang K, Ryan JF, Baxevanis AD, Martindale MQ (2011) Evolution of the TGF-β signaling pathway and its potential role in the ctenophore, *Mnemiopsis leidyi*. PLoS One 6(9):e24152

Petersen CP, Reddien PW (2009) Wnt signaling and the polarity of the primary body axis. Cell 139:1056–1068

Pires-daSilva A, Sommer RJ (2003) The evolution of signalling pathways in animal development. Nat Rev Genet 4:39–49

Rebscher N, Deichmann C, Sudhop S, Fritzenwanker JH, Green S, Hassel M (2009) Conserved intron positions in FGFR genes reflect the modular structure of FGFR and reveal stepwise addition of domains to an already complex ancestral FGFR. Dev Genes Evol 219(9–10):455–468

Rentzsch F, Fritzenwanker JH, Scholz CB, Technau U (2008) FGF signalling controls formation of the apical sensory organ in the cnidarian *nematostella vectensis*. Development 135(10):1761–1769

Richards GS, Degnan BM (2009) The dawn of developmental signaling in the metazoa. Cold Spring Harb Symp Quant Biol 74:81–90

Richards GS, Degnan BM (2012) The expression of Delta ligands in the sponge *Amphimedon queenslandica* suggests an ancient role for Notch signaling in metazoan development. Evodevo 3(1):15. doi:10.1186/2041-9139-3-15

Richards GS, Simionato E, Perron M, Adamska M, Vervoort M, Degnan BM (2008) Sponge genes provide new insight into the evolutionary origin of the neurogenic circuit. Curr Biol 18(15):1156–1161

Robbins DJ, Fei DL, Riobo NA (2012) The Hedgehog signal transduction network. Sci Signal 5(246):re6

Roettinger E, Dahlin P, Martindale MQ (2012) A framework for the establishment of a cnidarian gene regulatory network for "Endomesoderm" specification: the inputs of ß-catenin/TCF signaling. PLoS Genet 8:e1003164

Ryan JF, Pang K, Schnitzler CE, Nguyen AD, Moreland RT, Simmons DK, Koch BJ, Francis WR, Havlak P, NISC Comparative Sequencing Program, Smith SA, Putnam NH, Haddock SH, Dunn CW, Wolfsberg TG, Mullikin JC, Martindale MQ, Baxevanis AD (2013) The genome of the ctenophore *mnemiopsis leidyi* and its implications for cell type evolution. Science 342(6164):1242592

Saina M, Genikhovich G, Renfer E, Technau U (2009) BMPs and chordin regulate patterning of the directive axis in a sea anemone. Proc Natl Acad Sci U S A 106(44):18592–18597

Schierwater B (2005) My favorite animal, *trichoplax adhaerens*. Bioessays 27(12):1294–1302

Srivastava M, Begovic E, Chapman J, Putnam NH, Hellsten U, Kawashima T, Kuo A, Mitros T, Salamov A, Carpenter ML, Signorovitch AY, Moreno MA, Kamm K, Grimwood J, Schmutz J, Shapiro H, Grigoriev IV, Buss LW, Schierwater B, Dellaporta SL, Rokhsar DS (2008) The *trichoplax* genome and the nature of placozoans. Nature 454(7207):955–960

Srivastava M, Simakov O, Chapman J, Fahey B, Gauthier MEA, Mitros T, Richards GS, Conaco C, Dacre M, Hellsten U et al (2010) The *amphimedon queenslandica* genome and the evolution of animal complexity. Nature 466:720–726

Sudhop S, Coulier F, Bieller A, Vogt A, Hotz T, Hassel M (2004) Signalling by the FGFR-like tyrosine kinase, kringelchen, is essential for bud detachment in *hydra vulgaris*. Development 131(16):4001–4011

Taipale J, Beachy PA (2001) The Hedgehog and Wnt signalling pathways in cancer. Nature 411(6835):349–354

Technau U, Rudd S, Maxwell P, Gordon PM, Saina M, Grasso LC, Hayward DC, Sensen CW, Saint R, Holstein TW, Ball EE, Miller DJ (2005) Maintenance of ancestral complexity and non-metazoan genes in two basal cnidarians. Trends Genet 21(12):633–639

Thisse B, Thisse C (2005) Functions and regulations of fibroblast growth factor signaling during embryonic development. Dev Biol 287(2):390–402

Wagner GP (2007) The developmental genetics of homology. Nat Rev Genet 8(6):473–479

Windsor PJ, Leys SP (2010) Wnt signaling and induction in the sponge aquiferous system: evidence for an ancient origin of the organizer. Evol Dev 12:481–590

Xing Y, Clements WK, Kimelman D, Xu W (2003) Crystal structure of a beta-catenin/axin complex suggests a mechanism for the beta-catenin destruction complex. Genes Dev 17(22):2753–2764

Yu JK, Meulemans D, McKeown SJ, Bronner-Fraser M (2008) Insights from the amphioxus genome on the origin of vertebrate neural crest. Genome Res 18(7):1127–1132

The Evolution of Developmental Signalling in Dictyostelia from an Amoebozoan Stress Response

Yoshinori Kawabe, Christina Schilde, Zhi-hui Chen, Qingyou Du, Hajara Lawal and Pauline Schaap

Abstract Dictyostelid social amoebas represent one of several groups of genetically divergent lineages that display aggregative multicellularity. In this chapter, we describe the evolution of developmental complexity in Dictyostelia and discuss the signalling mechanisms that control the developmental programme of the model organism *Dictyostelium discoideum.* We also reconstruct the evolutionary history of these developmental control mechanisms from environmental sensing in the unicellular ancestors of Dictyostelia. Finally, we explore the parameters that define the boundary between uni- and multicellularity.

Keywords Dictyostelia · Colonial multicellularity · Evolution of phenotypic complexity · Evolution of cyclic AMP signalling · Encystation · Environmental stress

Introduction

In our macroscopic world, life forms such as animals, plants and fungi that develop through repeated divisions from a cellular spore or zygote are the most readily observed multicellular organisms. However, in both prokaryotes and eukaryotes, multicellularity evolved many times by aggregation of single cells. Currently,

P. Schaap (✉) · Y. Kawabe · C. Schilde · Z.-hui Chen · Q. Du · H. Lawal
College of Life Sciences, University of Dundee, Dundee, UK
e-mail: p.schaap@dundee.ac.uk

Y. Kawabe
e-mail: y.kawabe@dundee.ac.uk

C. Schilde
e-mail: c.schilde@dundee.ac.uk

Z.-hui Chen
e-mail: z.y.chen@dundee.ac.uk

Q. Du
e-mail: q.du@dundee.ac.uk

H. Lawal
e-mail: h.m.lawal@dundee.ac.uk

I. Ruiz-Trillo, A. M. Nedelcu (eds.), *Evolutionary Transitions to Multicellular Life,*
Advances in Marine Genomics 2, DOI 10.1007/978-94-017-9642-2_21

aggregative multicellularity has reached the highest level of complexity in the Dictyostelia, members of the eukaryote supergroup Amoebozoa. Up to a million dictyostelid amoebas can come together to form a multicellular organism that displays light-oriented migration and erection of a fruiting structure consisting of a spore mass with up to four different supporting cell types (Schaap 2007; Williams 2006; Yamada et al. 2010).

Within the otherwise unicellular Amoebozoa, the genus *Copromyxa* also forms multicellular fruiting structures by aggregation. Several unrelated amoeboid protists, such as *Acrasis* and *Pocheina* in the supergroup Excavata (Brown et al. 2012) and *Fonticula alba* in Opisthokonta (Brown et al. 2009) also form fruiting bodies by aggregation of single cells. Even a ciliate species, *Sorogena stoianovitchae* from the supergroup Chromalveolata aggregates by cell adhesion to form a fruiting structure with encysted cells (Olive 1978; Sugimoto and Endoh 2006).

For over 50 years, *Dictyostelium discoideum* has been a popular model organism, used to address various questions in cell, developmental and evolutionary biology, and a broad range of molecular genetic, biochemical and cell-biological techniques has been developed in this system. Consequently, our understanding of the molecular mechanisms that control the Dictyostelid multicellular life cycle is extensive. As most other species with aggregative multicellularity have mostly only been studied descriptively, in this chapter we will focus on the mechanisms that control Dictyostelid multicellularity and their evolutionary origins.

Dictyostelium Life Cycle Transitions

Dictyostelids are common inhabitants of a wide range of soil habitats, ranging from the arctic to the tropics and from desert to rainforest. They are most prevalent in tropical to temperate forests, where they feed on bacteria in the decaying leaf litter (Swanson et al. 1999). Individual species can utilise one of three survival strategies when food runs out (Fig. 1). Similar to their solitary amoebozoan ancestors, individual amoebas can encapsulate to form dormant cysts, called microcysts; dark and wet conditions that are not conducive to aggregation, or high solute or waste (ammonia) levels, favour the encystation pathway (Raper 1984). Under wet and dark conditions, amoebas can also fuse to form a zygote; this usually requires amoebas of opposite mating types, but some species also show homothallic mating. The zygote attracts and cannibalizes surrounding amoebas and uses their resources to build a thick-walled sphere, the macrocyst (O'Day and Keszei 2011). Lastly, amoebas can aggregate and form multicellular asexual fruiting structures.

Following food depletion, amoebas attract each other by secreting a chemoattractant. In the case of *D. discoideum*, the chemoattractant is cAMP, which is produced in an oscillatory manner by some cells in the population and relayed through cAMP-induced cAMP synthesis by the surrounding cells. Once aggregated, the tip of the cell mound continues to emit cAMP pulses, causing it to be pushed upward by the upward movement of the cells underneath, ultimately leading to the formation of the columnar sorogen or slug (Dormann and Weijer 2001). The slug topples over and, guided by

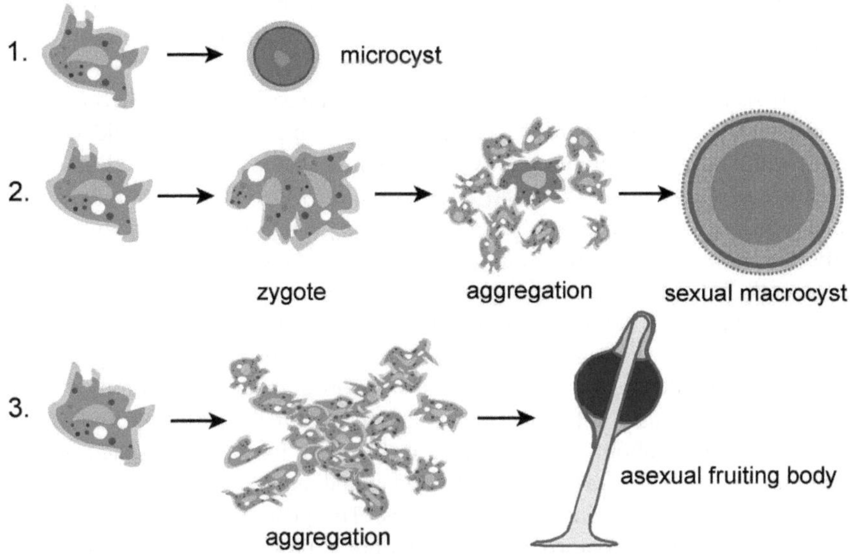

Fig. 1 The three *Dictyostelium* life cycles. Starving *Dictyostelium* amoebas can either encapsulate individually to form a microcyst, fuse to form a zygote (which attracts and cannibalizes other amoebas to form a sexual macrocyst) or aggregate and form a multicellular fruiting body in which cells specialize into dead stalk cells and viable spores

light and warmth, moves to the top layer of the soil to enable optimal spore dispersal. Meanwhile, the majority of the cells differentiate into spore-cell precursors, while separate smaller groups of cells enter pathways that lead to differentiation as stalk cells and other supporting cell types. These cells are first intermixed with prespore cells, but sort out to their respective positions due to selective chemotaxis and selective cell adhesion. The prestalk cells move towards the front of the slug, while the prebasal disk and upper cup cells move towards the rear and front of the prespore region to form support structures for the stalk and spore mass (Kessin 2001; Schaap 2007).

Upon initiation of fruiting body formation, the prestalk cells synthesize a central cellulose tube, move into the tube and differentiate into vacuolated stalk cells. The prespore cells follow the prestalk cells up the stalk. They mature into spores by exocytosing spore coat materials that are prepackaged in prespore vesicles and by *in situ* synthesis of a thick spore wall (West 2003).

Evolution of Phenotypic Complexity in Dictyostelia

Traditionally, social amoebas were classified according to fruiting body morphology into three genera: *Dictyostelium* (with unbranched or laterally branched fruiting bodies), *Polysphondylium* (with regular whorls of side branches) and *Acytostelium* (with acellular stalks) (Raper 1984). More recent phylogenetic inferences from single gene (SSU rDNA and α-tubulin) sequences and 32 concatenated proteins subdivide Dictyostelia in two branches that each consists of two major groups and some group-intermediate species (Fig. 2a; Schaap et al. 2006; Romeralo et al. 2011; Romeralo et al. 2013). The four groups do not coincide with the earlier genera. Groups 1, 3 and 4 consist only of *Dictyostelium* species, while group 2 can be subdivided into two clades with clade 2A containing only *Acytostelium* species and clade 2B comprising a mixture of *Polysphondylium* and *Dictyostelium*. More *Polysphondylium* species are present in a small clade positioned in between groups 3 and 4. Evidently, fruiting body morphology is not a reliable marker of relatedness between species.

Recently, about 30 morphological and behavioural characters were analysed across the hundred known social amoeba species and this data set was combined with the molecular phylogeny to reconstruct the evolution of phenotypes in Dictyostelia (Romeralo et al. 2013). The analysis showed that the last common ancestor (LCA) to all Dictyostelia formed several small unbranched fruiting bodies from a single aggregate of about 0.5 mm in height (Fig. 2b). Its stalks consisted of a cellulose tube that contained vacuolated cells and ended in a broadened tip, and its spores were elliptical and contained polar granules. The postulated LCA used cAMP to coordinate fruiting body morphogenesis, but not aggregation, which may have been mediated by the dipeptide glorin (Asghar et al. 2011). Except for the broadened stalk tip, these character states persisted into the LCAs of groups 1, 2 and 3. The LCA to group 2 gained pointed stalk tips, while the LCA to clade 2A lost cells from its stalk tube and granules from its spores. The LCA to clade 2B mostly formed a single fruiting body per aggregate, which it adorned with whorls of side branches. The LCA of group 4 formed very robust single fruiting bodies per aggregate and its sorogens acquired light-oriented migratory behaviour. It also used cAMP for aggregation and lost the granules from its spores. It is as yet unclear whether there is a causal relationship between use of cAMP as attractant, increased fruiting body robustness and acquisition of slug migration.

Developmental Signalling in *D. discoideum*

Until recently, *Dictyostelium* research was almost entirely focussed on the group 4 species *D. discoideum* and the major signals that control development in this species are summarized in Fig. 3. The most remarkable aspect of the development of *D. discoideum* is the dominant role of cAMP. As a secreted signal, cAMP controls cell movement and differentiation throughout the developmental programme. However, in its more common role as intracellular messenger acting on cAMP-dependent protein kinase or PKA, cAMP also regulates many developmental processes, such as the transition from growth to aggregation, the differentiation of prespore cells, the maturation of spores and stalk cells and the germination of the spores. In feeding cells, PKA translation is inhibited by the translational repressor PufA (Souza et al. 1999). This repression is relieved by the protein kinase YakA, which is considered to be up-regulated in response to the accumulation of a prestarvation factor, one of the two

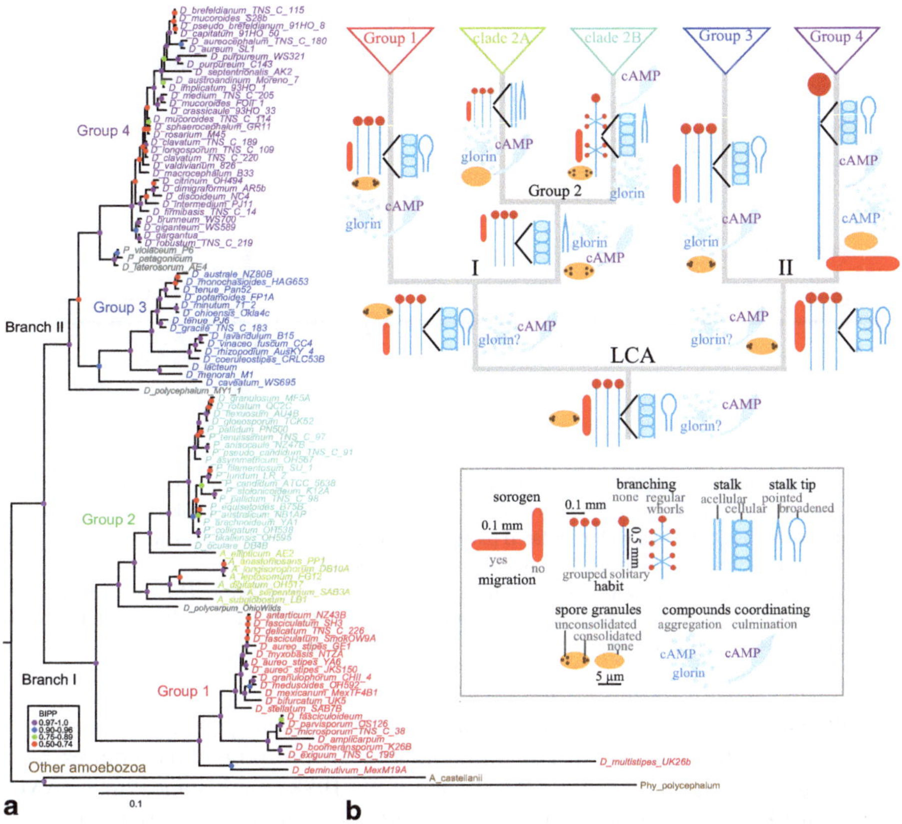

Fig. 2 Phylogeny and phenotypic evolution in Dictyostelia. **a** *Phylogeny.* A phylogenetic tree was constructed using SSU rDNA sequences for all Dictyostelia combined with 32 concatenated protein sequences, which were retrieved from the completed genomes of *D. discoideum*, *D. purpureum*, *D. lacteum*, *P. pallidum*, *A. subglobosum*, *D. fasciculatum*, *Acanthamoeba castellani* and *Physarum polycephalum,* which together represent all major groups and clades 2A and 2B of Dictyostelia and two amoebozoan outgroup taxa (Romeralo et al. 2013). Node probability values are indicated by *coloured dots*. **b** *Phenotypic evolution.* A range of 30 phenotypic characters was assessed for all Dictyostelia and the resulting character matrix was combined with the phylogenetic tree for inference of character evolution (Romeralo et al. 2013). For characters that showed a well defined character history, the ancestral state values at major nodes are plotted in *cartoon-style* onto a simplified phylogeny

quorum sensing factors that are required for entry into multicellular development (Souza et al. 1998; Clarke and Gomer 1995). PKA activity is required for basal expression of aggregation genes, such as the cAMP receptor (cAR1), the extracellular cAMP phosphodiesterase (PdsA), and the adenylate cyclase A (ACA) (Schulkes and Schaap 1995). These proteins are part of the biochemical network that generates the cAMP pulses that coordinate aggregation and morphogenesis (Sawai et al. 2005). The cAMP pulses in turn strongly stimulate the expression of aggregation genes (Gerisch et al. 1975).

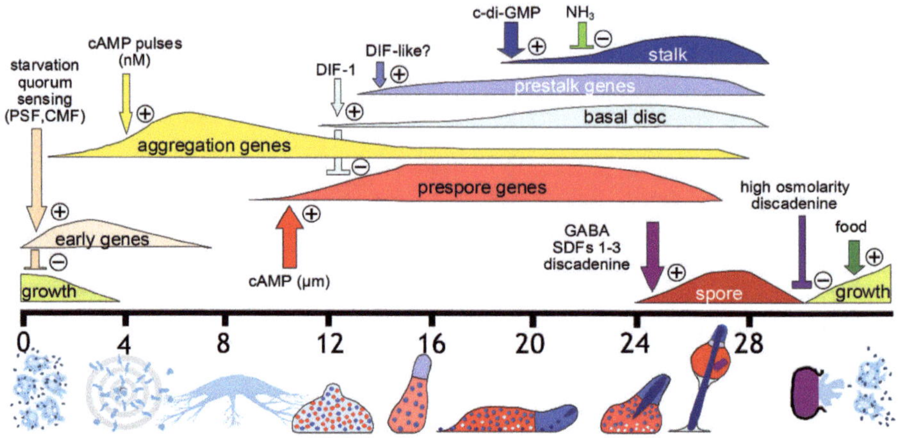

Fig. 3 Signals that regulate *D. discoideum* developmental gene expression. During their 28 h life cycle, starving *D. discoideum* cells aggregate and differentiate into spore-, stalk-, upper and lower cup-, and basal disk cells in a well regulated pattern. The changes in gene expression that cause cell differentiation are mostly controlled by secreted signal molecules.
Abbreviations: *PSF* prestarvation factor; *CMF* conditioned medium factor; *DIF-1* differentiation inducing factor 1; *cAMP* 3',5'-adenosine monophosphate; *c-di-GMP* 3',5'-cyclic diguanylic acid; *SDF* spore differentiation factor; *GABA* gamma-amino butyric acid

Following aggregation, cells start to differentiate into prespore cells and several prestalk-like cell populations. The differentiation of prespore cells requires both extracellular cAMP acting on cAR1 (Wang et al. 1988), and intracellular cAMP acting on PKA (Hopper et al. 1993). A second adenylate cyclase, ACG, is translationally upregulated in the emerging prespore cells and produces cAMP for both PKA and cAR1 activation (Alvarez-Curto et al. 2007). The prespore cells produce polyketide-based compounds that cause their redifferentiation into supporting cells. One of these compounds is DIF-1, which triggers differentiation of cells that will sort rearwards during fruiting body formation to form the basal disc and lower cup (Kay and Thompson 2001; Saito et al. 2008). The other polyketide-based compounds likely promote differentiation of cells that sort forward to form the stalk and upper cup of the fruiting body (Serafimidis and Kay 2005; Saito et al. 2006). In addition to diffusible molecules, cell differentiation in mounds also requires direct cell-cell interactions that are mediated by cell adhesion proteins, such as LagC and LagD (Dynes et al. 1994; Kibler et al. 2003). These proteins are members of a large family of membrane proteins with several immunoglobulin repeats. LagC and other members of this family of "Tiger" proteins also mediate kin recognition in Dictyostelia (Benabentos et al. 2009; Hirose et al. 2011).

The waste product ammonia has many signalling roles in *Dictyostelium*. Starving cells digest cellular proteins by autophagy (Otto et al. 2003), which yields substantial amounts of ammonia waste. The anterior prestalk cells display the most intensive autophagy (Schaap 1983) and during slug migration cells die off after having reached the tip. Ammonia, produced by anterior cells, normally prevents the forward sorting of prestalk-like cells into the anterior region. However, when the anterior becomes depleted by cell death, ammonia levels drop and forward sorting is resumed to replace the lost cells (Feit et al. 1990, 2001, 2007). Ammonia furthermore inhibits the transition from slug migration to fruiting body formation and the maturation of stalk cells (Schindler and Sussman 1977; Wang and Schaap 1989). Ammonia and several other signals that regulate terminal differentiation are detected by sensor-linked histidine kinases, which are common sensors for external stimuli in bacteria, fungi and plants, but not in animals (Thomason and Kay 2000). The histidine kinase moiety can also act as a histidine phosphatase, and ligand binding to the sensor domain triggers either the kinase or the phosphatase activity. Active kinase phosphorylates itself on a histidine residue, which starts a series of histidine-aspartate-histidine phosphoryl transfer reactions. In *D. discoideum*, the phosphoryl group is deposited via the phosphorelay intermediate RdeA on the response regulator of the intracellular cAMP phosphodiesterase RegA, thereby activating cAMP hydrolysis (Fig. 4a). When ligand activates the histidine phosphatase, phosphorelay runs in reverse and inactivates RegA. Ammonia activates the histidine kinase DhkC and consequently RegA (Singleton et al. 1998), resulting in cAMP degradation and preventing activation of PKA. PKA activation is essential for spore and stalk cell maturation, and ammonia activation of RegA therefore effectively blocks both processes (Harwood et al. 1992; Hopper et al. 1993). A third adenylate cyclase, termed ACB or ACR is expressed preferentially in prestalk cells (Alvarez-Curto et al. 2007); this adenylate cyclase is also the major cAMP source for PKA activation in both prestalk and prespore cells at early culmination (Soderbom et al. 1999). The ammonia transporter AmtC, which is expressed at the tip, facilitates loss of ammonia from the tip region, thus releasing the block on stalk cell maturation (Kirsten et al. 2005). A second transporter AmtA mediates DhkC activation by ammonia (Singleton et al. 2006).

However, release of inhibition cannot by itself induce stalk cell differentiation. The stalk-inducing-factor was recently shown to be cyclic-di-GMP (Chen and Schaap 2012). This molecule is produced by a diguanylate cyclase that is specifically expressed in the prestalk region of the slug. Cyclic-di-GMP causes the motile prestalk cells to differentiate into sessile stalk cells by secreting polysaccharide cell wall materials. Diguanylate cyclases were previously only found in bacteria, where cyclic-di-GMP is a major second messenger for external stimuli. Among other functions, it causes the loss of motility and the synthesis of adhesins and exopolysaccharide matrix components by bacteria, which mark their transition from a swarming motile phase to a sessile biofilm associated life style. In *Caulobacter crescentus* this transition also involves formation of a stalk (Hengge 2009; Jenal and Malone 2006). It is however more likely that its role in *Dictyostelium* stalk cell differentiation is the result of lateral gene transfer and convergent evolution, rather than being evolutionary derived from its role in bacteria.

The terminal stages of spore maturation are controlled by intensive communication between the prestalk and prespore cells to ensure that spores encapsulate (and thereby become immotile) only after they have moved to the top of the fruiting body. First, a steroid—SDF-3 (spore differentiation factor 3), is released, which triggers the production of GABA (gamma-amino butyric acid) by prespore cells

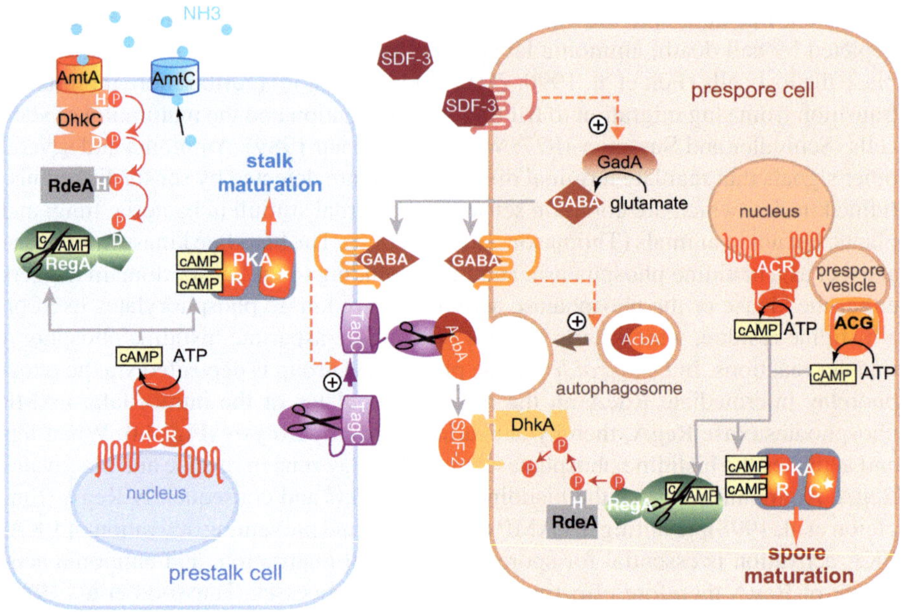

a

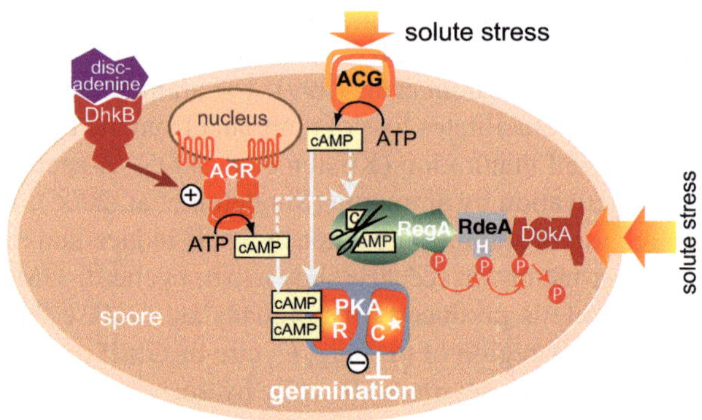

b

Fig. 4 Mechanisms that control terminal stalk and spore differentiation. The maturation of spores and stalk cells is triggered by PKA activation, and active PKA additionally keeps spores dormant, while still in the spore head. Spore and stalk cell maturation are under extensive spatio-temporal control, because these processes involve encapsulation and immobilization of amoebas, while at the same time the completion of morphogenesis requires cells to be motile. Spatio-temporal control is achieved by extensive signalling between the maturing stalk and spores cells as outlined in the main text. Most signals act mainly indirectly (via forward or reverse phosphorelay) to regulate the activity of the cAMP phosphodiesterase RegA and thereby PKA activity. In addition, high solute levels in the spore head also act directly on adenylate cyclase G to increase cAMP levels and to prevent precocious germination. Abbreviations: *AmtA, AmtC* ammonia transporters A and C; *DhkA, DhkB, DhkC, DokA* histidine kinases A, B, C and DokA; *ACR and ACG*: adenylate cyclases A and G; *RdeA* phosphoryl transfer protein; *RegA* cAMP phosphodiesterase with response regulator; *PKA* cAMP-dependent protein kinase; *TagC* ABC transporter with intrinsic serine protease moiety; *GadA* glutamate decarboxylase A; *GABA* gamma-amino butyric acid; *AcbA* acetyl coenzyme binding protein A; *SDF-2, SDF-3* spore differentiation factors 2 and 3

(Anjard et al. 2009). GABA induces secretion of AcbA (Acyl-CoA binding protein) from prespore cells. GABA also induces the ABC transporter TagC of prestalk cells to transport its intrinsic serine protease moiety to the cell surface. The protease moiety then cleaves secreted AcbA to form the peptide SDF-2 (Anjard and Loomis 2006). SDF-2 activates the histidine phosphatase DhkA of prespore cells, which dephosphorylates, and thereby inactivates, the cAMP phosphodiesterase RegA. cAMP levels increase, causing PKA activation and spore maturation (Wang et al. 1999).

Spore germination is also tightly regulated (Fig. 4b). High intracellular cAMP levels and active PKA maintain the dormant state of spores (Virdy et al. 1999; Van Es et al. 1996). The spore head contains large amounts of ammonium phosphate, which increases the osmolarity of the spore mass and inhibits spore germination (Cotter 1977; Cotter et al. 1999). High osmolarity or solute stress has two targets: it acts on the intrinsic osmosensor of ACG to activate cAMP synthesis (Van Es et al. 1996; Saran and Schaap 2004) and it inhibits cAMP degradation by RegA by triggering reverse phosphorelay through the histidine phosphatase DokA (Schuster et al. 1996). Another signal, the cytokinin discadenine also accumulates during fruiting body formation. Discadenine is considered to act via the histidine kinase DhkB to promote cAMP production by ACR, and causes both maturation of spores and inhibition of spore germination (Abe et al. 1981; Anjard and Loomis 2008).

Notwithstanding their complexity, these extensive cell communication systems have only a single target: the activity of PKA, which acts to bring cells into an encapsulated, and for spores—a dormant state, and to maintain dormancy until spores have been dispersed to novel feeding grounds.

Evolution of Developmental cAMP Signalling from Amoebozoan Encystation

Dictyostelium's complex development is an adaptation for the production of resilient dispersable spores in response to nutrient stress. The adenylate cyclase ACG plays a central role in spore formation, firstly by inducing the differentiation of prespore cells and secondly by regulating the process of spore germination (Van Es et al. 1996; Alvarez-Curto et al. 2007). ACG and its ability to be activated by high osmolarity appeared to be conserved throughout the Dictyostelid phylogeny, and high osmolarity was found to universally inhibit spore germination in species from all four taxon groups (Ritchie et al. 2008).

Many species in groups 1–3 have retained the ancestral stress survival strategy of encystation (Fig. 1). Similar to spore germination, the germination of cysts is also inhibited by high osmolarity. However, remarkably, high osmolarity actively triggers encystation, even when cells are still feeding. Solute stress-induced encystation is accompanied by increased cAMP levels and requires PKA, indicating that it is mediated by ACG acting on PKA. Nevertheless, unlike prespore differentiation, which also requires ACG acting on PKA (Alvarez-Curto et al. 2007), parallel activation of cAMP receptors by extracellular cAMP is not necessary for encystation (Ritchie et al. 2008).

The studies cited above and earlier work (Toama and Raper 1967) established high osmolarity as an independent trigger for encystation. Free-living soil amoebas are not only exposed to starvation, but also to drought. Increased solute stress resulting from raised mineral concentrations in drying soil is most likely a natural environmental trigger for encystation. The roles of ACG and PKA in prespore differentiation and spore germination are homologous to those in cyst formation and germination, and this strongly suggests that ACG and PKA's role in developmental sporulation is evolutionary derived from their regulatory role in encystation in response to drought.

In *D. discoideum* cAMP is also an important secreted signal. As a chemoattractant, it coordinates cell movement during aggregation and fruiting body morphogenesis, while secreted cAMP also induces expression of aggregation genes and prespore genes (Fig. 3). In *D. discoideum*, secreted cAMP is detected by four homologous cAMP receptors (cARs 1–4). Orthologs of cAR1 are present in the genomes of representative species of all four taxon groups, while independent duplications of cAR1 occurred in groups 1 and 2 (Alvarez-Curto et al. 2005; Kawabe et al. 2009). In *D.discoideum*, cAR1 is expressed from separate early and late promoters during aggregation and post-aggregative development, respectively (Louis et al. 1993), and *car1* null mutants loose the ability to produce cAMP pulses and to aggregate and form fruiting bodies (Sun and Devreotes 1991). Groups 1–3 do not use cAMP for aggregation (Fig. 2) and their cARs are mostly only expressed after aggregation. Ablation of cARs in the group 2 species *Polysphondylium pallidum* did not affect aggregation, but completely disrupted fruiting body morphogenesis. The stunted *P. pallidum car* null fruiting bodies contained a disorganized mass of stalk cells and cysts instead of spores. In addition, the *car* null mutant lost cAMP induction of prespore differentiation (Kawabe et al. 2009). These findings indicate that non-group 4 species use secreted cAMP to coordinate fruiting body morphogenesis and to induce spore formation.

When comparing the requirements for cyst and spore formation, it becomes clear why the *P.pallidum car* null mutant formed cysts in its fruiting bodies. Both spore formation and encystation require intracellular cAMP acting on PKA, and this condition was not altered in the mutant. However, spore formation also requires extracellular cAMP acting on cARs. Because this pathway was ablated in the *car* null mutant, the cells reverted to encystation. This points to what might have been the most ancestral role for secreted cAMP: the Dictyostelid ancestor already used intracellular cAMP to mediate stress-induced encystation. Dictyostelids secrete most of the cAMP that they produce, and accumulation of cAMP in aggregates may have acted to inform cells of their aggregated state and cause them to form spores and not cysts.

Oscillatory synthesis and secretion of cAMP probably evolved later to coordinate more sophisticated fruiting body morphogenesis. Group 4 species then brought extracellular cAMP forward in development by adding distal promoter elements to existing cAMP signalling genes, such as cAR1 and ACA, which enabled their expression before and during aggregation (Alvarez-Curto et al. 2005; Galardi-Castilla et al. 2010). Interestingly, the extracellular cAMP phosphodiesterase, PdsA, which is a non-selective, low affinity cyclic nucleotidase in groups 1–3, gained cAMP selectivity and a 200-fold increase in affinity in group 4. This adaptation enabled PdsA to effectively degrade the nanomolar cAMP concentrations that are produced by dispersed aggregating cells, as opposed to the much higher concentrations that accumulate in sorogens (Kawabe et al. 2012). In conclusion, the evolution of Dictyostelia presents us with an unfolding history of gradual elaboration of an environmental stress response into an extensive signalling network that regulates the cell-type specialization and coordinated cell movement, characteristic of multicellular development.

Defining the Boundary Between Uni- and Multicellularity

Colonial multicellularity appeared at least seven times in the course of eukaryote evolution and may, in its first manifestation, have been a simple adaptation that (i) prevented organisms from being ingested by a larger predator, (ii) improved their own access to food, and (iii) provided protection from environmental stress. Prokaryotes form large communities, called biofilms, where members of the same or a variety of species communicate and specialize to perform different functions (Chapter "Multicellularity in bacteria: from division of labor to biofilm formation"). These communities can also incorporate protozoa and are characterized by copious secretion of adhesive matrix and by extensive signalling between cells (Stoodley et al. 2002). The myxobacteria take this process further and, like Dictyostelia, form exquisitely shaped multicellular fruiting bodies from an aggregated cell mass (Kim et al. 1992), emphasizing that this form of multicellular morphogenesis has evolved many times independently.

Though sparsely explored, symbiotic, parasitic and predatory relationships amongst unicellular organisms and between unicellular and multicellular organisms in the rhizosphere are likely to be intense and dependent on extensive communication. The genomes of free-living protozoa such as *Naegleria gruberi* (Fritz-Laylin et al. 2010) and *Acanthamoeba castellani* (Clarke et al. 2013) show a repertoire and variety of signalling genes that is at least equal and sometimes greater than that of multicellular organisms such as *Dictyostelia* and early diverging Metazoa (Eichinger et al. 2005; Srivastava et al. 2010; Chapman et al. 2010). This supports the notion that the capacity for cell-cell communication and environmental sensing in unicellular organisms is vastly underestimated. Of particular interest is the large number of sensor histidine kinases and their known *Dictyostelium* target, the cAMP phosphodiesterase RegA, in both the *Acanthamoeba* and *Naegleria* genomes. This is quite remarkable, since *Naegleria* is not even a member of Amoebozoa. Furthermore, the presence of several adenylate cyclases and PKA in these genomes (Fritz-Laylin et al. 2010; Clarke et al. 2013) suggests that intracellular cAMP is widely used as a signalling intermediate in these organisms. The *Acanthamoeba* genome also contains an extensive and functionally complete repertoire of genes involved in receptor tyrosine kinase mediated signalling (Clarke et al. 2013). The receptor tyrosine kinase mediated signalling pathway in Metazoa is typically activated by peptide growth factors, and only some components of this pathway were previously detected in non-metazoan genomes (Chapter "Developmental signalling and emergence of animal multicellularity"). It was therefore assumed that possession of the full pathway was a hallmark of metazoan multicellularity (Jin and Pawson 2012), which is obviously not the case.

With such a large repertoire of genes in unicellular protists to mediate environmental sensing and cell-cell signalling, it is probably not surprising that colonial multicellularity has evolved many times, independently. This being said, it should be acknowledged that although cellular aggregation may have been a mechanistically facile transition to multicellularity, it is only zygotic multicellularity that has thus far given rise to truly macroscopic organisms with the highest levels of morphological and behavioural complexity.

One reason may be that the need for colonial organisms to accommodate varying cell numbers into a functional structure constrains the repertoire of patterning mechanisms that can further shape these structures. For instance, establishment of a bipartite or multipartite pattern by a gradient of a diffusible molecules, as is common in metazoan morphogenesis (Rogers and Schier 2011), is unlikely to function accurately when the size of multicellular structures within species can measure from 10 to 10,000 or more cell diameters across. Within species, the size of a zygote does not vary much and even when evolving into a larger form, its patterning mechanisms will have time to adapt. A critical step in the evolution of complex multicellularity may therefore have been the ability to keep cells together after the first division and to control their proliferation.

Summary

1. Colonial multicellularity evolved many times independently and is most thoroughly studied in Dictyostelia.
2. Dictyostelia can be subdivided into four major groups. Their last common ancestor formed very small aggregates and fruiting bodies, which increased most dramatically in size in the ancestor to group 4. This increase in size is correlated with the use of cAMP as chemoattractant for aggregation.
3. Dictyostelia have three alternative life cycles : (i) aggregation to form asexual fruiting bodies, (ii) zygotic fusion followed by cannibalism to generate a walled macrocyst, and (iii) encystation of individual amoebas. The latter strategy is broadly used by most amoeboid protists to survive environmental stress. It has however been lost from the group 4 Dictyostelia.
4. The group 4 species *D. discoideum* not only uses secreted cAMP as a chemoattractant but also as a signal for induction of the expression of aggregation genes and prespore genes. Cyclic di-GMP, a second messenger that triggers biofilm formation in bacteria, is secreted in the prestalk region by *Dictyostelium* cells to induce stalk cell differentiation.
5. A set of signals ranging from steroids, peptides, neurotransmitters and cytokinins to solute stress controls the timely maturation of stalk and spore cells and the germination of spores. All these signals act indirectly to control intracellular cAMP levels and the activation state of PKA. Fully active PKA triggers spore and stalk maturation and maintains spore dormancy.
6. In group 1–3, cyclic AMP receptors (cARs) are only expressed after aggregation. cARs and therefore cAMP signalling are essential for the coordination of

multicellular morphogenesis in group 1–3 species, but not for the aggregation process.

7. As in group 4, cARs are also required for prespore gene induction in groups 1–3. Uniquely ablation of *cAR* genes causes a group 2 species to form cysts instead of spores in its fruiting bodies. This suggests that accumulation of secreted cAMP initially served to signal the aggregated state and to induce starving cells to form spores and not cysts.
8. The elaborate signalling mechanisms that coordinate Dictyostelid spore and stalk maturation all converge on the core process of cAMP activating PKA, which mediates stress-induced encystation in their solitary ancestors. This strongly suggests that developmental cAMP signalling in Dictyostelia evolved from a stress-response in its unicellular ancestors.
9. Emerging genome sequencing projects reveal a suprisingly large repertoire of signalling genes in unicellular protists. This suggests that extensive inter- and intra species communication occurs in their natural habitats and that the threshold to convert these interactions into some form of colonial multicellularity may be relatively low.

Acknowledgements YK, CS and HL are supported by project grant BB/G020426/1 from the Biotechnology and Biological Sciences Research Council, and ZC and QD by project grant 090276 from the Wellcome Trust.

References

Abe H, Hashimoto K, Uchiyama M (1981) Discadenine distribution in cellular slime molds and its inhibitory activity on spore germination. Agric Biol Chem 45:1295–1296

Alvarez-Curto E, Rozen DE, Ritchie AV, Fouquet C, Baldauf SL, Schaap P (2005) Evolutionary origin of cAMP-based chemoattraction in the social amoebae. Proc Natl Acad Sci U S A 102(18):6385–6390

Alvarez-Curto E, Saran S, Meima M, Zobel J, Scott C, Schaap P (2007) cAMP production by adenylyl cyclase G induces prespore differentiation in Dictyostelium slugs. Development 134(5):959–966

Anjard C, Loomis WF (2006) GABA induces terminal differentiation of Dictyostelium through a GABA(B) receptor. Development 133(11):2253–2261

Anjard C, Loomis WF (2008) Cytokinins induce sporulation in Dictyostelium. Development 135:819–827

Anjard C, Su Y, Loomis WF (2009) Steroids initiate a signaling cascade that triggers rapid sporulation in Dictyostelium. Development 136(5):803–812. doi:10.1242/dev.032607 [doi] (dev.032607 [pii])

Asghar A, Groth M, Siol O, Gaube F, Enzensperger C, Glockner G, Winckler T (2011) Developmental gene regulation by an ancient intercellular communication system in social amoebae. Protist 163:25–37. doi:10.1016/j.protis.2010.12.002 (S1434-4610(11)00005-8 [pii])

Benabentos R, Hirose S, Sucgang R, Curk T, Katoh M, Ostrowski EA, Strassmann JE, Queller DC, Zupan B, Shaulsky G, Kuspa A (2009) Polymorphic members of the lag gene family mediate kin discrimination in Dictyostelium. Curr Biol 19(7):567–572. doi:10.1016/j.cub.2009.02.037 [doi] (S0960-9822(09)00747-7 [pii])

Brown MW, Spiegel FW, Silberman JD (2009) Phylogeny of the "forgotten" cellular slime mold, Fonticula alba, reveals a key evolutionary branch within Opisthokonta. Mol Biol Evol. doi:10.1093/molbev/msp185 [doi] (msp185 [pii])

Brown MW, Silberman JD, Spiegel FW (2012) A contemporary evaluation of the acrasids (Acrasidae, Heterolobosea, Excavata). Eur J Protistol. doi:10.1016/j.ejop.2011.10.001 (S0932-4739(11)00075-7 [pii])

Chapman JA, Kirkness EF, Simakov O, Hampson SE, Mitros T, Weinmaier T, Rattei T, Balasubramanian PG, Borman J, Busam D, Disbennett K, Pfannkoch C, Sumin N, Sutton GG, Viswanathan LD, Walenz B, Goodstein DM, Hellsten U, Kawashima T, Prochnik SE, Putnam NH, Shu S, Blumberg B, Dana CE, Gee L, Kibler DF, Law L, Lindgens D, Martinez DE, Peng J, Wigge PA, Bertulat B, Guder C, Nakamura Y, Ozbek S, Watanabe H, Khalturin K, Hemmrich G, Franke A, Augustin R, Fraune S, Hayakawa E, Hayakawa S, Hirose M, Hwang JS, Ikeo K, Nishimiya-Fujisawa C, Ogura A, Takahashi T, Steinmetz PR, Zhang X, Aufschnaiter R, Eder MK, Gorny AK, Salvenmoser W, Heimberg AM, Wheeler BM, Peterson KJ, Bottger A, Tischler P, Wolf A, Gojobori T, Remington KA, Strausberg RL, Venter JC, Technau U, Hobmayer B, Bosch TC, Holstein TW, Fujisawa T, Bode HR, David CN, Rokhsar DS, Steele RE (2010) The dynamic genome of Hydra. Nature 464(7288):592–596. doi:10.1038/nature08830

Chen ZH, Schaap P (2012) The prokaryote messenger c-di-GMP triggers stalk cell differentiation in *Dictyostelium*. Nature 488:680–683. doi:10.1038/nature11313

Clarke M, Gomer RH (1995) PSF and CMF, autocrine factors that regulate gene expression during growth and early development of Dictyostelium. Experientia 51:1124–1134

Clarke M, Lohan AJ, Liu B, Lagkouvardos I, Roy S, Zafar N, Bertelli C, Schilde C, Kianianmomeni A, Burglin TR, Frech C, Turcotte B, Kopec KO, Synnott JM, Choo C, Paponov I, Finkler A, Soon Heng Tan C, Hutchins AP, Weinmeier T, Rattei T, Chu JS, Gimenez G, Irimia M, Rigden DJ, Fitzpatrick DA, Lorenzo-Morales J, Bateman A, Chiu CH, Tang P, Hegemann P, Fromm H, Raoult D, Greub G, Miranda-Saavedra D, Chen N, Nash P, Ginger ML, Horn M, Schaap P, Caler L, Loftus B (2013) Genome of Acanthamoeba castellanii highlights extensive lateral gene transfer and early evolution of tyrosine kinase signaling. Genome Biol 14(2):R11. doi:10.1186/gb-2013-14-2-r11

Cotter DA (1977) The effects of osmotic pressure changes on the germination of *Dictyostelium discoideum* spores. Can J Microbiol 23(9):1170–1177

Cotter DA, Dunbar AJ, Buconjic SD, Wheldrake JF (1999) Ammonium phosphate in sori of Dictyostelium discoideum promotes spore dormancy through stimulation of the osmosensor ACG. Microbiology-Uk 145(Pt8):1891–1901

Dormann D, Weijer CJ (2001) Propagating chemoattractant waves coordinate periodic cell movement in *Dictyostelium* slugs. Development 128(22):4535–4543

Dynes JL, Clark AM, Shaulsky G, Kuspa A, Loomis WF, Firtel RA (1994) LagC is required for cell-cell interactions that are essential for cell-type differentiation in Dictyostelium. Genes Dev 8:948–958

Eichinger L, Pachebat JA, Glockner G, Rajandream MA, Sucgang R, Berriman M, Song J, Olsen R, Szafranski K, Xu Q, Tunggal B, Kummerfeld S, Madera M, Konfortov BA, Rivero F, Bankier AT, Lehmann R, Hamlin N, Davies R, Gaudet P, Fey P, Pilcher K, Chen G, Saunders D, Sodergren E, Davis P, Kerhornou A, Nie X, Hall N, Anjard C, Hemphill L, Bason N, Farbrother P, Desany B, Just E, Morio T, Rost R, Churcher C, Cooper J, Haydock S, van Driessche N, Cronin A, Goodhead I, Muzny D, Mourier T, Pain A, Lu M, Harper D, Lindsay R, Hauser H, James K, Quiles M, Madan Babu M, Saito T, Buchrieser C, Wardroper A, Felder M, Thangavelu M, Johnson D, Knights A, Loulseged H, Mungall K, Oliver K, Price C, Quail MA, Urushihara H, Hernandez J, Rabbinowitsch E, Steffen D, Sanders M, Ma J, Kohara Y, Sharp S, Simmonds M, Spiegler S, Tivey A, Sugano S, White B, Walker D, Woodward J, Winckler T, Tanaka Y, Shaulsky G, Schleicher M, Weinstock G, Rosenthal A, Cox EC, Chisholm RL, Gibbs R, Loomis WF, Platzer M, Kay RR, Williams J, Dear PH, Noegel AA, Barrell B, Kuspa A (2005) The genome of the social amoeba *Dictyostelium discoideum*. Nature 435(7038):43–57

Feit IN, Bonner JT, Suthers HB (1990) Regulation of the anterior-like cell state by ammonia in Dictyostelium discoideum. Dev Genetics 11:442–446

Feit IN, Medynski EJ, Rothrock MJ (2001) Ammonia differentially suppresses the cAMP chemotaxis of anterior-like cells and prestalk cells in Dictyostelium discoideum. J Biosci 26(2):157–166

Feit IN, Pawlikowski J, Zawilski C (2007) A model for cell type localization in the migrating slug of Dictyostelium discoideum based on differential chemotactic sensitivity to cANIP and differential sensitivity to suppression of chemotaxis by ammonia. J Biosci 32(2):329–338

Fritz-Laylin LK, Prochnik SE, Ginger ML, Dacks JB, Carpenter ML, Field MC, Kuo A, Paredez A, Chapman J, Pham J, Shu S, Neupane R, Cipriano M, Mancuso J, Tu H, Salamov A, Lindquist E, Shapiro H, Lucas S, Grigoriev IV, Cande WZ, Fulton C, Rokhsar DS, Dawson SC (2010) The genome of *Naegleria gruberi* illuminates early eukaryotic versatility. Cell 140(5):631–642. doi:10.1016/j.cell.2010.01.032

Galardi-Castilla M, Garciandia A, Suarez T, Sastre L (2010) The *Dictyostelium discoideum acaA* gene is transcribed from alternative promoters during aggregation and multicellular development. PLoS One 5(10):e13286. doi:10.1371/journal.pone.0013286

Gerisch G, Fromm H, Huesgen A, Wick U (1975) Control of cell-contact sites by cyclic AMP pulses in differentiating Dictyostelium cells. Nature 255:547–549

Harwood AJ, Hopper NA, Simon M-N, Driscoll DM, Veron M, Williams JG (1992) Culmination in *Dictyostelium* is regulated by the cAMP-dependent protein kinase. Cell 69:615–624

Hengge R (2009) Principles of c-di-GMP signalling in bacteria. Nat Rev Micro 7(4):263–273

Hirose S, Benabentos R, Ho H-I, Kuspa A, Shaulsky G (2011) Self-recognition in social amoebae is mediated by allelic pairs of tiger genes. Science 333(6041):467–470. doi:10.1126/science.1203903

Hopper NA, Harwood AJ, Bouzid S, Véron M, Williams JG (1993) Activation of the prespore and spore cell pathway of *Dictyostelium* differentiation by cAMP-dependent protein kinase and evidence for its upstream regulation by ammonia. EMBO J 12:2459–2466

Jenal U, Malone J (2006) Mechanisms of cyclic-di-GMP signaling in bacteria. Annu Rev Genet 40:385–407. doi:10.1146/annurev.genet.40.110405.090423

Jin J, Pawson T (2012) Modular evolution of phosphorylation-based signalling systems. Philos Trans R Soc Lond B Biol Sci 367 (1602):2540–2555. doi:10.1098/rstb.2012.0106

Kawabe Y, Morio T, James JL, Prescott AR, Tanaka Y, Schaap P (2009) Activated cAMP receptors switch encystation into sporulation. Proc Natl Acad Sci U S A 106(17):7089–7094. doi:10.1073/pnas.0901617106 (0901617106 [pii])

Kawabe Y, Weening KE, Marquay-Markiewicz J, Schaap P (2012) Evolution of self-organisation in Dictyostelia by adaptation of a non-selective phosphodiesterase and a matrix component for regulated cAMP degradation. Development 139(7):1336–1345. doi:10.1242/dev.077099

Kay RR, Thompson CR (2001) Cross-induction of cell types in Dictyostelium: evidence that DIF-1 is made by prespore cells. Development 128(24):4959–4966

Kessin RH (2001) *Dictyostelium*: Evolution, cell biology and the development of multicellularity. Developmental and cell biology series, vol 38. Cambridge University Press, Cambridge

Kibler K, Svetz J, Nguyen TL, Shaw C, Shaulsky G (2003) A cell-adhesion pathway regulates intercellular communication during Dictyostelium development. Dev Biol 264 (2):506–521

Kim SK, Kaiser D, Kuspa A (1992) Control of cell density and pattern by intercellular signaling in *Myxococcus* development. Annu Rev Microbiol 46:117–139. doi:10.1146/annurev.mi.46.100192.001001

Kirsten JH, Xiong YH, Dunbar AJ, Rai M, Singleton CK (2005) Ammonium transporter C of Dictyostelium discoideum is required for correct prestalk gene expression and for regulating the choice between slug migration and culmination. Developmental Biology 287(1):146–156

Louis JM, Saxe CL III, Kimmel AR (1993) Two transmembrane signaling mechanisms control expression of the cAMP receptor gene cAR1 during *Dictyostelium* development. Proc Natl Acad Sci U S A 90:5969–5973

O'Day DH, Keszei A (2011) Signalling and sex in the social amoebozoans. Biol Rev Cambridge Phil Soc. doi:10.1111/j.1469-185X.2011.00200.x

Olive LS (1978) Sorocarp development by a newly discovered ciliate. Science 202(4367):530–532. doi:10.1126/science.202.4367.530 (202/4367/530 [pii])

Otto GP, Wu MY, Kazgan N, Anderson OR, Kessin RH (2003) Macroautophagy is required for multicellular development of the social amoeba Dictyostelium discoideum. J Biol Chem 278(20):17636–17645

Raper KB (1984) The Dictyostelids. Princeton University Press, Princeton

Ritchie AV, van Es S, Fouquet C, Schaap P (2008) From drought sensing to developmental control: evolution of cyclic AMP signaling in social amoebas. Mol Biol Evol 25(10):2109–2118. doi:10.1093/molbev/msn156 [doi] (msn156 [pii])

Rogers KW, Schier AF (2011) Morphogen gradients: from generation to interpretation. Annu Rev Cell Dev Biol 27:377–407. doi:10.1146/annurev-cellbio-092910-154148

Romeralo M, Cavender JC, Landolt JC, Stephenson SL, Baldauf SL (2011) An expanded phylogeny of social amoebas (Dictyostelia) shows increasing diversity and new morphological patterns. BMC Evol Biol 11:84. doi:10.1186/1471-2148-11-84 (1471-2148-11-84 [pii])

Romeralo M, Skiba A, Schilde C, Gonzalez-Voyer A, Lawal H, Kedziora S, Cavender JC, Gloeckner G, Urushihara H, Schaap P (2013) Detailed mapping of phenotypic evolution in social amoebas highlights developmental plasticity as a consequence of colonial multicellularity. Proc Biol Sci 280(1764):20130976

Saito T, Taylor GW, Yang JC, Neuhaus D, Stetsenko D, Kato A, Kay RR (2006) Identification of new differentiation inducing factors from Dictyostelium discoideum. Biochim Biophys Acta 1760(5):754–761. doi:10.1016/j.bbagen.2005.12.006 (S0304-4165(05)00392-2 [pii])

Saito T, Kato A, Kay RR (2008) DIF-1 induces the basal disc of the *Dictyostelium* fruiting body. Dev Biol 317(2):444–453. doi:10.1016/j.ydbio.2008.02.036 [doi] (S0012-1606(08)00150-4 [pii])

Saran S, Schaap P (2004) Adenylyl cyclase G is activated by an intramolecular osmosensor. Mol Biol Cell 15(3):1479–1486

Sawai S, Thomason PA, Cox EC (2005) An autoregulatory circuit for long-range self-organization in Dictyostelium cell populations. Nature 433(7023):323–326

Schaap P (1983) Quantitative analysis of the spatial distribution of ultrastructural differentiation markers during development of Dictyostelium discoideum. Roux's Arch Dev Biol 192:86–94

Schaap P (2007) The evolution of size and pattern in the social amoebas. BioEssays 29:635–644

Schaap P, Winckler T, Nelson M, Alvarez-Curto E, Elgie B, Hagiwara H, Cavender J, Milano-Curto A, Rozen DE, Dingermann T, Mutzel R, Baldauf SL (2006) Molecular phylogeny and evolution of morphology in the social amoebas. Science 314(5799):661–663

Schindler J, Sussman M (1977) Ammonia determines the choice of morphogenetic pathways in Dictyostelium discoideum. JMolBiol 116:161–169

Schulkes C, Schaap P (1995) cAMP-dependent protein kinase activity is essential for preaggegative gene expression in *Dictyostelium*. FEBS Lett 368:381–384

Schuster SC, Noegel AA, Oehme F, Gerisch G, Simon MI (1996) The hybrid histidine kinase DokA is part of the osmotic response system of *Dictyostelium*. EMBO J 15:3880–3889

Serafimidis I, Kay RR (2005) New prestalk and prespore inducing signals in *Dictyostelium*. Dev Biol 282(2):432–441

Singleton CK, Zinda MJ, Mykytka B, Yang P (1998) The histidine kinase dhkC regulates the choice between migrating slugs and terminal differentiation in *Dictyostelium discoideum*. Dev Biol 203:345–357

Singleton CK, Kirsten JH, Dinsmore CJ (2006) Function of ammonium transporter A in the initiation of culmination of development in Dictyostelium discoideum. Eukaryot Cell 5(7):991–996

Soderbom F, Anjard C, Iranfar N, Fuller D, Loomis WF (1999) An adenylyl cyclase that functions during late development of *Dictyostelium*. Development 126:5463–5471

Souza GM, Lu SJ, Kuspa A (1998) Yaka, a protein kinase required for the transition from growth to development in dictyostelium. Development 125:2291–2302

Souza GM, daSilva AM, Kuspa A (1999) Starvation promotes Dictyostelium development by relieving PufA inhibition of PKA translation through the YakA kinase pathway. Development 126 (14):3263–3274

Srivastava M, Simakov O, Chapman J, Fahey B, Gauthier ME, Mitros T, Richards GS, Conaco C, Dacre M, Hellsten U, Larroux C, Putnam NH, Stanke M, Adamska M, Darling A, Degnan

SM, Oakley TH, Plachetzki DC, Zhai Y, Adamski M, Calcino A, Cummins SF, Goodstein DM, Harris C, Jackson DJ, Leys SP, Shu S, Woodcroft BJ, Vervoort M, Kosik KS, Manning G, Degnan BM, Rokhsar DS (2010) The Amphimedon queenslandica genome and the evolution of animal complexity. Nature 466(7307):720–726. doi:10.1038/nature09201

Stoodley P, Sauer K, Davies DG, Costerton JW (2002) Biofilms as complex differentiated communities. Annu Rev Microbiol 56:187–209. doi:10.1146/annurev.micro.56.012302.160705

Sugimoto H, Endoh H (2006) Analysis of fruiting body development in the aggregative ciliate Sorogena stoianovitchae (Ciliophora, Colpodea). J Eukaryot Microbiol 53(2):96–102. doi:10.1111/j.1550-7408.2005.00077.x (JEU77 [pii])

Sun TJ, Devreotes PN (1991) Gene targeting of the aggregation stage cAMP receptor cAR1 in Dictyostelium. Genes Dev 5:572–582

Swanson AR, Vadell EM, Cavender JC (1999) Global distribution of forest soil dictyostelids. J Biogeogr 26(1):133–148

Thomason P, Kay R (2000) Eukaryotic signal transduction via histidine-aspartate phosphorelay. J Cell Sci 113(Pt 18):3141–3150

Toama MA, Raper KB (1967) Microcysts of the cellular slime mold *Polysphondylium pallidum*. I. Factors influencing microcyst formation. J Bacteriol 94(4):1143–1149

Van Es S, Virdy KJ, Pitt GS, Meima M, Sands TW, Devreotes PN, Cotter DA, Schaap P (1996) Adenylyl cyclase G, an osmosensor controlling germination of *Dictyostelium* spores. J Biol Chem 271:23623–23625

Virdy KJ, Sands TW, Kopko SH, van Es S, Meima M, Schaap P, Cotter DA (1999) High cAMP in spores of *Dictyostelium discoideum*: association with spore dormancy and inhibition of germination. Microbiology 145(Pt8):1883–1890

Wang M, Schaap P (1989) Ammonia depletion and DIF trigger stalk cell differentiation in intact Dictyostelium discoideum slugs. Development 105:569–574

Wang M, Van Driel R, Schaap P (1988) Cyclic AMP-phosphodiesterase induces dedifferentiation of prespore cells in Dictyostelium discoideum slugs: evidence that cyclic AMP is the morphogenetic signal for prespore differentiation. Development 103:611–618

Wang N, Soderbom F, Anjard C, Shaulsky G, Loomis WF (1999) SDF-2 induction of terminal differentiation in *Dictyostelium discoideum* is mediated by the membrane-spanning sensor kinase DhkA. Mol Cell Biol 19(7):4750–4756

West CM (2003) Comparative analysis of spore coat formation, structure, and function in Dictyostelium. Int Rev Cytol 222:237–293

Williams JG (2006) Transcriptional regulation of *Dictyostelium* pattern formation. EMBO Rep 7(7):694–698

Yamada Y, Kay RR, Bloomfield G, Ross S, Ivens A, Williams JG (2010) A new *Dictyostelium* prestalk cell sub-type. Dev Biol 339(2):390–397. doi:10.1016/j.ydbio.2009.12.045 (S0012-1606(10)00005-9 [pii])

Signaling in Swarming and Aggregating Myxobacteria

Dale Kaiser

Abstract Myxobacteria are renowned for the ability to sporulate within multicellular fruiting bodies whose shapes are species-specific. That capacity arises from the ability of *M. xanthus* to organize high cell density swarms, in which the cells are aligned with each other while constantly moving. The head-tail polarity of rod-shaped cells lays the foundation, and each cell uses two polarized engines for gliding on surfaces, including on the surface of other cells. Regularly periodic reversals of the gliding direction were found to be required for swarming. Such reversals are generated by a G-protein switch that is driven by an oscillator tuned by protein modification. Developmental gene expression is regulated by a network of 2-component systems, which senses the approach of starvation and regulates the transitions between phases.

Keywords Type IV pili · Timer · Pattern formation · Cell polarity · Synchrony · Polysaccharide fibrils

Origin of the Myxobacteria

Multicellular cyanobacteria with differentiated nitrogen-fixing cells arose early in the history of life on earth (Schirrmeister et al. 2011; also see Chapter "Multicellular Life Cycles as an Emergent Property in Filamentous Bacteria"). Myxobacteria evolved later as a branch from the delta proteobacteria (Goldman et al. 2006). They are renowned for their ability to build structurally complex multicellular structures, such as fruiting bodies filled with differentiated myxospores. It appears that the capacity for cell-to-cell signaling accompanies the evolution of multicellularity with cell differentiation. More than 40 species of myxobacteria can be recognized by their 16S RNA sequence and reliably distinguished from each other by distinctive fruiting body shapes (Sproer et al. 1999). Hans Reichenbach has documented the species

D. Kaiser (✉)
Departments of Biochemistry and of Developmental Biology, Stanford University School of Medicine Stanford, Beckman Center, B300, 279 Campus Drive, 94305-5329 Stanford, CA, USA
Tel.: 650/723-6165
e-mail: adkaiser@stanford.edu

I. Ruiz-Trillo, A. M. Nedelcu (eds.), *Evolutionary Transitions to Multicellular Life*, Advances in Marine Genomics 2, DOI 10.1007/978-94-017-9642-2_22

Table 1 Myxobacterial genomes sequenced to date

Species	Genome size (Mb)	Chromosome	G+C (%)	Proteins	GenBank Ref
Myxococcus xanthus 1622	9.14	1 circle	68.9	7388	CP000113
Myxococcus stipitatus	10.35	1 circle	69.2	7361	CP004025
Stigmatella aurantiaca	10.26	1 circle	67.5	8352	CP002271
Coralllococcus coralloides	10.08	1 circle	69.9	8033	CP003389
Chondromyces apiculatus	11.58	1 circle	70.3		Framework GCA_000 601485.1
Myxococcus fulvus HW1	9.004	1 circle	70.6	7361	CP002830

similarities and differences in swarming and fruiting body morphogenesis in a remarkable series of annotated time-lapse movies for *Myxococcus* and *Chondromyces* (Kuhlwein and Reichenbach 1968; Reichenbach et al. 1965a, 1965b; Reichenbach 1966, 1968, 1974, 1984). In addition to the 2006 genome sequence of *Myxococcus xanthus*, the complete genome sequences of several of its close relatives have been determined (Table 1). They have remarkably similar sized circular genomes. But their genes are arranged in different sequences, reflecting different regulatory patterns used for their distinctive swarm and fruiting body developmental patterns. Despite the different gene sequences, there is a high degree of local synteny with *M. xanthus* in COG motility category N, which suggests the presence of A- and S-motility (Schneiker 2007). It may also suggest a similar use of signals. A myxobacterial swarm resembles a large school of fish that has no leader to give them directions. Rather, every myxobacterial swarm cell is both leader and follower, giving and taking directions from the movements of neighboring cells.

Cell Locomotion

Myxobacteria cannot swim in liquid, instead they glide over the solid surface of agar, glass, plastic or soil particles (Hartzell et al. 2008). Jonathan Hodgkin investigated many motility mutants derived from the same genetically characterized strain of *Myxococcus xanthus*. Comparisons between mutants revealed two strikingly different swarm patterns, indicative of two different gliding engines called engine A and engine S (Hodgkin and Kaiser 1979a, 1979b). The swarm patterns are known as A-motility and S-motility. The similarity of genome size and gene content shown in Table 1 suggests that those species will share A and S engines.

Table 2 Type IV pilus genes and the function and localization of their encoded proteins

Gene	Clustered[a]	Conserved[b]	Protein function	Cellular localization
PilA	+	+	Pilin, monomer unit of the pilus filament	Assembled into the pilus fiber stored in the inner membrane
PilB	+	+	Pilus extension	Inner membrane
PilT	+	+	Pilus retraction	Inner membrane
PilC	+	+	Unknown	Inner membrane
PilD	+	+	PilA leader peptidase	Inner membrane
PilG	+		ABC transporter	Periplasm
PilH	+		ABC transporter	Periplasm
PilI	+		ABC transporter	Periplasm
PilM	+	+	ATPase	Inner membrane
PilN	+	+	Unknown	Periplasm
PilO	+	+	Unknown	Periplasm
PilP	+	+	Unknown	Anchored in outer membrane
PilQ	+	+	Secretin	Outer membrane
PilR	+	–	Regulates transcription of pilA	
PilS	+	–	Two-component sensor for pilR	
PilR1	+	–	Transcriptional regulator	
PilS1	+	–	Two-component sensor for pilR1	
Tgl	–	–	Secretin assembly factor	Outer membrane

[a] A set of 16 contiguous genes in *M. xanthus*

[b] Conserved in *M. xanthus*, *P. aeruginosa*, *N. gonorrhoeae*, *N. meningitigis* and *Synechocystis* PCC 6803

S-Motility

There is general agreement that the S engines, which occupy the leading end of the cell, are polar type IV pili (Hartzell et al. 2008). Their structure and ability to retract with great force were reviewed by Nudleman and Kaiser (2004) and Hartzell et al. (2008). Table 2 shows the functions of the pilus proteins.

A-Motility

A-motility is regulated by the same reversal switch as the type IV pili (Kaiser 2008), and is used for swarming (Kaiser 2007) and to build multicellular fruiting bodies (Hodgkin and Kaiser 1979a). Three conceptually different motors have been proposed for A-motility. One proposal involves focal adhesions, discovered by Mignot (Mignot et al. 2005, 2007) and revealed by fluorescently labeled clusters of AglZ, a protein necessary but not sufficient for A-motility (Mauriello et al. 2009b). The clusters, found along the sides of cells are considered to be molecular motors (Mignot et al. 2007). A second proposal considers the deformation of the cell envelope generated by the proton motive force to propel cells in the direction of their long axis (Nan and Zusman 2011). Finally, the secretion of polysaccharide slime from nozzles located at the trailing pole of each cell has been proposed to push each cell forward. Slime trails are evident in Fig. 1, and *M. xanthus* cells are always found on a trail. Hodgkin isolated mutants specifically deficient in A-motile gliding, and established that A motility involved a different set of genes from the type IV pilus-dependent, S motility. The trail elongates at exactly the rate at which the cell that moves forward is depositing the trail. The trail begins at the lagging end of each cell, and is observed to lengthen at the same rate as the leading end of the cell advances. High resolution transmission electron microscopy (EM) revealed long, narrow, amorphous filaments, which bind fluorescent dyes like extruded polysaccharide slime, emerging from one end of the cell (Wolgemuth et al. 2002). Moreover, by phase contrast visible light microscopy filaments were seen to emerge only from one end of each cell (Yu and Kaiser 2007). At any particular moment, both EM and light microscopy clearly show that slime filaments are present only at one of the two cell poles (Wolgemuth et al. 2002; Yu and Kaiser 2007). Unipolarity of the extruded filaments of slime matches the instantaneous unidirectional movement of cells.

As mentioned, both EM and light microscopy showed that a single (united) filament emerges from one end of a cell, while the opposite end has no filament. Yu and Kaiser isolated transposon insertions (null mutations) within several classes of A- motility genes that link them to slime secretion (Yu and Kaiser 2007). Half of the mutants were found to secrete slime from both ends simultaneously, and such mutations were found in *agmK, agnA, agmX, agnB, agnC, aglU,* and *mglA*. The *mglAB* null mutants remained in place while rapidly reversing their gliding direction, at least 10-times more frequently than wild type (Spormann and Kaiser 1999). Eighteen mutants were knockouts of glycosyl transferase genes or regulatory genes that seem to change the sequence of sugars in the repeat unit polysaccharide, resulting in a reduced rate of swarm expansion (Yu and Kaiser 2007). No mutants failing to produce any propulsive slime have been isolated, despite extensive searches because the slime serves as the capsule that protects the cells from lysis by the extracellular digestive enzymes (Cuthbertson et al. 2009), which is the vital function of the capsule in all Gram-negative bacteria (Whitfield 2006).

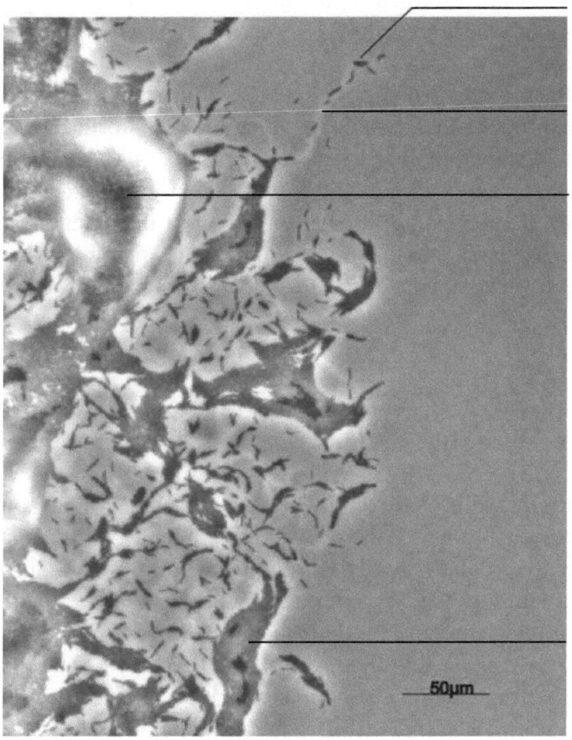

Fig. 1 A phase contrast image of cells at the edge of a DK1622 swarm on 1 % CTT, 1 % agar plate. The swarm is expanding in the radial direction, which is to the *right* in this image, of a small section of the swarm (Scale bar, 50 μm). Photographed with a 20X phase contrast objective. A side-by-side cluster of 5 cells, a slime trail, a multicellular mound with 5 layers, and a large multicellular raft are identified. The inner edge of the annulus of exponential growth reaches 0.517 μm from the outer edge of the swarm

Fibrils

Along with type IV pilus fibers (Kaiser 1979), a high molecular weight repeat unit polysaccharide, the fibril polysaccharide, is required for S-motility. Fibrils are found in the extra-cellular matrix (ECM) (Behmlander and Dworkin 1994a). ECM extracts contain equal amounts of protein and a repeat unit polysaccharide that contains galactose, glucosamine, glucose, rhamnose, and xylose (Behmlander and Dworkin 1991, 1994a, 1994b). Fibrils make up an elastic meshwork of strands that bundle neighboring cells close together, which have been revealed in scanning electron microscope (SEM) images (Kearns and Shimkets 2001). Because the fibril-deficient *dsp/dif* mutants, which grow dispersed in liquid culture, can accept fibrils from an extra-cellular polysaccharide fraction of normal cells (Lu et al. 2005) and regain their S-motility, we infer that the tips of *M. xanthus* pili attach firmly enough to fibrils to withstand the full force of pilus retraction, about 100 pN without letting go (Maier et al. 2002). *Pseudomonas aeruginosa* PAK type IV pili, which are structurally similar to *M. xanthus* pili, have been shown to bind the polysaccharide asialo-GM1 (Giltner et al. 2006) offering a precedent for a specific polysaccharide binding to type IV pili. We suggest that proper attachment of *M. xanthus* pili to fibrils rapidly initiates a strong retraction. The observation that 15 nm diameter fibrils were most frequently found on the cell surface, while larger fibrils were most often seen farther from the cell surface (Dworkin 1993) suggests that the smallest-diameter fibrils may be secreted by each cell in a cluster, and that those small fibrils assemble extracellularly with other small fibrils from adjacent cells to form the 30 nm diameter fibrils that are observed in the SEM.

Swarming

Within *M. xanthus* colonies, cells move only when they elongate, divide, and the two daughters separate from each other. By contrast, cells in an *M. xanthus* swarm use their A- and S-motility to circulate rapidly within the swarm. Individual rod-shaped cells are constantly moving, transiently interacting with one another, and independently reversing their gliding direction. Independent reversal is, in fact, essential for creating a swarm (Wu et al. 2009). Gliding reversals are timed by a feed-back oscillator involving three *frz* genes, called the pacemaker. The pacemaker circuit shown in Fig. 2 was confirmed experimentally by the observation that in-frame deletions of a gene controlling any one of the three sectors of the negative feedback loop generates the same swarming phenotype, quantitatively (Kaiser and Warrick 2011). Deleting *frz* genes or significant parts of *frz* genes outside that loop generated different phenotypes, some stronger, some weaker but none identical to those without feedback. For example, reversal does not appear to depend on the Frz methylesterase.

The pacemaker drives a small G-protein to switch the gliding direction, as shown in Fig. 2. All three deletion mutants lacking individual proteins that constitute the pacemaker are still able to swarm (but at a low rate) and they all have the same long reversal period of 34 min (Bustamante et al. 2004). Mutants lacking the MglA protein are the only motility mutants that completely lack the ability to swarm (Hodgkin and Kaiser 1979a, 1979b; Kroos et al. 1988). The adjacent *mglB* gene, when deleted, renders cells partially motile (Yu and Kaiser 2007), because the deletion destabilizes the MglA protein (Hartzell and Kaiser 1991). The *mglA* gene encodes a small Ras-like protein with a G-loop (Stephens et al. 1989). Missense mutations within the G-loop of MglA prevent swarming (Fremgen et al. 2010), strongly suggesting that MglA switches between GDP-bound and GTP-bound states (Bourne et al. 1990). Leonardy et al. have provided biochemical evidence that MglB could be the cognate GAP protein for MglA that stimulates its GTPase activity (Leonardy et al. 2010). Leonardy et al. also provided evidence that the Frz proteins lie upstream of the MglAB switch to govern the frequency of switching (Leonardy et al. 2010), consistent with the swarming data.

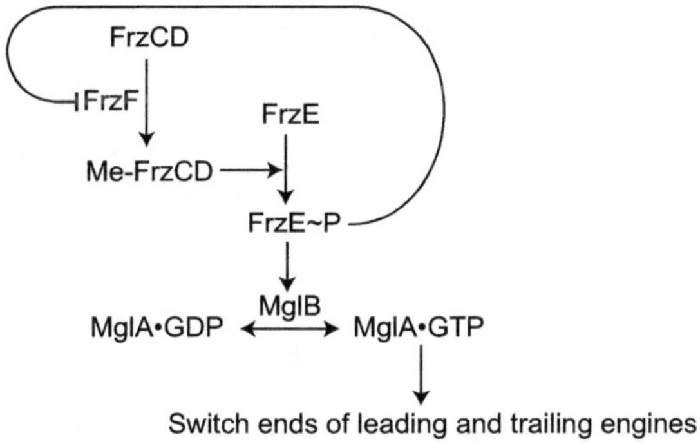

Fig. 2 Feedback-induced oscillator that drives the MglAB reversal switch as demonstrated by the phenotypes of gene deletion mutants. *Arrows* indicate reactions; –| FrzF indicates that the action of the FrzF methyltransferase is inhibited by FrzE ∼ P. FrzCD and FrzE constitute a two-component system. A high level of FrzE ∼ P activates the formation of MglA · GTP which causes both the A and the S engines to switch from one cell end to the other

That there is a connection between pulses of FrzE ∼ P and the MglAB switch is supported by an abundance of experimental evidence, but the identity of proteins that make the connection remains to be clarified. Apparently the pulses of FrzE ∼ P are conveyed by a protein that acts when a certain threshold level of FrzE ∼ P is reached. Since FrzE protein has both a kinase/response regulator domain and an additional CheY domain, FrzE protein alone might be capable of sensing the threshold. Alternatively, the connection might involve FrzZ that has two CheY domains and has been proposed as the output of the Frz system (Inclan et al. 2007). Moreover, Leonardy et al. showed that the dynamic polar localization of RomR, which is required for A-motility, depends on the Mgl switch (Leonardy et al. 2007, 2010). These authors also showed that the dynamic polar localization of PilT, which is required for S-motility (Nudleman and Kaiser 2004) is downstream of the MglAB switch as well (Leonardy et al. 2010). Together, these data support the circuit, diagrammed in Fig. 2 (Wu et al. 2009) proposed that the circuit evolved to drive GDP/GTP oscillations to a shorter period, one of 8–9 min that would maximize the swarm expansion rate of cells gliding at a speed averaging 4 μm/min (Kaiser and Crosby 1983; Wu et al. 2009).

Signaling in the Swarm

Evidence for a signal that can synchronize reversals in adjacent cells was found in the behavior of two sorts of highly organized, multicellular structures seen near the edge of an expanding swarm and indicated in Fig. 1: the rectangular rafts and the round, multi-layered mounds. Rafts are dense, single-layered, rectangular assemblies of a 100 cells that have their long axes in parallel with each other. Although raft cells are packed together, adjacent cells can slide smoothly along side each other because their polysaccharide capsules mix and lubricate the sliding. The capacity to slide enables raft cells to reverse independently of each other. Mounds are multi-layered assemblies of a thousand cells. The layers are nested one on top of another, and adjacent layers are separated by an organized mesh-work of fibrils. Mound cells can move rapidly from one layer to the next, either up or down using their S-motility. Recall that tips of pili avidly bind fibrils. Cells can move more slowly to a new position within the same layer using A-motility based on slime secretion. Since mounds are built from the bottom up, cells in the top layer will have been mound residents longer than any other cells in the mound.

Because each cell has its own pacemaker that causes the cell to reverse its gliding direction regularly, the newly-found synchronizing signal is proposed to bring the pacemakers in a pair of cells, which are signaling to each other, to the same phase of their reversal cycle, and thus synchronized with each other. Moreover, once synchronized by signal exchange more signaling keeps the pacemakers synchronized. The signal was discovered when viewing time-lapse movies of cell movement at the swarm edge. After a 1 h delay, all cells in the top (fifth) layer of a mound "exploded". Suddenly they increased their speed of movement more than 3-fold to 12 μm/min, then within a minute decreased their speed 3-fold back to their long term average speed of 4 μm/min. The timing signal is thought to spread by transient aligned contact between pairs of adjacent cells. Accordingly, the hour's delay is the time required to spread the signal among all cells in the mound's top layer.

The signal is proposed to be transferred from one cell to another by multi-protein structures termed "focal adhesions" (Mignot et al. 2005, 2007). Focal adhesions are found as a series of discrete foci running along one side of a cell from its leading end to just forward of mid-cell. Although focal adhesions were initially proposed to be complexes of motor proteins (Nan et al. 2011; Sun et al. 2011), the 15 or more proteins that characterize an adhesion seem more suited for signaling than for motorizing cells. One reason for proposing them as signaling proteins is their capacity to bind each other in pairs (Nan et al. 2010). Another reason is their apparent localization in different membrane-bound compartments of *M. xanthus* cells. The synchronizing circuit proposed is shown in Fig. 3. Proceeding inward, the compartments are defined by the outer membrane, the periplasm, the outer surface of the peptidoglycan sacculus, the inner membrane, and finally the cytoplasm of these Gram-negative bacteria. Accordingly, the signal could link oscillating methylated-FrzCD in the pacemaker of one cell mechanically through pairs of protein 1 transiently bound to protein 2 links to CglB—one of the 15 motility proteins found in an adhesion—on the surface of that cell. It is proposed that CglB on the surface of the first cell would assemble with CglB on the surface of the second cell. From CglB the signal would link through the same series of protein 1 · protein 2 pairs until it reached FrzCD in the pacemaker of the second cell. When completed, the series of signal links would bring the pacemakers of both cells to the same, average phase of their oscillatory cycles. Two focal adhesions seem to bind each other weakly and transiently as pictured (Mauriello et al. 2009a).

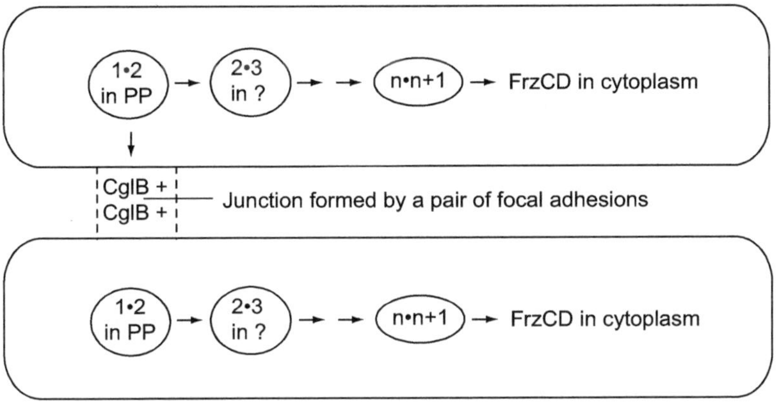

Fig. 3 Path followed by the signal postulated to synchronize the pacemakers of the cell pair shown. Signaling is proposed to be initiated by the formation of a junction between the CglB proteins in the outer membrane of two cells that are in physical contact with each other as shown. That contact induces protein 1 and protein 2 located in the periplasm to bind each other. $1 \cdot 2$ binding then induces $2 \cdot 3$ binding in the next compartment, etc. in both cells. *Arrows* in the figure point toward the next pair of numbered A-motility proteins to bind together, as described in the text. The proteins listed in Table 3 are represented in this figure by a number that indicates their position in the sequence of pairwise binding steps; unless, like CglB and FrzCD, their location is established and they are named. $1 \cdot 2$ is the first pair of proteins to bind, $2 \cdot 3$ is the second pair, and $n \cdot n \cdot n \cdot n+1$ is the next to last pair. The last pair is $n+1 \cdot$ FrzCalthouFrzCD is a methylated regulatory protein and not an A-motility protein. Once the signal has reached FrzCD in the cytoplasm of both cells, the phases of their oscillations are reset to the average value of phase they had prior to this connection. The two cells shown are joined for a short time, long enough to complete the whole binding series and average phases. Then the two cells disjoin before moving on to pair with other cells, spreading the signal to other cells nearby

Responding to Starvation and Building a Multicellular Fruiting Body

When growth begins to outrun the food supply, a swarm begins to allocate its remaining resources. Strict allocation is necessary because the cells are constantly circulating to access oxygen for efficient aerobic ATP production (Kaiser and Warrick 2011). Cell behavior changes abruptly in response to starvation: the swarm stops expanding outward, instead the cells migrate inward to build their fruiting bodies, containing stress-resistant spores. In addition to allocating ATP for cell movement, which is necessary throughout development (Kroos et al. 1988), ATP must be allocated to DNA replication so that each spore will contain two complete copies of the genome (Tzeng and Singer 2005). *M. xanthus* has a single origin of bidirectional replication located near the *dnaA* locus on its circular genome (Goldman et al. 2006). Using flow cytometry, it was shown that exponentially growing *M. xanthus* cells contain 1–2 copies of the chromosome, indicating a single DNA replication initiation event per cell cycle (Tzeng and Singer 2005). As cells depleted their nutrient sources, no new rounds of DNA replication were initiated. Instead, there was a

Table 3 A-motility proteins found in focal adhesions

Protein	Localization	Predicted function
CglB	In outer membrane	Binding to next protein in legend to Fig. 3
AglT	Associated with inner membrane	Binding to next protein in sequence
AglU	Periplasm	Binding to next protein in sequence
AglW	Inside surface of outer membrane	Binding to next protein in sequence
AglZ	Cytoplasm	Binding to next protein in sequence
AgmK	Periplasm, crosses inner membrane	Binding to next protein in sequence
AgmX	Periplasm, crosses inner membrane	Binding to next protein in sequence
AgmU	Cytoplasmic side of inner membrane	Binding to next protein in sequence
AgmV	Unknown	Binding to next protein in sequence
AgnB	Unknown	Binding to next protein in sequence
AgnC	Unknown	Binding to next protein in sequence
AgnK	Unknown	Binding to next protein in sequence
MXAN_4864	Unknown	Binding to next protein in sequence
MXAN_4868	Unknown	Binding to next protein in sequence
Pgl I	Associated with inner membrane	Binding to next protein in sequence

Proteins listed in this table are located in different cellular compartments, CglB for example is a lipo-protein embedded in the outer surface of the cell's outer membrane. Each of these proteins binds and signals to the next protein also listed in this table. However, the order of protein- protein signaling is yet to be worked out. Signaling is proposed to be initiated by the formation of a junction between the CglB proteins in the outer membrane of two cells that come into physical contact with each other. That contact induces protein 1 (Cgl B) and protein 2 (there are several candidates: Agl U, Agm K, Agm V, and is called protein 2) located in the periplasm to bind each other. 1 · 2 binding then induces 2 · 3 binding in the next compartment, etc. In general the number indicates their position in the sequence of pairwise binding steps, unless like CglB and FrzCD their location is established and they are named. 1 · 2 is the first pair of proteins to bind, 2 · 3 is the second pair, and n · n+1 is the next to last pair. The last pair is n+1 · FrzCD, although FrzCD is a methylated regulatory protein. Once the signal has reached FrzCD in the cytoplasm of both cells, the phases of their oscillations are reset to the average value of phase they had prior to this connection. Each pair of cells shown are joined for a short time, less than a minute. After the two phases are averaged the cells disjoin, and move on to pair with other cells, to spread the signal to other cells nearby

stringent response (Cashel et al. 1996) with the accumulation of (p)ppGpp. Indeed, it was shown that (p)ppGpp synthesis was both necessary and sufficient to trigger fruiting body development (Singer and Kaiser 1995). Although *B. subtilis* and *E. coli* arrest DNA synthesis immediately with their stringent responses, *M. xanthus* chromosome replication continues. Continuation ensures that each myxospore contains two complete copies of the genome (Tzeng and Singer 2005). Evidently, *M. xanthus* has a different survival strategy for its sporulation than bacilli that may be related to the threat posed by long exposure of myxospores in the top soil to bright sunlight. A more detailed description of starvation and sporulation can be found in Diodati et al. (2008).

Additional ATP must be allocated for protein synthesis on ribosomes. More than 30 new proteins must be made as spore-filled fruiting bodies are under construction (Inouye et al. 1979; Dahl et al. 2007). Consequently, the swarm must initiate its program for fruiting body development before any nutrient essential for protein synthesis has been eliminated. This explains why growth limiting concentrations of any amino acid, of usable carbon sources, or of phosphate induce *M. xanthus* to initiate fruiting body development (Manoil and Kaiser 1980a, 1980b). By contrast, neither the lack of oxygen, or the lack of purines or of pyrimidines, which *M. xanthus* scavenges by digesting its prey's nucleic acids, will induce development (Kimsey and Kaiser 1991). These observations point to deficiencies of any amino-acylated tRNA leading the swarm to initiate fruiting body development. In *M. xanthus* as in many other bacteria, the absence or shortage of any one of the charged tRNAs leads a ribosome, sensing with a codon that lacks its cognate amino-acylated tRNA, to synthesize guanosine tetra (and penta) -phosphate, (p)ppGpp, in a reaction catalyzed by the *relA* ppGpp synthase.

Then, instead of a cascade of sigma factors like that used by *B. subtilis* for sporulation (Kroos et al. 1999; Errington 2003), *M. xanthus* has a cascade of enhancer-binding proteins (EBPs). The cascade organizes the transition from exponential growth through the staged development of multicellular fruiting bodies (Caberoy et al. 2003). A cascade of several EBPs, each with its own metabolic sensor also replaces an early commitment to sporulation found in *B. subtilis* with the possibility of responding to newly found nutrient, post-starvation, by restarting growth. *M. xanthus* appears not to commit to sporulation until it has begun to differentiate the spores (Licking et al. 2000).

Though ordinarily considered an alternative sigma factor, sigma-54 is essential for *M. xanthus* growth and development (Keseler and Kaiser 1997). Currently, the myxobacteria have more EBPs than any other taxonomic group of sequenced bacterial genomes (Goldman et al. 2006; Ronning and Nierman 2008). EBPs are specific transcriptional activators that work in conjunction with sigma-54 RNA polymerase to activate transcription at designated sigma-54 promoters (Caberoy et al. 2003). In response to an activating signal such as phosphorylation by a histidine kinase sensor protein, EBPs use the energy from ATP hydrolysis to form a transcription-competent open promoter complex. Cascade EBPs appear to be regulated for reliability (Caberoy et al. 2003). First, the expression of a downstream EBP is activated at the proper time by a preceding EBP in the cascade. Second, Nla4 and Nla18 are important for (p)ppGpp production. Third, Nla6, Nla28, and ActB (Gronewold and Kaiser 2001, 2002) positively regulate their own expression. Since EBPs typically activate gene expression in response to a specific interaction with a signal transduction partner that detects a particular environmental cue (Studholme and Dixon 2003), it is suggested that the cascade's sensor kinases measure the level of metabolites that inform a cell whether those levels render fruiting body development an outcome to be sought, despite the death of many cells that accompanies development. Early detection of approaching starvation seems to be limiting spore formation because no more than 1 % of the cells initiating fruiting body development ever become spores (Harvey et al. 2013). It is thought that 99 % of cells are cannibalized to sustain the continued movement of the surviving 1 %. After (p)ppGpp formation, Nla6 and Nla28 help to manage the subsequent pre-aggregation stage, and the ActB and MXAN4899 regulate gene expression during the aggregation of fruiting bodies.

In parallel with a starvation-induced cascade of EBPs, the transition from growth to development is guided by a diffusible cell-to-cell signal, the A-signal. A-signal molecules, purified from medium conditioned by developing cells, is a set of amino acids and peptides containing those amino acids (Kuspa et al. 1986, 1992a). Each developing *Myxococcus* cell releases a small quantity of A-signal about two hours into development. Consequently, the extracellular concentration of A-signal is directly proportional to the density of *M. xanthus* cells that are beginning to develop (Kuspa et al. 1992b; Kaplan and Plamann 1996). Cells respond to A-signal only if its concentration is above a certain threshold, which may reflect the number of cells necessary to produce at least one spore-filled cyst. In *M. xanthus,* the cyst is a single fruiting body, and individual fruiting bodies are very similar in diameter and spherical shape, reflecting inheritance of the threshold value. The threshold number of cells constitute a quorum, and the A-signal is a quorum sensor.

Genes for the synthesis of secondary metabolites account for 17 % of the *M. xanthus* genome. After aggregation, *M. xanthus* cells express a unique set of A-signal-dependent genes (Kroos and Inouye 2008), including *csgA,* the gene for C-signal. Expression of the *csgA* gene ensures that C-signaling comes after A-signaling.

C-Signaling

C-signaling is essential for *M. xanthus* fruiting body morphogenesis. Unlike diffusible small molecules, C-signaling requires cell movement for its transmission from one cell to another (Kroos et al. 1988). C-signal deficient mutants (*csgA*) were found to grow and swarm normally but they failed to aggregate or to sporulate (Hagen et al. 1978; Kim and Kaiser 1990b; Shimkets et al. 1983). Active C-signal was found to be a 17 kDa cell-surface-bound protein that communicates when pairs of cells make an end-to-end contact with each other (Kim and Kaiser 1990a; Sager and Kaiser 1994). CsgA protein is 25 kD and is secreted to the cell surface where it is cleaved to the active 17 kDa signal by a membrane protease (Lobedanz and Søgaard-Andersen 2003; Rolbetski et al. 2008). No receptors for the C-signal have been found on either the upstream or the downstream cell (Søgaard-Andersen 2008), while subsequent fruiting body development and the activation of FruA, a developmentally important response regulator (Ellehauge et al. 1998), clearly demonstrate that C-signaling has occurred. C-signal transfer appears to require forceful collisions between pairs of aligned cells actively moving into end-to-end contact by A motility (Kim and Kaiser 1990a). C-signal transfer closely parallels the stimulation of CglB, as described (Nudleman et al. 2005). CglB stimulation was shown to result in the equal sharing of outer membrane CglB protein between pairs of colliding cells (Nudleman et al. 2005; Wall and Kaiser 1998; Wall et al. 1998; Wei et al. 2011).

When development starts, there are few C-signal molecules per cell. However, expression of the C-signal increases rapidly due to a positive feedback loop involving the five proteins of the *act* operon (Gronewold and Kaiser 2001). Jelsbak and Søgaard-Andersen (1999) found that cells in an aggregation stream continue to reverse their gliding direction, at the period set by the pacemaker (within measurement error) (Jelsbak and Søgaard-Andersen 1999). Whenever a cell reverses in response to its pacemaker, it would come into end-to-end contact with the cell immediately behind it that is still moving toward a nascent fruiting body. Such forceful contact should allow the two cells to exchange the C-signal. Each time the C-signal is exchanged between cells in an aggregation stream, the positive feedback loop would increase expression of *csgA* and elevate the number of signal molecules on both signaling cells (Gronewold and Kaiser 2001).

Spore differentiation is likely to be the final step in fruiting body development because spores have lost their poles and without their polar engines would no longer be able to propel themselves and to raise the level of C-signal further. Consequently, sporulation should be triggered only after cells had been signaling each other long enough for the level of the C-signal to have reached some elevated threshold, to ensure that spores would form inside the nascent fruiting body and not prematurely in an aggregation stream. The patchy spatial distribution of spores within a fruiting body fits a C-signal rise to some threshold (Harvey et al. 2013).

Summary

1. Myxobacteria are renowned for the ability to sporulate within multicellular fruiting bodies whose shapes are species-specific.
2. Their capacity to sporulate arises from the ability of *M. xanthus* to organize high cell density swarms, in which the cells are aligned with each other while constantly moving.
3. The head-tail polarity of rod-shaped cells lays the foundation, and each cell uses two polarized engines for gliding on surfaces, including on the surface of other cells.
4. Regularly periodic reversal of the gliding direction was found to be required for swarming. Reversals are generated by a G-protein switch, driven by an oscillator that is tuned by protein modification.
5. Developmental gene expression is regulated by a network of two-component systems that senses the approach of starvation and regulates the transitions between phases.
6. Each of the signaling pathways described in this chapter is presented as a specific model that is detailed enough to be tested experimentally and improved upon on.

Acknowledgement I thank Dr. Marianne Bromme Powell, Stanford University, for her critique of the text.

References

Behmlander RM, Dworkin M (1991) Extracellular fibrils and contact-mediated cell interactions in *Myxococcus xanthus*. J Bacteriol 173(24):7810–7821

Behmlander RM, Dworkin M (1994a) Biochemical and structural analyses of the extracellular matrix fibrils of *Myxococcus xanthus*. J Bacteriol 176(20):6295–6303

Behmlander RM, Dworkin M (1994b) Integral proteins of the extracellular matrix fibrils of *Myxococcus xanthus*. J Bacteriol 176(20):6304–6311

Bourne HR, Sanders DA, McCormick F (1990) The GTPase superfamily: a conserved switch for diverse cell functions. Nature 348:125–132

Bustamante VH, Martínez-Flores I, Vlamakis HC, Zusman D (2004) Analsis of the Frz signal transduction system of *Myxococcus xanthus* shows the importance of the conserved C-terminal region of the cytoplasmic chemoreceptor FrzCD in sensing signals. Mol Microbiol 53(5):1501–1513

Caberoy NB, Welch RD, Jakobsen JS, Slater SC, Garza AG (2003) Global mutational analysis of NtrC-like activators in *Myxococcus xanthus*: identifying activator mutants defective for motility and fruiting body devleopment. J Bacteriol 185(20):6083–6094

Cashel M, Gentry DR, Hernandez VJ, Vinella D (1996) The stringent response. In: Neidhardt F (ed) *Escherichia coli and Salmonella*, 2nd edn. ASM Press, Washington, DC, pp 1458–1496

Cuthbertson L, Mainprize I, Naismith J, Whitfield C (2009) Pivotal roles of the outer membrane polysaccharide export and polysaccharide copolymerase protein families in export of extracellular polysaccharides in gram-negative bacteria. Microbiol Mol Biol Rev 73(1):155–177

Dahl JL, Tengra F, Dutton D, Yan J, Andacht T, Coyne L, Windell V, Garza AG (2007) Identification of major sporulation proteins of *Myxococcus xanthus* using a proteome approach. J Bacteriol 189:3187–3197

Diodati M, Gill R, Plamann L, Singer M (2008) Initiation and early developmental events (Chapter 3). In: Whitworth DE (ed) Myxobacteria, multicellularity and differentiation. ASM Press, Washington DC, pp 43–76

Dworkin M (1993) Cell surfaces and appendages. In: Dworkin M, Kaiser D (eds) Myxobacteria II. American Society of Microbiology, Washington, DC, pp 63–83

Ellehauge E, Norregaard-Madsen M, Søgaard-Andersen L (1998) The FruA signal transduction protein provides a checkpoint for the temporal coordination of intercellular signals in *Myxococcus xanthus* development. Mol Microbiol 30:807–813

Errington J (2003) Regulation of endospore formation in *Bacillus subtilis*. Nat Rev Microbiol 1:117–126

Fremgen S, Burke N, Hartzell P (2010) Effects of site-directed mutagenesis of *mglA* on motility and swarming of *Myxococcus xanthus*. BMC Microbiol 10:295

Giltner CL, van Schaik EJ, Audette GF, Kao D, Hodges RS, Hassett DJ, Irvin RT (2006) The *Pseudomonas aeruginosa* type IV pilin receptor binding domain functions as an adhesin for both biotic and abiotic surfaces. Mol Microbiol 59(4):1083–1096

Goldman BS, Nierman WC, Kaiser D et al (2006) Evolution of sensorycomplexity recorded in a myxobacterial genome. Proc Natl Acad Sci U S A 103:15200–15205

Gronewold TMA, Kaiser D (2001) The *act* operon controls the level and time of C-signal production for *M. xanthus* development. Mol Microbiol 40:744–756

Gronewold TMA, Kaiser D (2002) The act operon control of developmental gene expression in *Myxococcus xanthus*. J Bacteriol 184:1172–1179

Hagen DC, Bretscher AP, Kaiser D (1978) Synergism between morphogenetic mutants of *Myxococcus xanthus*. Dev Biol 64:284–296

Hartzell P, Kaiser D (1991) Upstream gene of the mgl operon controls the level of mglA protein in *Myxococcus xanthus*. J Bacteriol 173:7625–7635

Hartzell P, Shi W, Youdarian, P. (2008) Gliding motility of *Myxococcus xanthus*. In: Whitworth DE (ed) Myxobacteria, multicellularity and differentiation. ASM Press, Washington, D.C., pp 103–132

Harvey C, Du H, Xu Z, Kaiser D, Aranson I, Alber MS (2013) Interconnected cavernous structure of bacterial fruiting bodies. PLoS Comput Biol 8(12):e1002850

Hodgkin J, Kaiser D (1979a) Genetics of gliding motility in *M. xanthus* (Myxobacterales): genes controlling movement of single cells. Mol Gen Genet 171:167–176

Hodgkin J, Kaiser D (1979b) Genetics of gliding motility in *M. xanthus* (Myxobacterales): two gene systems control movement. Mol Gen Genet 171:177–191

Inclan Y, Vlamakis H, Zusman D (2007) FrzZ, a dual cheY-like response regulator, functions as an output for the Frz chemosensory pathway of *Myxococcus xanthus*. Mol Microbiol 65(1):90–102

Inouye M, Inouye S, Zusman D (1979) Gene expression during development of *Myxococcus xanthus*: pattern of protein synthesis. Dev Biol 68:579–591

Jelsbak L, Søgaard-Andersen L (1999) The cell-surface associated C-signal induces behavioral changes in individual *M. xanthus* cells during fruiting body morphogenesis. Proc Natl Acad Sci U S A 96:5031–5036

Kaiser AD (1979) Social gliding is correlated with the presence of pili in *Myxococcus xanthus*. Proc Natl Acad Sci U S A 76:5952–5956

Kaiser D (2007) Bacterial swarming, a re-examination of cell movement patterns. Curr Biol 17(14):R561–R570

Kaiser D (2008) Chapter 5. Reversing *Myxococcus xanthus* polarity. In: Whitworth DE (ed) Myxobacteria, multicellularity and differentiation. ASM Press, Washington, DC, pp 93–102

Kaiser AD, Crosby C (1983) Cell movement and its coordination in swarms of *Myxococcus xanthus*. Cell Motil 3:227–245

Kaiser D, Warrick H (2011) *M. xanthus* swarms are driven by growth and regulated by a pacemaker. J Bacteriol 193(21):5898–5904

Kaplan HB, Plamann L (1996) A *Myxococcus xanthus* cell density-sensing system required for multicellular development. FEMS Microbiol Lett 139:89–95

Kearns DB, Shimkets LJ (2001) Lipid chemotaxis and signal transduction in *Myxococcus xanthus*. Trends Microbiol 9:126–129

Keseler IM, Kaiser D (1997) Sigma-54, a vital protein for *Myxococcus xanthus*. Proc Natl Acad Sci U S A 94:1979–1984

Kim SK, Kaiser D (1990a) Cell alignment required in differentiation of *Myxococcus xanthus*. Science 249:926–928

Kim SK, Kaiser D (1990b) Purification and properties of *Myxococcus xanthus* C-factor, an intercellular signaling protein. Proc Natl Acad Sci U S A 87:3635–3639

Kimsey HH, Kaiser D (1991) Targeted disruption of the *Myxococcus xanthus* orotidine 5'-monophosphate decarboxylase gene: effects on growth and fruiting-body development. J Bacteriol 173:6790–6797

Kroos L, Inouye S (2008) Transcriptional regulatory mechanisms during *Myxococcus xanthus* development. In: Whitworth DE (ed) Myxobacteria, multicellularity and differentiation. ASM Press, Washington, DC, pp 149–168

Kroos L, Hartzell P, Stephens K, Kaiser D (1988) A link between cell movement and gene expression argues that motility is required for cell-cell signalling during fruiting body development. Genes Dev 2:1677–1685

Kroos L, Zhang B, Ichikawa H, Yu U-TN (1999) Control of sigma factor activity during *Bacillus subtilis* sporulation. Mol Microbiol 31(5):1285–1294

Kuhlwein H, Reichenbach H (1968) Schwarmentwicklung und Morphogenese bei Myxobacterien—Archangium, Myxococcus, Chondrococcus, Chondromyces. Film C893. Encyclopedia Cinematographica. Institut fur den Wissenschaftlichen Film, Gottingen

Kuspa A, Kroos L, Kaiser D (1986) Intercellular signaling is required for developmental gene expression in *Myxococcus xanthus*. Dev Biol 117:267–276

Kuspa A, Plamann L, Kaiser D (1992a) Identification of heat-stable A-factor from *Myxococcus xanthus*. J Bacteriol 174:3319–3326

Kuspa A, Plamann L, Kaiser D (1992b) A-signalling and the cell density requirement for *Myxococcus xanthus* development. J Bacteriol 174:7360–7369

Leonardy S, Freymark G, Hebener S, Ellehauge E, Søgaard-Andersen L (2007) Coupling of protein localization and cell movement by a dynamically localized response regulator in *Myxococcus xanthus*. EMBO J 26(21):4433–4444

Leonardy S, Miertzschke M, Bulyha I, Sperling E, Wittinghofer A, Søgaard-Andersen L (2010) Regulation of dynamic polarity switching in bacteria by a Ras-like G-protein and its cognate GAP. EMBO J 29:1–14

Licking E, Gorski L, Kaiser D (2000) A common step for changing the cell shape in fruiting body and starvation-independent sporulation of *Myxococcus xanthus*. J Bacteriol 182:3553–3558

Lobedanz S, Søgaard-Andersen L (2003) Identification of the C-signal, a contact-dependent morphogen coordinating multiple developmental responses in *Myxococcus xanthus*. Genes Dev 17:2151–2161

Lu A, Cho K, Black WP, Duan X-y, Lux R, Yang Z, Kaplan HB, Zusman D, Shi W (2005) Exopolysaccharide biosynthesis genes required for social motility in *Myxococcus xanthus*. Mol Microbiol 55:206–220

Maier B, Potter L, So M, Seifert HS, Sheetz MP (2002) Single pilus motor forces exceed 100 pN. Proc Natl Acad Sci U S A 99(25):16012–16017

Manoil C, Kaiser D (1980a) Accumulation of guanosine tetraphosphate and guanosine pentaphosphate in *Myxococcus xanthus* during starvation and myxospore formation. J Bacteriol 141:297–304

Manoil C, Kaiser D (1980b) Guanosine pentaphosphate and guanosine tetraphosphate accumulation and induction of *Myxococcus xanthus* fruiting body development. J Bacteriol 141:305–315

Mauriello E, Astling DP, Sliusarenko O, Zusman D (2009a) Localization of a bacterial cytoplasmic receptor is dynamic and changes with cell-cell contacts. Proc Natl Acad Sci U S A 106:4852–5857

Mauriello E, Nan B, Zusman D (2009b) AglZ regulates adventurous [A-] motility in *Myxococcus xanthus* through its interaction with the cytoplasmic receptor, FrzCD. Mol Microbiol 72:964–977

Mignot T, Merlie JP, Zusman D (2005) Regulated pole-to-pole oscillations of a bacterial gliding motility protein Science 310:855–857

Mignot T, Shaevitz J, Hartzell P, Zusman D (2007) Evidence that focal adhesions power bacterial gliding motility. Science 315:853–856

Nan B, Zusman DR (2011) Uncovering the mystery of gliding motility in the myxobacteria. Annu Rev Genet 45:21–39

Nan B, Mauriello E, Sun I-H, Wong A, Zusman D (2010) A multi-protein complex from *Myxococcus xanthus* required for bacterial gliding motility. Mol Microbiol 76 (6):1539–1554

Nan B, Chen J, Neu J, Berry R, Oster G, Zusman D (2011) Myxobacteria gliding motility requires cytoskeleton rotation powered by proton motive force. Proc Natl Acad Sci U S A 108(6):2498–2503

Nudleman E, Kaiser D (2004) Pulling together with type IV pili. J Mol Microbiol Biotechnol 7:52–62

Nudleman E, Wall D, Kaiser D (2005) Cell-to-cell transfer of bacterial outer-membrane lipoproteins. Science 309:125–127

Reichenbach H (1966) *Myxococcus* spp. (Myxobacterales) Schwarmentwicklung und Bildung von Protocysten. vol E778/1965. Institut fur den Wissenschaftlichen Film, Gottingen

Reichenbach H (1968) *Archangium violaceum* (Myxobacteriales) Schwarmentwicklung und bildung von protocysten Film E 777/1965. Encyclopedia Cinematographica INST WISS FILM. Institut fur den Wissenschaftlichen Film, Gottingen

Reichenbach H (1974) *Chondromyces apiculatus* (Myxobacteriales) Schwarmentwicklung und morphogenese. Film E 779/1965, Encyclopedia Cinematographica. Institut fur den Wissenschaftlichen film, Gottingen

Reichenbach H (1984) Myxobacteria: a most peculiar group of social prokaryotes. In: Rosenberg E (ed) Myxobacteria. Springer-Verlag, New York, pp 1–50, p 34

Reichenbach H, Heunert HH, Kuczka H (1965a) *Archangium violaceum* (Myxobacterales) -Schwarmentwicklung und Bildung von Protocysten. Film E777. Encyclopedia Cinematographica. Institut fur den Wissenschaftlichen film, Gottingen
Reichenbach H, Heunert HH, Kuczka H (1965b) *Chondromyces apiculatus* (Myxobacterales)-Schwarmentwicklung und Morphogenese. Film E779. Schwarmentwicklung und Morphogenese. Institut fur den Wissenschaftlichen film, Gottingen
Rolbetski A, Ammon M, Jakovljevic V, Konovalova A, Søgaard-Andersen L (2008) Regulated secretion of a protease activity activates intercellular signaling during fruiting body formation in *M. xanthus*. Dev Cell 15:627–634
Ronning CM, Nierman WC (2008) The genomes of *Myxococcus xanthus* and *Stigmatella aurantiaca*. In: Whitworth DE (ed) Myxobacteria: multicellularity and differentiation. ASM Press, Washington, DC, pp 285–298
Sager B, Kaiser D (1994) Intercellular C-signaling and the traveling waves of *Myxococcus*. Genes Dev 8:2793–2804
Schneiker S. et al (2007) Complete genome sequence of the myxobacterium *Sorangium cellulosum*. Nat Biotechnol 25(11):1281–1289
Schirrmeister BE, Antonelli A, Bagheri HC (2011) The origin of multicellularity in Cyanobacteria. BMC Evol Biol 11:45
Shimkets LJ, Gill RE, Kaiser D (1983) Developmental cell interactions in *Myxococcus xanthus* and the spoC locus. Proc Natl Acad Sci U S A 80:1406–1410
Singer M, Kaiser D (1995) Ectopic production of guanosine pent-and tetraphosphate can initiate early developmental gene expression in *Myxococcus xanthus*. Genes Dev 9:1633–1644
Søgaard-Andersen L (2008) Contact-dependent signaling in *Myxococcus xanthu*s: the function of the C-signal in fruiting body morphogenesis. In: Whitworth DE (ed) Myxobacteria. ASM Press, Washington, DC, pp 77–91
Spormann AM, Kaiser D (1999) Gliding mutants of *Myxococcus xanthus* with high reversal frequencies and small displacements. J Bacteriol 181:2593–2601
Sproer C, Reichenbach H, Stackebrandt E (1999) Correlation between morphological and phylogenetic classification of myxobacteria. Int J Syst Bacteriol 49:1255–1262
Stephens K, Hartzell P, Kaiser D (1989) Gliding motility in *Myxococcus xanthus*: the *mgl* locus, its RNA and predicted protein products. J Bacteriol 171:819–830.
Studholme DJ, Dixon R (2003) Minireview: domain architectures of s^{54}-dependent transcriptional activators. J Bacteriol 185 (6):1757–1767
Sun M, Wartel M, Cascales E, Shaevitz J, Mignot T (2011) Motor-driven intracellular transport powers bacterial gliding motility. Proc Natl Acad Sci U S A 108:7559–7564
Tzeng L-F, Singer M (2005) DNA replication during sporulation in *Myxococcus xanthus* fruiting bodies. Proc Natl Acad Sci U S A 102:14428–14433
Wall D, Kaiser D (1998) Alignment enhances the cell-to-cell transfer of pilus phenotype. Proc Natl Acad Sci U S A 95:3054–3058
Wall D, Wu SS, Kaiser D (1998) Contact stimulation of Tgl and type IV pili in *Myxococcus xanthus*. J Bacteriol 180:759–761
Wei X, Pathak D, Wall D (2011) Heterologous protein transfer within structured myxobacteria biofilms. Mol Microbiol 81:315–326. doi:10.1111/j.1365–2958.2011.07710.x
Whitfield C (2006) Biosynthesis and assembly of capsular polysaccharides in *Escherichia coli*. Annu Rev Biochem 75:39–68
Wolgemuth C, Hoiczyk E, Kaiser D, Oster G (2002) How myxobacteria glide. Curr Biol 12:369–377
Wu Y, Kaiser D, Jiang Y, Alber M (2009) Periodic reversal of direction allows myxobacteria to swarm. Proc Natl Acad Sci U S A 106 (4):1222–1227
Yu R, Kaiser D (2007) Gliding motility and polarized slime secretion. Mol Microbiol 63:454–467

Index

I. Ruiz-Trillo, A. M. Nedelcu (eds.), *Evolutionary Transitions to Multicellular Life*, Advances in Marine Genomics 2, DOI 10.1007/978-94-017-9642-2

MIX
Papier aus verantwortungsvollen Quellen
Paper from responsible sources
FSC® C105338

If you have any concerns about our products,
you can contact us on
ProductSafety@springernature.com

In case Publisher is established outside the EU,
the EU authorized representative is:
Springer Nature Customer Service Center GmbH
Europaplatz 3, 69115 Heidelberg, Germany

Printed by Libri Plureos GmbH
in Hamburg, Germany